Multiphase Flows with Droplets and Particles

Multiphase Flows with Droplets and Particles provides an organized, pedagogical study of multiphase flows with particles and droplets. This revised edition presents new information on particle interactions, particle collisions, thermophoresis and Brownian movement, computational techniques and codes, and the treatment of irregularly shaped particles. An entire chapter is devoted to the flow of nanoparticles and applications of nanofluids.

FEATURES

- Discusses the modelling and analysis of nanoparticles.
- Covers all fundamental aspects of particle and droplet flows.
- Includes heat and mass transfer processes.
- Features new and updated sections throughout the text.
- Includes chapter exercises and a Solutions Manual for adopting instructors.

Designed to complement a graduate course in multiphase flows, the book can also serve as a supplement in short courses for engineers or as a stand-alone reference for engineers and scientists who work in this area.

Multiphase Flows with Droplets and Particles

Third Edition

Efstathios E. Michaelides,
Martin Sommerfeld and
Berend van Wachem

CRC Press
Taylor & Francis Group
Boca Raton London New York

CRC Press is an imprint of the
Taylor & Francis Group, an **informa** business

Designed cover image: Efstathios E. Michaelides.

Third edition published 2023
by CRC Press
6000 Broken Sound Parkway NW, Suite 300, Boca Raton, FL 33487–2742

and by CRC Press
4 Park Square, Milton Park, Abingdon, Oxon, OX14 4RN

CRC Press is an imprint of Taylor & Francis Group, LLC

© 2023 Taylor & Francis Group, LLC

ISBN: 978-0-367-54431-7 (hbk)
ISBN: 978-0-367-54434-8 (pbk)
ISBN: 978-1-003-08927-8 (ebk)

DOI: 10.1201/9781003089278

Typeset in Times New Roman
by Apex CoVantage, LLC

Contents

Nomenclature

This is a list of the commonly used symbols in this book. When certain subjects of *multiphase flows* utilize their own nomenclature, the commonly used symbols are used in the pertinent equations with a short description, immediately following.

LATIN SYMBOLS

a	particle radius, radius of curvature [m]
A	projected area [m^2], Hamaker constant [J]
A_i	cross-sectional area at station i [m^2]
Ac	acceleration parameter [–]
Ar	Archimedes number
B	transfer number [–]
c	speed of sound [m/s], particle circularity [–]
c_d	specific heat of dispersed phase [J/kg-K]
c_p	specific heat of continuous phase at constant pressure [J/kg-K]
C	mass concentration [–]
C_B	coefficient for Basset force equation [–]
C_D	drag coefficient [–]
C_s	thermal slip coefficient [–]
C_S	Smagorinsky coefficient [–]
C_{LR}	lift coefficient due to particle rotation [–]
C_t	thermal exchange coefficient [–]
C_{vm}	coefficient for virtual mass force equation [–]
C_μ	coefficient for turbulence effective viscosity [–]
$C_{\varepsilon 1}$	coefficient for production of dissipation [–]
$C_{\varepsilon 2}$	coefficient for dissipation of dissipation [–]
$C_{\varepsilon 3}$	coefficient for particle production of dissipation [–]
d_f	fringe spacing [m]
D	Particle diameter [m]
D_A	area-equivalent sphere diameter [m]
D_M	median diameter [m]
D_V	volume-equivalent sphere diameter [m]
D_T	throat diameter [m]
$\mathcal{D}$	diffusivity [m^2/s]
D_ε	diffusion of dissipation [kg/m · s^3]
$\bar{D}$	average diameter [m]
D_{32}	Sauter mean diameter [m]
e	coefficient of restitution [–], charge on an electron [C]
E	energy [J], Youngs modulus [Pa]

E_i	electric field intensity [N/C]
E_0	electric field strength [N/C]
f	drag factor [–], friction factor [–], frequency [Hz]
$\hat{f}$	focal length [m]
f_i	force per unit volume due to particles [N/m³]
$f(D)$	continuous frequency distribution [–]
$\tilde{f}(D)$	discrete frequency distribution [–]
$\tilde{F}(D)$	discrete cumulative distribution [–]
$F(D)$	continuous cumulative distribution [–]
F_d	drag force [N]
F_i, $\mathbf{F}$	force vector [N]
F_T	thermophoretic force [N]
g_i	gravitational acceleration [m/s²]
g_0	distribution function [–]
$G(x_i)$	filter function [–]
$\mathbf{G}$	relative velocity vector [m/s]
Ga	Galileo number [–]
h	specific enthalpy [J/kg]
h	convective heat transfer coefficient [W/m²K]
h_L	latent heat [J/kg]
h_m	mass transfer coefficient [m/s]
H	heat of combustion [J/kg], shear modulus [Pa]
I	moment of inertia [kg-m²]
I_r	relative turbulence intensity [–], radiation intensity [W/m²]
J	radiosity [W/m²]
$\mathbf{J}$	impulsive force [N]
k	stiffness [N/m], turbulence kinetic energy [m²/s²]
k'_{eff}	effective thermal conductivity [W/mK]
k'	thermal conductivity [W/mK]
K_a	absorption coefficient [–]
K_s	scattering coefficient [–]
Kn	Knudsen number [–]
K_{wall}	wall drag multiplier
ℓ	inter-particle spacing [m], Eulerian length scale [m]
l	length scale [m]
L	volume side dimension [m], characteristic length [m]
L_f	fractal dimension
L_{mf}	mean free path of molecules [m]
m	particle mass [kg], refractive index ratio [–]

$\dot{m}$	mass flow rate [kg/s]
m''	mass flux [kg/m^{2}s]
M	molecular weight [kg/mole]
Ma	Mach number [-]
n	number density (m^{-3})
N	total number [−]
N_R	spin factor [-]
Nu	Nusselt number [−]
n_i, $\mathbf{n}$	unit outward normal vector [−]
p	pressure [Pa]
p_A	partial pressure of species A [Pa]
P	perimeter [m]
P_{ij}	pressure strain tensor [Pa]
P_{pl}	yield pressure [Pa]
Pr	Prandtl number [−]
P_ε	production of dissipation [kg/ms^3]
q	charge [C]
$\hat{\dot{q}}$	average heat transfer rate over surface [W/m^2]
$\dot{q}_c$	heat flux through the continuous phase [W/m^2]
$\dot{Q}_d$	heat transfer rate to particle [W]
Q_s	scattering efficiency factor [−]
r	radial coordinate [m], particle radius [m]
r_i	radial vector [m]
Re	Reynolds number [−]
Re_r	Reynolds number based on relative velocity [-]
Re_G	shear Reynolds number [−]
R_{ij}	volume-averaged Reynolds stress [m^2/s^2]
R_E	Eulerian correlation function [−]
R_{xx}	cross-correlation function [−]
s	source per unit volume
S	surface area [m^2]
St	Stokes number [−], Stanton number [−]
Sh	Sherwood number [−]
Sc	Schmidt number [−]
t	time [s]
T	temperature [K]
T_i	torque vector [N · m]
T_L	Lagrangian timescale [s]
u	continuous phase velocity [m/s]

u_i, $\mathbf{u}$	continuous phase velocity vector [m/s]
U	free stream velocity [m/s], superficial velocity [m/s]
v	particle velocity [m/s]
v_i, $\mathbf{v}$	particle velocity vector [m/s]
v_{crit}	critical velocity [m/s]
v_T	terminal velocity [m/s]
V	volume [m³]
w_i	velocity of gases at particle surface with respect to surface [m/s]
w_i'	velocity of gases at particle surface with respect to droplet center [m/s]
$W(v_i, x_i, t)$	phase space density [–]
$\dot{W}$	Work rate [W]
x	wetness [–]
x_i	spatial coordinate [m]
Y	Mass fraction of the dispersed phase [–]
z	local loading [–]
Z	overall loading [–]

GREEK SYMBOLS

α	volume fraction [–]
α_c	thermal diffusivity of continuous phase [m²/s]
β	thermal expansion coefficient [1/K]
β_V	parameter for momentum coupling [Ns/m⁴]
γ	flow shear [s⁻¹]; dissipation of granular temperature [kg/ms³]; ratio of specific heats [–]
ε	emissivity [–]
ε_o	permittivity [–]
ε	dissipation [m²/s³]
η	damping coefficient, Kolmogorov length scale [m]
η_i	impact efficiency [–]
Θ	granular temperature [m²/s²]
κ	coefficient for conductivity of granular temperature [kg/ms]
λ	mean free path [m], evaporation constant [m²/s], burning rate [m²/s];
λ_t	viscosity ratio [–].
	Thermal slip parameter []
μ	mean value, viscosity [N · s/m²]
μ_s	solids phase shear viscosity [N · s/m²]
$\bar{\mu}$	viscosity ratio [–]
v	kinematic viscosity [m²/s]
v_s	subgrid kinematic viscosity [m²/s]
v_T	turbulent kinematic viscosity [m²/s]
Ξ	dimensionless distance related to boundary [-]

Π	coupling parameter [−]
ρ	density [kg/m³], Poisson ratio [−], charge density [C/m²]
$\bar{\rho}$	bulk density [kg/m³]
σ	standard deviation, capillary force [N/m], charge density [C/m²], effective Schmidt number [−], Stephan-Boltzmann constant [W/m²K⁴]
σ_2	Variance
τ	response time (s), solid stress [N/m²]
τ_{ij}	shear stress tensor [Pa]
τ_η	Kolmogorov timescale [s]
ϕ	velocity ratio [−], cross-wise sphericity [−], potential function
ψ	elevation angle in PDA [°]
Ψ	sphericity [−]
φ, ϕ	angle, scattering angle in PDA [°]
φ_B	Brewster angle [°]
φ_R	rainbow angle [°]
ω_A	mass fraction of species A [−]
Ω	relative angular velocity vector [rad/s]
ω	angular velocity vector [rad/s]

SUBSCRIPTS

c	continuous phase, carrier phase
cl	center line
C	Collision
d	dispersed phase, droplet phase
e	effective
E	Emitter
II	Parallel
m	pertains to mass; pertains to film coefficients
n	Number
p	pressure, computational particle
r	radiation, receiver
R	rotation
s	particle surface
t	Tangential
tp	transverse, perpendicular
T	thermophoresis
SS	steady state
∞	far from the particle surface

OPERATORS

$\overset{=}{()}$ time average

$\overline{()}$ volume average, superficial

$\langle\ \rangle$ phase average, interstitial

$\tilde{()}$ mass average, filtered

$\widehat{()}$ number averaged

Foreword

Since the publication of the first and second editions of *Multiphase Flow with Droplets and Particles*, in 1998 and 2012 respectively, there have been numerous and important advances in the science and technology of dispersed multiphase flows. The intent of the third edition is to include these advances, while retaining the organized and pedagogical approach of the first two editions. All chapters have been modified to reflect the new knowledge and applications in the field of *multiphase flows*. Chapter 1 was enhanced to include several examples of current and emerging technologies. Chapters 2 and 3 have been revamped to include more rigorous approaches to the continuum hypothesis; the thermodynamic and transport properties of the phases; and the characteristics of the dispersed phase, especially their size characterization. The chapter on phase interactions—mass, momentum, and energy—was split in two, Chapter 4 and Chapter 5, to better elucidate the several types of forces that act on the two phases and to present the mass and energy transfer in a holistic manner. Chapters 6 and 7 were expanded to include all the modern research findings on the inter-particle and particle-wall interactions as well as the analytical and numerical modelling of these interactions, including fundamental aspects of particle agglomeration and wall deposition with adhesion. Chapter 8 has been completely rewritten to include the significant progress and advances in single-phase and multiphase numerical modelling of the last 15 years. All the important numerical methods are introduced ranging from particle-resolved simulations to point-mass-based approaches, as applied in engineering practice. The introduced theory is underpinned by illustrative examples. With 42 original figures and more than 230 references, this chapter is of the size of a research monograph. Chapter 9 was also revamped to include all the modern experimental methods and instruments. This chapter offers a great deal of heuristics and practical suggestions for reliable multiphase measurements. Chapter 10 was added—on nanofluids—to cover the salient characteristics and properties of these dispersed mixtures of base fluids with nano-size particles that are increasingly becoming important in several applications ranging from drug delivery to advanced energy conversion systems.

About the Authors

Efstathios E. (Stathis) Michaelides, PhD, is currently the holder of W.A. (Tex) Moncrief Chair of Engineering at Texas Christian University (TCU) and also a Fellow of the Ralf Lowe Energy Institute at TCU. He is recognized as a leading scholar in the areas of multiphase flows and energy conversion, where he has authored seven monographs. He has published more than 170 journal papers and has contributed more than 250 presentations in national and international conferences. He chaired the Fourth International Conference on Multiphase Flows (New Orleans May 27 to June 1, 2001). He was awarded an honorary M.A. degree from Oxford University (1983); the ASEE Centennial Award for Exceptional Contributions to the Profession of Engineering (1993); the Lee H. Johnson award for Teaching Excellence (1995); the Senior Fulbright Fellowship (1997); the ASME Freeman Scholar award (2002); the Outstanding Researcher award at Tulane University (2003); the ASME Outstanding Service award (2007); the ASME Fluids Engineering award (2014); the ASME-FED Ninetieth Anniversary Medal (2016); and the ASME Edwin F. Church Medal for "eminent service in increasing the value, importance and attractiveness of mechanical engineering education."

Martin Sommerfeld, Dr.-Ing., studied at the Technical University of Aachen, where he received his doctorate in 1984. Thereafter, he conducted research at the Department of Aeronautical Engineering, Kyoto University, Japan, with a research fellowship obtained from the Japan Society for Promotion of Science and the Alexander von Humboldt Foundation. At the Institute of Fluid Mechanics of the University of Erlangen, he headed a research group on two-phase flow from 1986 to 1994 and completed his habilitation. In October 1994 he was appointed as Professor of Mechanical Process Engineering at the Martin-Luther-University of Halle-Wittenberg. In 1997 he received the DECHEMA Award 1996 for his contributions to multiphase flow measurements, modelling, and numerical prediction. Since 2017 he is Professor at the Otto-von-Guericke University Magdeburg in the Faculty Process and Systems Technology, leading the Institute Multiphase Flow Systems. Professor Sommerfeld has organized a continuing series of workshops on two-phase flow predictions, ASME symposia, the International Conference on Multiphase Flow (ICMF 2007), and several other international conferences. For 15 years Professor Sommerfeld has organized jointly with ERCOFTAC the "Best Practice Guidance Seminar CFD for Dispersed Multi-Phase Flows." He has published 220 journal publications and more than 210 conference papers as well as 40 contributions to monographs. His present research activities on multiphase flows are related to modelling of particulate flows through experimental analysis using modern optical methods and direct numerical simulation for extending Euler/Lagrange numerical methods with applications range from particulate flows in transport lines over spraying systems and reactive bubbly flows to particle-laden flows in bio-medical systems.

Berend van Wachem, PhD, is currently Chair of Mechanical Process Engineering at the Otto-von-Guericke University in Magdeburg, Germany. His main research areas comprise multiphase flow, particle technology, numerical methods, and computational fluid dynamics. He has published more than 100 scientific journal papers on these topics and has contributed to more than 150 presentations at conferences on these topics. He is also the main developer of the software package that may be accessed at www.multiflow.org, to predict the behavior of multiphase flows. He holds a PhD degree from Delft University of Technology and was Professor at the Imperial College London prior to moving to Magdeburg, Germany. In his free time, he enjoys rowing, and he is the proud father of twin daughters.

Acknowledgments

First and foremost, the authors wish to acknowledge the enormous contribution of the late Professor Clayton T. Crowe, who was the principal author in the first two editions of the book. Professor Crowe had the vision to write the first version of the book in a pedagogical way to inspire, educate, and train graduate students in the area of multiphase flow. The authors also thank Ms. Kyra Lindholm and Mr. Kendall Bartels of CRC for their continuous guidance and patience with the submission of the manuscript. Professor Michaelides acknowledges the support of the W.A. (Tex) Moncrief Chair of Engineering at TCU, which enables him to undertake such long-term projects and the assistance of several former students—particularly of Professor Zhigang Feng of the University of Texas at San Antonio—who contributed to his group's research on multiphase flows and taught him a great deal about the flows of particles and drops.

Professor Sommerfeld acknowledges the research work of former PhD students and postdocs (i.e. Dr. S. Blei, Dr. C.-U. Böttner, Dr. Y. Cui, Dipl.-Ing. M. Dietzel, Dr. M. Ernst, Dr. C.-A. Ho, Dr. A. Hoelzer, Dr. J. Kussin, Dr. S. Lain, Dr. J. Lipowsky, Dr. G.A. Novelletto, Dipl.-Ing. S. Schmalfuss, Dr. O. Sgrott, Dr. S. Sübing, and Dr. M.A. Taborda), which is partly included in the book and is mostly based on financial support of the Deutsche Forschungsgemeinschaft (DFG), German Research Foundation. Professor van Wachem would like to thank the numerous PhD students that have graduated under his supervision and postdocs who have assisted with building the knowledge of his research group and developed theories and codes over the years, of which some of the results can be found in this book.

Last but not least, all the authors acknowledge the support of their families, who patiently indulged their writing of this book.

Efstathios E. (Stathis) Michaelides
Martin Sommerfeld
Berend van Wachem

1 Introduction

The flow of particles and droplets in fluids is a category of multiphase flows, often involving multiple components. A *component* is a chemical species such as nitrogen, oxygen, water, iron, carbon, or Freon-134a. A *phase* refers to the solid, liquid, and vapor state of the matter. The flow of multicomponent, multiphase mixtures covers a wide spectrum of applications and flow conditions with some examples of single and multicomponent, multiphase flows listed in Table 1.1.

Although air is composed of several components, it is common practice to consider the ambient air at temperatures below 2,000 K as a single component (79% nitrogen and 21% oxygen) and assign to it unique thermodynamic (e.g. enthalpy, entropy, specific heat capacity, etc.) and transport (e.g. viscosity, thermal conductivity, diffusivity, etc.) properties. Air is treated as a multicomponent mixture at higher temperatures, where dissociation occurs, and at cryogenic temperatures, where some of its constituent species may condense.

The flow of mixtures of liquids is also an important industrial application. For example, water and supercritical carbon dioxide are often used to flush petroleum from wells (secondary and tertiary oil recovery), a process that gives rise to a multicomponent single- or multiphase flow. If the two liquids are miscible (e.g. water and ethyl alcohol), then the homogeneous mixture is commonly treated as a single-phase with modified properties. If the liquids are immiscible (e.g. oil and water), then the liquid cannot be regarded as homogeneous, and treatment of the flow problem becomes much more complex. In this situation, one may have "globs," or drops of oil in the water, or (at high oil content) "globs," or drops of water carried in a continuous matrix of oil. The mixtures of two liquids are generally referred to as emulsions.

Single-component, multiphase flows are typically the flow of a liquid with its vapor. The most common example is steam-water flows, which are found in a wide variety of industries. Another example of single-component, multiphase flows are refrigerants in a refrigeration system.

The flow of fluids of a single phase has occupied the attention of scientists and engineers for many years. The equations for the motion and thermal properties of single-phase fluids are well accepted (Navier-Stokes equations), and closed-form

TABLE 1.1

Examples of Single and Multicomponent, Multiphase Flows

	Single component	Multicomponent
Single-phase	Water flow	Combustion gases
	Nitrogen flow	Flow of oil-water emulsions
Multiphase	Steam-water flow	Air-water flow
	Freon-Freon liquid-vapor flow	Coal slurry flow

DOI: 10.1201/9781003089278-1

solutions for specific cases are well documented. A major difficulty is the model-ling and quantification of turbulence and its influence on mass, momentum, and energy transfer. The state-of-the art for multiphase flow models is considerably more primitive in that the correct formulation of the governing equations is still subject to debate. For this reason, the study of multiphase flows represents a challenging and potentially fruitful area of endeavor for the scientist or engineer.

Multiphase flows can be subdivided into four categories: Gas-liquid, gas-solid, liquid-solid, and three-phase flows. Examples of these four categories are shown in Table 1.2. A gas-liquid flow can assume several different configurations. For exam-ple, the motion of bubbles in a liquid, in which the liquid is the continuous phase is a gas-liquid flow. On the other hand, the motion of liquid droplets in a gas is also a gas-liquid flow. In this case, the gas is the continuous phase. Also, a separated flow, in which the liquid moves along the bottom of a pipe and the gas along the top is also a gas-liquid flow. In this situation, both phases are continuous. The first two examples, bubbles in a liquid and droplets in a gas, are known as dispersed phase flows since one phase is dispersed, and the other is continuous. By definition, one can pass from one point to another in the continuous phase while remaining in the same medium. One cannot pass from one droplet to another without going through the gas.

Gas-solid flows are usually considered to be a gas with dispersed and suspended solid particles. This category includes pneumatic transport as well as fluidized beds. Another example of a gas-solid flow is the motion of particles down a chute or inclined plane. These are known as granular flows. Particle-particle and particle-wall interactions are much more important than the forces due to the interstitial gas. If the particles become motionless, the problem reduces to flow through a porous medium in which the viscous force on the particle surfaces is the primary mechanism affect-ing the gas flow. An example is a pebble-bed heat exchanger. It is not appropriate to refer to flow in a porous medium as a gas-solid flow since the solid phase is not in motion. Gas-solid flow is another example of a dispersed phase flow since the par-ticles constitute the dispersed phase, and the gas is the continuous phase.

Liquid-solid flows consist of flows in which solid particles are carried by the liquid and are referred to as slurry flows. Slurry flows cover a wide spectrum of

TABLE 1.2
Categories of Multiphase Flows

Gas-liquid flows	Bubbly flows
	Separated flows
	Gas-droplet flows
Gas-solid flows	Gas-particle flows
	Pneumatic transport
	Fluidized beds
Liquid-solid flows	Slurry flows
	Hydrotransport
	Sediment transport
Three-phase flows	Bubbles in a slurry flow
	Droplets/particles in gaseous flows

applications from the transport of coals and ores to the flow of mud. These flows can also be classified as dispersed phase flows and are the focus of considerable interest in engineering research. Once again it is not appropriate to refer to the motion of liquid through a porous medium as a liquid-solid flow since the solid phase is not in motion.

Three-phase flows are also encountered in engineering problems. For example, bubbles in a slurry flow give rise to the presence of three phases flowing together. There is little work reported in the literature on three-phase flows.

The subject of this book is the flow of particles or droplets in a fluid, specifically the flow of particles and/or droplets in a conveying gas as well as particles in a conveying liquid. The other area of dispersed phase flows, namely, bubbly flows, will not be addressed here. Particulate flows—a term that includes the flows of particles and drops in fluids—has a wide application in industrial, energy generation, and environmental processes: The removal of particulate material from exhaust gases is essential to the control of pollutants generated by power plants fired by fossil fuels. The efficient combustion of droplets and coal particles in a furnace depends on the multiple interactions of particles or droplets with air. Some food production depends on the drying of liquid droplets to powders in high-temperature gas streams. The transport of powders in pipes is common to many chemical and processing industries.

For many years, the design of systems with particles and droplets was based primarily on empiricism. Since the 1980s, more sophisticated measurement techniques have led to improved process control and evaluation of fundamental parameters. Increased computational capability has enabled the development of numerical models that can be used to complement engineering system design. The improved understanding of this engineering subject is a rapidly growing field of technology, which has far-reaching benefits in upgrading the operation and efficiency of processes, in system optimization, and in supporting the development of new and innovative approaches. A great deal more material in multiphase flows and the current status of multiphase flow technology in industrial applications can be found in the second edition of the *Multiphase Flow Handbook* (Michaelides *et al.*, 2017).

1.1 INDUSTRIAL APPLICATIONS

The objective of this book is to provide a background in this important area of physics and engineering, to assist new researchers in the field, and to provide a resource to those actively involved in the design and development of particulate systems. In this chapter, examples of multiphase flows in industrial, energy conversion, and environmental processes are outlined to illustrate the wide application of this technology and the field of science that supports it.

1.1.1 SPRAY DRYING

Many products such as foods, detergents, and pharmaceuticals are produced using spray drying (Masters, 1972). This is a process in which a liquid-solid mixture is atomized as droplets that include the solid particles and is injected in a stream of hot

Drying gases out

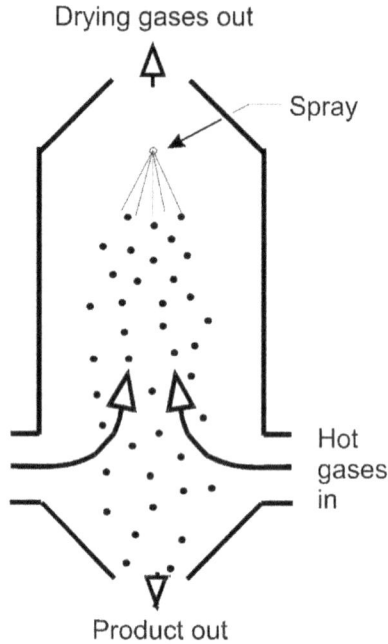

Spray

Hot
gases
in

Product out

FIGURE 1.1   Countercurrent flow spray dryer.

gases. The droplets are dried, and their solid content forms a powder. The general configuration of a countercurrent flow spray dryer is shown in Figure 1.1. A slurry or concentrated mixture is introduced at the top of the dryer and atomized into droplets. Hot gases are fed into the bottom with a swirl component and move upward through the dryer. The droplets are dried as they fall through the hot rising gases to form a powder, which is collected by precipitation at the bottom and removed as the final product.

The gas-droplet (particle) flow within the dryer is very complex. The swirling motion of the gases transports the particles toward the wall, and this may lead to impingement and accumulation. The temperature distribution in the dryer will depend on the local concentration of the droplets as they fall through the dryer. High local concentrations will depress the local gas temperature and lead to less effective drying. The result may be a non-uniformly dried product reducing product quality.

In practical driers, accumulation of the dried product on the wall is to be avoided because of uncontrolled drying, material degradation, and the possibility of fire. Also, in the case of food production, the product cannot become too hot to avoid altering any proteins or other ingredients and jeopardizing its nutritional value or its taste.

Even though spray drying technology has been continuously improved through the years, it is still difficult to scale up models to prototype operation. It is also difficult to determine, without actual testing, how modifying the design of a conventional dryer will affect performance. There has been significant progress

(Verdurmen *et al.*, 2004, 2005) in the development of numerical and analytic tools that adequately simulate the gas-droplet flow field in the dryer. Such models and the pertinent analyses, which require sophisticated modelling techniques, could be effectively used to improve the efficiency of current driers, optimize their design, predict off-design performance, and serve as a tool for scale-up of promising bench-scale designs to prototype operation.

1.1.2 Materials Transport Systems

Materials are transported by either gases or liquids. The transport of materials by air is known as pneumatic transport, while that by liquids (usually water) is known as slurry transport. More information and comprehensive reviews of pneumatic and slurry transport are given in the *Multiphase Flow Handbook* (Michaelides *et al.*, 2017).

1.1.2.1 Pneumatic Transport

Pneumatic transport is used widely in industry for the transport of cement, grains, metal powders, ores, plastic pellets, coal, etc. The major advantage over a conveyer belt is continuous operation, flexibility of location, and the ability to tap into the pipe at arbitrary locations. Moreover, such conveying systems are completely closed, and therefore fine particle emission is avoided during powder handling. Pneumatic transport has been particularly useful in layouts where obstacles prohibit straight-line transport or systems that require tapping the line (e.g. for sampling and quality control) at arbitrary locations.

Flow patterns will depend on many factors, such as solids loading, Reynolds number, and particle properties. Vertical pneumatic transport corresponds to gas flow velocities exceeding the fast fluidization velocity. The following regimes that are illustrated in Figure 1.2 have been identified for horizontal, gas-particle flows: In homogeneous flow, the gas velocity is sufficiently high that the particles are well mixed and maintained in a nearly homogeneous state by turbulent mixing, as illustrated in Figure 1.2a. As the gas velocity is reduced, the particles begin to settle out and collect on the bottom of the pipe, as shown in Figure 1.2b. The velocity at which deposition begins to occur in the pipe is called *saltation velocity*. After a solids layer builds up, ripples begin to form due to the gas flow. These ripples resemble "dunes." As the solid particles continue to fill the pipe, there are alternate regions where particles have settled and where they are still in suspension, as shown in Figure 1.2c. This is called slug flow. Finally, at even lower gas velocities, the solid particles completely fill the pipe, and the flow of gas represents flow through a packed bed depicted by Figure 1.2d. At this point, very few solids exit the pipeline.

Pneumatic conveying systems are generally designated as dilute- or dense-phase transport. Dilute-phase transport is represented by Figure 1.2a. These systems normally operate on low-pressure differences with low solids loading and high velocity (higher than the saltation velocity). This implies, even for dilute-phase conveying, a high risk of wall erosion due to high-velocity particle impacts in conveying elements used for flow deflection or splitting, such as bends and T-sections. Dense-phase transport is represented by Figure 1.2c, in which the pressure drop (pressure

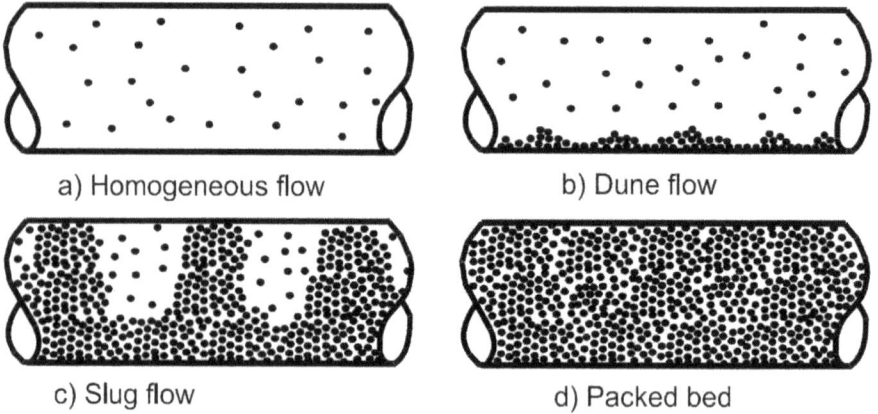

a) Homogeneous flow

b) Dune flow

c) Slug flow

d) Packed bed

FIGURE 1.2 Horizontal pneumatic transport.

loss) and the solids loading are higher. The lower velocity leads to less material degradation and line erosion.

Many studies on pressure drop in pneumatic transport have been reported, but there are considerable discrepancies in the reported data. The discrepancies are primarily due to the different nature of the solid particles and the particle-wall collisions, which are affected by the particles' inertia, turbulence, particle shapes, and wall roughness (Michaelides and Roy, 1987; Michaelides, 1987; Sommerfeld and Lain, 2015). Extensive experience with the design, installation, and operation of pneumatic transport systems has given rise to design criteria, which ensure a functional system. In dense-phase transport, the pressure drop is proportional to the square of the length of the slug, so various schemes have been devised to achieve this end. Still, there are situations where extensive experience is insufficient. One such case is the transport of wet particles, for which little information is available on the tendency of the particles to stick together, to accumulate on the wall or in bends, and to plug the pipe. Even more fundamental, there is essentially no information on the pressure drop associated with the conveying of the different types of wet solids. Hence, even simple conveying of particles comprises numerous elementary particle-scale processes, which need to be considered, such as fluid forces on non-spherical particles, inter-particle collisions and long-term particle contacts, wall collisions associated with possible particle degradation, erosion, and deposition.

1.1.3 SLURRY TRANSPORT

The transport of particles in liquids is identified as slurry flow (Shook and Roco, 1991). The term "hydrotransport" is also used for the transport of large particles like rock or chunks of coal. The flow of mud is regarded as a slurry flow. Considerable effort has been devoted to the development of coal-water slurries, which could be substituted for fuel oils.

As with pneumatic transport, slurries are classified as homogeneous, heterogeneous, moving bed, and stationary bed. Homogeneous slurries normally consist of small particles, which are kept in suspension by the turbulence of the carrier fluid. On the other hand, heterogeneous slurries are generally composed of coarser particles, which tend to settle on the bottom of the pipe. The velocity at which the particles settle out is the *deposition velocity*, which is equivalent to the saltation velocity in pneumatic transport. Of course, because of particle settling, no slurry flow is completely homogeneous. The rule of thumb is that the slurry is homogeneous if the variation in particle concentration from the top to bottom of the pipe is less than 20%, but there are no well-established empirical rules that predict whether a slurry will be homogeneous or not. The moving bed regime occurs when the particles settle on the bottom of the pipe and move along as a bed. In this case, the flow rate is considerably reduced because the bed moves more slowly than the fluid above the bed. Finally, when the particles fill the duct and no further motion is possible, the flow configuration becomes a stationary bed. This flow is now analogous to the flow through a porous medium.

The fluid mechanics theory of liquid-solid flows is complex because of the particle-particle and fluid-particle interaction. Usually, the homogeneous slurry is treated as a single-phase fluid with modified properties, which depend on solids loading. Experience shows that the various correlations for pressure drop, which have been developed for slurry flows can only be used with confidence for slurries with properties identical to those for which the correlations have been obtained. Extrapolation of the correlations to other slurries may lead to significant errors in pressure drop predictions.

1.1.4 Manufacturing and Material Processing

The flow of droplets and particles in gases is important to many manufacturing and material processing methods, a few of which are described next.

1.1.4.1 Spray Forming

An often-used process in the manufacturing industry is spray forming or spray casting. During this process, molten metals are atomized into fine droplets, transported by a high-temperature carrier gas, and deposited on a substrate, as shown in Figure 1.3. This casting technique has several advantages, including (a) the rapid solidification of the small metal droplets generates a fine grain structure; (b) improved material properties of the newly solidified and deposited material; (c) by moving the substrate, it is possible to produce shapes close to the final product, which minimizes material waste.

The importance of gas-particle flow is evident in spray forming. The state of the droplet upon impact with the substrate is also important. A completely solidified droplet will have to be melted to form a homogeneous deposit. A liquid droplet may splatter, complicating the deposition pattern. Also, the energy associated with the latent heat of the droplet will have to be removed through the substrate, and this may slow the cooling of the material on the surface. The cooling rate of the droplets depends on the droplet size and the local temperature in the spray. Droplets on the

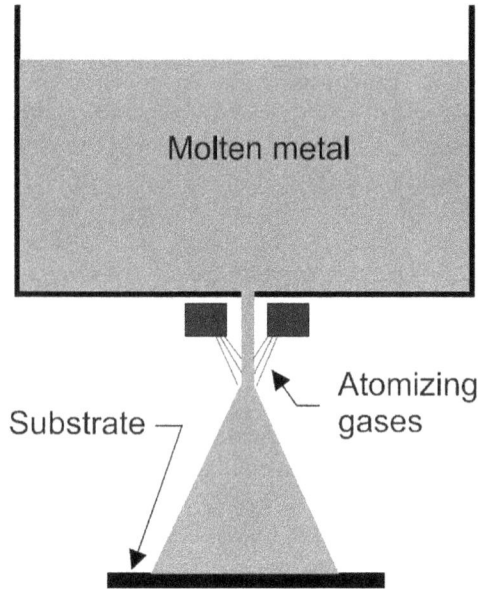

FIGURE 1.3   Spray forming process.

edge of the spray will cool faster than those near the core. The understanding and ability to predict the movement and thermal behavior of the droplets in the spray is important to the continued development and optimization of the spray forming process.

1.1.4.2   Plasma Spray Coating

Another important area of gas-particle flows in manufacturing is plasma coating. The typical plasma torch consists of a chamber where a plasma is produced. Particles, introduced into the plasma flow, are melted and advected toward the substrate, as shown in Figure 1.4. The heat transfer and drag on a particle in a plasma is fundamental to the operation of the torch. At present, plasma torches are designed primarily by experience and empirical data. The improvement of efficiency, operation in off-design conditions, optimization, and scale-up will follow the development of improved multiphase flow models for particles in plasma.

1.1.4.3   Abrasive Water-Jet Cutting

The rapid and accurate cutting of various materials through the use of high-velocity water jets with entrained abrasive materials is another application of multiphase flows. A typical abrasive water jet is shown in Figure 1.5. Water issues through a small orifice from a high-pressure source. Some systems operate at 60,000 psi (400 MPa) with jet velocities exceeding 900 m/sec. The material is moved while the jet remains stationary to cut out the desired shapes. The inclusion of an abrasive

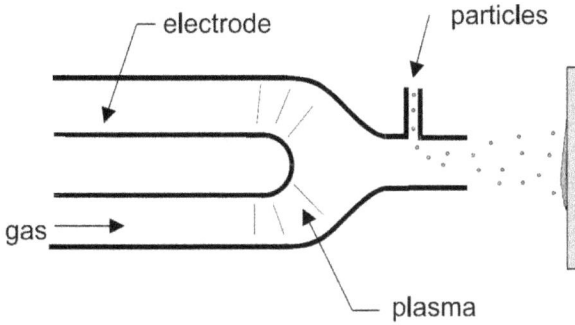

FIGURE 1.4   Plasma spray coating.

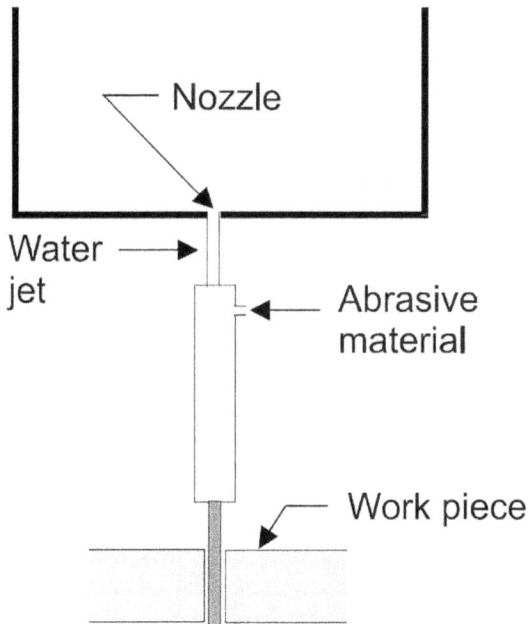

FIGURE 1.5   Abrasive water jet cutting.

material (usually garnet) in the jet enables cutting of hard materials, such as concrete and glass. The effectiveness of the abrasive water jet depends on the speed with which the abrasive material impacts the surface. The particles of the abrasive material are accelerated by the drag of the fluid, which is an intriguing multiphase flow problem.

1.1.4.4   Synthesis of Nanophase Materials

An emerging area of multiphase flow applications in materials processing is the generation of nanophase materials and nanofluids. These materials have grain sizes less than 200 nm and can be produced by the compaction of noncrystalline materials. One approach that has been investigated to produce nanoclusters is gas phase synthesis. A second technique is the injection of precursor materials into diffusion or premixed flames, where the formed particle sizes are in the range 1 to 500 nanometers. The wide range of sizes results from the lack of control of the steep thermal gradients. A third approach is with thermal reactors, in which the precursor materials are introduced into the furnace in the form of an aerosol. Chemical reactions occur as the multiphase mixture passes through the furnace. The control of particle size is highly dependent on the regulation of temperature and flow velocity. This represents an important application of multiphase flows, in which thermal coupling between the gaseous and particulate phases is critical for the quality of the formed material. Spray pyrolysis is also being used in connection with furnace reactors, in which a precursor is atomized and advected through the furnace. The solvent evaporates and reactions occur between the multitude of particles to form the desired material. The last approach is promising because of the possibility of making multicomponent materials. It is important in this approach to control the temperature, the velocity, and the residence time of the particles in the furnace, a challenging problem in multiphase flow technology.

1.2 ENERGY CONVERSION AND PROPULSION

There are many examples of droplet and particle flows in energy conversion and propulsion systems ranging from coal-fired or oil-fired furnaces to rocket propulsion. A very important application in this area is the formation of sprays (atomization) as well as the droplet interactions inside the spray, such as droplet breakup and inter-droplet collisions. Moreover, the wall interaction of sprays is an important elementary process, causing further atomization and enhancing evaporation.

1.2.1 PULVERIZED-COAL-FIRED FURNACES

Furnaces fired by pulverized coal operate by blowing a coal particle-air mixture from the corner of a furnace, as shown in Figure 1.6. The corner-fired furnace produces a swirling flow in the furnace, which enhances mixing and the combustion reactions. When the particles enter the furnace, the radiative and convective heat transfer heats the coal particles, and the volatiles (mostly methane) are released. These gases serve as the primary fuel for combustion. Ultimately, the remaining char burns, but at a lower rate. Obviously, the effective mixing of the volatiles and the gas is important for efficient combustion with minimum pollutant production. The gas-particle flow and heat transfer in the furnace is complex because of the interaction of heat transfer, combustion reactions, and particle dynamics.

1.2.2 FLUIDIZED BEDS

The fluidized bed is another example of an important industrial operation involving multiphase flows. A fluidized bed consists of a vertical cylinder containing particles where gas is introduced through the holes of the distributor in the bottom of the cylinder, as shown in Figure 1.7. The gas rising through the bed fluidizes and suspends

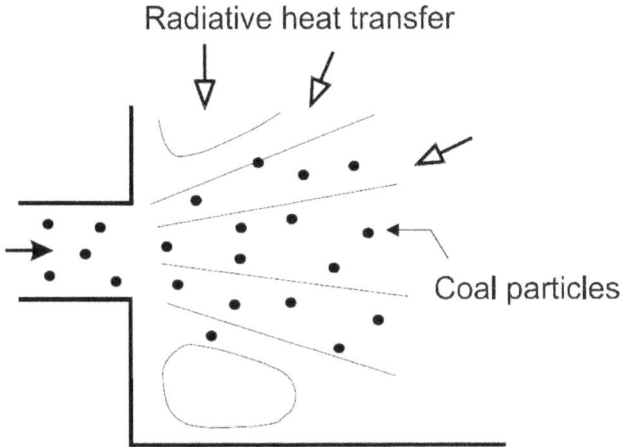

FIGURE 1.6   Corner-fired furnace with pulverized coal.

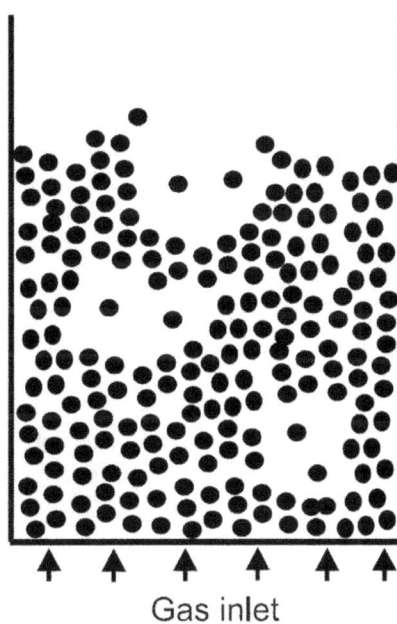

FIGURE 1.7   Fluidized bed.

most of the particles. At a given flow rate, "bubbles"—regions of low particle density—appear in the fluidized bed and rise through the bed, thus intensifying the mixing within the bed. Fluidized beds are used for many chemical processes, such as coal gasification, combustion, liquefaction, as well as the disposal of organic, biological, and toxic wastes.

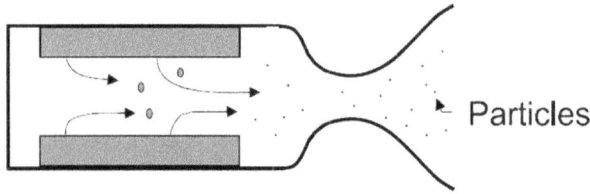

FIGURE 1.8   Solid propellant rocket motor.

Currently the design and operation of a fluidized bed is the result of many years of experience in building, modifying, and testing to achieve the best performance possible. The mechanics of the multiphase flow in a fluidized bed has been, and continues to be, a challenge to the scientist and practicing engineer. Some have chosen to treat this flow as a homogeneous fluid with a modified viscosity and thermal conductivity. Others have modeled the problem using the discrete particle approach, in which the motion of each particle is considered. The former approach depends on developing appropriate relationships for the transport properties of the particulate phase. The latter approach requires extensive computer capability to include a sufficient number of particles to simulate the heterogeneous system (Michaelides, 2013). Even though the numerical models for multiphase flow in a fluidized bed appear promising, there are still many issues that have to be included, such as the cohesiveness of particles, the sticking probability of wet particles, as well as particle-wall interactions.

1.2.3 Solid Propellant Rockets

An example of a gas-particle flow in a propulsion system is the solid propellant rocket shown in Figure 1.8. The fuel of solid propellant rocket can be aluminum powder, or another solid powder. When the fine powder particles burn, small droplets—about a micrometer in diameter—are produced and advected out the nozzle in the exhaust gases. The presence of these particles lowers the specific impulse of the rocket. The principles of gas-particle flows and mass transfer are used to design nozzles to achieve the best specific impulse possible within the design constraints of the system.

1.3   ENVIRONMENTAL APPLICATIONS

1.3.1 Pollution Control

The removal of particles and droplets from industrial effluents is a very important application of gas-particle and droplet flows (Jorgensen and Johnsen, 1981). Several devices are used to separate particles or droplets from gases.

1.3.1.1   Cyclone Separators

If the particles are sufficiently large (greater than 50 microns), a settling chamber can be used in which the condensed phase simply drops out of the flowing gas and is

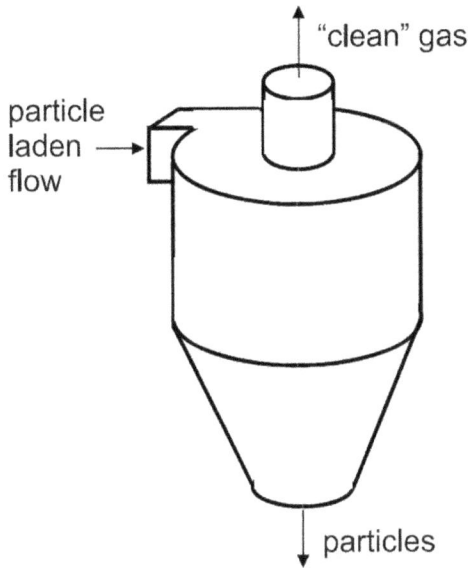

FIGURE 1.9 Cyclone separator.

collected. For very fine particles (less than 50 microns), the cyclone separator, which is schematically shown in Figure 1.9, is used. The gas-particle flow enters the device in a tangential direction at the top of the separator. The resulting vortex motion and the centrifugal force in the separator causes the particles to migrate toward the wall. Gravity and secondary flows drive the particles to the bottom of the cyclone, where they are collected in a bin. The gases converge toward the center and form a vortex flow, which exits through a pipe at the top of the cyclone body. The performance of the cyclone is quantified by the particle "cut size"—the particle diameter above which all the particles are collected. This value may be obtained through a simple model based on comparisons of the centrifugal force with the inward directed radial drag force at the imaginary surface of a cylinder extending between outlet pipe and cyclone bottom. Years of experience in cyclone separator design have resulted in "standard" designs that, under normal operating conditions, have predictable performance. Numerical modelling and other analytical approaches are needed to design cyclones for special applications, such as hot-gas cleanup. Particle behavior in cyclone separators is strongly affected by wall collisions, inter-particle collisions, and possible agglomeration. Figure 1.10 shows the results of a numerical simulation using the Euler/Lagrange approach, which considers four different modelling levels. A snapshot of the near-wall particle concentration is shown exhibiting the typical spiral downward transport of the particle in strands or ropes. Two-way coupling (2-WC) implied passive transport of the particles and four-way coupling includes the influence of the particle phase on the fluid flow. The right two graphs are results obtained by different particle agglomeration models (sphere model implies that new agglomerates are treated as volume equivalent spheres, and history models account

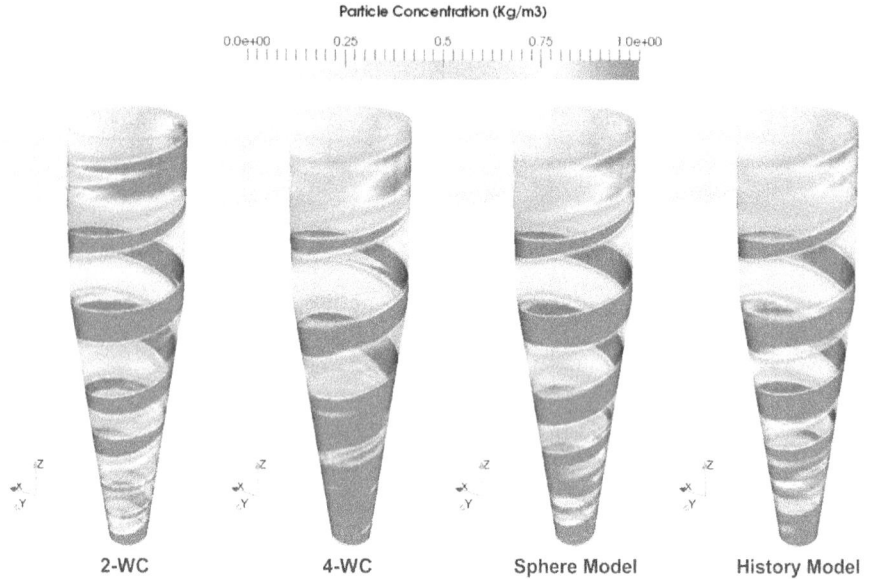

FIGURE 1.10 Distribution of particle concentration within the cyclone body near the walls, obtained by numerical Euler/Lagrange simulations.

for the temporal evolution of porosity during the agglomeration process). It is obvious that the near wall particle concentration and the rope development is completely different in the four modelling cases (Sgrott *et al.*, 2019).

1.3.1.2 Electrostatic Precipitators

The particles issuing from the furnaces of fossil fuel power plants are on the order of a micron in diameter. To filter out the particles in these applications, an electrostatic precipitator is generally used. The top view of a conventional electrostatic precipitator is shown in Figure 1.11. High voltage (tens of kilovolts) applied to the wires creates a corona discharge (mostly a negative corona is used) whereby electrons are emitted. The surrounding gas molecules are ionized by the so-called impact ionization. The ions travel along the electric lines of force to the particles, accumulate on the particle surface, and electrically charge the particles. This interaction between ions and particles occurs through convective (field charging) or diffusional transport, depending on the particle size (for particles larger about 0.2 μm field charging is dominant). The resulting negatively charged particles are transported toward the positively charged wall by Coulomb forces and deposited on the wall. A porous dust cake grows on the collection plates to several centimeters thickness. Periodically, the plates are vibrated (rapped), and the filter cake falls downward by gravity into a collection bin.

The fluid mechanics of the electrostatic precipitator are quite complex mainly due to the coupling between fluid flow and electric field as illustrated in the three parts of Figure 1.12 (from Böttner and Sommerfeld, 2003). The particle-fluid interaction

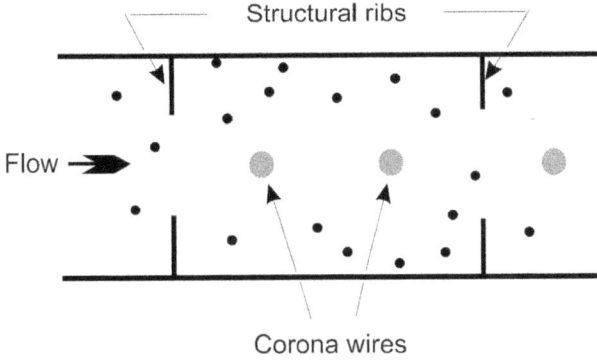

FIGURE 1.11 Schematic diagram of an electrostatic precipitator.

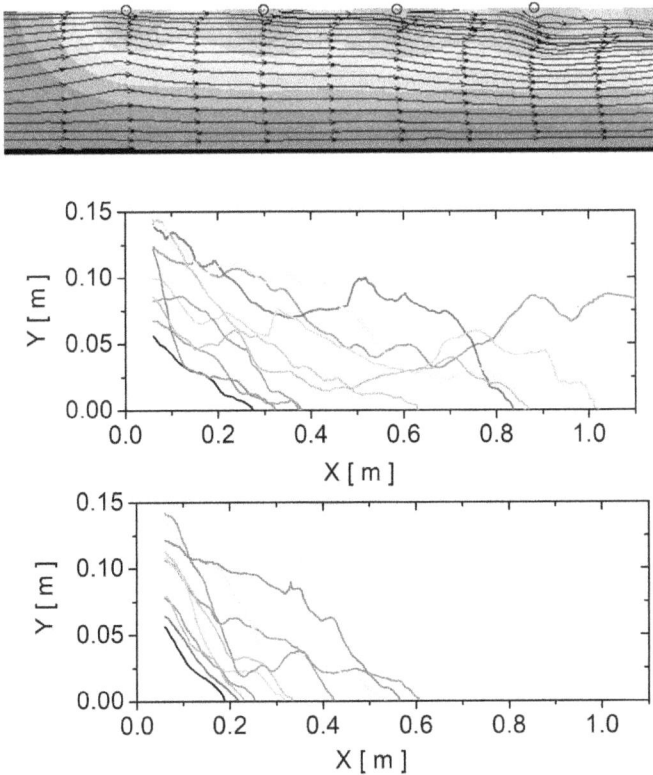

FIGURE 1.12 Numerically simulated flow field (coupled flow and electric field) and particle motion through an electrostatic precipitator (a) flow field through the channel in the initial part showing streamlines and electric field strength as gray scale; (b) trajectories of particles with diameters between 0.9 and 1.5 μm; (c) trajectories of particles with diameters between 12 and 20 μm. The inflow velocity is 0.5 m/s, the applied voltage 60 kV, negative corona, initial particle charge 66% of full charge.

obviously influences the particle concentration and the charge density. These, in turn, affect the electric field. Flow turbulence is also introduced by the structural ribs in the system and the corona discharge. The effects of turbulence are visible in the calculated particle trajectories (parts b and c), where the smaller particles exhibit stronger fluctuations. The larger particles are faster separated due to their higher level of charging, as shown in part c of the figure.

1.3.1.3   Scrubbers

Another pollution control device is the wet gas scrubber, which is designed to remove particulate as well as gaseous pollutants. Scrubbers come in many configurations and the venturi scrubber, shown in Figure 1.13, represents one of the simple designs. Liquid (atomized by the airstream) or droplets are introduced upstream of the venturi or in the narrow throat, and then the particles carried in the flue gas are captured by the droplets. The process of particle-droplet collisions is very important at this stage. The collision rate depends mainly on the droplet and particle concentrations, as well as the relative velocity between them. During this interaction, the impact efficiency is important since the particles are normally much smaller than the droplets and may bypass the droplet with the relative flow yielding no collision. Therefore, the size ratio needs to be properly adapted in order to provide sufficiently high collision rates. The droplets, being much larger than the particles, can be more easily separated from the flow.

Sulfur dioxide gas is also removed from the combustion products in coal power plants by using a "shower" of water droplets mixed with lime. The sulfur dioxide is absorbed on the surface of the droplets and reacts with the lime to form $CaSO_3$, which is a solid. The droplets are collected in tanks by precipitation, and then the solid $CaSO_3$ particles also precipitate in the aqueous solution and are removed. The remaining water is reused to generate droplets in the scrubber.

1.3.2   FIRE SUPPRESSION AND CONTROL

Fire suppression systems in buildings usually consist of nozzles located in ceilings that are activated in the event of a fire, as shown in Figure 1.14. Usually, the high temperatures produced by the fire melt a wax seal in the nozzle, and this allows the water to flow. The suppression systems are designed to deliver the amount of water

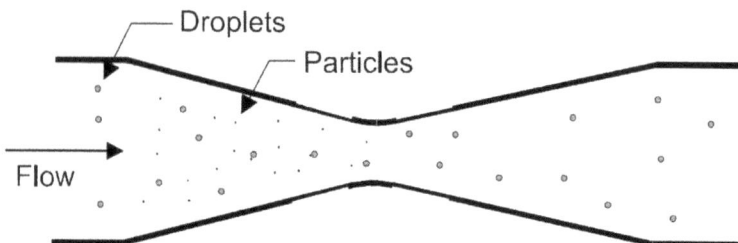

FIGURE 1.13   Schematic diagram of a venturi scrubber.

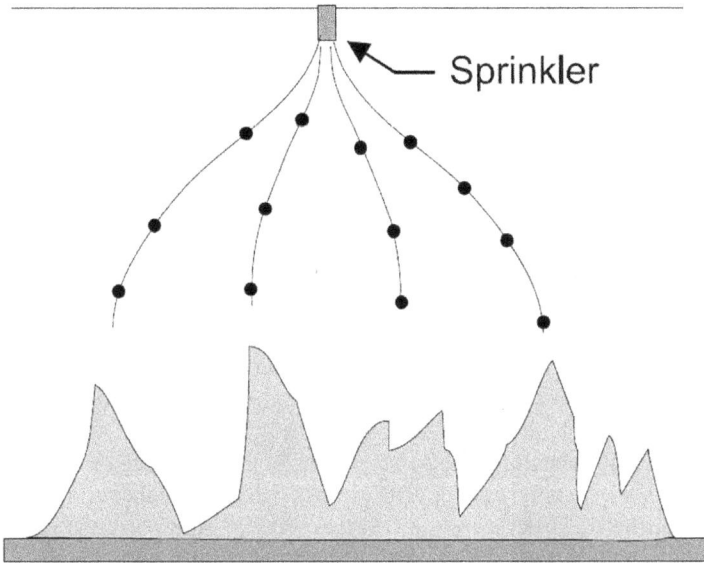

FIGURE 1.14 Ceiling sprinkler fire suppression system.

flux per unit area (required delivery density) to extinguish the fire. This design criterion is generally established by experiment. The phenomena associated with the spray are complex. As the droplets are projected toward the fire, they are evaporated by the hot gases and may not penetrate to reach the location of the fire. The evaporating droplets also cool the gases and reduce the radiative thermal feedback to the flame. Modelling fire suppression by droplets is an active area of research.

1.4 BIO-MEDICAL APPLICATIONS

Forms and regimes of multiphase flows are also met in biological and medical applications. Most important are the airflow in the respiratory system and the blood flow in the cardiovascular system. The flow of red blood cells in the human arteries is rather complex due to the high number density of platelets, their biconcave shape, and their size with respect to the arterial capillaries (red blood cell dimensions: diameter about 7–8 μm and their thickness is about 2.5 μm). Particles entering the human respiratory system are typically relatively small with respect to the airway dimensions. However, both the blood vessels as well as the airways are flexible and may deform.

1.4.1 DRY POWDER INHALERS

Inhaler devices are used for drug delivery directly to the human lungs and airways in the therapy of asthma and other chronicle pulmonary diseases. Three main types of inhaler devices are currently in use: (a) the pressurized metered dose inhaler (pMDI); (b) the dry powder inhaler (DPI); and (c) the soft mist inhaler (SMI) operating with a pressurized spray. Dry powder inhalers are low-cost, cheap, and flexible devices, which are easy to use for delivering particulate drugs directly to the lung. However, in DPI applications, only about 30%–50% of the loaded dose reaches deep into the lung and is effectively deposited. The active pharmaceutical ingredient (API) particles that can be administered via dry powder inhalers have to exhibit aerodynamic particle diameters in the size range up to about 5 μm, as only particles of this size range can reach the tiny airways of the deep lung. The performance and efficiency of DPIs is strongly related to the way of drug delivery and the API formulations being used (Daniher and Zhu, 2008).

There are three kinds of dry powder inhalers that are classified based on the drug powder delivery, namely, single-dose inhalers, multi-unit dose inhalers, and reservoir-based systems. In the case of single-dose inhalers, the required amount of API is loaded into capsules or blisters, which then, before the inhalation, are placed into the inhaler and then pricked by needles or broken in order to release the drug formulation. Very specific ways of drug powder delivery is done for multi-unit-dose devices by a rotating disk or multiple blisters. Finally, reservoir inhaler types have a powder storage with hundreds of doses. Just prior to inhalation, one dose is transported into a small pre-chamber, wherefrom the powder dose is withdrawn by the breathing airstream (Daniher and Zhu, 2008).

The most frequently found drug formulations in dry powder inhalers are carrier-based systems (i.e. the drug powder is blended with larger carrier particles in a mixer, normally with carrier diameters between about 50 μm to 200 μm, so that the fine drug particles are homogeneously distributed on the carrier surface). On the other hand, a fine API powder may be used directly.

The inhaler should perform in such a way that the drugs are detached from the carrier, or if drug agglomerates exist, they are destroyed (also called aerosolization). These processes essentially determine the inhaler efficiency, characterized by the fine particle fraction (FPF) being eventually delivered to the lung airways. Hence, in order to deliver a high amount of fine drug particles, a proper balancing of the adhesive forces between carrier and drug particles and the removal forces during the inhalation process is necessary. This also applies to the adhesion among drug particles in situations where they form clusters and agglomerates. Thus, adhesive interactions between API particles and between carrier particles and API play a key role in these kind of formulations (Cui *et al.*, 2014; Zellnitz *et al.*, 2015).

The flow structure inside inhalers is mostly very complex and highly turbulent depending on the design, and there are numerous designs available on the market (Islam and Cleary, 2012). The typical maximum inhalation flow rate is about 70 L/min, depending on the health conditions and the age of the patient. Aerosolization of the drugs may be realized by fluid stresses induced by the breathing airflow or collisions with device walls. The fluid dynamic detachment of drug powder from a carrier (or the breakage of agglomerates) in the very complex airflow of inhalers is realized by acceleration/deceleration of the clusters (i.e. inertial forces), flow shear gradients, and turbulent stresses (Telko and Hickey, 2005). Drug detachment or agglomerate breakage during wall impact is mainly caused by inertial effects, which depend on impact velocity and angle (Yang *et al.*, 2015; Ariane *et al.*, 2018).

If aerosolization does not take place, the drug particles sticking on the carrier surface or existing in agglomerated form will deposit together in the mouth and throat region.

The deposition of fine drug powder on the inhaler walls may constitute a problem and reduce the emitted fine particle dose. For this reason, and for the understanding of the processes within inhalers as a basis for optimization, CFD (computational fluid dynamics) has been applied to dry powder inhalers in the last fifteen years.

A typical capsule-based swirl-type inhaler is shown in Figure 1.15 together with a steady-state RANS (Reynolds-Averaged-Navier-Stokes) solution of the complex flow field within the inhaler at a flow rate of 70 L/min (Sommerfeld et al., 2019). In this single-dose inhaler, the capsule with the required dose of API is initially placed in the capsule chamber and, just before breathing, is pricked from the side faces with needles to allow drug release. When breathing starts, the flow enters from the outside through two tangential inlets into the swirl chamber, inducing a strongly swirling flow that can easily reach 50 m/s. With this flow, the capsule is lifted and begins to rotate, releasing the powder through the holes in the caps of the capsule. These particles will, depending on their size, frequently collide with the walls (Sommerfeld et al., 2019) and are carried through the grid and the mouthpiece, eventually entering the oral cavity and lung airways of a human. The grid causes particle-wall collisions inducing dispersion and daps the swirling motion to avoid particle-wall collisions in the mouthpiece and have a pressure recovery. Consequently, the flow in inhalers is very complex, and particle transport is affected by numerous elementary processes, including wall collisions, inter-particle collisions, de-agglomeration, and a possible wall deposition.

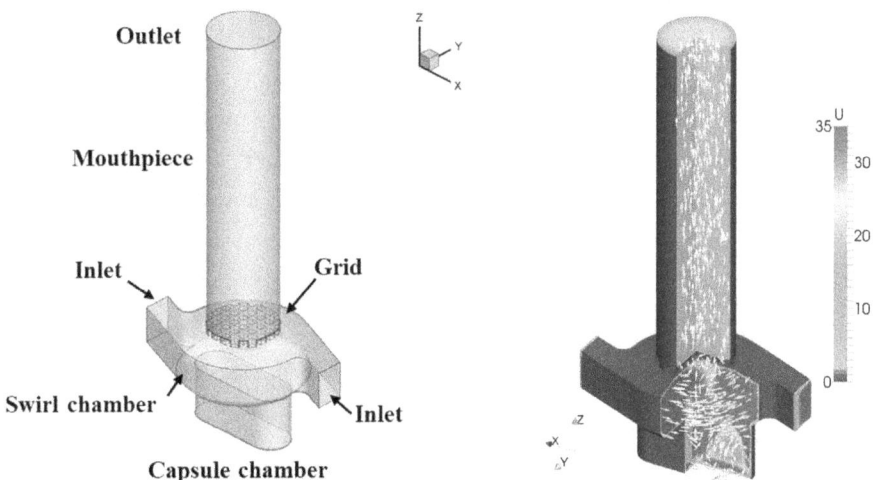

FIGURE 1.15 Configuration of a typical capsule-based swirl-type inhaler, brand name Cyclohaler® (left), with a typical flow field at 70 L/min (right).

1.4.2 Airway Deposition

The understanding of fine particle transport and deposition in the upper respiratory tract of the lung is important for aerosolized delivery of drug powders as well as unwanted exposure to environmental aerosols. Moreover, the developed airflow within the lung is rather complex and strongly time dependent. In addition, the airway walls expand and contract during inhalation and exhalation, respectively. The temporal variation of inhalation and exhalation flow rate, as well as the continuous variation of the airway cross-section (due to the continuous branching) is naturally coupled with transitions from turbulent to laminar flow and vice versa. The major parts of the human lung system, which are shown in Figure 1.16, are the extrathoracic region with the nasal and oral cavities and the pharynx; and the tracheobronchial region with the trachea, the bronchi, and the bronchioles, which are also called conduction zone. Finally, there is the alveolar region, also named pulmonary

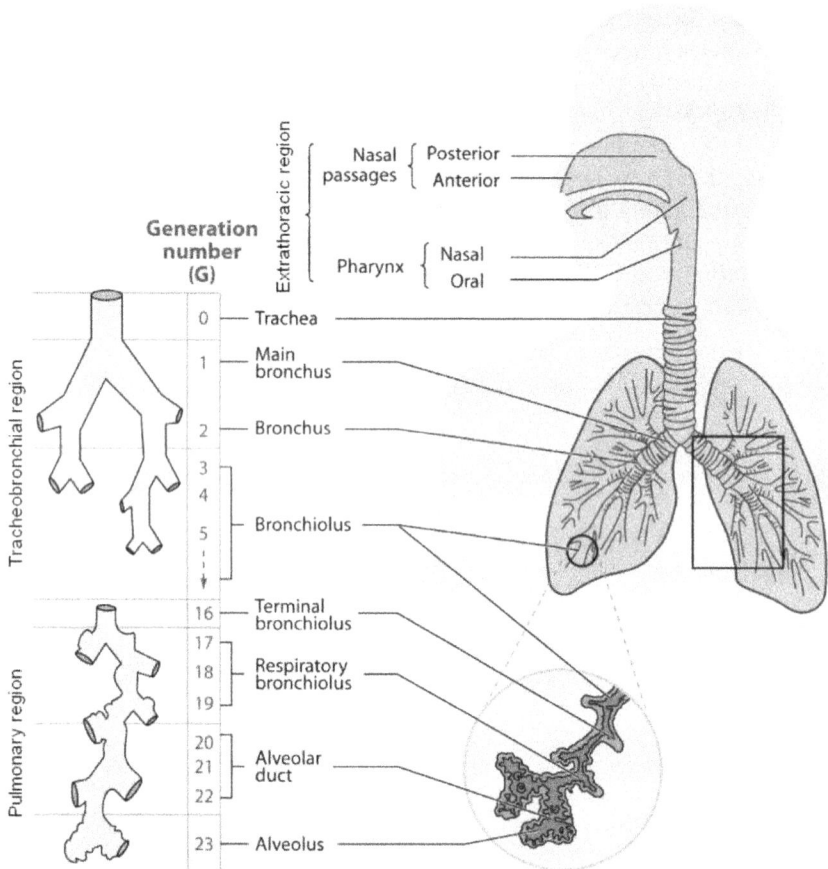

FIGURE 1.16 Anatomical regions of the complete human pulmonary system and airway generation model.

region. Additionally, in Figure 1.16, the generation number is specified, indicating the number of branching. The particles entering the lung airways during an inhalation process are typically less than about 10 μm, but those reaching deep into the lung are smaller than 5 μm. The location of deposition depends on the particle size and the local flow conditions, that is, laminar or turbulent. The mechanisms for particle deposition are numerous, namely, inertia or impaction, interception, sedimentation (gravity effect), and Brownian diffusion. In the extrathoracic region and the upper airways, up to the bronchus region, particle inertia and interception are the dominant mechanism. In the pulmonary region with narrow airway passages and low Reynolds numbers, Brownian diffusion is the most relevant mechanism for deposition (Hussain *et al.*, 2011). Due to the sticky mucus layer, the particles will always stick to the lung walls.

Fine particle transport in lung airways is governed by the drag force with slip correction, the shear-induced lift forces, and the additional wall effects in drag and lift (Sommerfeld *et al.*, 2021). Another issue influencing aerosol transport is the humid environment in the lung system. Hence, liquid droplets may be subject to evaporation/condensation and solid particles will swell (i.e. increase their size) upon absorption of water vapor.

Computational fluid dynamics (CFD) is an efficient analysis tool for fine particle deposition in lungs. The Euler/Lagrange approach, which is explained in detail on Chapter 8 (Sommerfeld *et al.*, 2008), is the most common choice to simulate the airflow and aerosol transport in the human airways (Kolanjiyil and Kleinstreuer, 2016; Longest *et al.*, 2019; Islam *et al.*, 2020) since it can provide high resolution and size-dependent information of localized deposition. A complete inhalation period is of course an unsteady process and, therefore, numerically very expensive. Most of the studies are done for distinct typical flow rates. Direct numerical simulations (DNS) is rarely applied and only usable for small sections of the airway system. More common to analyze aerosol lung deposition are large eddy simulations (LES) and RANS (Reynolds-Averaged Navier-Stokes) approaches in combination with various available turbulence closures (Koullapis *et al.*, 2016; Islam *et al.*, 2020; Sommerfeld *et al.*, 2021).

In doing such numerical computations with simplifications in the boundary conditions, such as stationary flow and rigid geometry (i.e. no flexible airways), one has to make first the choice for the region of interest, which is determined by the numerical method applied and the objective of the planned studies. This is also a consequence of medical imaging resolution and computational cost constraints. It should be noted that lung geometries are very much person specific and also depend on the health condition of the person.

1.5   SUMMARY AND OBJECTIVES OF THIS BOOK

There are many other applications of the flow of particles and droplets in fluids that have not been addressed here. Those discussed in the previous sections illustrate the technological significance of this important area of fluid mechanics.

The objective of this book is to present the fundamental concepts and approaches to address fluid particle flows. Chapter 2 provides important definitions to characterize the flow. The various parameters used to define particle and droplet size are presented in Chapter 3. The interaction between the carrier fluid and the particle and droplet phase are addressed in Chapters 4 and 5. The equations describing particle-particle and particle-wall interactions are discussed in Chapter 6. The volume-averaged form of the continuous phase equations and the modelling techniques of the continuous phase are presented in Chapter 7. The techniques for the modelling of particulate and droplet flows are discussed in Chapter 8. Chapter 9 describes the currently available methods, instruments, and measurement techniques for fluid-particle systems. Finally, Chapter 10 is devoted to the subject of nanofluids, an emerging technological area.

REFERENCES

Ariane, M., Sommerfeld, M. and Alexiadis, A., 2018, Wall collision and drug-carrier detachment in dry powder inhalers: Using DEM to devise a sub-scale model for CFD calculations, *Powder Technology*, **334**, 65–75.

Böttner, C.-U. and Sommerfeld, M., 2003, Messung und numerische Berechnung der Partikelbewegung im Elektroabscheider, *Chemie Ingenieur Technik, Jahrg*, **75**, 188–194.

Cui, Y., Schmalfuß, S., Zellnitz, S., Sommerfeld, M. and Urbanetz, N., 2014, Towards the optimisation and adaptation of dry powder inhalers, *Intern. J. of Pharmaceutics*, **470**, 120–132.

Daniher, D.I. and Zhu, J., 2008, Dry powder platform for pulmonary drug delivery, *Particuology*, **6**, 225–238.

Hussain, M., Madl, P. and Khan, A., 2011, Lung deposition predictions of airborne particles and the emergence of contemporary diseases Part-I, *The Health*, **2**(2), 51–59.

Islam, M.S., Paul, G., Ong, H.X., Young, P.M., Gu, Y.T. and Saha, S.C., 2020, A review of respiratory anatomical development, air flow characterization and particle deposition, *Int. J. Environ. Res. Public Health*, **17**, 380.

Islam, N. and Cleary, M.J., 2012, Developing an efficient and reliable dry powder inhaler for pulmonary drug delivery: A review for multidisciplinary researchers, *Medical Engineering & Physics*, **24**, 409–427.

Jorgensen, S.E. and Johnsen, I., 1981, *Principles of Environmental Science and Technology*, Elsevier Scientific Publishing Co., New York, NY.

Kolanjiyil, A.V. and Kleinstreuer, C., 2016, Computationally efficient analysis of particle transport and deposition in a human whole-lung-airway model. Part I: Theory and model validation, *Computers in Biology and Medicine*, **79**, 193–204.

Koullapis, P.G., Kassinos, S., Bivolarova, M.P. and Melikov, A.K., 2016, Particle deposition in a realistic geometry of the human conducting airways: Effects of inlet velocity profile, inhalation flowrate and electrostatic charge, *J of Biomechanics*, **49**, 2201–2212.

Longest, P.W., Bass, K., Dutta, R., Rani, V., Thomas, M.L., El-Achwah, A. and Hindle, M., 2019, Use of computational fluid dynamics deposition modeling in respiratory drug delivery, *Expert Opinion Drug Delivery*, **16**, 7–26.

Masters, K., 1972, *Spray Drying: An Introduction to Principles, Operational Practice and Applications*, Leonard Hill Books, London.

Michaelides, E.E., 1987, Motion of particles in gases: Average velocity and pressure loss, *J. Fluids Engineering*, **109**, 172.

Michaelides, E.E., 2013, *Heat and Mass Transfer in Particulate Suspensions*, Springer, New York.

Michaelides, E.E., Crowe, C.T. and Schwarzkopf, J.D. (eds.), 2017, *Multiphase Flow Handbook*, 2nd edition, CRC Press, Boca Raton.

Michaelides, E.E. and Roy, I., 1987, An evaluation of several correlations used for the prediction of pressure drop in particulate flows, *Int. J. of Multiphase Flows*, **13**, 433.

Sgrott Junior, O.L. and Sommerfeld, M., 2019, Influence of inter-particle collisions and agglomeration on cyclone performance and collection efficiency, *Canadian J. Chemical Engineering*, **97**, 511–522.

Shook, C.A. and Roco, M.C., 1991, *Slurry Flow: Principles and Practice*, Butterworth-Heinemann, Boston, MA.

Sommerfeld, M., Cui, Y. and Schmalfuß, S., 2019, Potentials and constraints for the application of CFD combined with lagrangian particle tracking to dry powder inhalers, *European J. of Pharmaceutical Sciences*, **128**, 299–324.

Sommerfeld, M. and Lain, S., 2015, Parameters influencing dilute-phase pneumatic conveying through pipe systems: A computational study by the Euler/Lagrange approach, *Canadian J. of Chemical Engineering*, **93**, 1–17.

Sommerfeld, M., Sgrott Jr., O.L., Taborda, M.A., Koullapis, P., Bauer, K. and Kassinos, S., 2021, Analysis of flow field and turbulence predictions in a lung model applying RANS and implications for particle deposition, *European J. of Pharmaceutical Sciences*, **166**, 105959.

Sommerfeld, M., van Wachem, B. and Oliemans, R., 2008, Best practice guidelines for computational fluid dynamics of dispersed multiphase flows. *ERCOFTAC (European Research Community on Flow, Turbulence and Combustion)*, ISBN 978-91-633-3564-8.

Telko, M.J. and Hickey, A.J., 2005, Dry powder inhaler formulation, *Respiratory Care*, **50**, 1209–1227.

Verdurmen, R.E.M., Menn, P., Ritzert, J., Blei, S., Nhamaio, G.C.S., Sonne Sørensen, T., Gunsing, M., Straatsma, J., Verschueren, M., Sineijn, M., Schulte, G., Fritsching, U., Bauckhage, K., Tropea, C., Sommerfeld, M., Watkins, A.P., Yule, A.J. and Schonfeldt, H., 2004, Simulation of agglomeration in spray drying installations: The EDECAD Project, *Drying Tech.*, **22**, 1403.

Verdurmen, R.E.M., Verschuren, M., Gunsing, M., Straatmans, H., Blei, S. and Sommerfeld, M., 2005, Simulation of agglomeration in spray dryers: The EDECAD project, *Le Lait*, **85**, 1–9.

Yang, J., Wu, C.-Y. and Adams, M., 2015, DEM analysis of the effect of particle-wall impact on the dispersion performance in carrier-based dry powder inhalers, *Intern. J. of Pharmaceutics*, **487**, 32–38.

Zellnitz, S., Schroettner, H. and Urbanetz, N.A., 2015, Influence of surface characteristics of modified glass beads as model carriers in dry powder inhalers (DPIs) on the aerosolization performance, *Drug Development and Industrial Pharmacy*, **41**, 1710–1717.

2 Properties of Dispersed Phase Flows

2.1 THE CONTINUUM HYPOTHESIS

The *continuum hypothesis* is a fundamental concept, central to the description and modelling of most systems. Using the continuum concept, one may apply the principles and methodology of geometry and calculus—two abstract mathematical subjects—to actual materials composed of discrete atoms and molecules. The properties of a continuum are assumed to be defined at every geometric point and to vary continuously from point to point. The continuum theory is based on several implicit assumptions pertaining to the meaning of limit mathematical operators, such as derivatives and integrals, that have implications on the definitions of the local thermodynamic and transport properties.

Consider a heterogeneous mixture of dispersed solid particles in a matrix of base fluid as depicted in Figure 2.1. The total volume of the mixture, V, is occupied either by the dispersed solid phase, V_d, or by the continuous fluid phase, V_c. The total mass enclosed within the volume, V, may be calculated by the integral

$$m = \int_V \rho(x, y, z)dV, \tag{2.1}$$

where $\rho(x,y,z)$ denotes the density function of the mixture at the point (x,y,z). A glance at Figure 2.1 proves that the density function of heterogeneous mixtures is

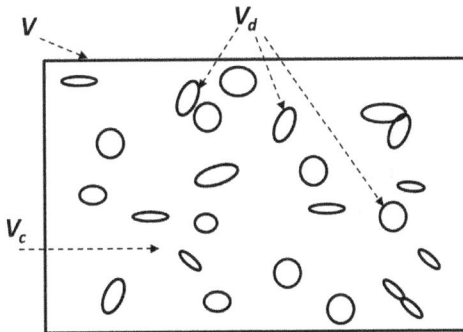

FIGURE 2.1 A heterogeneous mixture of gas and particles.

DOI: 10.1201/9781003089278-2

non-uniform. According to the rules of calculus, the integration operation in Eq. (2.1) may be performed under the condition that the density function of the heterogeneous material, $\rho(x,y,z)$, exists and is well-defined at every point within the volume, V. It must be noted that the integration operation does not require that the density function be continuous.

It is recalled that the local density function is defined mathematically by another limit operation, that of differentiation

$$\rho(x, y, z) = \lim_{\Delta V \to 0} \frac{\Delta m}{\Delta V} . \tag{2.2}$$

Δm is the mass of the material contained within a small volume, ΔV. The mathematical limit operation of differentiation implies that the volume, ΔV, becomes very small and shrinks to the point (x,y,z). This is typically referred to in mathematical terminology as "the volume, ΔV, becomes arbitrarily small."

The limit operations of differentiation and integration to actual materials present us with a conundrum: All matter is composed of atoms, which are composed of subatomic particles. When the volume ΔV is continuously reduced to a geometric point in space by becoming "arbitrarily small," an atom or a part of an atom may or may not exist within this small volume. If the "arbitrarily small" volume, ΔV, were to be of subatomic dimensions, there would be very low probability that an atom or part of an atom existed within this small volume, and the local density of the material, $\rho(x,y,z)$, would most likely be zero. However, if part of an atom existed in the volume, ΔV, the density, as defined by Eq. (2.2), would be extremely large. It is apparent that when we consider the "arbitrarily small" volume, ΔV, the density function becomes highly non-uniform and, because atoms are in continuous movement, time dependent. In addition, the uncertainty principle and quantum mechanics dictate that the existence of matter within an arbitrarily small volume, ΔV, would be given by a probability function. As a result, when the volume ΔV is allowed to become "arbitrarily small," the function, $\rho(x,y,z)$, for all materials must be defined in a probabilistic manner. This makes the calculation of the mass, m, within the volume, V, as defined by Eq. (2.1), a very formidable task.

A moment's reflection proves that, for the calculation of the integral of Eq. (2.1), an alternative definition for the density function, $\rho(x,y,z)$, is needed, and this is achieved by the modelling framework that is often referred to as the *continuum model*. According to this model, the elemental volume, ΔV, is defined to be large enough to contain a sufficiently large number of atoms or molecules, for example, several thousand. When the volume ΔV is sufficiently large, individual molecules that enter or leave the volume have very little effect on the mass, Δm, inside the volume, ΔV. Consequently, the ratio $\Delta m/\Delta V$ is almost constant. Therefore, if the elemental volume, ΔV, can be large enough and contain a sufficiently large number of molecules, the density function defined by Eq. (2.2) converges to a well-defined limit, which is defined as the operational *density* of the material.

Figure 2.2 demonstrates the operational definition of density. The left part shows schematically the volume, ΔV, and the number of molecules of the material(s) inside this volume. The right part depicts the function $f(\Delta V)=\Delta m/\Delta V$ as a function of the

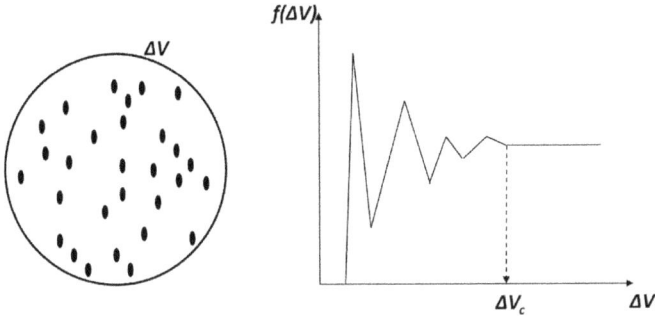

FIGURE 2.2 The operational definition of density in a continuum.

size of the volume ΔV. When ΔV is extremely small (less than the size of an atom) the magnitude of the function $f(\Delta V)$ is either very small or very large depending on whether it contains vacuum or part of an atom. As ΔV increases, the fluctuations of the function $f(\Delta V)$ decrease continuously, and when $\Delta V > \Delta V_c$, the volume contains a sufficient number of atoms for the function $f(\Delta V) = \Delta m / \Delta V$ to converge to a stable limit, which is defined as the *density of the material* at the point (x,y,z), the center of the volume, ΔV

$$\rho(x, y, z) = \lim_{\Delta V \to \Delta V_c} \left(\frac{\Delta m}{\Delta V} \right). \tag{2.3}$$

This definition of the operational local density applies to all the geometric points, (x,y,z), within a material. The definition resolves the original conundrum and defines the property density of a material in a meaningful way that enables the performance of limit mathematical operations. Following this procedure, it is possible to assign a single value for the density function $\rho(x,y,z)$ to every point within a material, whether it is homogeneous or heterogeneous. With the mathematical function $\rho(x,y,z)$ defined at every point within the larger volume, V, the integration denoted by Eq. (2.1) is easily performed, and the mass of the material is determined. It must be emphasized that, for this integration to be performed, the density function $\rho(x,y,z)$ only needs to be properly defined at every point, (x,y,z), within the volume, V, and it does not need to be uniform, continuous, or differentiable.

Functions for the other local thermodynamic and transport properties of any (homogeneous or heterogeneous) material—such as the specific enthalpy, $h(x,y,z)$; the specific energy, $e(x,y,z)$; the dynamic viscosity, $\mu(x,y,z)$; and the thermal conductivity, $k(x,y,z)$—are defined in a similar manner at every point, (x,y,z). When these functions are well-defined, all the mathematical operations that appear in the governing equations may be performed. By using this procedure for the definition of all the local properties, useful and accurate results for the behavior of homogeneous and heterogeneous materials are obtained.

It is generally accepted that the property fluctuations are less than 1% when the volume, ΔV_c, encloses approximately 10^5 atoms or molecules. Given that 22.4 m^3

of a gas at standard conditions contain $6.023*10^{26}$ molecules (1 kmol), the 10^5 molecules are enclosed in a spherical gaseous volume with diameter approximately 200 nm. For the much denser solids and liquids, the size of ΔV_c is significantly less; for example, for liquid water that has density approximately 1,000 kg/m³, the 10^5 molecules are contained in a sphere of diameter approximately 18 nm. Such dimensions for the elemental volume, ΔV_c, are low enough for homogeneous materials to be regarded as continua in most engineering calculations. One must be cognizant, however, that when the dispersed phase in a heterogeneous mixture is assumed to be a continuum—as is the case with the volume-averaged Eulerian models or two-fluid models of multiphase mixtures—the point volume of the computations should be commensurately greater. For example, a particulate flow system with fine spherical particles of 200 μm diameters at 10% volumetric fraction would contain 10^5 fine particles in a sphere of approximately 2 cm. This would be the minimum length scale for any meaningful results based on volume-averaged Eulerian models. It must be noted that the continuum hypothesis is not necessary in models based on the probability density function (PDF), which are based on ensemble averages. However, the closure equations used with PDF models have often been derived for continua.

It must be noted that the validity of the continuum hypothesis does not stem from a physical principle or a mathematical proof. It is inferred from the fact that the resulting "continuum description" of materials does not conflict with any empirical observations. The resulting *continuum theory*, which is applied in many branches of science and engineering, is supported by all the available empirical data for the physical systems that contain a large enough number of molecules or particles to be considered as continua. For systems that contain a lesser number of such elements, these elements must be treated as discrete in modelling methods known as *molecular models* (for materials) and *discrete particle models* (for particulate and droplet mixtures).

2.2   DENSITY AND VOLUME FRACTION OF DISPERSED FLOWS

Dispersed multiphase flows are heterogeneous mixtures of a carrier fluid and one or more dispersed phases, which are composed of solid particles or liquid drops. The fluid—gas or liquid—is always considered to be a continuum, while each dispersed phase may be considered either discrete or continuum. Figure 2.3 depicts such a heterogeneous mixture composed of a fluid and several dispersed spherical particles. Using the same arguments as in section 2.1 for the molecules in a sampling volume, the heterogeneous mixture sampling volume, ΔV_m, must contain a sufficient number of the dispersed-phase elements to ensure insignificant variations. Consequently, all the properties of the mixture at the point (x,y,z), the center of the volume, ΔV_m, may be defined in terms of the properties of the two phases.

The *number density* of the dispersed phase is defined as the number of particles per unit volume

$$n = \frac{\Delta N}{\Delta V_m}, \tag{2.4}$$

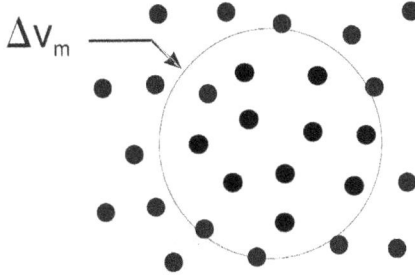

FIGURE 2.3 Characterization of number density, volume fraction, and related properties in fluid-particle systems.

where ΔN denotes the total number of particles inside the sampling volume, ΔV_m. For the definition of the number density and the other properties of the mixture, fractions of particles enclosed within the volume, ΔV_m, are counted as fractional numbers, for example, 2/10, 73/100, etc. The number density is useful in the characterization of particles in air, especially of pollutants, and is measured as particles per unit volume—$58*10^6$ aerosol particles/m^3, 58 particles/cm^3, etc.

The *volume fraction* of the dispersed phase is defined as

$$\alpha_d = \frac{\Delta V_d}{\Delta V_m},$$ (2.5)

where ΔV_d is the volume occupied by the dispersed phase within the entire sampling volume, ΔV_m. Oftentimes, the volume fraction of the particles or drops is referred to as the *volumetric concentration* of the mixture.

The volume fraction of the continuous phase is similarly defined as

$$\alpha_c = \frac{\Delta V_c}{\Delta V_m},$$ (2.6)

where ΔV_c is the volume occupied by the continuous phase within the total sampling volume ΔV_m. In gas-solid mixtures this volume fraction is sometimes referred to as the *void fraction*.

By the conservation of space principle (every part of the volume, ΔV_m, is either occupied by a solid or by the carrier fluid), the sum of the volumes occupied by the two phases is equal to the sampling volume, ΔV_m, and therefore the sum of the two volume fractions is equal to one

$$\alpha_c + \alpha_d = 1$$ (2.7)

The *bulk density* (or *apparent density*) of the dispersed phase is the total mass of the dispersed phase per unit volume of mixture

$$\bar{\rho}_d = \frac{\Delta m_d}{\Delta V_m},$$ (2.8)

where ΔM_d is the total mass of the dispersed phase within the volume, ΔV_m. If all the particles in a volume have the same mass, m, the bulk density is proportional to the number density

$$\bar{\rho}_d = nm \tag{2.9}$$

The bulk density of the continuous phase (the carrier fluid) is defined by an equation similar to Eq. (2.8). The sum of the bulk densities for the dispersed and continuous phases is the *mixture density*

$$\bar{\rho}_d + \bar{\rho}_c = \rho_m \tag{2.10}$$

It must be noted that the bulk densities of the two phases are different than the *material densities* (or *actual* densities, or simply, *densities*) of the solids in the dispersed phase, ρ_d, and the fluid of the carrier phase, ρ_c. For example, if the dispersed phase were water droplets in air, ρ_d would be the density of liquid water (approximately 1,000 kg/m³) and ρ_c the density of air (approximately 1.2 kg/m³).

The bulk density of the heterogeneous mixture is related to the two volume fractions and material densities. Since the mass of the dispersed phase within the sampling volume, ΔV_m, may be calculated as the product of the actual density of the solid phase and the volume occupied by the solid phase

$$\Delta M_d = \rho_d \Delta V_d, \tag{2.11}$$

a combination of Eqs. (2.5) and (2.8) yields

$$\bar{\rho}_d = \alpha_d \frac{\Delta M_d}{\Delta V_d} = \alpha_d \rho_d. \tag{2.12}$$

Similarly, the bulk density of the continuous phase may be obtained as

$$\bar{\rho}_c = \alpha_c \rho_c. \tag{2.13}$$

Therefore, the mixture density may be written in terms of the densities of the two materials as

$$\rho_m = \alpha_c \rho_c + \alpha_d \rho_d. \tag{2.14}$$

If the mixture is composed of several, n, dispersed phases, 1, 2, 3 . . . n, with corresponding volume fractions, $\alpha_{d1}, \alpha_{d2} \ldots \alpha_{dn}$, and densities, $\rho_1, \rho_2 \ldots \rho_n$, the mixture density is given by the sum

$$\rho_m = \alpha_c \rho_c + \sum_{i=1}^{n} \alpha_{di} \rho_i. \tag{2.15}$$

The *mass ratio* of the dispersed phase to that of the continuous phase is often used with multiphase mixtures

$$C = \frac{\bar{\rho}_d}{\bar{\rho}_c} = \frac{\alpha_d}{\alpha_c} \frac{\rho_d}{\rho_c}. \tag{2.16}$$

Similar to the volume fraction of the heterogeneous mixtures, the mass fraction of the phases is defined as the total mass of a given phase to the total mass of the mixture within the sampling volume. Thus, the *mass fraction*, *Y*, of the dispersed phase depicted in Figure 2.3 is

$$Y = \frac{M_d}{M_d + M_c} = \frac{\alpha_d \rho_d}{\alpha_d \rho_d + \alpha_c \rho_c} = \frac{\alpha_d \rho_d}{\alpha_d \rho_d + (1 - \alpha_d)\rho_c} = \frac{C}{1 + C}. \tag{2.17}$$

The mass fraction of the carrier fluid is equal to *1-Y*. The mass fraction, *Y* is sometimes referred to as the *mass concentration* of the mixture.

A useful parameter for gas-solids mixtures in multiphase flows is the local *loading* of the mixture, defined as the ratio of the mass flux of the dispersed phase to that of the continuous phase

$$z = \frac{\dot{m}_d}{\dot{m}_c} = \frac{\bar{\rho}_d v}{\bar{\rho}_c u}, \tag{2.18}$$

where v and u are the local velocities of the dispersed and continuous phases, respectively. For example, in a gas-droplet spray, the local loading, z, is the local mass flux ratio, which is finite and varies throughout the spray and is zero outside the spray envelope. The total or overall loading, Z, is the ratio of the total mass flux of the dispersed phase to the overall mass flux of the continuous phase within a flow domain, for example, a pipeline

$$Z = \frac{\bar{\rho}_d \bar{v}}{\bar{\rho}_c \bar{u}} = \frac{\alpha_d}{\alpha_c} \frac{\rho_d}{\rho_c} \frac{\bar{v}}{\bar{u}}. \tag{2.19}$$

In a gas-droplet spray, the overall loading, Z, would be the ratio of total liquid mass flow rate supplied to the atomizer to the total gas flow rate.

2.3   INTER-PARTICLE DISTANCE—DILUTE AND DENSE FLOWS

All the characteristics of dispersed flows, including phase couplings and interactions, depend on the average distance between the dispersed particles or drops. Consider the dispersed elements shown in Figure 2.4 that are enclosed in cubes with side ℓ, the average distance between the element centers. The volume fraction of the dispersed phase is

$$\alpha_d = \frac{\pi D^3}{6 \ell^3}. \tag{2.20}$$

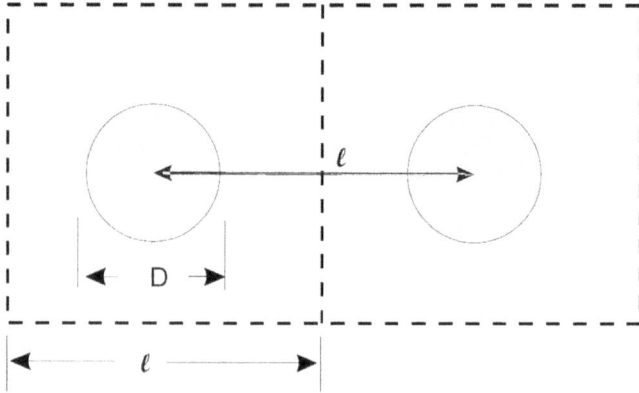

FIGURE 2.4   Inter-particle spacing in a cubic arrangement.

If the particles were in contact in this cubic configuration, the volume fraction of the dispersed phase would be $\pi/6 = 0.52$. It must be noted that the cubic arrangement of particles is not the arrangement with the highest particle density. This is the triangular configuration, which results in the *close packing of equal spheres* with maximum density 0.74.

One may rewrite Eq. (2.20) to obtain the average particle or droplet spacing in terms of the volume fraction

$$\frac{\ell}{D} = \left(\frac{\pi}{6\alpha_d}\right)^{1/3}. \qquad (2.21)$$

It follows that for a volume fraction of 10% for the dispersed phase, the average spacing is 1.7 diameters. This implies that the dispersed phase elements are close and that any local mass, momentum, and energy interactions are influenced by the neighboring elements. If one requires that the dispersed elements are on average at least one diameter apart ($l = 2D$), the volume fraction becomes $\alpha_d = 6.5\%$.

Eq. (2.21) may also be used to assess the size of the mixture sampling volume that would have insignificant variations of the mixture properties. With the stipulation that the particles are arranged in a cubic matrix and that $N = 10^5$ maintains the effect of the fluctuations less than 1%, one obtains

$$\frac{\ell_m}{D} = \left(\frac{N\pi}{6\alpha_d}\right)^{1/3} \Rightarrow \ell_m = \frac{37D}{(\alpha_d)^{1/3}}. \qquad (2.22)$$

For a dispersed system with 100 μm particles and volume fraction 1%, the size of the sampling volume, ΔV_m, would be a cube of 17 mm side.

The volume fraction of the dispersed phase may also be expressed in terms of the mass ratio, C, of the dispersed phase and the densities of the constituent materials in the mixture, as in Eq. (2.15). Since for a binary mixture $\alpha_d = 1-\alpha_c$, Eq (2.15) may be rewritten to express the average spacing in terms of the mass ratio and the densities of the materials that constitute the dispersed multiphase mixture

$$\frac{\ell}{D} = \left[\frac{\pi}{6} \frac{1 + C\left(\frac{\rho_c}{\rho_d}\right)}{C\left(\frac{\rho_c}{\rho_d}\right)} \right]^{1/3}. \tag{2.23}$$

For most gas-particle and gas-droplet flows, the ratio of material densities, ρ_c/ρ_d, is on the order of 10^{-3}. For such mixtures, the inter-particle spacing of flows with equal masses of the dispersed and carrier phases, ($Y = 0.5$, or $C = 1$), the average inter-particle distance is on the order of 10 diameters. In this case, individual particles or droplets may be treated as isolated with negligible influence by their neighbors. It must be noted, however, that in most engineering applications, the mass ratio of solid particles is large, and the average inter-particle distance is significantly less than 10 diameters. For liquid-particle (slurry) flows, the ratio of the densities is on the order of 1. In such flows, the average inter-particle distance is low, and the particles should not be treated as isolated from their neighbors, as the following example shows.

Example: A slurry composed of sand and water has mass fraction $Y = 0.3$. Determine the particle volume fraction and the inter-particle spacing.

Solution: The density ratio of sand to water is $1000/2700 = 1/2.7$. For $Y = 0.3$, $C = 0.43$, and Eq. (2.23) yields $l = 1.56D$. This is a very short inter-particle distance for the particles to be considered as isolated. It must also be noted that this value must be regarded as an estimate since the particles in the slurry are not arranged in a perfect cubic lattice configuration, which is an implicit assumption in Eq. (2.23).

2.4 RESPONSE TIMES, THE STOKES NUMBER, COLLISIONS

The response time of particles and droplets to the carrier fluid velocity and temperature variations are important in establishing non-dimensional parameters to characterize the flow and heat transfer processes. The momentum response time, τ_v, pertains to the time required for a particle or droplet to respond to changes in flow velocity and is derived from the equation of motion of a sphere in a fluid under Stokesian flow conditions (see Chapter 4)

$$m\frac{dv}{dt} = 3\pi D\mu_c (u-v) \Rightarrow \frac{\pi D^3 \rho_d}{6}\frac{dv}{dt} = 3\pi D\mu_c (u-v), \tag{2.24}$$

where v is the particle velocity, u is the gas velocity, and μ_c is the viscosity of the continuous phase. Rearranging the terms of the last equation, one obtains

$$\frac{D^2 \rho_d}{18\mu_c}\frac{dv}{dt} = (u-v) \Rightarrow \tau_v \frac{dv}{dt} = (u-v). \tag{2.25}$$

The solution to this differential equation for constant u and an initial particle velocity of zero is

$$v = u\left(1 - e^{-t/\tau_v}\right). \tag{2.26}$$

The timescale, τ_v, which is equal to $D^2 \rho_d / 18\mu_c$, is referred to as the *characteristic time*, the *momentum (or velocity) response time*, or simply *the response time* of the particles or drops and is based on the fluid viscosity. Essentially this is the time required for a particle released from rest to accelerate to 63% of the free stream velocity in the Stokesian flow regime. An extension of response times to non-Stokesian flows is typically achieved using a pertinent expression of the drag coefficient from section 4.1.2. For example, the use of the empirical drag force factor, f, of Eq. (4.26) in Eqs. (2.24) and (2.25) yields the following expressions

$$m\frac{dv}{dt} = 3\pi D\mu_c \left(u - v\right)\left(1 + 0.15 + \mathrm{Re}_r^{0.687}\right)$$

$$\Rightarrow \frac{\pi D^3 \rho_d}{6}\frac{dv}{dt} = 3\pi D\mu_c \left(u - v\right)\left(1 + 0.15 + \mathrm{Re}_r^{0.687}\right), \tag{2.27}$$

and the pertinent response time becomes

$$\tau_v = \frac{D^2 \rho_d}{\left(18\mu_c \left(1 + 0.15 + \mathrm{Re}_r^{0.687}\right)\right)} = \frac{D^2 \rho_d}{18\mu_c f}. \tag{2.28}$$

In the case of non-Stokesian flows, and because the pertinent drag is non-linear with respect to the velocity, the corresponding response time does not yield the evolution of the particle velocity, as in Eq. (2.26).

The particle rotational response time is similarly defined for rotating particles. The rotational equation of particle motion may be written as follows in Stokesian flow

$$\frac{d\vec{\omega}_p}{dt} = -\frac{60\mu_c}{D^2 \rho_d}\vec{\Omega} = -\frac{1}{\tau_R}\vec{\Omega} \tag{2.29}$$

where $\vec{\omega}_p$ and $\vec{\Omega}$ denote the rotational velocities of the particle and the fluid respectively. It is apparent from a comparison of Eqs. (2.28) and (2.29) that the ratio of the rotational to the translational response time in the Stokes regime is 18/60 or 3/10. As with the response time for the translational velocity, the extension of τ_R to arbitrary particle Reynolds numbers may be done by introducing an empirical relationship for the rotational drag coefficient. More details on this subject may be found in section 4.6.

Similarly, the *thermal response time* pertains to the response of particles and droplets to temperature changes and is derived from the thermal balance equation during heat conduction

$$mc_d \frac{dT_d}{dt} = 2\pi Dk_c \left(T_c - T_d\right) \Rightarrow \frac{\rho_d c_d D^2}{12 k_c}\frac{dT_d}{dt} = \left(T_c - T_d\right), \tag{2.30}$$

where c_d is the specific heat capacity of the spherical droplet or particle, and k_c is the thermal conductivity of the carrier fluid. It is apparent that the factor, $\rho_d c_d D^2 / 12 k_c$, has the units of time. This factor is called the *thermal response time* and is denoted by the symbol τ_T. As with the momentum response time, it represents the time for the particles and drops to reach 63% of the temperature difference with the carrier fluid during heating or cooling by conduction. It is apparent from the last two equations that the momentum and thermal response times are very sensitive to the sizes of particles and drops and that they diminish for very small diameters. This implies that the smaller particles and droplets respond faster to the carrier fluid fluctuations of velocity and temperature.

It is easy to prove that the ratio of the thermal to the momentum response times is proportional to the Prandtl number of the carrier fluid, Pr, and the ratio of the two specific heat capacities

$$\frac{\tau_T}{\tau_v} = \frac{3}{2} Pr \frac{c_d}{c_c}. \tag{2.31}$$

2.4.1 THE STOKES NUMBER

An important characteristic number of fluid-particle flows is the *Stokes number*, which is defined as the ratio of the momentum response time of particles (or drops) and the characteristic time of the flow

$$St = \frac{\tau_v}{\tau_f}. \tag{2.32}$$

The characteristic time of the flow is typically defined as the ratio of a characteristic velocity, U, and a characteristic dimension, L, of the flow domain. For example, the characteristic dimension for the flow through a venturi is the diameter at the throat of the venturi, D_T, and the characteristic velocity is the average velocity, U. Therefore, the Stokes number of particles or drops with diameter, D, in venturi flow is

$$St = \frac{\tau_v U}{D_T} = \frac{D^2 \rho_d U}{18 \mu_c D_T}. \tag{2.33}$$

If $St \ll 1$, the response time of particles is much less than that of the fluid, and the particles have sufficient time to respond to flow velocity changes and fluctuations. As a result, the particles' and the fluid velocities will be nearly equal (velocity equilibrium), and the particles will follow the flow. On the other hand, if $St \gg 1$, the particles do not have sufficient time to respond to the fluid velocity fluctuations and changes. In this case, the fluid and particles' velocities differ significantly, and the particles do not follow the flow.

The Stokes number establishes an approximate relationship for the particle-to-fluid velocity under the "constant lag" assumption. According to this, the velocity ratio of the two phases is expressed as $\phi = v/u$ and is assumed to vary much more slowly than the fluid velocity, that is, $d\phi/dt \ll du/dt$. The substitution of the particle velocity, v, in terms of ϕ in the equation of motion, Eq. (2.25), yields

$$\phi \frac{du}{dt} = \frac{u}{\tau_v}(1-\phi). \tag{2.34}$$

Since the carrier fluid acceleration may be approximated by the ratio u/τ_f, one obtains the following algebraic equation for the relationship between the (constant) velocity ratio and the Stokes number

$$\phi\frac{u}{\tau_f} = \frac{u}{\tau_v}\left(1-\phi\right) \Rightarrow \phi = \frac{v}{u} = \frac{1}{1+St}. \tag{2.35}$$

The last equation indicates that when $St \to 0$—a condition satisfied by very small or very light particles—the particle velocity approaches the carrier phase velocity; also, when $St \to \infty$—a condition satisfied by large or very heavy particles—the particle velocity diminishes. In the latter case, the particle velocity is unaffected by the fluid.

2.4.2   DILUTE FLOWS AND DENSE FLOWS

In dilute dispersed phase flows, the dispersed elements (particles or drops) are far apart so that the elements' movement is controlled by the hydrodynamic forces (drag and lift) and not by collisions with other elements. In dense flows, the dispersed elements' movement is primarily determined by collisions or contact forces. The average time between collisions, τ_C, and the momentum timescale, τ_v, of the dispersed elements are two important factors in the determination of whether a multiphase flow is dilute or dense. The flow is considered *dilute* if

$$\frac{\tau_v}{\tau_C} \ll 1. \tag{2.36}$$

This happens because the dispersed elements have sufficient time to respond to the local hydrodynamic forces before the next collision occurs.

On the other hand, when

$$\frac{\tau_v}{\tau_C} \gg 1, \tag{2.37}$$

the dispersed phase does not have enough time to respond to the hydrodynamic forces before the next collision, the collisions dominate the dynamics of the flow, and the flow is defined as *dense*. The flow is considered *intermediate* when the two timescales are of the same order, $\tau_C \sim \tau_v$.

Because of the complexity of dispersed flows, there is not a single and definitive scaling parameter that defines the boundary between dilute and dense flows.

The time between collisions may be estimated from the classic equations for collision frequency that emanate from the kinetic theory of gases: Consider a group of spherical particles, each of diameter, D, as shown in Figure 2.5. One of the particles is moving with a relative velocity, v_r, with respect to the other particles. During a time, δt, this particle will intercept and collide with all the particles in a tube with radius $2D$ and length, $v_r\delta t$. The number of particles in this tube, which is equal to the number of collisions, is

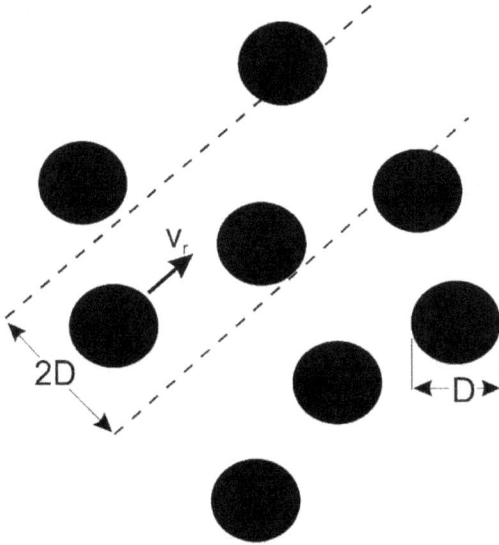

FIGURE 2.5 Collisions between an array of particles and a single particle with relative velocity, v_r.

$$\delta N = n\left(\pi D^2\right)v_r \delta t. \tag{2.38}$$

Therefore, the collision frequency is

$$f_C = n\left(\pi D^2\right)v_r, \tag{2.39}$$

and the average time between collisions (the characteristic time of collisions) is

$$\tau_C = \frac{1}{f_C} = \frac{1}{n\left(\pi D^2\right)v_r}. \tag{2.40}$$

In the various types of dispersed multiphase flows, there are several mechanisms responsible for the finite relative velocity between particles, such as turbulence, shear flow, carrier-flow velocity changes, particle-particle and particle-wall impact, etc. For solid particles in turbulent flows with mean fluctuation velocity, v', Abrahamson (1975) suggests that the particles' collision frequency is

$$f_C = 2\sqrt{\pi}\,nD^2 v'. \tag{2.41}$$

With this collision frequency, the ratio of response times is

$$\frac{\tau_v}{\tau_C} = \tau_v f_C = \frac{4n\sqrt{\pi}\,D^4 \rho_d v'}{18\mu_c}. \tag{2.42}$$

Bearing in mind that the volume fraction of the dispersed phase is $\alpha_d = \pi n D^3/6$, Eq. (2.42) yields an expression for the particle diameter

$$D = \frac{3\sqrt{\pi}\,\mu_c}{4\rho_d v' \alpha_d} \frac{\tau_v}{\tau_C}.$$

(2.43)

If dilute flows are defined by the relationship $\tau_v/\tau_C < 0.1$ and dense flows as $\tau_v/\tau_C > 10$, Figure 2.6 shows the relationship between the diameter of the elements of the dispersed phase and the volume fraction of the dispersed phase for velocity fluctuations equal to 1 m/s. One observes in this case that for dispersed element diameters of 10 µm, the dispersed volume fraction will have to be less than 10^{-4} (0.01%) for the flow to be considered dilute. The volume fraction of the dispersed case would have to be less than 0.1% if the velocity fluctuations were $v' = 0.1$ m/s.

Because there are many mechanisms that are responsible for particle-particle collisions, and the collisions' modelling is approximate, it is difficult to establish exact boundaries for dilute and dense flows. As it becomes apparent from the preceding example, the volume fraction of the dispersed phase provides a useful indication for the dilute and dense flow regimes, as it is graphically depicted in Figure 2.7. For particle volume fractions 0.001 or smaller, the flow may be considered to be dilute. The dense flow region, at the right side of the figure, is separated into two distinct regimes. The *collision-dominated* regime is a flow, where particles predominantly collide and rebound following a different trajectory. The time during contact is small compared with the time between collisions. The *contact-dominated* regime is a flow where particles predominantly are in continuous contact, and the contact forces are principally responsible for the movement of the particles. Examples of

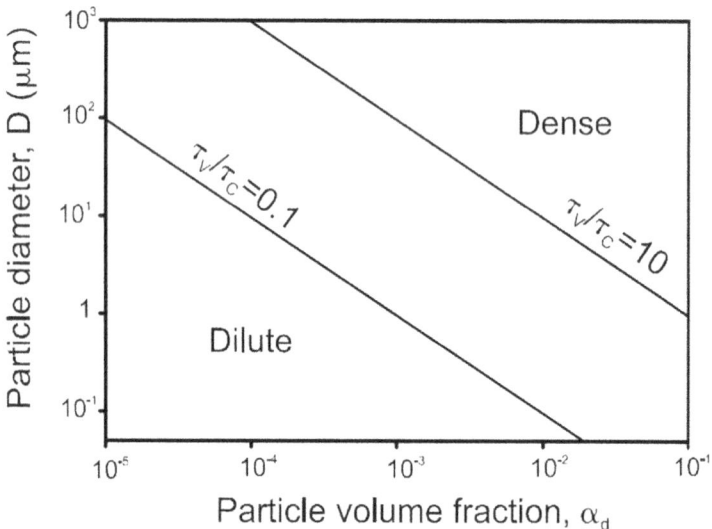

FIGURE 2.6 Domains for dilute and dense flows for glass particles in air with a fluctuating velocity magnitude of 1 m/s.

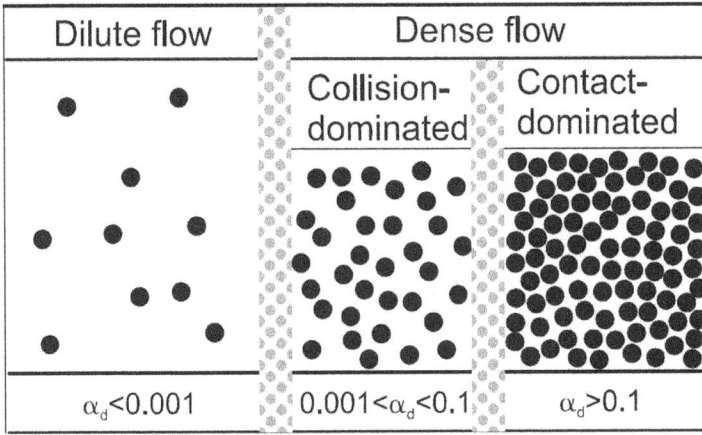

FIGURE 2.7 Flow regimes for dilute and dense flows.

dilute particulate flows are the gas-particle flows in electrostatic precipitators, dilute pneumatic conveying, and cyclone separators. Collision-dominated flows occur in fluidized beds, where the particles are suspended by the fluid and frequently collide with each other as well as with the walls of the vessel. The particle volume fraction for collision-dominated flows is typically in the range 0.001 to 0.1. Examples of contact-dominated flows are gravity flows, such as snow avalanches, mud flows, and dune flows. The particle volume fractions in contact-dominated flows are typically higher than 0.1. It is apparent that since different physical mechanisms are responsible for the particle movement, each one of these particle-flow regimes requires a different modelling approach for the dispersed phase.

2.5 THERMODYNAMIC AND TRANSPORT PROPERTIES

The equilibrium thermodynamic properties of heterogeneous mixtures are important variables that are used in all engineering applications. The thermodynamic properties of such mixtures may be obtained using the properties of the constituent homogeneous materials and the theory of heterogeneous mixtures (Gibbs, 1878).

The specific properties of a homogeneous material—water, nitrogen, air, copper—in thermodynamic equilibrium are functions of any two independent properties, usually the intensive variables pressure, P, and temperature, T, which are typically easiest to measure. The functional relationships between any specific property and the two independent variables are obtained either by explicit equations, for example, $s(T,P)$; or by thermodynamic tables, for example, as steam tables; or by software that calculate numerically the thermodynamic properties from equations of state. The extensive thermodynamic properties of heterogeneous mixtures—volume, enthalpy, entropy, etc.—are equal to the product of the corresponding specific properties and the mass of the heterogeneous mixture (Michaelides, 2014).

The mixing of homogeneous materials to produce a heterogeneous mixture includes a step change in the corresponding extensive property of the mixture. An extensive thermodynamic property of a heterogeneous mixture composed of N constituent materials and denoted by Φ_m,[1] is obtained, in general, as the sum of the corresponding extensive properties of the materials, Φ_i, and a step change of the mixture, $\Delta\Phi_{mix}$

$$\Phi_m = \sum_{i=1}^{N} \Phi_i + \Delta\Phi_{mix}. \tag{2.44}$$

The change due to the mixing, $\Delta\Phi_{mix}$, is associated with physicochemical processes that may occur during the mixing process and is typically a function of the temperature of the mixture, that is: $\Delta\Phi_{mix}(T)$.

Typical particle and droplet multiphase flows are composed of binary mixtures ($N = 2$) of a carrier fluid and the particles or droplets. The step change of the corresponding property in such mixtures is expressed in terms of the mass of the dispersed phase, m_d. Thus, the explicit equations for the volume of the mixture, V_m; the internal energy, U_m; the enthalpy, H_m; and the entropy, S_m, of the heterogeneous mixture are written as follows (Michaelides, 2021)

$$V_m = m_c v_c + m_d v_d + m_d \Delta v_{mix}, \tag{2.45}$$

$$U_m = m_c u_c + m_d u_d + m_d \Delta u_{mix}, \tag{2.46}$$

$$H_m = m_c h_c + m_d h_d + m_d \Delta h_{mix}, \tag{2.47}$$

and

$$S_m = m_c s_c + m_d s_d + m_d \Delta s_{mix}, \tag{2.48}$$

Δv_{mix}, Δu_{mix}, and Δh_{mix}, are related to the change of volume and the release or absorption of energy during the mixing process. Energy release of absorption may occur as the result of ionization processes or chemical reactions during the mixing process and the formation of new molecules or complex ions that are typically different from those of the original materials. For most particulate and droplet systems, the mixed materials are inert, and the quantities Δv_{mix}, Δu_{mix}, and Δh_{mix} vanish. The entropy of mixing, Δs_{mix}, is not equal to zero because all mixing processes are irreversible and occur with an entropy increase. However, in most common particulate and droplet mixtures with inert materials (where no chemical reactions occur upon mixing), the entropy of mixing is significantly less than the entropy of the constituent materials and may be neglected. Hence, the corresponding specific properties (extensive properties divided by the total mas of the mixture) may be written in terms of the mass fraction, Y, and the specific properties of the constituent materials

$$v_m = (1 - Y)v_c + Y v_d, \tag{2.49}$$

$$u_m = (1 - Y)u_c + Y u_d, \tag{2.50}$$

$$h_m = (1-Y)h_c + Yh_d, \qquad (2.51)$$

and

$$s_m = (1-Y)s_c + Ys_d. \qquad (2.52)$$

The specific heat capacities of the mixture at constant volume and at constant pressure may be similarly expressed as

$$c_{vm} = \left(\frac{\partial u_m}{\partial T}\right)_v = (1-Y)c_{vc} + Yc_{vd} \qquad (2.53)$$

and

$$c_{Pm} = \left(\frac{\partial h_m}{\partial T}\right)_P = (1-Y)c_{Pc} + Yc_{Pd}. \qquad (2.54)$$

The last two properties are useful in thermodynamic calculations because they enable us to express the differences of enthalpies and internal energies in terms of temperatures that are easily measured.

Experimental evidence has shown that the transport properties of a heterogeneous mixture depend on the volumetric concentration of the dispersed phase as well as on the shape and orientation of the particles that constitute the dispersed phase, as illustrated in Figures 2.8a and 2.8b. Figure 2.8a depicts the flow of a heterogeneous mixture composed of ellipsoidal particles in a narrow tube, for example, a capillary viscometer, driven by the pressure difference, $P_1 - P_2$. In part A of the figure, the particles are aligned at the center of the tube, and only the fluid is in contact with the tube walls. In part B, the ellipsoidal particles are randomly distributed; they collide with the walls and affect the velocity of the carrier fluid close to the walls. As a result, the velocity in part B is lower, $u_A > u_B$, and the viscometer will indicate that the viscosity of the mixture in part B is higher, $\mu_A < \mu_B$. Figure 2.8b similarly depicts the heat transfer process in two heterogeneous particulate mixtures with ellipsoidal particles that have higher thermal conductivity than the carrier fluid. Heat transfer occurs in the transverse direction because of the temperature difference, $T_1 > T_2$. Because the particles have higher thermal conductivity and are aligned in all directions, there is higher rate of heat transfer in part B, indicating that the thermal conductivity of the mixture in part B is higher, $k_A < k_B$ (Michaelides, 2014).

When the particles are spherical and uniformly distributed in the carrier fluid domain, the transport properties depend on the volume fraction of the dispersed phase alone. Analytical expressions for the transport properties of such fluid-particle systems have been derived as functions of the volume fraction. Einstein (1906, corrigendum, 1911) derived the following expression for the effective viscosity of a fluid-solid mixture, composed of spherical particles, in terms of the carrier fluid viscosity

$$\mu_m = \mu_c(1 + 2.5\alpha_d). \qquad (2.55)$$

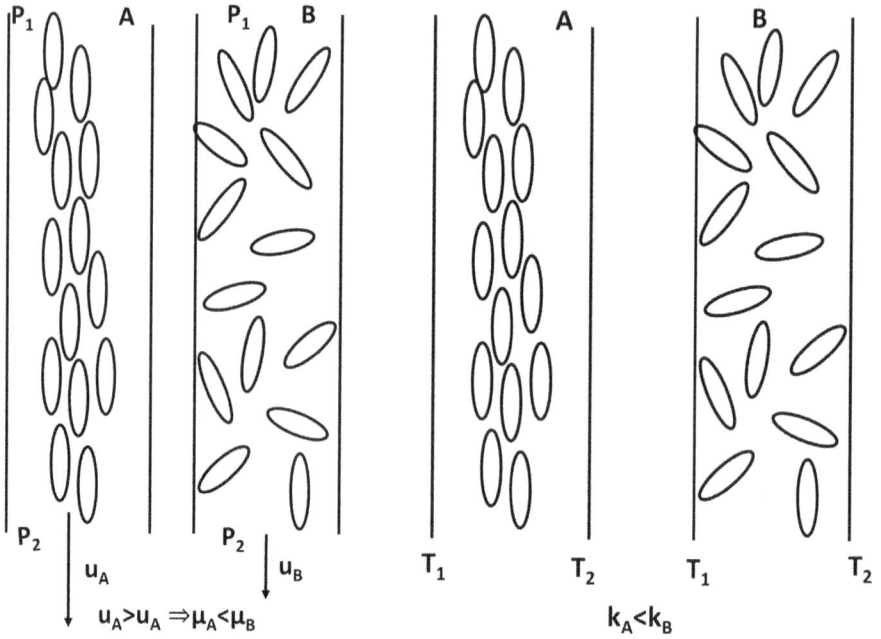

FIGURES 2.8 Effect of particle position, shape, and orientation on the viscosity and thermal conductivity of a fluid-solids heterogeneous mixture.

The expression by Maxwell (1881), which was derived for the electrical conductivity of mixtures with uniformly distributed spherical particles, is also applicable for the analogous thermal conductivity of a fluid-solids mixture

$$k_m = k_c \left[1 + \frac{3(k_d - k_c)\alpha_d}{(k_d + 2k_c) - (k_d - k_c)\alpha_d} \right]. \tag{2.56}$$

In the case of particles with much higher conductivity than the fluid, $k_d \gg k_c$, this equation is approximated as

$$k_m = k_c \left(1 + \frac{3\alpha_d}{1 - \alpha_d} \right) \approx k_c (1 + 3\alpha_d). \tag{2.57}$$

For the diffusivity of heterogeneous mixtures composed of uniformly distributed spherical particles, Cussler (2009) determined analytically that the effective diffusivity of the mixture is

$$\mathcal{D}_m = \mathcal{D}_c \frac{1 - \alpha_d}{1 + 0.5\alpha_d}. \tag{2.58}$$

It must be noted that the last four equations for the transport properties of heterogeneous mixtures apply strictly to dilute mixtures with uniformly distributed spherical particles. Analytical expressions that pertain to different shapes of particles and second-order corrections of these equations are important in nanofluidic technology and abound in the scientific literature (Michaelides, 2014).

2.6 PHASE INTERACTIONS—COUPLING

An important subject in the analysis of all multiphase flows is the way distinct phases interact and how these interactions affect the exchange of mass, momentum, and energy among the constituent phases. This is called the *coupling* among the phases. For two phases in the multiphase mixture—for example, solid particles in a gas—if one phase affects the other phase, while there is no significant reverse effect on the first phase, the mixture is said to be *one-way coupled*. If the interactions result in mutual effects between the flows of both phases, then the flow is *two-way coupled*.

A schematic diagram of gas-solids coupling is shown in Figure 2.9: The state of the carrier phase is described by the density, temperature, pressure, velocity field, and the concentrations of the gaseous species in the carrier gas. The particle phase is described by the volumetric concentration (volume fraction), the temperature, the velocity field, the size, and the loading of the particles. Interactions between the phases (coupling) may take place through mass, momentum, and energy transfer between the phases. Mass coupling occurs with the transfer of mass from one phase to another through evaporation, sublimation, or condensation. Momentum coupling happens by two mechanisms: (a) the action of the drag forces on the dispersed and continuous phase; and (b) momentum exchange due to the mass transfer. Energy coupling occurs also by two mechanisms, the heat transfer between the phases and the energy (both thermal and kinetic) transferred as a result of the mass transfer.

It is apparent that analyses of multiphase systems based on one-way coupling are simpler. They require lesser sophistication, less parameters to be defined, and lesser time for the solution. For this reason, it is important to estimate—using experience or order of magnitude—if two-way coupling is necessary or one-way coupling would be sufficient. The following example illustrates this process: Assume that hot particles were injected into a cool gas flowing in a horizontal channel, as shown in

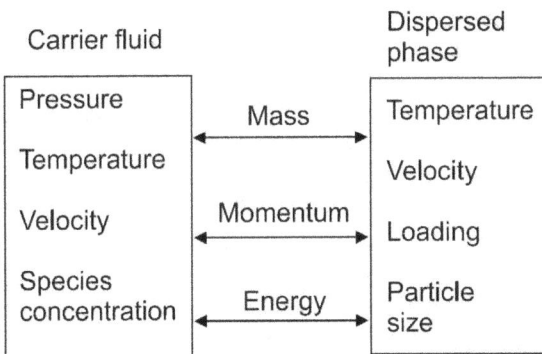

FIGURE 2.9 Schematic diagram of phase interactions and coupling effects.

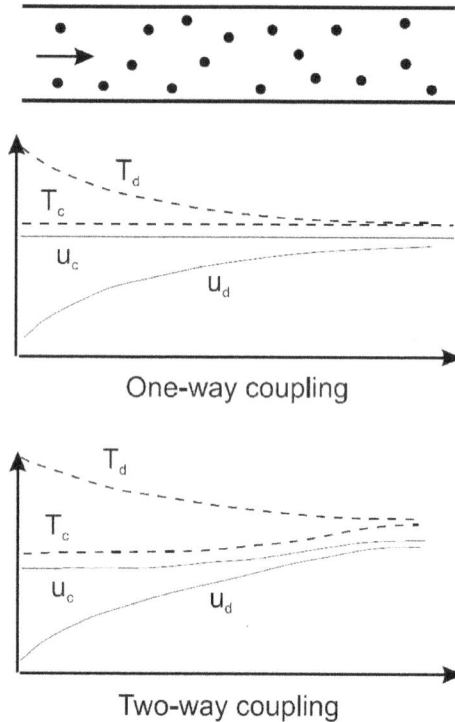

FIGURE 2.10   Illustration of one- and two-way coupling for the cooling of solid particles in a channel.

Figure 2.10. The volumetric fraction of the solids, α_d, is very low, and the fraction of the carrier gas, α_c, is near one. One may calculate the trajectory and the corresponding temperature evolution of the particles using the local velocity and temperature of the gas (Michaelides and Lasek, 1987). The particle temperature continuously decreases and, if the channel is long enough, it will asymptotically approach the temperature of the gas. Simultaneously, the injected particle velocity will asymptotically approach the carrier gas velocity. One-way coupling implies that the particles do not affect the gas flow velocity and temperature. Hence, the gas velocity and temperature would remain constant during this process.

Two-way coupling takes into account the effect the hot injected particles have on the gas. Because of heat and momentum exchanges with the particles, the gas temperature increases, the carrier gas density decreases, and the gas velocity increases to accommodate these changes as well as the momentum exchange with the particles. As a result, the particles would be accelerated to a higher velocity, and the temperature difference between the particles and the gas would decrease. The heat transfer (rate of cooling) adjusts accordingly to accommodate the local temperature difference and the local velocity difference between the phases. It is apparent that the analysis with two-way coupling would require the knowledge of more parameters and entails significantly more work than the simpler one-way coupling. The coupling parameters provide an indication of the importance coupling effects and guidance on the type of coupling needed in the various multiphase flow processes (Crowe, 1991).

2.6.1 MASS COUPLING

We consider a gas-droplet system where n droplets per unit volume are present in a cubic domain with side L, as shown in Figure 2.11. Each droplet is evaporating at a rate $\dot{m}$, and therefore, the mass of the gaseous phase increases by a rate

$$\dot{M}_{ev} = \dot{m} n L^3. \tag{2.59}$$

The mass of the droplets decreases by the same amount. When the flow of the gas-droplet system is considered, the mass flux of the gaseous phase through this volume is

$$\dot{M}_c = u \bar{\rho}_c L^2. \tag{2.60}$$

In this case, a *mass coupling parameter* is defined as

$$\Pi_{mass} = \frac{\dot{M}_{ev}}{\dot{M}_c} = \frac{\dot{m} n L^3}{u \bar{\rho}_c L^2} = \frac{L \dot{m}}{u m} \frac{\bar{\rho}_d}{\bar{\rho}_c}. \tag{2.61}$$

If $\Pi_{mass} \ll 1$, then the effect of mass addition to the continuous phase would be insignificant, and the mass coupling could be neglected.

The ratio $\dot{m}/m$ has the dimensions of inverse time and is sometimes denoted as $(\tau_m)^{-1}$. Since the last ratio in Eq. (2.61) represents the mass ratio of the dispersed phase, C, the mass coupling parameter may be written as

$$\Pi_{mass} = C \frac{L}{u \tau_m}. \tag{2.62}$$

Bearing in mind that Π_{mass} is a measure of the instantaneous evaporation, a second parameter that must be examined in this case is the loading, Z, of the mixture, the

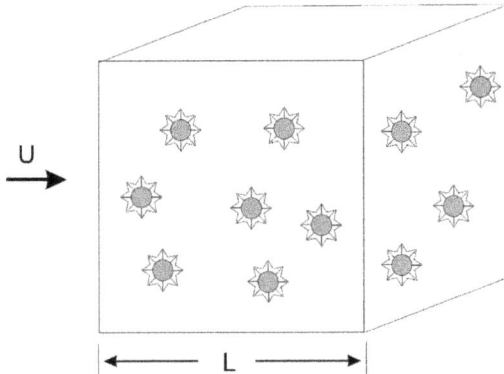

FIGURE 2.11 Droplets evaporating in a cube of side L.

ratio of the mass of the dispersed phase to the mass of the carrier fluid. If $Z\sim1$ and the evaporation is allowed to occur for a long time, all the droplets will evaporate, the total mass of the carrier phase would significantly increase, and the carrier fluid velocity would increase commensurably. Actually, in cases where $\rho_d \gg \rho_c$, for example, with water droplet in air or steam, even a very low value of Z may cause significant velocity effects if total evaporation is allowed, as the following example demonstrates.

Example: A mixture of liquid water droplets and steam flows in a circular pipe of 5 cm diameter, D_{pipe}, with common average velocity, $u = 7$ m/s. At the entrance of the pipe, the volume fraction of the carrier phase (the steam) is $\alpha_c = 99\%$. The mixture is at the thermodynamic saturated conditions of P = 15.54 bar and T = 200°C. Heat transfer causes the evaporation of the droplets at a rate $6*10^{-3}$ kg/s. Determine whether or not the mixture may be modelled with one- or two-way coupling in the two cases: (a) the length of the pipe is 1 m; (b) the length of the pipe is 50 m.

Solution: From thermodynamic tables, we obtain the densities of the steam and the water droplets at 200°C: $\rho_c = 12.7$ kg/m³ and $\rho_d = 639$ kg/m³. The mass flow rate of the steam at the entrance of the pipe is $\pi(D_{pipe}/2)^2\rho_c u \alpha_c = 0.1727$ kg/s and that of the droplets: $\pi(D_{pipe}/2)^2\rho_d u(1-\alpha_c) = 0.0878$ kg/s. Hence, $Z = 0.51$ and the mass coupling parameter is $\Pi_{mass} = 0.034 \ll 1$.

   When the pipe length is short, 1 m, it takes 0.143 s for the mixture to exit. At the exit of the pipe, the evaporation has increased the mass flow rate of the carrier phase by $0.143*6*10^{-5}$ kg/s $= 8.571*10^{-4}$ kg/s. This increase is insignificant (less than 5%) in comparison to the mass flow rate of the carrier fluid ($172.7*10^{-4}$ kg/s), and its effect on the velocity would be insignificant too. In this case, the mixture may be modelled with one-way coupling.
   With the longer pipe length of 50 m, the time spent in the pipe is $50/7 = 7.14$ s, and the additional mass flow rate of the carrier fluid at the exit of the pipe is: $7.14*6*10^{-3}$ $= 0.0428$ kg/s. This addition increases the mass flow rate of the carrier phase from 0.1727 kg/s to 0.2155 kg/s. In this case $\Pi_{mass} = 0.027$, and still $\Pi_{mass} \ll 1$. However, the evaporation has caused a significant increase of the carrier mass flow rate, which causes the velocity at the exit to increase to 8.73 m/s. In this case, the mixture must be modelled with two-way coupling.
   Note that, if the pipe is long enough and the entire mass of droplets evaporates (as it happens in boilers and steam generators), the mass flow rate of the steam would increase to 0.2605 kg/s, and the velocity of the carrier phase in the pipe would increase to 10.6 m/s.
   It becomes apparent from this example that whether or not a process may be modelled with one-way coupling depends on the parameter Π_{mass} as well as the other details of the process.

2.6.2 MOMENTUM COUPLING

The drag forces determine the relative velocity between the fluid and the dispersed phase. The carrier fluid exerts a drag force on the elements of the dispersed phase, and the dispersed phase exerts an equal and opposite force on the fluid. One-way momentum coupling considers this interaction on the disperse phase alone. Accordingly, it is only the velocity of the dispersed phase that is affected. When the drag on the carrier fluid is significant, two-way momentum coupling must be applied. The importance of momentum coupling is assessed by comparing the drag force exerted by the dispersed phase with the momentum flux of the carrier phase. Thus, a *momentum coupling parameter* is defined as the ratio of the drag on the particles, which is equal in magnitude to the force exerted on the fluid, and the momentum of the carrier fluid. Referring to the particle-fluid mixture depicted in Figure 2.11 and assuming Stokesian flow conditions, the drag force on the particles is

$$F_D = 3\pi n L^3 \mu_c D (u - v). \tag{2.63}$$

The momentum flux of the fluid is

$$\dot{M}_{mom} = \alpha_c \rho_c L^2 u^2. \tag{2.64}$$

Hence, the momentum coupling parameter may be expressed as

$$\Pi_{mom} = \frac{3\pi n L \mu_c D}{\alpha_c \rho_c u} \left(1 - \frac{v}{u}\right). \tag{2.65}$$

Substituting the number density of particles for the density of the dispersed phase, we obtain

$$\Pi_{mom} = \frac{\bar{\rho}_d}{\bar{\rho}_c} \left(\frac{18\mu_c}{D^2 \rho_d}\right) \frac{L}{u} \left(1 - \frac{v}{u}\right) = C \frac{L}{\tau_v u} \left(1 - \frac{v}{u}\right). \tag{2.66}$$

Where C is the dispersed phase mass ratio, defined in Eq. (2.16). When the particles are very small, τ_v approaches zero; the velocity ratio of the two phases, v/u, approaches 1; and the last term in parenthesis approaches zero too. In this case, the momentum coupling parameter approaches asymptotically the value

$$\Pi_{mom} = \frac{C}{1 + \left(\dfrac{\tau_v u}{L}\right)}. \tag{2.67}$$

Sometimes the term in parenthesis in the denominator of the last expression is called the *Stokes number for momentum transfer*.

Example: Coal particles, 100 μm in equivalent diameter and density 1,200 kg/m³, flow in an air stream into a venturi section with a 2 cm diameter throat. The velocity at the throat is 10 m/s, the mass ratio is 1, and the air viscosity is 1.8×10^{-5} Ns/m². Evaluate the magnitude of the momentum coupling parameter.

Solution: The velocity response time of the coal particles is $\tau_v = 0.037$ s. Using the throat diameter as the characteristic distance, L, the Stokes number for momentum transfer is calculated to be 18.5. This yields the value 0.051 for the momentum coupling parameter, which implies that the momentum coupling effects are not important.

2.6.3 ENERGY COUPLING

When the temperatures of the two phases are different, there is heat transfer between the dispersed phase and the carrier fluid. The significance of this energy coupling is assessed by comparing the rate of heat transfer between the phases and the rate of energy transported by the carrier phase. Referring to the schematic of Figure 2.11, the rate of heat transfer by conduction associated with the dispersed phase elements in the volume, L^3, is

$$\dot{Q}_d = 2\pi n L^3 k_c D (T_d - T_c). \tag{2.68}$$

The rate of thermal energy associated with the transport of the continuous phase is

$$\dot{E}_c = \alpha_c \rho_c u L^2 c_d T_c, \tag{2.69}$$

where c_d is the specific heat of the carrier fluid at constant pressure. Therefore, the *energy or thermal coupling parameter* may be expressed as

$$\Pi_{en} = \frac{2\pi n L k_c D (T_d - T_c)}{\alpha_c \rho_c c_d u T_c} = \frac{\bar{\rho}_d}{\bar{\rho}_c} \left(\frac{12 k_c}{c_d D^2 \rho_d} \frac{L}{u} \right) \left(\frac{T_d - T_c}{T_c} \right). \tag{2.70}$$

One easily recognizes that the first ratio in the first parenthesis of Eq. (2.70) is the inverse of the thermal response time, τ_T, of the dispersed phase. The inverse of the entire term in the parenthesis, $(u^*\tau_T)/L$, is a dimensionless number that is sometimes referred to as the *thermal Stokes number*. As with the momentum transfer, when the elements of the dispersed phase are very small, the thermal Stokes number vanishes, and the temperature of the dispersed phase, T_d, approaches the temperature of the carrier fluid, T_c. In this case, the asymptotic value of the fraction that defines the thermal coupling parameter becomes

$$\Pi_{en} = \frac{C}{1 + \left(\dfrac{u\tau_T}{L} \right)}, \tag{2.71}$$

where C is the mass ratio defined in Eq. (2.16). For dispersed flows with gaseous carriers, the momentum and thermal response times are of the same order of magnitude and

$$\Pi_{mom} \sim \Pi_{en}. \tag{2.72}$$

This suggests that the justification of one-way coupling for the analysis of momentum transfer justifies the one-way coupling for energy transfer. It must be recalled, however, that the thermal energy coupling parameter has been defined in terms of the sensible heat transfer (e.g. conduction) and not in terms of phase change, which is characteristic of evaporation, sublimation, and condensation processes and involves latent heat. When most of the energy transfer in a multiphase system is associated with the latent heat of the dispersed phase, the rate of heat transfer associated with the phase change in the volume, L^3, is

$$\dot{Q}_d = nL^3 \dot{m}_{fg} h_{fg},\tag{2.73}$$

where $\dot{m}_{fg}$ is the rate of the dispersed mass that undergoes phase change, and h_{fg} is the latent heat associate with this phase change (evaporation, sublimation, or condensation). Therefore, the thermal coupling parameter becomes

$$\Pi_{fg} = \frac{nL^3 \dot{m}_{fg} h_{fg}}{\alpha_c \rho_c u L^2 c_d T_c} = C\left(\frac{L}{\tau_m u}\right)\left(\frac{h_{fg}}{c_d T_c}\right) = \Pi_m \left(\frac{h_{fg}}{c_d T_c}\right),\tag{2.74}$$

the product of the mass coupling parameter, and the factor, $h_{fg}/c_p T_c$. Because typically the latent heat per unit mass is significantly higher than the sensible heat per unit mass, this factor is very large, sometimes on the order of 1,000. Consequently, two-way coupling for the modelling of energy processes may be needed in cases when one-way coupling is sufficient for the modelling of mass transfer and the momentum transfer.

SUMMARY

The continuum hypothesis, which is invariably applied to multiphase flow systems, entails the implicit assumption that all the flow variables are defined not in geometric points but at the centers of sampling volumes, which have finite size. The discrete nature of a dispersed phase flow is facilitated by the use of volume-averaged properties such as bulk density and mass-averaged properties, such as enthalpy and entropy. Because of the interactions of the phases, apart from the parameters that characterize the single-phase flows, additional parameters are necessary to characterize and model the multiphase flow processes. Such parameters are the response times, the Stokes numbers, the volumetric ratio, and the loading. Dilute and dense flows refer to the importance of the interactions between particles. Coupling refers to the interactions between phases. One-way coupling implies that the carrier phase affects the dispersed phase, but the dispersed phase has negligible influence on the continuous carrier phase. Two-way coupling implies that there are significant effects on both the carrier and the dispersed phase. The significance of coupling may be quantified by the respective coupling parameters.

NOTE

1 Following the convention of thermodynamics, extensive properties are denoted with capital letters, specific properties with the corresponding low case letters, and intensive properties (only P and T are of interest) with capital letters.

REFERENCES

Abrahamson, J., 1975, Collision rates of small particls in a vigorously turbulent fluid, *Chem. Eng. Sci.*, **30**, 1371–1378.

Crowe, C.T., 1991, The state-of-the-art in the development of numerical models for dispersed phase flows, *Proc. First Intl. Conf. on Multiphase Flows—'91-Tsukuba*, **3**, 49.

Cussler, E.L., 2009, *Diffusion: Mass Transfer in Fluid Systems*, 3rd edition, Cambridge University Press, New York.

Einstein, A., 1906, Eine neue bestimung der molekuldimensionen, *Annalen der Physik*, **19**, 289–306.

Einstein, A., 1911, Berichtigung zu meiner Arbeit: "Eine neue bestimung der molekuldimensionen", *Annalen der Physik*, **34**, 591–592.

Gibbs, J.W., 1878, On the equilibrium of heterogeneous substances, in *The Collective Works of J. Willard Gibbs*, Longmans, New York, 1928.

Maxwell. J.C., 1881, *A Treatise on Electricity and Magnetism*, 2nd edition, Clarendon Press, Oxford.

Michaelides, E.E., 2014, *Nanofluidics: Thermodynamic and Transport Properties*, Springer, Heidelberg.

Michaelides, E.E., 2021, *Exergy and the Conversion of Energy*, Cambridge University Press, Cambridge.

Michaelides, E.E. and Lasek, A., 1987, Fluid-solids flow with thermal and hydrodynamic non-equilibrium, *Int. J. Heat and Mass Transfer*, **30**, 2263.

PROBLEMS

2.1.   A sand-water slurry has a solids volume fraction of 40%. Determine the ratio of the average inter-particle spacing to the particle diameter.

2.2.   The bulk density of water droplets in air at standard conditions (25°C and 1 atm) is 0.5 kg/m^3. The droplets are monodisperse (uniform size) with diameter D = 100 μm. Determine the number density, void fraction, and mass fraction.

2.3.   Particles of uniform size are stacked in a packed bed such that each particle sits in the "pocket" of the four particles adjacent to it. This is a "body-centered" cubic configuration and represents the minimum void fraction. Determine the solids volume fraction.

2.4.   Droplets are released in a hot air stream and evaporate as they are advected with the flow, as shown in the figure in the next section. The mass coupling parameter is small, while the latent heat coupling parameter is large. Sketch the variation of the gas temperature, gas velocity, gas density, and droplet velocity for both one-way and two-way coupling in the duct.

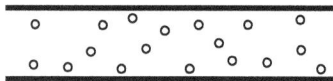

PROBLEM 2.4

2.5.   A pneumatic conveying system consists of particles transported by an air stream in a circular duct with diameter D_p. Ten percent of the particle mass impinges on the wall per diameter of length. The rebound velocity is 1/2 the impingement velocity. The gas velocity is u, and the average particle velocity is v. The loading is Z. The skin friction coefficient is C_f. Derive an expression for the pipe friction factor as a function of C_f, Z, v, and u. Assume a void fraction of approximately one.

2.6.   A gas-particle flow accelerates through a venturi as shown. Sketch the pressure distribution through the venturi for a small momentum and for a large momentum coupling parameter and explain the rationale for your sketches.

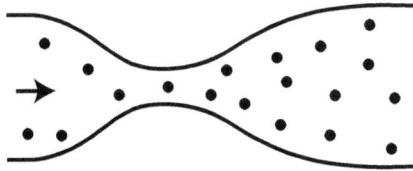

PROBLEM 2.6

2.7.   Water droplets are evaporating in an airflow at standard pressure and temperature. The airflow rate is 1 kg/s, and the water flow rate is 0.01 kg/s. The droplet velocity is 30 m/s, and the droplet diameter is 100 microns. The pipe diameter is 5 cm. The evaporation time for a water droplet is expressed as

$$\tau_e = \frac{D^2}{\lambda}$$

where D is the droplet diameter, and λ is the evaporation constant equal to 0.02 cm²/s for water droplets. Evaluate the mass coupling and the energy coupling parameters. Explain whether or not these processes may be treated with one- or two-way coupling.

2.8.   Ice particles flow in a duct with air at 60°F. The ice particles are not completely melted at the end of the duct. The ice particles and air are in dynamic equilibrium. Sketch how air temperature, particle temperature, gas velocity, and pressure would change along the duct assuming one-way and two-way coupling. Provide physical arguments explaining your results.

3 Distributions and Statistics of Particles and Droplets

Size is a very important characteristic that governs the interactions of particles and drops with the carrier fluid. For this reason, it is useful to determine what is meant by the "size" of particles and drops and to have a basic knowledge of the statistical parameters relating to size distributions. At first, one must recognize that the surface tension of the liquid keeps the shape of small drops spherical. In this case, the diameter of the drops determines their "size." However, the vast majority of solid particles have non-spherical shapes—their shapes are conical, prismatic, cylindrical, polygonal, irregular, etc. One must define what is the best parameter (the equivalent diameter) to quantify the "size" of the particles in the governing equations of mass, momentum, and energy interactions with the carrier fluid.

Experience shows that several multiphase systems are not composed of droplets and particles, where the characterization of these droplets and particles can be merely described by a single equivalent diameter. In *monodisperse* systems, the particles or drops are very close to a single size, while *polydisperse* systems are composed of particles and drops that have a wide range of sizes (equivalent diameters). An approximate definition of a monodisperse distribution is one for which the standard deviation of the equivalent diameters is less than 10% of the mean diameter.

3.1 THE "SIZE" OF PARTICLES

While small drops are spherical and large drops are spheroidal, most solid particles have irregular shapes that may not be described by one or two easily measurable dimensions. The *size* of spherical particles and drops are equal to their diameters. Figure 3.1 depicts particles that have non-spherical, regular, and irregular shapes and for which the choice of the size is not so apparent and straightforward. The "size"—a dominant parameter in all governing equations—of non-spherical particles must be well-defined and not be left to interpretation. Following the practice of spherical particles, for which most of the analytical and experimental work has been performed, an *equivalent diameter* may be defined for non-spherical particles. The practical usefulness of the equivalent diameter concept is that one may correlate the transport coefficients of particles, e.g. drag coefficients and heat/mass transfer coefficients, with the known transport coefficients for spheres. Several equivalent diameters have been proposed for non-spherical particles, including the diameter of a sphere that would have the same volume, V, which is denoted as D_V; the diameter of a sphere that would have the same area, A, denoted as D_A; the diameter of a

DOI: 10.1201/9781003089278-3

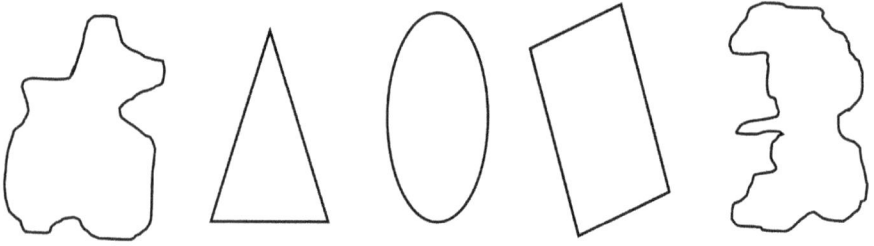

FIGURE 3.1   What is the "size" of these particles?

sphere that would have the same projected area, D_{PA}; and the diameter of a sphere that would have the same perimeter, P, projected in the direction of the motion of the non-spherical particle, denoted as D_P. These three equivalent diameters are defined as follows

$$D_V = \sqrt[3]{\frac{6V}{\pi}}, \quad D_A = \sqrt{\frac{A}{\pi}}, \quad D_{PA} = \sqrt{\frac{4A_P}{\pi}} \quad and \quad D_P = \frac{P}{\pi}. \tag{3.1}$$

For spherical particles and drops, all three equivalent diameters are the same and equal to the actual diameter, D. A fourth equivalent diameter, which is frequently used with irregular particles and aggregates, is the diameter of the minimum sphere, in which the irregular particle will fit in. Typically, this is the longest dimension of the particle, D_L. While D_P depends on the direction of the particle movement and its magnitude may vary in an arbitrary way, for all the other measures of the "size" of particles, the inequality $D_V \leq D_A \leq D_L$ applies, with the equal sign applying to spheres only (Michaelides, 2014). Figure 3.2 shows schematically the last three diameters, the candidates for the quantification of "size," for a particle that appears as an elongated parallelepiped. It is observed in this figure that the three equivalent diameters vary significantly in magnitude. Because of its significant variability for elongated and irregularly shaped particles, a precise definition, or the measurement of the "size" of particles, must include how this "size" has been defined or measured.

Another definition of an equivalent diameter, which is extensively used with coal, sediments, and liquid suspensions of particles, is the "*sieve diameter*." This is obtained from a sieve mesh analysis and is defined as the maximum standard sieve mesh size (or minimum sieve aperture) through which the particles may pass through (Rajapakse, 2016). Since the standard sieves do not extend to the micrometer and nanometer sizes, this method is only applicable to coarser particles.

Regarding shapes, the *Corey shape factor*, which has been widely used in the past for ellipsoidal particles, is defined as the ratio of the shortest principal axis of the particle to the square root of the product of the longest two principal axes. The Corey factor is not related to volume or area calculations, which are important for the determination of the transport coefficients of particles. Also, it is difficult to apply this factor to irregular particles, where the principal axes are not well-defined.

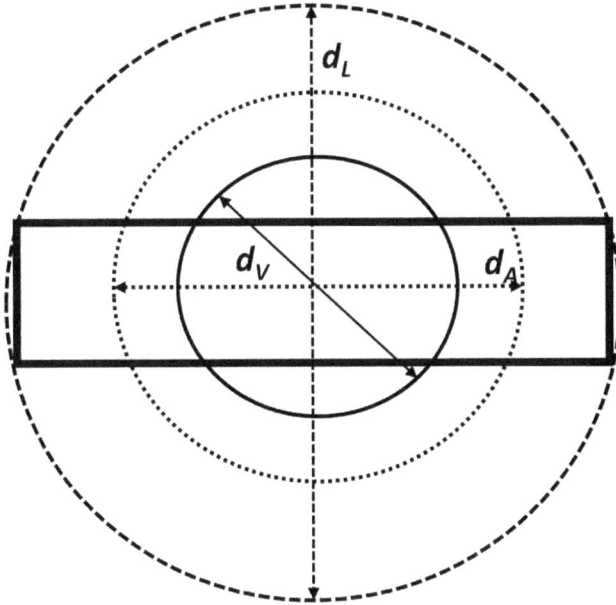

FIGURE 3.2   Volume equivalent diameter, area equivalent diameter, and longest dimension of an elongated parallelepiped.

3.1.1   FRACTAL DIMENSION

Fractal geometry is a recent tool, often used to analyze the structure of irregular patterns of lines, surfaces, and volumes. Fractal shapes are composed of self-similar parts when viewed or measured at different length scales. Photographs and diagrams of fractal objects at different scales show that the shapes of the objects remain the same as the scale of view changes. However, the actual lengths and areas of fractal shapes are different and increase when the scale of measurement decreases. A classic example of a fractal shape is the coastline of a country (Mandelbrot, 1967): One may measure the length of the coastline using a map, using aerial photographs, using a one-meter ruler on the ground, or using a smaller ruler whose length is one grain of sand. The length of the coastline increases as the unit of measurement becomes smaller, more details of the coastline are revealed, and these details are measured and counted. Figure 3.3 illustrates an example of the fractal concept using the so-called *Koch curve*. The curve is a complex, self-similar shape that evolves from a single straight segment according to the following rule: at each stage of evolution, every straight segment of the curve is substituted by four other straight segments of length 1/3 the length of the old segment. Two of the new segments span the ends of the old segment. The other two segments, which occupy 1/3 of the old segment's length, form the sides of an equilateral triangle and make up the inner part of the new segment. The process of evolution of the curve may produce an infinite number of stages, three of which are shown in Figure 3.3. The end points of all the stages, *A* and *B*, are at the same distance apart in all the stages of the pattern evolution.

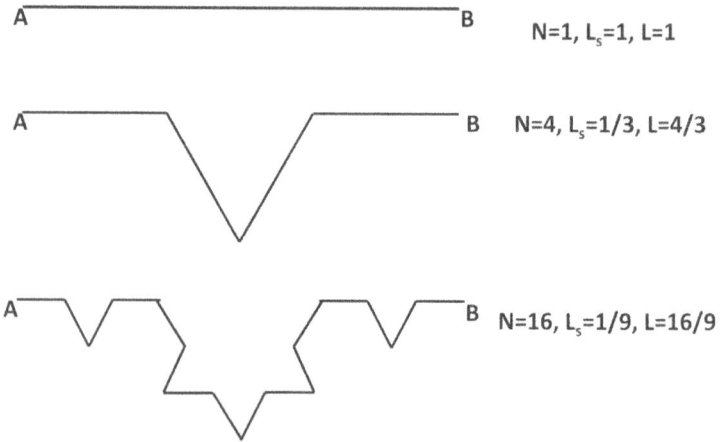

FIGURE 3.3   The Koch curve, a self-similar, fractal shape.

Without loss of generality, we may stipulate that the segment AB is of unit length. The measure of the length in the first stage of the curve is, $N = 1$. In the second stage, the length scale is $L_s = 1/3$, and the number of segments, $N = 4$. In the third stage, the length scale is $L_s = 1/9$, and the number of segments, $N = 16$. In terms of the original units, the total length of the curve, L, is $4/3$ units in the second stage, $16/9$ in the third stage, and so on. It quickly becomes apparent that the total length increases at every stage of the evolution.

For all self-similar objects, such as the Koch curve, there is a fundamental dimension, the *fractal dimension*, L_f, which does not change with the scale of measurement. This fundamental dimension is related to the fundamental equation of self-similar patterns by the relationship

$$N = \left(L_s \right)^{-L_f},$$ (3.2)

where N is the number of straight-line segments of the scale length, L_s, that span the entire length of the curve. From the definition of the fractal shapes, N is a strong function of the scale length. The value of L_f for a straight line is 1 and for a sphere 3. All other shapes have fractal dimensions between 1 and 3. For the Koch curve in Figure 3.3, $L_f = 1.2619$.

It must be noted that the fractal dimension describes shapes and surfaces in very general terms. It does not define shapes and patterns and does not contain enough information to construct the shapes. In the case of the Koch curve, there are several other patterns that have the same fractal dimension but are dramatically different in shape and complexity.

Several types of particles have irregular but not self-similar shapes. Because of this, the fractal dimension, L_f, of single particles is meaningless. Aggregates of primary particles, and especially very large aggregates of spheres, have been observed to have complex shapes, surfaces, and perimeters, which are often approximated as self-similar surfaces and lengths. In these cases, a fractal dimension of the aggregates, L_f, may be defined using the limit of Eq. (3.2) as the length scale, L_s, vanishes

$$L_f = -Lim\frac{\ln(N)}{\ln(L_s)} \quad as \quad L_s \to 0. \tag{3.3}$$

For aggregates of particles, L_f is typically calculated by taking microscopic images of the aggregates at different scales and using Eq. (3.3) with the measuring length scale, L_s, taking several different values. Optical image software may perform this task automatically (Wang and Chau, 2009). A simpler, albeit not as accurate method, is to measure the fractal dimension from light scattering (Bushell *et al.*, 2002).

The fractal dimension of the aggregates is sometimes used to give an estimate of the porosity, ε_p, of an aggregate of several smaller (primary) particles (Li and Logan, 2001)

$$\varepsilon_p = 1 - \left(\frac{D_a}{D}\right)^{3-L_f}, \tag{3.4}$$

where D is the diameter of the primary particles. and D_a is the "diameter" of the aggregate. The latter is typically assumed to be equal to the longest dimension of the aggregate, that is: $D_a = D_L$.

Since the 1990s several studies have used the fractal dimension to describe the structure and morphology of single particles and aggregates of particles. It is apparent from the previous example that two conditions must be explicitly satisfied when one uses the fractal dimension in a quantitative manner:

a) The original object must have a self-similar shape to make the fractal dimension, L_f, meaningful.
b) Since all particles are three dimensional, the two-dimensional microscopic images or the light scattering method must be adequate for the determination of the fractal dimension.

Because the second condition is not always satisfied, it has been observed that measurements of L_f by two-dimensional photographic images are laden with high measurement error, which becomes significantly high if the "measured" fractal dimension is close to or greater than 2 (Vicsek, 1999).

3.2   DISCRETE SIZE DISTRIBUTIONS

3.2.1   FREQUENCY DISTRIBUTION

Polydisperse multiphase flow mixtures are composed of drops and particles with a large range of sizes, which are typically characterized by an equivalent diameter. For any calculations on the interactions between the carrier fluid and the dispersed phase(s) it is necessary to have a knowledge of the range and distribution of the sizes of the particles. Typically, this is accomplished by determining the size distribution of a sample of particles using a measuring technique, such as photography, light scattering, etc. The measured sample must be large enough to ensure that a sufficiently large number of particles will populate every size interval, ΔD_i, of the proposed distribution. The size of the interval, ΔD_i, is decided by the detail needed for the characterization of the multiphase mixture or for the calculations and does not have to be the same for all intervals. Subsequently, the number of particles in each size interval are counted, recorded, and divided by the total number of particles in the sample. The results may be plotted in the form of a histogram (bar chart), as shown in part (a) of Figure 3.4. This is identified as the *discrete number frequency distribution* for the particle size, which is also referred to as the *probability density function*, or *pdf*, for the particle size. The ordinate corresponding to each size interval is known as the *number frequency*, $\hat{f}_n(D_i)$ of the particles. The sum of each bar and of all numbers $\hat{f}_n(D_i)$ is 1, because the number in each size category has been "normalized" by dividing each number with the total number of particles in the sample

$$\sum_{i=1}^{N} \hat{f}_n(D_i) = 1,$$ (3.5)

where N is the total number of intervals that span the entire range of diameters. The representative size for the interval, D_i, is typically the diameter corresponding to the midpoint of the interval, ΔD_i. The number-averaged particle diameter (or mean diameter) of this distribution is defined as

$$\bar{D}_n = \sum_{i=1}^{N} D_i \hat{f}_n(D_i),$$ (3.6)

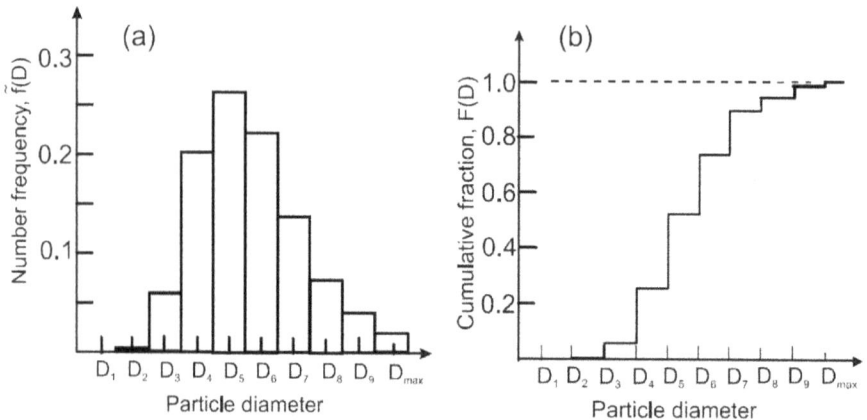

FIGURE 3.4 Particle size distribution: (a) Discrete number frequency distribution and (b) discrete cumulative number distribution.

and the number variance is

$$\sigma_n^2 = \frac{N}{N-1} \sum_{i=1}^{N} \left(D_i - \bar{D}_n\right)^2 \hat{f}_n(D_i) = \frac{N}{N-1} \sum_{i=1}^{N} D_i^2 \hat{f}_n(D_i) - \bar{D}_n^2. \quad (3.7)$$

The standard deviation, σ_n, is the square root of the variance. The ratio of the standard deviation to the average diameter is used as a measure of the *spread* of the distribution, with narrow distributions having low ratios. A mixture is, in general, characterized as monodisperse when $\sigma_n/\bar{D}_n < 0.1$.

Two other methods are used to describe particle distributions using the mass and the volume instead of the number of particles in each interval. For the mass distribution, the mass of each particle is measured or calculated from the size measurements, and the fraction of mass associated with each size interval is used to construct the distribution. This is known as the discrete mass-frequency distribution, $\hat{f}_m$. Thus, one may calculate the mass-averaged particle diameter and mass-averaged variance as follows

$$\bar{D}_m = \sum_{i=1}^{N} D_i \hat{f}_m(D_i) \quad (3.8)$$

and

$$\sigma_m^2 = \frac{N}{N-1} \sum_{i=1}^{N} \left(D_i - \bar{D}_m\right)^2 \hat{f}_m(D_i) = \frac{N}{N-1} \sum_{i=1}^{N} D_i^2 \hat{f}_m(D_i) - \bar{D}_m^2. \quad (3.9)$$

The volume-averaged particle diameter and volume-averaged variance are calculated in a similar manner. If the material density of the particles is constant, the volume- and mass-averaged distributions are identical.

It must be noted that, in order to achieve a reasonably smooth and meaningful frequency distribution function, a very large number of particles must be counted. Several modern optical experimental techniques make this task feasible with relatively low effort.

3.2.2 Cumulative Distribution

Another commonly used method to quantify the particle size is the *cumulative* distribution, which is the sum of the components of the frequency distribution. The cumulative number distribution associated with size D_k is

$$\hat{F}_n(D_k) = \sum_{i=1}^{k} \hat{f}_n(D_i). \quad (3.10)$$

The numerical value of $\hat{F}_n$ is the fraction of particles with sizes less than D_k. It is apparent that the cumulative number distribution is close to 0 at the very fine sizes of particles and close to 1 at the higher end of the sizes. Part (b) in Figure 3.4 shows the cumulative number distribution that corresponds to the number frequency distribution of part (a). Both the cumulative number and cumulative mass distributions may be determined from the data of the corresponding frequency distribution.

The value of the cumulative distribution represents the fraction of particles with sizes less than or equal to D_k. Thus, $\hat{F}(100\ \mu m) = 0.4$ implies that 40% of the distribution has particles of 100 microns or less. By its definition, the value of the cumulative distribution is one for the largest particle size, provided the frequency distribution, $\hat{f}$, has been normalized as in Eq. (3.5). All three—the cumulative number, the cumulative volume, and the cumulative mass distributions—may be generated from the corresponding frequency distribution, the shape, and material properties of the particles.

3.3   CONTINUOUS SIZE DISTRIBUTIONS

If the size intervals in a discrete distribution are made progressively smaller, in the limit as ΔD approaches zero, one obtains the continuous number frequency function

$$f_n(D) = \lim_{\Delta D \to 0} \left(\frac{\hat{f}_n(D)}{\Delta D} \right). \tag{3.11}$$

The number fraction of particles with diameters within the infinitesimally small interval between D and $D + d(D)$ is given by the differential $f_n(D)d(D)$. The function of the continuous frequency distribution is a continuous function of the equivalent diameter, as shown in part (a) of Figure 3.5. If the continuous distribution is normalized, as

$$\int_0^{D_{max}} f_n(D)dD = 1, \tag{3.12}$$

where D_{max} is the maximum equivalent diameter, then the area under the continuous frequency distribution curve is equal to 1.

It must be noted that measurements of size data for particles generate discrete distributions. Continuous distributions are obtained for convenience in the calculations by fitting empirical functions (correlations) to the discrete size data.

Using the continuous distribution function, $f_n(D)$, and the density of the dispersed phase, or the discrete mass frequency distribution, $\hat{f}_m$, one obtains the continuous mass distribution function, $f_m(D)$. In this case, the differential quantity, $f_m(D)dD$, is the fraction of particle mass associated with sizes between D and $D + dD$.

The *continuous cumulative distribution*, $F_n(D)$, is obtained from the integral of the continuous frequency distribution as follows

$$F_n(D) = \int_0^D f_n(\xi)d\xi. \tag{3.13}$$

The cumulative distribution function associated with all continuous distribution functions is an S-shaped curve, as depicted in part (b) of Figure 3.5. All cumulative distribution functions approach the value 1, as the particle size approaches the

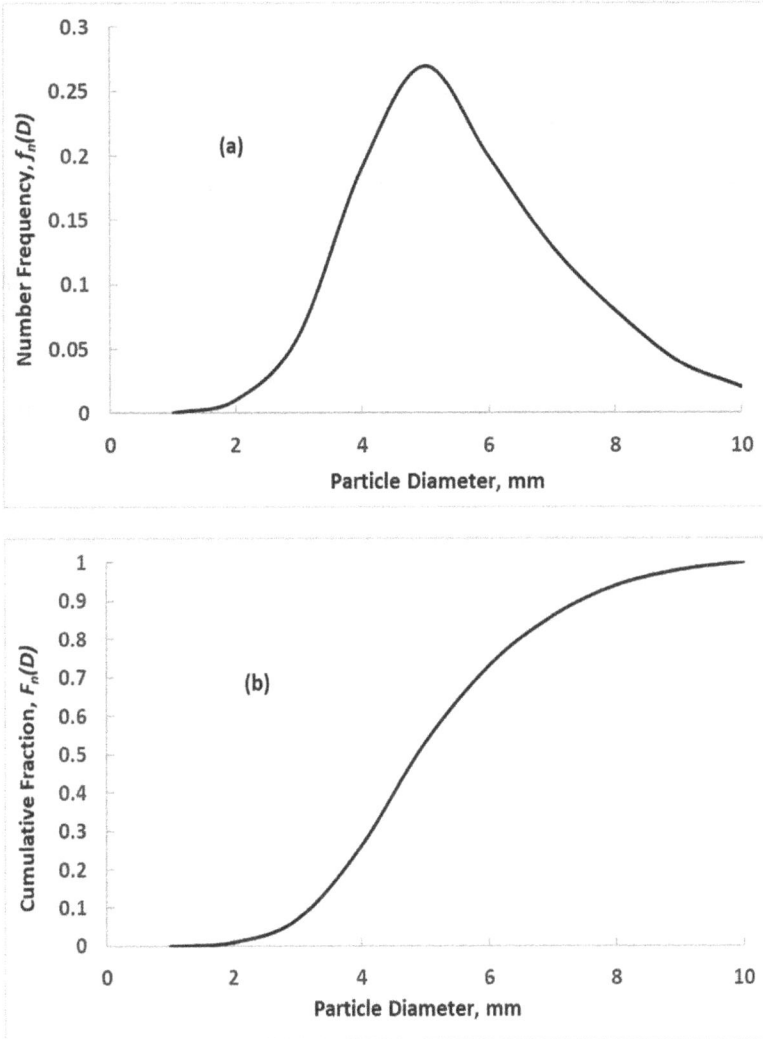

FIGURE 3.5 Continuous distributions of particle sizes: (a) Frequency size distribution and (b) cumulative size distribution.

maximum size. At the limit of a uniform continuous distribution, when $f_n(D)$ = constant, the cumulative distribution is a straight line.

3.4 STATISTICAL PARAMETERS

Several parameters are used to quantify the distribution functions. Those used most commonly for dispersed multiphase flows are presented in the following subsections. Typically, these parameters have different values for the number, mass, and volume distributions.

3.4.1 Mode, Mean, and Median

The mode is the size that corresponds to the maximum of a continuous frequency function, obtained by the expression: $df_n/dD = 0$. Thus, in Figure 3.5 part (a), the mode is the
maximum of the curve, at approximately $D = 5$ mm, with number frequency $f_n = 0.27$.
The corresponding mode for a discrete size distribution is the size that corresponds to
the highest value in the frequency histogram. For example, $d_5 \approx 0.27$ is the mode of the
discrete distribution depicted in Figure 3.4. A distribution that has two local maxima
is referred to as a bimodal distribution, and it has two modes, not necessarily of equal
value. The number frequency curve and the cumulative fraction curve of a bimodal distribution are depicted in Figure 3.6, parts (a) and (b). Similarly, trimodal and multimodal
distributions may be defined, but they are seldom met in dispersed multiphase flows.

The mean and variance of discrete distributions have been defined in Eqs. (3.6)
and (3.7). The mean of a continuous distribution is obtained from the integral of the
equivalent diameter, D, multiplied by the pertinent frequency distribution. Thus, for
the continuous number distribution function, the corresponding mean is

$$\mu_n = \int_0^{D_{max}} D f_n d(D). \tag{3.14}$$

A similar expression applies to the continuous mass distribution mean, μ_m.
Sometimes this mean is called the arithmetic mean diameter and denoted by d_{AM}.
Two other mean diameters are useful in the modelling of dispersed multiphase flows:
The surface mean diameter

$$D_{SM} = \left[\int_0^{D_{max}} D^2 f_n d(D) \right]^{1/2}, \tag{3.15}$$

and the volume mean diameter

$$D_{VM} = \left[\int_0^{D_{max}} D^3 f_n d(D) \right]^{1/3}. \tag{3.16}$$

Another statistical mean, the *Sauter Mean Diameter* (*SMD*), is frequently encountered in the spray and atomization literature

$$D_{32} = \frac{\int_0^{D_{max}} D^3 f_n(D) d(D)}{\int_0^{D_{max}} D^2 f_n(D) d(D)}. \tag{3.17}$$

The SMD is a measure of the volume to the surface area ratio for drops.

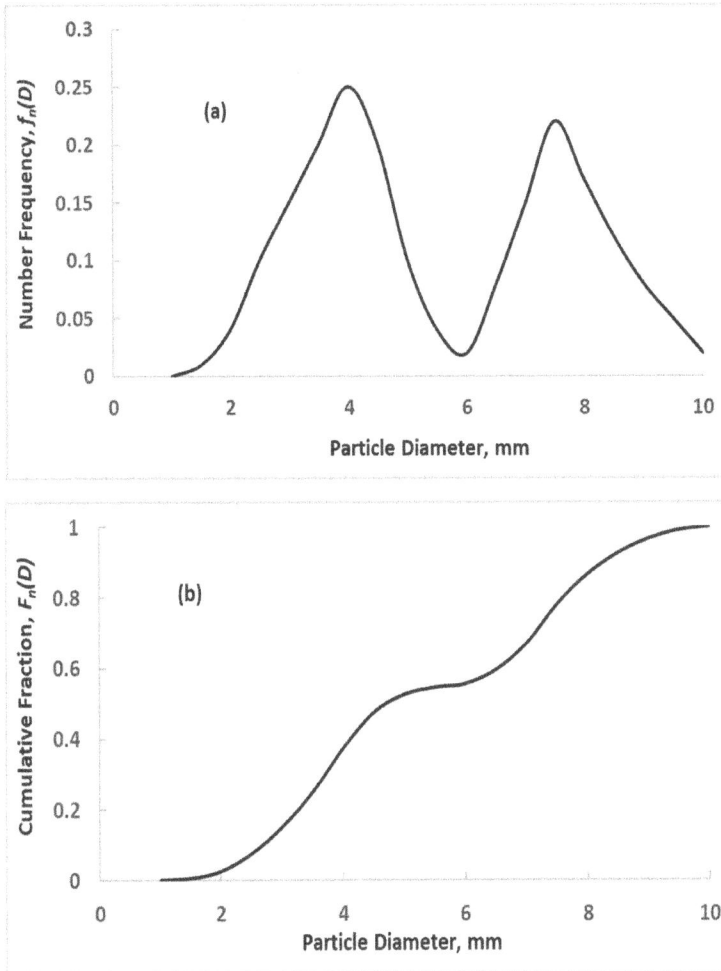

FIGURE 3.6 (a) Frequency size distribution and (b) cumulative size distribution for a bimodal size distribution.

The median diameter, D_M, corresponds to the diameter for which the cumulative distribution has the value 0.5 and is typically different than the mean, μ. Effectively, half of the particles have equivalent diameters shorter than the median and the other half of the particles have diameters longer than the median. In the number distributions depicted in Figures (3.5) and (3.6) the median diameter is approximately 4.8 mm. The analogous values for the mass median diameter, D_{mM}, and the volume median diameter, D_{VM}, may be determined from the cumulative mass and volume distribution functions respectively.

3.4.2 VARIANCE AND STANDARD DEVIATION

The variance of the continuous distribution with mean μ, as defined in Eq. (3.14), is calculated from the expression

$$\sigma^2 = \int_0^{dD} (D-\mu)^2 \, f(D)d(D) = \int_0^{D_{max}} D^2 f(D)d(D) - \mu^2 \qquad (3.18)$$

As with the mean, a variance may be defined for the number distribution, the mass distribution, and the volume distribution of particle sizes. The variance and its square root, the standard deviation, are measures of the spread of the distribution.

3.5 ANALYTICAL SIZE DISTRIBUTIONS

While there are several analytical distribution functions in the literature of statistics, the following are statistical distributions that are commonly used in the literature of dispersed multiphase flows.

3.5.1 LOG-NORMAL DISTRIBUTION

The log-normal distribution is frequently used to represent the size of solid particles. This distribution derives from the normal or Gaussian distribution which is defined as

$$f(x) = \frac{1}{\sqrt{2\pi}\sigma} \exp\left[-\frac{1}{2}\left(\frac{x-\mu}{\sigma}\right)^2\right], \qquad (3.19)$$

where σ is the standard deviation, and μ is the mean of the distribution. The Gaussian distribution corresponds to the well-known "bell-shaped" curve, shown in Figure 3.7a. The curve is symmetric about the mean and trails off to plus and minus infinity. The normal distribution is normalized so the area under the curve is one

$$\int_0^\infty f(x)dx = 1. \qquad (3.20)$$

The log-normal distribution is obtained from the normal distribution by replacing the independent variable with the logarithm of the particle diameter. Thus, the number of particles with diameters between D and $D + dD$ is

$$dN = f_n(D)dD = \frac{1}{\sqrt{2\pi}\sigma_0} \exp\left[-\frac{1}{2}\left(\frac{\ln(D)-\mu_0}{\sigma_0}\right)^2\right]\frac{dD}{D}, \qquad (3.21)$$

where σ_o and μ_o are the standard deviation and the mean of the log-normal distribution. The shape of this distribution is shown in Figure 3.7b, where it is observed that its limits range from zero to infinity.[1]

The cumulative number distribution of the log-normal distribution is given by the expression

$$F_n(D) = \frac{1}{\sqrt{2\pi}\sigma_0} \int_0^D \frac{1}{D} \exp\left[-\frac{1}{2}\left(\frac{\ln(D) - \mu_0}{\sigma_0} \right)^2 \right] dD. \tag{3.22}$$

This integral may be expressed in terms of the error function, which is defined as follows

$$erf(t) = \frac{2}{\sqrt{\pi}} \int_0^t \exp\left(-\xi^2\right) d\xi. \tag{3.23}$$

Hence, the expression for the cumulative distribution becomes

$$F_n(D) = \frac{1}{2}\left[1 + erf\left(\frac{\ln D - \mu_0}{\sqrt{2}\sigma_0} \right) \right]. \tag{3.24}$$

When the error function is equal to zero, the value of the cumulative distribution is 0.5, and this corresponds to the median of the distribution. Since the value of the error function is zero when its argument is zero, this implies that the mean of the log-normal distribution is the logarithm of the number median diameter, that is: $\mu_o = \ln D_M$. Therefore, the log-normal number frequency distribution may be conveniently expressed as

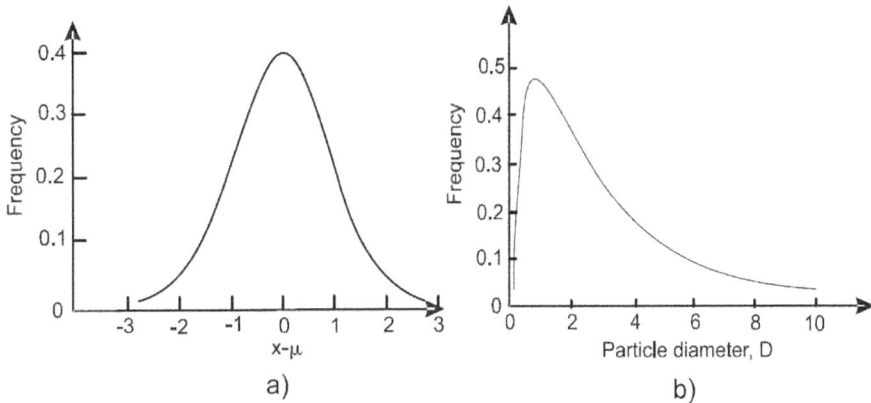

FIGURE 3.7   Frequency distributions: (a) normal (Gaussian) and (b) log-normal.

[1] The independent variable ($\ln D$) approaches $-\infty$ as D approaches zero.

$$f_n(D) = \frac{1}{\sqrt{2\pi} D\sigma_0} \exp\left[-\frac{1}{2}\left(\frac{\ln(D)-\ln(D_M)}{\sigma_0}\right)^2\right].$$ (3.25)

The corresponding mean for the log-normal mass frequency distribution is the logarithm of the mass median diameter, $\ln D_{mM}$

$$f_m(D) = \frac{1}{\sqrt{2\pi} D\sigma_0} \exp\left[-\frac{1}{2}\left(\frac{\ln(D)-\ln(D_{mM})}{\sigma_0}\right)^2\right].$$ (3.26)

The variance of the log-normal distribution, σ_0, is usually determined by plotting the cumulative distribution on a semi-logarithmic graph ($\ln D$ vs. frequency). In this graph, the distribution appears as a straight line, as it is shown in Figure 3.8. From this graph one obtains the values that correspond to the 84th percentile and the 50th percentile, $D_{84\%}$ and $D_{50\%}$. At $D_{50\%}$ the error function is zero, and the diameter is equal to the median diameter, D_M, while at $D_{84\%}$ the value of the error function is one. Therefore, the following relationship applies

$$\sigma_0 = \ln\left(\frac{D_{84\%}}{D_{50\%}}\right).$$ (3.27)

For particles and drops that have constant density, the value for σ_0 is the same for both the number and the mass distributions.

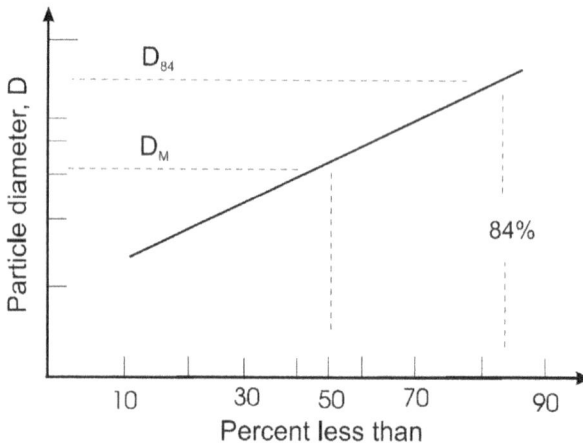

FIGURE 3.8 Determination of $D_{84\%}$ and $D_{50\%}$ of a log-normal distribution.

For the calculation of the several statistical parameters in the log-normal distribution, a useful mathematical relationship is

$$D_M^k \exp\left(\frac{k^2\sigma_0^2}{2}\right) = \frac{1}{\sqrt{2\pi}\sigma_0} \int_0^\infty D^k \exp\left[-\frac{1}{2}\left(\frac{\ln(D)-\ln(D_{mM})}{\sigma_0}\right)^2\right]\frac{d(D)}{D}, \quad (3.28)$$

or

$$D_M^k \exp\left(\frac{k^2\sigma_0^2}{2}\right) = \int_0^\infty D^k f(D)d(D). \quad (3.29)$$

These relationships may be readily used to calculate the several statistical parameters of the distribution. An example illustrating the use of this relationship is the determination of the number mean of the distribution, defined as

$$\mu_n = \int_0^\infty D f_n(D)dD. \quad (3.30)$$

In this case, $k = 1$ in Eq. (3.29), and the number mean diameter is simply

$$\mu_n = D_{nM} \exp\left(\frac{\sigma_0^2}{2}\right). \quad (3.31)$$

One may also derive the following relationship for the number median diameter and the mass median diameter

$$D_{nM} = D_{mM} \exp\left(\frac{3\sigma_0^2}{2}\right). \quad (3.32)$$

Therefore, one only needs the information on one of the median diameters and σ_0 to determine all the other statistical parameters of the log-normal distribution.

Example: The mass median diameter of a log-normal distribution with $\sigma_0 = 0.3$ was measured to be 200 microns. Determine the standard deviation of the mass distribution.

Solution: From Eq. (3.18), the variance of this distribution is

$$\sigma_m^2 = \int_0^\infty D^2 f(D)dD - \mu_m^2.$$

From Eq. (3.31)

$$\mu_m^2 = D_{mM}^2 \exp\left(\sigma_0^2\right),$$

and from Eq. (3.29)

$$D_{mM}^2 \exp\left(\frac{4\sigma_0^2}{2}\right) = \int_0^\infty D^2 f(D)d(D).$$

Therefore, the mass variance is

$$\sigma_m^2 = D_{mM}^2 \left[\exp\left(2\sigma_0^2\right) - \exp\left(\sigma_0^2\right) \right],$$

and this implies that the standard deviation is 200 $(1.197 - 1.094)^{1/2} = 64.2$ microns.

3.5.2   UPPER-LIMIT LOG-NORMAL DISTRIBUTION

The upper-limit log-normal distribution is frequently used to model droplet data from spray nozzles. The distribution is an off-shoot of the log-normal distribution and is designed to set the maximum particle diameter as the upper limit of the distribution (Mugele and Evans, 1951). This is accomplished by replacing ln D in Eqs. (3.25) and (3.26) with the independent variable

$$\ln\left(\frac{\alpha D}{D_{max} - D}\right),$$

where D_{max} is the maximum diameter and a is a constant, which is selected to normalize the distribution. The modified independent variable approaches $-\infty$ as D approaches zero and $+\infty$ as D approaches D_{max}.

Therefore, the number frequency function in this distribution is defined as

$$f_n(D) = \frac{1}{\sqrt{2\pi}D\sigma_0}\exp\left[-\frac{1}{2}\left(\frac{\ln(\alpha D) - \ln\left(D_{max} - D_{nM}\right)}{\sigma_0}\right)^2\right], \qquad (3.33)$$

where D_{max} is the maximum diameter in the distribution. The constant a that normalizes the distribution is

$$\alpha = \frac{D_{max}}{D_{nM}} - 1. \qquad (3.34)$$

As with the log-normal distribution, the standard deviation may be calculated from the slope of the logarithmic plot of the cumulative frequency data.

3.5.3   SQUARE-ROOT NORMAL DISTRIBUTION

The number frequency function of the square-root normal distribution is given by the expression (Tate and Marshall, 1953)

$$f_n(D) = \frac{1}{2\sqrt{2\pi\sigma}\,D} \exp\left[-\left(\frac{\left(\sqrt{D}-\sqrt{\bar{D}}\right)}{2\sigma}\right)^2\right], \tag{3.35}$$

where σ is the variance of the square-root normal distribution, and $\bar{D}$ is the mean of the distribution. This distribution appears to fit well with empirical data from vane-type atomizers. Its principal advantage is that it does not have the long end "tails" of a normal distribution that extend to infinity.

3.5.4 ROSIN-RAMMLER DISTRIBUTION

The Rosin-Rammler distribution (Mugele and Evans, 1951) is frequently used to model droplet size distributions in sprays. It is expressed in terms of the cumulative mass distribution

$$F_m(D) = 1 - \exp\left[-\left(\frac{D}{\delta}\right)^n\right]. \tag{3.36}$$

δ and n are two empirical constants, which are determined by plotting the cumulative distribution on logarithmic coordinates. By taking twice the logarithm of the last equation, one obtains

$$\ln\left[-\ln\left(1 - F_m(D)\right)\right] = n\left(\ln(D) - \ln(\delta)\right). \tag{3.37}$$

Therefore, n, is the slope of this line.

The parameter δ is obtained from the expression for the mass median of the distribution. Since $F_m(D_{mM}) = \frac{1}{2}$ and $\ln\frac{1}{2} = -0.693$, the following equation yields δ in terms of the mass median diameter

$$\delta = \frac{D_{mM}}{0.693^{1/n}}. \tag{3.38}$$

The mass frequency function of this distribution is obtained from the derivative of the cumulative distribution of Eq. (3.36)

$$f_m(D) = \frac{dF_m(D)}{d(D)} = \frac{n}{\delta}\left(\frac{D}{\delta}\right)^{n-1} \exp\left[-\left(\frac{D}{\delta}\right)^n\right]. \tag{3.39}$$

As with the log-normal distribution and the error function, the gamma function

$$\Gamma(x) = \int_0^\infty \xi^{x-1} \exp(-\xi)d\xi, \tag{3.40}$$

is a useful tool to calculate the moments of the Rosin-Ramler distribution because the following expression holds

$$\int_0^\infty nD^\alpha \left(\frac{D}{\delta}\right)^{n-1} \exp\left[-\left(\frac{D}{\delta}\right)^n\right] d\left(\frac{D}{\delta}\right) = \delta^\alpha \Gamma\left(\frac{\alpha}{n}+1\right). \qquad (3.41)$$

This expression helps obtain the relationship for the statistical parameters of this distribution

$$\int_0^\infty D^\alpha f_m(D)dD = \delta^\alpha \Gamma\left(\frac{\alpha}{n}+1\right). \qquad (3.42)$$

For example, the mass mean diameter, which is defined as

$$\mu_m = \int_0^\infty Df_m(D)dD = \int_0^\infty D\frac{n}{\delta}\left(\frac{D}{\delta}\right)^{n-1} \exp\left[-\left(\frac{D}{\delta}\right)^n\right], \qquad (3.43)$$

is obtained when $\alpha = 1$. In this case, Eq. (3.42) yields

$$\mu_m = \delta\Gamma\left(1+\frac{1}{n}\right). \qquad (3.44)$$

With a given value for n, the mass mean diameter may be obtained from tables of the gamma function, which are found in most mathematical texts.

Example: A Rosin-Rammler distribution has a mass median diameter of 120 microns with $n = 2.0$. Determine the mass mean of the distribution.

Solution: At first, the parameter δ is calculated from Eq. (3.38) to be 144 μm. The mass mean is calculated from Eq. (3.44) as $144\Gamma(1.5) = 128$ μm.

3.5.5 NUKIYAMA-TANASAWA DISTRIBUTION

The Rosin-Rammler distribution is a special case of the more general Nukiyama-Tanasawa distribution. The frequency function of this distribution is

$$f_n(D) = BD^2 \exp\left(-CD^q\right), \qquad (3.45)$$

where B, C, and q are empirically determined parameters. For spray distributions, q typically varies from 0.166 to 2.0 with the higher values corresponding to narrower distributions.

3.5.6 LOG-HYPERBOLIC DISTRIBUTION

Several measurements on spray droplets showed that neither the log-normal nor the Rosin—Rammler distribution fit the data sufficiently well. Two droplet-size distributions fitted with the Rosin-Rammler distribution may have the same parameters, but the actual distributions may be very different. This happens because of a significant shortcoming of both the log-normal and Rosin-Rammler distributions: The inadequate representation of the tails of the distributions, that is, the accuracy of the values of the frequency function for very large and very small sizes.

Researchers have determined that a better fit for the tails of the distributions is the log-hyperbolic distribution proposed by Barndorff-Nielsen (1977). The form of the number frequency function for this distribution is

$$f_n(D) = A \exp\left[-\alpha\sqrt{\delta^2 - (\ln(D) - \mu)^2} + \beta(D - \mu)\right], \qquad (3.46)$$

where α, β, δ and μ are empirical parameters. It is apparent that the random variable of this distribution is the logarithm of the particle diameter. The coefficient A is a normalization factor related to the other parameters by the expression

$$A = \frac{\sqrt{(\alpha^2 - \beta^2)}}{2\alpha\delta K_1\left(\delta\sqrt{(\alpha^2 - \beta^2)}\right)}, \qquad (3.47)$$

where K_1 is the third-order Bessel function of the third kind. The logarithm of the frequency function yields the equation of a hyperbola

$$\ln\left[f_n(D)\right] = \ln(A) - \alpha\sqrt{\delta^2 - (\ln(D) - \mu)^2} + \beta(D - \mu). \qquad (3.48)$$

For $[ln(D) - \mu]/\delta < 0$, the slope of the asymptote is $\alpha + \beta$, while for $[ln(D) - \mu]/\delta > 0$, the slope of the asymptote is $-\alpha + \beta$. Thus, when the logarithm of the frequency is plotted versus $ln(D)$, the slopes of the two asymptotes may be measured to determine the parameters α and β. This procedure is shown in Figure 3.9. The parameter

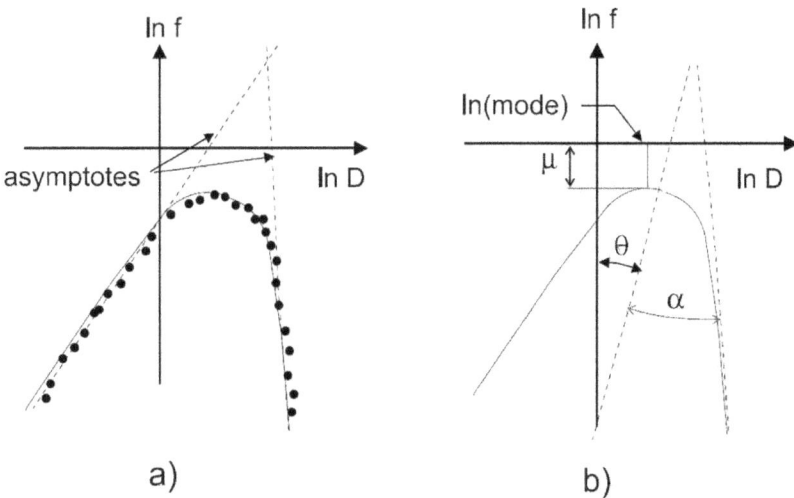

FIGURE 3.9   The two asymptotes of the log-hyperbolic distribution and the determination of the parameters of the distribution.

μ is the mode of the distribution, and the value for δ is determined from a curve-fitting procedure.

One of the problems with the log-hyperbolic distribution, as well as of other distributions, is the uncertainty of the parameters. If the tails of the distribution are not long enough to yield accurate values of slopes of the asymptotes, there is high uncertainty in the determination of α and β. This implies that the parameters are not unique and that the same distribution may be matched by various combinations of the parameters. A method to circumvent this problem has been proposed by Xu *et al.* (1993): The data are fitted with a hyperbola, which is rotated at an angle θ with respect to the y-axis. as shown in part (b) of Figure 3.9. This transformation makes the dissector of the angle of the hyperbola the abscissa of the graph. Now the asymptotes are symmetric and more accurately determined. The mean becomes the vertex of the hyperbola. This is referred to as the "three-parameter log-hyperbolic distribution."

SUMMARY

Since particle "size" plays an important role in the multiphase particulate flows, several quantitative measures for the size of particles are introduced, including the fractal dimension of irregularly shaped particles. The distribution of particle and droplet sizes are formulated as discrete or continuous in terms of particle number or particle mass. Parameters that describe these distributions are the mean, the median, the mode, the standard deviation, and the variance. Numerous analytical distributions are available in the literature to quantify particle and droplet size distributions. Among these distribution functions, the log-normal, the upper limit log-normal, the square-root normal, the Rosin-Rammler, the Nukiyama-Tanasawa, and the log-hyperbolic distributions, which are primarily used for droplets, are succinctly presented, and their properties are examined. For more details of these and other distributions, the reader is referred to the abundant literature in this area.

REFERENCES

Barndorff-Nielsen, O., 1977, Exponentially decreasing distributions of the logarithm of particle size, *Proc. Res. Soc. Lond. A.*, **353**, 401.

Bushell, G.C., Yan, Y.D., Woodfield, D., Raper, J. and Amal, R., 2002, On techniques for the measurement of the mass fractal dimension of aggregates, *Adv. in Colloid and Interface Sci.*, **95** (1), 1–50.

Li, X.-Y. and Logan, B. E., 2001, Permeability of fractal aggregates, *Water Research*, **35** (14), 3373–3380.

Mandelbrot, B. B., 1967, How long is the coast of Britain?, *Science*, **156**, 636–638.

Michaelides, E.E., 2014, *Nanofluidics: Thermodynamic and Transport Properties*, Springer, Heidelberg.

Mugele, R.A. and Evans, H.D., 1951, Droplet size distribution in sprays, *Ind. Eng. Chem.*, **43**, 1317–1324.

Rajapakse, R., 2016, *Geotechnical Engineering Calculations and Rules of Thumb*, 2nd edition, Elsevier, Amsterdam.

Tate, R.W. and Marshall, Jr., W.R., 1953, Atomization by centrifugal pressure nozzles (Part II), *Ind. Eng. Prog.*, **49**, 226–234.

Vicsek, T., 1999, *Fractal Growth Phenomena*, 2nd edition, World Scientific, Singapore.

Wang, W. and Chau, Y., 2009, Self-assembled peptide nanorods as building blocks of fractal patterns, *Soft Matter*, **5** (24), 4893–4898.

Xu, T.-H., Durst, F. and Tropea, C., 1993, The three-parameter log-hyperbolic distribution and its application to particle sizing, *Atom. and Sprays*, **3**, 109.

PROBLEMS

3.1   Prove that for a sphere, all the equivalent diameters are equal to the actual diameter, D.

3.2   The three axes of an ellipsoidal particle are $L_1 = 2$ mm, $L_2 = 2.8$ mm, and $L_3 = 6$ mm. Determine the equivalent diameters D_V, D_A, and D_L.

3.3   A measurement of a particle size distribution indicates that the distribution is log-normal with a mass median diameter of 60 μm and a slope (σ_0) of 0.3. Determine the following parameters:

a) Number median diameter
b) Sauter mean diameter
c) Mass mean diameter
d) Mode of mass distribution

3.4   Data obtained for a size distribution of droplets produced by an atomizer indicate that the data fit the Rosin-Rammler distribution with $n = 1.8$ and a mass median diameter of 120 μm. Evaluate the following statistical parameters:

a) The parameter δ for the distribution
b) The mass mean diameter and the mass variance for the distribution
c) The Sauter mean diameter

3.5   The continuous number frequency uniform distribution is as shown. The maximum diameter is 100 μm.

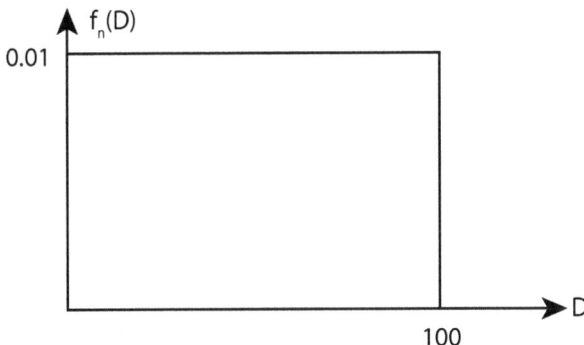

PROBLEM 3.5

Determine the mean, standard deviation, and the Sauter mean diameter, as well as the following:

a) Derive an expression for the cumulative distribution and the number mean diameter

b) Derive an expression for the mass frequency distribution and find the mode, mean and standard deviation

c) Obtain an expression for the mass cumulative distribution and find the mass median diameter

3.6   The following data were obtained for the cumulative mass distribution of a particle sample.

% less than	D (microns)
94	50
90	40
76	32
60	25.4
42	20
26	16
14	12.7
7	10
3	8

Carry out the following procedures:

a) Plot the data on log-probability paper and determine D_{mM} and σ_0

b) Calculate D_{nM} and SMD (Sauter mean diameter)

Calculate the standard deviation for both the mass and number distribution. Is the distribution monodisperse?

3.7   The data for a Rosin-Rammler distribution show that $n = 2$ and the mass median diameter is 20 microns.

a) Show that

$$\int_0^\infty nD^\alpha \left(\frac{D}{\delta}\right)^{n-1} \exp\left(-\frac{D}{\delta}\right)^n d\left(\frac{D}{\delta}\right) = \delta^\alpha \Gamma\left(\frac{\alpha}{n} + 1\right)$$

b) Using the expression generated in the previous section, find an expression for the mass variance of the Rosin-Rammler distribution in the form $\sigma_m^2 = \delta^2 f[\Gamma(n)]$ and evaluate for the distribution parameters.

c) Given that

$$f_n(D) = \frac{\dfrac{f_m(D)}{D^3}}{\dfrac{\displaystyle\int_0^\infty f_m(D)\,dD}{D^3}}$$

show that

$$f_n(D) = \frac{n}{\delta}\left(\frac{D}{\delta}\right)^{-3}\frac{\exp\left[-\left(\dfrac{D}{\delta}\right)^n\right]}{\Gamma\left(1-\dfrac{3}{n}\right)}\left(\frac{D}{\delta}\right)^{n-1}$$

for the Roslin-Rammler distribution.

Also, obtain an equation for the Sauter mean diameter in the form $SMD = \delta$ $f[\Gamma(n)]$ and evaluate for the preceding parameters.

3.8   A bimodal distribution consists of two peaks as shown. Forty percent of the mass of the particles is associated with 60 microns, and the remaining 60% of the mass is associated with 30 micron particles.

Determine the number average and the mass average diameters and the corresponding standard deviations.

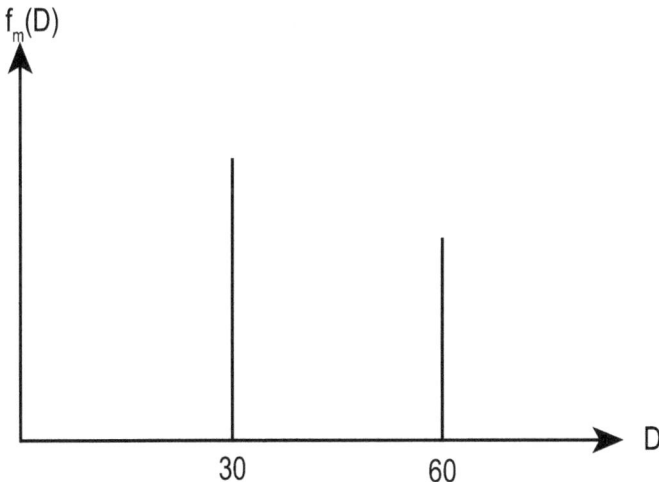

PROBLEM 3.8

3.9   The continuous number frequency distribution of a particle sample is shown in the following figure.

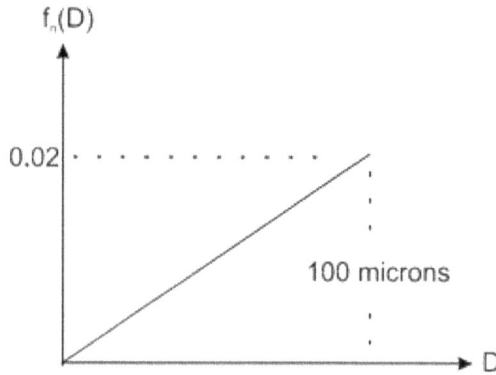

PROBLEM 3.9

Determine the following:

a) The number mean diameter, Sauter mean diameter, and the number variance
b) The mass mean and standard deviation
c) The number median and mass median diameter

3.10 Show that

$$D_m^k \exp\left(\frac{\sigma_0^2 k^2}{2}\right) = \frac{1}{\sqrt{2\pi}\sigma_0} \int_0^\infty D^k \exp\left[-\frac{1}{2}\left(\frac{\ln D - \ln D_m}{\sigma_0}\right)^2\right] \frac{dD}{D}$$

3.11 Using the definition

$$f_n(D) = \frac{D^3 f_n(D)}{\int_0^\infty D^3 f_n(D) dD}$$

obtain the relationship

$$\frac{D_{nm}}{D_{mm}} = f(\sigma_0)$$

3.12 The results of a digital image analysis of a particle sample yield the frequency distribution of the particle cross-sectional area. The following results for a sample were obtained:

f(D)	A, m²
0.2	0.286×10^{-8}
0.3	0.785×10^{-8}
0.5	1.13×10^{-8}

Assume each area is the projected area of a spherical particle and determine the following:

a) The number averaged diameter
b) The mass averaged diameter
c) The Sauter mean diameter

3.13 The following data were obtained with a Coulter Counter. The population refers to the number of particles of the given diameter.

Bin No.	Diameter (μm)	Population
1	1.75	0
2	2.21	3797
3	2.78	3119
4	3.51	2522
5	4.42	1878
6	5.75	1396
7	7.02	850
8	8.58	523
9	11.15	297
10	14.0	179
11	17.7	93
12	22.3	18
13	28.1	2

Form the cumulative number distribution, plot it on log-probability paper and calculate the following:

a) The number median diameter
b) The mass median diameter
c) The Sauter mean diameter

3.14 The following data are obtained from the measurement of a particle size distribution. The particles are spherical.

D (microns)	Number
10	20
20	50
30	73
40	62
50	33
60	18
70	5

Determine the mass- and number-averaged diameters. Estimate the number and mass median diameters.

3.15   Show for the log-hyperbolic distribution that the mode of the distribution is given by the expression

$$x = \ln D = \mu + \frac{\beta\delta}{\sqrt{\alpha^2 - \beta^2}}$$

and the value of the frequency function at the mode is

$$f = A\exp\left(-\delta\sqrt{\alpha^2 - \beta^2}\right)$$

4 Forces on Single Particles and Drops

The movement of single particles and drops is governed by Newton's second law, which may be written succinctly

$$m_p \frac{d\vec{v}}{dt} = \sum_i \vec{F}_i, \tag{4.1}$$

where m_p represents the mass of the particle;[1] $\vec{v}$ is the velocity vector of the particle; and the right-hand side represents the summation of all the forces, $\vec{F}_i$, acting on the particle. These forces may be hydrodynamic—resulting from the interactions between the particle and the carrier fluid—electrostatic and magnetic forces, and body forces, such as gravity. Of these forces, the modelling of hydrodynamic drag and lift can be very complex because of the contribution of several variables pertaining to both the particle and the carrier fluid. This chapter elucidates the origins and nature of the most common forces acting on particles and presents methods and closure equations for their modelling.

4.1 STEADY DRAG ON SPHERICAL PARTICLES AND DROPS

When the relative velocity between the particle and the carrier fluid is constant, the fluid exerts on the particle a steady drag force, which is, in general, a function of the Reynolds number of the particle, Re_r, defined in terms of the relative velocity of the particle

$$Re_r = \frac{D|\vec{u} - \vec{v}|\rho_c}{\mu_c} = \frac{2\alpha|\vec{u} - \vec{v}|\rho_c}{\mu_c}. \tag{4.2}$$

It is convenient to express the steady drag force on particles in terms of a dimensionless variable, C_D, the drag coefficient

$$\vec{F}_D = \frac{1}{2} C_D A_{cs} \rho_c |\vec{u} - \vec{v}|(\vec{u} - \vec{v}) = \frac{\pi}{2} C_D \alpha^2 \rho_c |\vec{u} - \vec{v}|(\vec{u} - \vec{v}), \tag{4.3}$$

where A_{cs} is the cross-sectional surface area of the particle, $\pi\alpha^2 = \pi D^2/4$. Eq. (4.3) implies that the drag force acts in the direction of the relative velocity of the particle. It accelerates the particles when the velocity of the continuous medium—the fluid—is higher and decelerates the particles when the velocity of the fluid is less than that of the particle. In general, the drag coefficient is a function of the Reynolds number.

---

[1] For brevity, the word "particle" will be used in this chapter to denote both solid particles and liquid drops. When this generalization does not apply, clarifications will be given as "solid particles" and "liquid drops" or just "drops."

DOI: 10.1201/9781003089278-4

4.1.1 DRAG AT VERY SMALL REYNOLDS NUMBERS—CREEPING OR STOKES FLOW

At very low Reynolds numbers, $Re_r \ll 1$, the effects of the fluid inertia on the particle vanish, and the flow around the particle is referred to as *creeping flow*, or *Stokes flow*. In this case there is fore and aft symmetry in the flow field around a spherical particle, and it is possible to derive analytical solutions for the steady drag on particles. Stokes (1851) was the first to derive an expression for the steady drag for solid spherical particles, while Hadamard (1911) and Rybczynski (1911) independently derived analytical expressions for the drag on viscous spheres (drops). Since the latter approach is more general, it will be followed to yield a general expression for the drag coefficients of viscous spheres. The drag coefficient of sold spheres (solid particles) may be easily derived as a partial case of the latter, when the viscosity of the sphere becomes very high.

Consider the flow field around a viscous, fluid sphere at rest, as shown in Figure 4.1, inside a fluid of different viscosity at $Re_r \ll 1$. The carrier fluid velocity is uniform, and its magnitude is U, far from the sphere. The no-slip velocity condition applies at the surface of the sphere and implies that the local velocity of the carrier fluid is equal to the local velocity of the viscous sphere. The center of coordinates is coincident with the center of the sphere, and the flow domain is much larger than the diameter of the sphere. This is referred to sometimes as an *infinite domain*. Under these conditions, the stream functions inside and outside the sphere are (Hadamard, 1911; Rybczynski, 1911)

$$\psi_i = \frac{Ur^2(a^2 - r^2)\sin^2\theta}{4(\lambda + 1)a^2} \tag{4.4}$$

and

$$\psi_o = \frac{Ur^2\sin^2\theta}{2}\left[1 - \frac{(3\lambda + 2)a}{2(\lambda + 1)r} + \frac{\lambda a^3}{2(\lambda + 1)r^3}\right], \tag{4.5}$$

where λ is the ratio of the dynamic viscosities of the sphere and the carrier fluid, μ_d/μ_c. The case of a solid sphere is given at the limit $\lambda \to \infty$, and the case of an inviscid sphere (a bubble) is given at the limit $\lambda \to 0$. Liquid drops are characterized by values of λ between the two limits. The presence of the sphere creates a disturbance in the carrier fluid velocity field, which is determined (in the radial and transverse directions) from the stream function of Eq. (4.5)

$$u_r = U\cos\theta\left[1 - \frac{(3\lambda + 2)a}{2(\lambda + 1)r} + \frac{\lambda a^3}{2(\lambda + 1)r^3}\right] \tag{4.6}$$

and

$$u_\theta = -U\sin\theta\left[1 - \frac{(3\lambda + 2)a}{4(\lambda + 1)r} - \frac{\lambda a^3}{4(\lambda + 1)r^3}\right]. \tag{4.7}$$

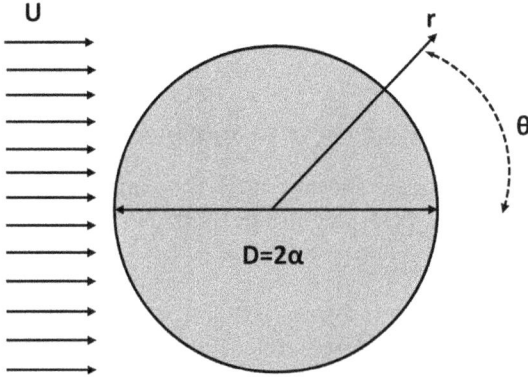

FIGURE 4.1   A solid or fluid sphere in a carrier fluid with uniform velocity U.

The velocity field described by the last two equations yields $u_r = 0$ on the surface of the sphere and $u = U$ far away from the sphere. The analytical expression of the velocity field may be used to determine the pressure field and the fluid stress tensor. Integration of the normal and shear stresses on the surface of the fluid sphere yields the total hydrodynamic force exerted by the fluid on the viscous sphere as follows

$$\vec{F}_D = \pi D \vec{U} \mu_c \frac{3\lambda + 2}{\lambda + 1}. \tag{4.8}$$

Since the viscous sphere is at rest, the fluid velocity, U, is the relative velocity of the sphere with respect to the fluid and, hence, $Re_r = DU\rho_c/\mu_c$ and $U = u - v$. For the more general case when the viscous spheres are not stationary, their scalar drag coefficient is defined as

$$C_D = \frac{|\vec{F}_D|}{\frac{1}{2}\pi a^2 \rho_c |\vec{u} - \vec{v}|^2} = \frac{8(3\lambda + 2)}{Re_r(\lambda + 1)}. \tag{4.9}$$

At the limit $\lambda \to \infty$, Eqs. (4.8) and (4.9) yield the so-called "Stokes drag" for solid spheres

$$\vec{F}_D = 6\pi a (\vec{u} - \vec{v}) \mu_c = 3\pi D (\vec{u} - \vec{v}) \mu_c \Rightarrow C_D = \frac{24}{Re_r}. \tag{4.10}$$

The corresponding drag coefficient for a spherical inviscid sphere (e.g. a bubble) is $C_D = 16/Re_r$ and the magnitude of the drag force on the inviscid sphere is $F_D = 2\pi D(u - v)\mu_c$. The latter is sometimes referred to as the *form drag* or *pressure drag*, while the difference of the two expressions, which is equal to $\pi D(u - v)\mu_c$, is referred to as the *friction drag*. The form drag is due to the pressure distribution over the surface of the sphere, and the friction drag is the result of the viscous stress field at the surface of the sphere. While some publications make this distinction between the two parts of the drag force, it must be noted that the drag force is a single entity that arises from the hydrodynamic interactions between the fluid and the sphere and not two different forces.

4.1.2   STEADY DRAG ON SPHERICAL AT FINITE REYNOLDS NUMBERS

At finite Re_r, the fluid has significant inertia that cannot be neglected, the fore-aft symmetry of the flow field around the sphere breaks down, and the stream function and the corresponding velocity field cannot be determined analytically. Experimental observations for both solid spheres and viscous spheres have proven that, even at low Re_r, a wake is formed behind the sphere. This is a steady wake that becomes stronger as Re_r increases and the inertia of the flow around the sphere overcomes the viscosity effects.

4.1.2.1   The Flow Field Around the Solid Sphere

Early experimental observations on solid spheres by Taneda (1956), Achenbach (1974), and Seeley *et al.* (1975), as well as recent numerical computations, indicate that the following flow descriptions or *regimes* are observed around solid spheres:

1. *Attached flow.* In the range $0 < Re_r < 20$, the flow is attached to the sphere, and there is no visible recirculation behind it. There is a *velocity defect* region behind the sphere, which is the characteristic of a weak wake. The fore-aft asymmetry of the flow becomes progressively more pronounced as Re_r increases.
2. *Steady wake.* Separation of the flow aft of the sphere commences at approximately $Re_r = 20$, and a weak recirculating wake becomes visible. The volume of the wake is very small and the wake is attached to the sphere. As Re_r increases, the wake becomes wider and longer, and its point of attachment on the sphere moves forward. The wake is steady up to approximately $Re_r = 130$. At this flow regime, the separation angle is a monotonically decreasing function of Re_r, while the wake length and volume increase with Re_r. An approximate correlation for the length of the wake in this flow regime is

$$L_W = D[0.0203(Re_r - 20) + 0.00012(Re_r - 20)^2].\qquad(4.11)$$

3. *Unsteady wake, laminar flow.* The onset of the wake instability occurs in the range $130 < Re_r < 150$. At these Reynolds numbers, a weak, long-period, laminar oscillation is formed at the tip of the wake with its amplitude increasing with Re_r. Pockets of vorticity begin to shed from the tip of the sphere and influence the flow field downstream. This flow regime has been observed in experiments in the range $(130 \text{ to} 150) < Re_r < 270$.
4. *High subcritical range.* This regime covers the range $270 < Re_r < 3*10^5$. As Re_r increases, stronger vortices are shed regularly from alternate sides of the sphere. At the lower end of this regime, the Strouhal number, Sl, of the

vortices is a monotonic function of Re_r and ranges from 0.1 at $Re_r = 400$ to approximately 2 at $Re_r = 6,000$. The following correlation describes the relationship between Sl and Re_r in the range $400 < Re_r < 4,000$

$$Sl = -3.252 + 2.874 \ln \left(Re_r^{1.43} \right) . \tag{4.12}$$

In the range $4,000 < Re_r < 6,000$, there is a great deal of scatter in the experimental data, with Sl appearing to level at the value 0.21. For values of Re_r above 6,000 wake separation occurs at a point that rotates around the sphere with frequency equal to the shedding frequency of the vortices. As a result of the wake separation (at approximately $Re_r = 6,000$), the value of Sl drastically reduces from 0.21 to 0.125. Subsequently, it gradually increases from 0.125 to 0.18 in the range $6,000 < Re_r < 3*10^4$, while in the range $3*10^4 < Re_r < 2*10^5$ Sl, it only rises to 0.19 (Achenbach, 1974). All the studies indicate that, in this flow regime, the wake is periodic but not turbulent. The trends and the relationship between the Strouhal number of the shed vortices and the Reynolds number are depicted in Figure 4.2. Additional observations by Seeley *et al.* (1975) have shown that for $Re_r > 1,300$, small jets and eddies appear, which signifies three-dimensional rotation of the flow. These observations confirm that, in the subcritical range, the flow separation point moves forward, and the angle where separation occurs is given by the correlation

$$\theta_S = 83 + 660 Re_r^{-0.5} . \tag{4.13}$$

5. *Supercritical flow.* The onset of the transition to a turbulent wake occurs at $Re_r = 2*10^5$, and the transition is completed at $Re_r = 3.7*10^5$. Significant changes in the flow pattern around the sphere occur in this range and are referred to as *critical transition*. At $Re_r > 3.7*10^5$, the wake separation point

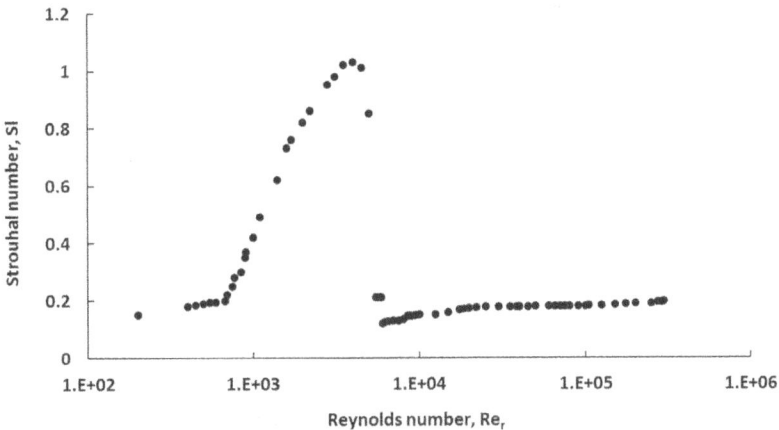

FIGURE 4.2 The relationship between Strouhal number and Reynolds number for spheres.

begins to move downstream, and fluctuations in the position of the separation point become evident. The free shear layer becomes turbulent and attaches to the surface of the sphere. The most evident result of the laminar to turbulent transition is a sharp drop of the drag coefficient from approximately 0.44 to 0.07. The transition to a turbulent boundary layer is sensitive to the intensity of the free-stream turbulence and may be accelerated by "tripping" the flow with a thin wire.

4.1.2.2 Steady Drag on Solid Spheres

The first accurate determination of the hydrodynamic force on a sphere at finite Reynolds numbers is attributed to Oseen (1913), who used a simple perturbation method to calculate a first order correction for the steady flow drag coefficient. The expression he obtained is known as the *Oseen equation* and may be written in terms of the drag coefficient as follows

$$C_D = \frac{24}{Re_r}\left(1+\frac{3}{16}Re_r\right). \tag{4.14}$$

This expression is valid in the range $Re_r < 0.45$, which covers many practical applications in the chemical industry, where particles are very small, and the carrier fluids have high viscosity.

Proudman and Pearson (1956) used an asymptotic method to calculate the velocity field around a solid sphere and extended Oseen's result to calculate the steady drag coefficient to the order $O(Re_r^2\ln Re_r)$

$$C_D = \frac{24}{Re_r}\left[1+\frac{3}{16}Re_r+\frac{9}{160}Re_r^2\ln\left(\frac{Re_r}{2}\right)+O\left(Re_r^2\right)\right]. \tag{4.15}$$

Eq. (4.15) may be used with accuracy in applications of finite but small Re_r in the range from $0 < Re_r < 1.5$. At higher values of Re_r, it is advisable to use one of the empirical or semi-empirical expressions for the steady drag coefficient that abound in the literature. One such expression that is accurate and very frequently used is the Schiller and Nauman (1933) correlation, which is recommended to be used in the range $1 < Re_r < 800$

$$C_D = \frac{24}{Re_r}\left(1+0.15Re_r^{0.687}\right) \tag{4.16}$$

Empirical correlations and numerical results for solid spheres have generated the *standard steady drag coefficient curve*, which is depicted in Figure 4.3 and extends to $Re_r = 10^7$. It is apparent in this figure that the steady drag coefficient of a sphere becomes almost constant and approximately equal to 0.445 in the range $10^3 < Re_r < 3*10^5$. In this high-inertia range, the friction part of the drag force is insignificant, and all the contribution comes from the form drag.

A key determinant of the magnitude of the drag coefficient is the type of the boundary layer formed outside the sphere and the separation point of the wake behind the sphere. At $Re_r = 3*10^5$ the boundary layer becomes turbulent, and the

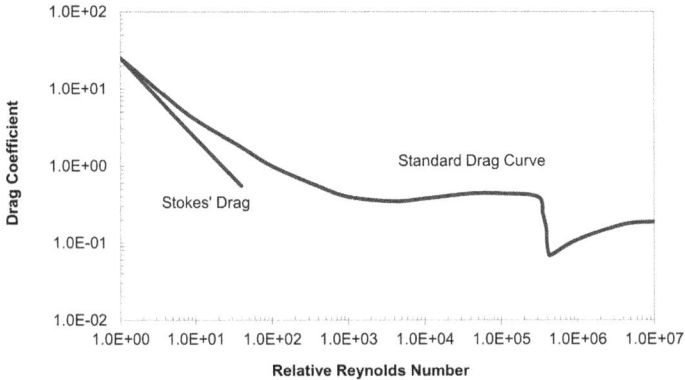

FIGURE 4.3   The standard, steady flow drag coefficient curve for solid spheres.

separation point of the flow moves downstream. The result is a sharp reduction of the form drag and, hence, a drastic reduction of the drag coefficient to the value *0.07.* This kind of drag reduction is typical of laminar-to-turbulent transitions. At these high values of Re_r, the flow becomes inherently transient. The standard hydrodynamic drag force on a sphere may be used (instead of the more complex transient expressions that are presented in section 4.11) only if the timescale of the sphere $(\tau_v = D^2\rho_c/18\mu_c)$ is much smaller than the timescale of the transients in the carrier fluid, $\tau_v \ll \tau_c$, a condition that implies $St \ll 1$.

4.1.2.3   Steady Drag on Liquid Spheres

When Re_r is finite, the external flow induces an internal circulation that generates a flow field inside the viscous spheres. The circulation modifies the external flow field and affects the drag coefficients that are markedly different from those of solid spheres. Figure 4.4, obtained from Feng and Michaelides (2001), depicts the streamlines (top semicircles) and the vorticity (bottom semicircles) of three flow fields for viscosity ratio $\lambda = 7$ and three Reynolds numbers, $Re_r = 10, 100,$ and *500.* Three significant observations may be made in Figure 4.4:

1. It is evident in the depiction of the streamlines that there is a fore-aft asymmetry of the flow, which is also present with rigid spheres. Even though at $Re_r = 10$ the recirculation region behind the sphere has not been formed, there is a clearly visible velocity-defect region. At $Re_r = 100$ there is a well-defined recirculation region and at $Re_r = 500$ this region expands considerably and extends downstream to more than one sphere diameter.
2. At $Re_r = 500$, there are sharp vorticity gradients in the outer field of the sphere, indicating the formation of a boundary layer. This is a laminar boundary layer, and its thickness is on the order of $Re_r^{-1/2}$.
3. There are two counter-rotating vortices inside the sphere—sometimes referred to as *Hill's vortices*—whose strength increases with Re_r. An internal recirculation region at the aft of the sphere appears at the higher values of Re_r or at lower values of λ.

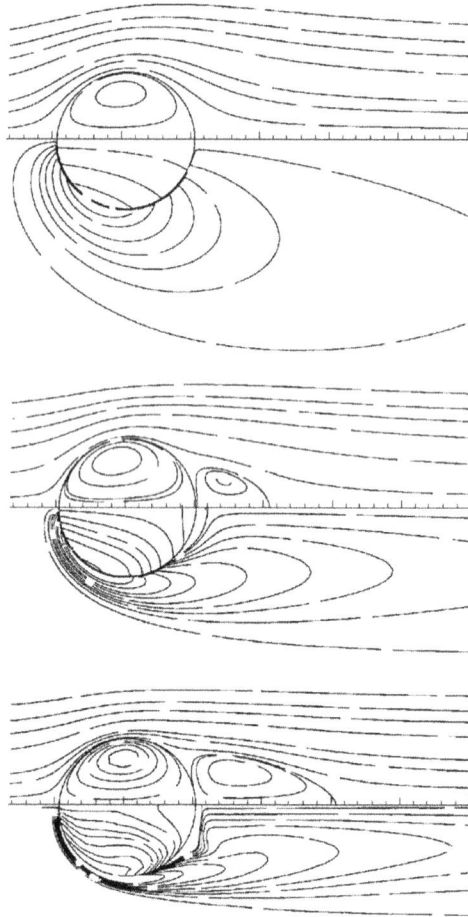

FIGURE 4.4 Streamlines (top semicircles) and vorticity (bottom semicircles) inside and around fluid spheres: (a) $\lambda = 7$, $Re_r = 10$; (b) $l = 7$, $Re_r = 100$; (c) $\lambda = 7$, $Re_r = 500$.

The internal flow in viscous spheres modifies the entire velocity field and results in lower drag. Accurate data on the drag coefficients of viscous spheres were numerically obtained by Feng and Michaelides (2001), who used a two-layer concept for the computational grid and performed computations that extend to values of Re_r up to 1,000 and for values of the viscosity ratio in the range $0 < \lambda < \infty$. The lower limit of the viscosity ratio applies to bubbles and the higher limit to solid particles. Engineering correlations for the drag coefficients of the viscous spheres in terms of the two principal variables, Re_r and λ, were developed

$$C_D\left(Re_r,\lambda\right) = \frac{2-\lambda}{2}C_D\left(Re_r,0\right) + \frac{4\lambda}{6+\lambda}C_D\left(Re_r,2\right) \quad for \quad 0 \leq \lambda \leq 2,$$

$$and \quad 5 < Re_r \leq 1,000 \tag{4.17}$$

and

$$C_D\left(Re_r,\lambda\right)=\frac{4}{\lambda+2}C_D\left(Re_r,2\right)+\frac{\lambda-2}{\lambda+2}C_D\left(Re_r,\infty\right)\quad for\quad 2\leq\lambda\leq\infty,$$

$$and\quad 5< Re_r\leq1,000 \tag{4.18}$$

The functions $C_D(Re_r,0)$, $C_D(Re_r,2)$, and $C_D(Re_r,\infty)$ are the drag coefficient for inviscid bubbles ($\lambda=2$), for viscous drops with $\lambda=2$, and for solid spheres ($\lambda=\infty$), respectively. The following functions for these drag coefficients are recommended to be used with Eqs. (4.17) and (4.18)

$$C_D\left(Re_r,0\right)=\frac{48}{Re_r}\left(1+\frac{2.21}{\sqrt{Re_r}}-\frac{2.14}{Re_r}\right), \tag{4.19}$$

$$C_D\left(Re_r,2\right)=17.0Re_r^{-2/3}, \tag{4.20}$$

and

$$C_D\left(Re_r,\infty\right)=\frac{24}{Re_r}\left(1+\frac{1}{6}Re_r^{2/3}\right). \tag{4.21}$$

The last expression is an approximate representation of the Schiller and Nauman expression, Eq. (4.16).

In the lower range of Re_r, which is not covered by the correlations in Eqs. (4.17) and (4.18), the following expression is recommended (Feng and Michaelides, 2001)

$$C_D\left(Re_r,\lambda\right)=\frac{8}{Re_r}\frac{3\lambda+2}{\lambda+1}(1+0.05\frac{3\lambda+2}{\lambda+1}Re_r)$$

$$-0.01\frac{3\lambda+2}{\lambda+1}Re_r\ln(Re_r)\qquad 0\leq Re_r\leq5. \tag{4.22}$$

Eq. (4.22) was derived from the results of the numerical computations in a way that it reduces asymptotically to the Hadamard (1911) and Rybczynski (1911) solutions at $Re_r\rightarrow0$ and to the Oliver and Chung (1987) expression for small but finite values of Re_r.

All the experimental data and numerical results show that the ratio of the density of the viscous spheres to the density of the carrier fluid, ρ_d/ρ_c has very small influence on the drag coefficients (less than 2%). At the higher values of Re_r, the viscous spheres become elongated spheroids, and their drag coefficient is a function of Re_r and λ as well as of a geometric variable that characterizes the deformation of the sphere, the elongation, or the eccentricity. In general, viscous spheres remain spherical when the dimensionless Bond number, Bo, is less than or equal to 0.2

$$Bo=\frac{We}{Fr}=\frac{gD^2\left|\rho_d-\rho_c\right|}{\sigma}\leq0.2. \tag{4.23}$$

When the Bond number is higher, the correlation equations for spheroids and ellipsoids should be used for higher accuracy (section 4.3.1). Water droplets in air will maintain their spherical shape at values of Re_r up to 470. For drops with high surface tension (e.g. liquid metals), the corresponding Re_r is significantly larger (up to $Re_r = 1,150$ for mercury droplets in air).

It must be noted that bubbles and drops in several industrial applications do not exactly follow the predictions of the analyses for viscous spheres, because of the presence of impurities in the carrier fluid. The impurities concentrate at the drop-carrier fluid interface and act as surfactants that cover the interface with a solids-like layer. This layer partly separates the external from the internal flow, which becomes weaker. The drag on the viscous spheres increases, and in several cases, the viscous spheres behave as solid spheres. When bubbles and drops are in motion, the surfactants are swept to the aft, leaving the forward surface relatively uncontaminated. This establishes a concentration gradient on the surface of the viscous spheres, which generates a surface tension gradient and, subsequently, a tangential stress that opposes the motion of the surface. The net result of this phenomenon is an increase of the total drag and is manifested in the retardation of the terminal velocity of viscous spheres (Levich, 1962). For this reason, one must consider the drag given by Eqs. (4.9), (4.17), and (4.18) as the lower limit of the hydrodynamic force, achieved under conditions of exceptional purity of the carrier fluid. The upper limit is the drag on solid spheres. Experiments have shown that, regardless of impurities, viscous spheres behave as solid spheres at creeping flow conditions ($Re_r \ll 1$) when the Eötvos number based on the density difference is less than 4

$$Eo = \frac{(\rho_d - \rho_c)gD^2}{\sigma} < 4. \tag{4.24}$$

4.1.2.4 The Drag Factor, f

It is oftentimes convenient to express the drag coefficient in terms of a *drag factor, f*, which is the ratio of the actual drag coefficient to the Stokes drag.

$$f = \frac{C_D}{24 / Re_r} = \frac{C_D D \rho_c |u - v|}{24 \mu_c}. \tag{4.25}$$

Obviously, $f = 1$ in creeping (Stokes) flow. The drag factor for the Schiller and Nauman expression, which is often used in calculations, becomes

$$f = 1 + 0.15 Re_r^{0.687}. \tag{4.26}$$

The variation of the drag factor with Reynolds number is shown in Figure 4.5 for the flow range $1 < Re_r < 10^5$. This range does not include the transition to turbulence.

4.1.3 STEADY DRAG WITH VELOCITY SLIP AT THE INTERFACE

The results and expressions for the drag force in the previous sections pertain to applications where the no-slip boundary condition applies at the interface of the

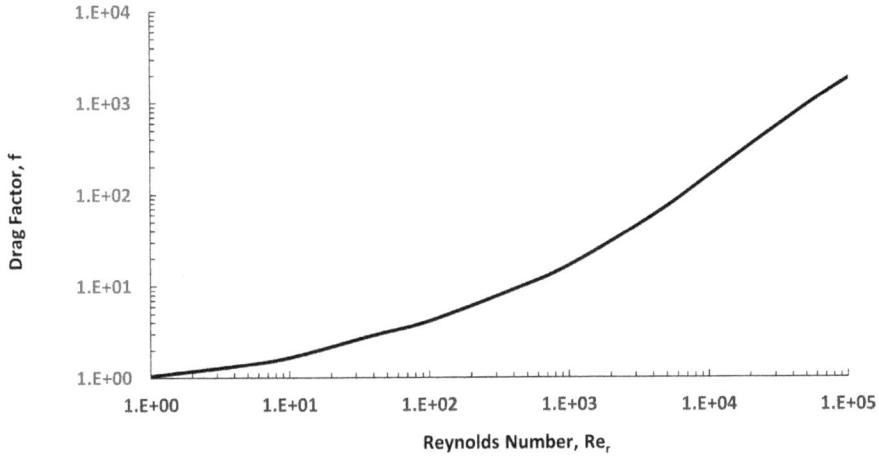

FIGURE 4.5 Variation of drag factor with the Reynolds number.

carrier fluid and the particles. Experimental observations have shown that, when the particle diameter is less than 1,000 nm or when the pressure of the carrier fluid is low enough to induce rare faction effects, there is velocity slip at the interface that significantly affects the drag coefficients.

In one of the first attempts to quantify the effect of interfacial velocity slip, Basset (1888b) stipulated that the shear stress on the sphere is proportional to the tangential slip, Δu_t, that is, $\tau = \beta \Delta u_t$. The coefficient β is referred to as the *slip parameter*. Thus, a dimensionless *slip ratio* may be defined for a sphere of radius α as follows

$$Sp = \frac{\mu_c}{\alpha\beta} = \frac{2\mu_c}{D\beta}. \tag{4.27}$$

In the case of the very small particles (nanoparticles and nanodroplets), the pertinent Reynolds number is very small and, in most cases, significantly less than one. Feng *et al.* (2012) performed an asymptotic analysis of the movement of a viscous sphere with interface slip at low Reynolds numbers, $Re_r < 1$, and derived the following expression for the steady flow drag factor

$$f = \frac{F_D}{3\pi D \mu_c U} = \frac{1}{3}\frac{2+3\lambda+6\lambda Sp}{1+\lambda+3\lambda Sp} + \frac{1}{24}\left(\frac{2+3\lambda+6\lambda Sp}{1+\lambda+3\lambda Sp}\right)^2 \frac{Re_r}{2} +$$
$$+ \frac{1}{120}\left(\frac{2+3\lambda+6\lambda Sp}{1+\lambda+3\lambda Sp}\right)^3 \frac{Re_r^2}{4}\ln\left(\frac{Re_r}{2}\right) + O\left(Re_r^2\right) \tag{4.28}$$

For a solid sphere, $\lambda \rightarrow \infty$, the drag factor becomes

$$f = \frac{1+2Sp}{1+3Sp} + \frac{3}{8}\left(\frac{1+2Sp}{1+3Sp}\right)^2 \frac{Re_r}{2} + \frac{9}{40}\left(\frac{1+2Sp}{1+3Sp}\right)^3 \frac{Re_r^2}{4}\ln\left(\frac{Re_r}{2}\right) + O\left(Re_r^2\right). \tag{4.29}$$

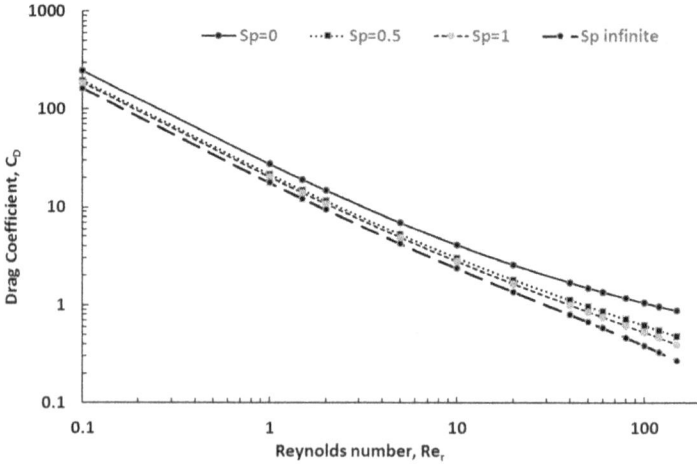

FIGURE 4.6   Effect of the slip ratio on the drag coefficient of solid spheres.

When the slip at the interface vanishes, $Sp \to 0$, and Eq. (4.29) reduces to the expression of Eq. (4.15) derived by Proudman and Pearson (1956).

Feng (2010) performed numerical simulations and derived a correlation for the drag factor on solid spheres with interfacial slip at Reynolds numbers up to 150

$$f = \frac{24}{Re_r}\frac{1+2Sp}{1+3Sp}\left(1+0.1509\sqrt{\frac{1+2Sp}{1+3Sp}}Re_r^{0.678}-0.0254\frac{Sp}{1+3Sp}Re_r^{1.104}\right). \quad (4.30)$$

This correlation applies to all values of the slip ratio, Sp, between 0 and ∞. The effect of the interfacial slip on the drag coefficients, and the standard drag curve is shown in Figure 4.6.

4.2   COMPRESSIBILITY AND RAREFACTION EFFECTS

Two phenomena, associated with compressibility and rarefaction, affect the hydrodynamic drag on particles:

1. The formation of shock waves on particles. Shock waves are formed when the Mach number, Ma, is higher than 0.6, which is referred to as the *critical Mach number*.
2. Velocity slip on the surface of the particles.

The *Mach number* and the *Knudsen number* are the dimensionless numbers that define the effects of compressibility and rarefaction on particles

$$Ma = \frac{|\vec{u}-\vec{v}|}{c} \quad and \quad Kn = \frac{L_{mf}}{D}, \quad (4.31)$$

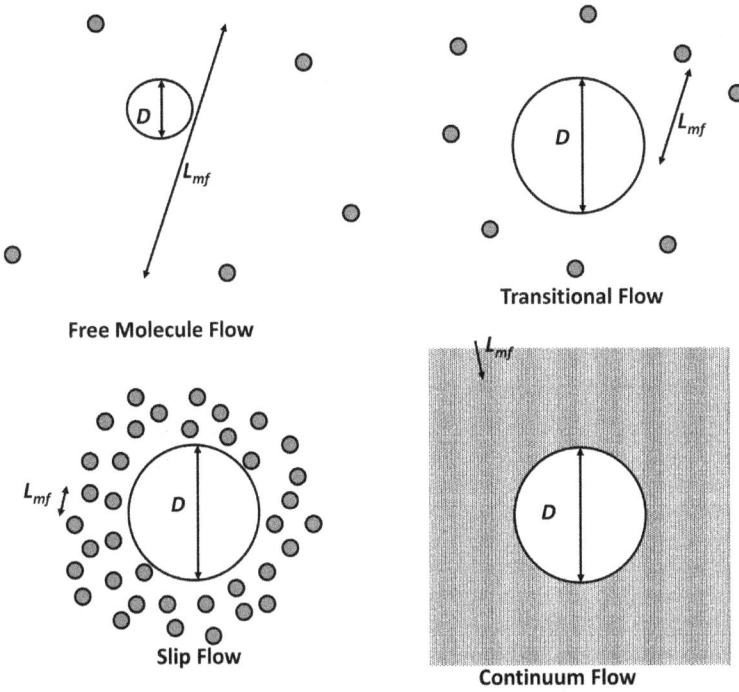

FIGURE 4.7   Schematic diagram of the flow regimes in rarefaction flows.

where c is the velocity of sound, and L_{mf} is the mean free path of the carrier fluid molecules. Large values of Kn imply that the carrier fluid may not be considered as a continuum, while the continuum postulation implies that $Kn \ll 1$. Since the molecular free path of a gas is proportional to the ratio of the viscosity divided by the product of the speed of sound and the gas density ($L_{mf} \sim \mu_c / c \rho_c$), it follows that Kn is proportional to the ratio of the Mach number and the Reynolds number ($Kn \sim Ma/Re$).

Four flow regimes, which are depicted schematically in Figure 4.7, are identified to characterize the rarefaction flows (Schaaf and Chambré, 1958):

1. *Free molecule flow* is defined by the conditions $Kn > 10$ and $Ma > 3Re_r$. In this regime, there are no interactions among the molecules. Discrete molecules collide with the particles and exchange momentum. The drag coefficient of a sphere is computed analytically to be $C_D = 2$, regardless of the value of the Reynolds number, Re_r.
2. *Transitional flow* happens in the ranges $0.25 < Kn < 10$, $0.1Re_r^{1/2} < Ma < 3Re_r$. In this regime, the size of the particles is comparable to the size of the mean free path, there are collisions among the molecules of the carrier fluid, and the interactions between particles and molecules create a flow field around the particles.
3. *Slip flow* occurs in the ranges $0.001 < Kn < 0.25$, and $0.01Re_r^{1/2} < Ma < 0.1Re_r^{1/2}$. The flow field around the particles becomes distinct; the temperature and

velocity at the surface of the particle are different from the temperature and velocity on the side of the gas close to the surface (presence of temperature and velocity slip). When $Kn < 0.01$, the flow field around the particle is obtained from the solution of the Navier-Stokes equations with the stipulation of a suitable closure equation for the slip, and the drag coefficient is determined as demonstrated in section 4.1.3.

4. *Continuum flow* occurs at $Kn < 0.001$ and $Ma < 0.1\ Re_r^{1/2}$. The no-slip boundary condition applies to the surface of the particles, and the flow field is obtained by the solution of the Navier-Stokes equations.

4.2.1 THE CUNNINGHAM CORRECTION FACTOR

Cunningham (1910) recognized that the effect of rarefaction and interfacial slip on small particles is to reduce the Stokesian drag ($C_D = 24/Re_r$) and suggested the following expression for the drag coefficient in the slip flow regime

$$C_D = \frac{24}{Re_r}\frac{1}{1+2AKn}. \qquad (4.32)$$

The coefficient A is referred to as the *Cunningham factor*, or the *Cunningham correction factor*, and is a function of Kn. Following Millikan (1923), who derived a suitable functional relationship in the form $A = C_1 + C_2 exp(-C_3/Kn)$, several authors have expressed the effects of rarefaction on the drag coefficients in terms of the Cunningham factor using the same functional form. Table 4.1 shows the results of seven such experimental studies in chronological order (Michaelides, 2014).

It is observed in Table 4.1 that the numerical values of the coefficients in all the expressions agree fairly well for the several classes of particles and drops used in the experiments. The following approximate values of the three constants are suggested to be used in calculations: $C_1 \approx 1.15$, $C_2 \approx 0.5$, and $C_3 \approx 0.5$. It must be noted that all the expressions for the Cunningham factor apply to low Reynolds and Mach numbers.

TABLE 4.1
Parameters for the Cunningham Factor in the form $A = C_1 + C_2 exp(-C_3/Kn)$

Author(s)	Particles	C_1	C_2	C_3
Knudsen and Weber (1911)	Glass spheres	1.034	0.536	0.610
Millikan (1923)	Oil droplets	1.209	0.406	0.447
Allen and Raabe (1982)	Oil droplets	1.155	0.474	0.298
Allen and Raabe (1985)	Latex spheres	1.142	0.558	0.500
Rader (1990)	Oil droplets	1.209	0.441	0.440
Hutchins *et al.* (1995)	Latex spheres	1.234	0.469	0.589
Kim *et al.* (2005)	Latex spheres	1.165	0.483	0.449

4.2.2 EFFECTS OF THE MACH NUMBER

Several authors have proposed correlations for the drag coefficient of particles in rarefied flows as functions of Ma and Re_r. The general trends of these correlations are depicted in the three-dimensional diagram of Figure 4.8. The standard, steady flow drag curve for particles corresponds to the vertical plane at $Ma = 0$. It is observed in this figure that, at the lower range of Reynolds numbers, the drag coefficient decreases with increasing Mach numbers, an observation that is congruent with all the studies shown in Table 4.1 for slip flow. At the higher range of Re_r and $Ma > 2$, the drag coefficient increases with increasing Reynolds numbers.

Given the significant variation of the size of droplets during combustion and evaporation, Crowe *et al.* (1969) showed that burning particles and droplets in rockets are subjected to all four flow regimes in rarefied flows. Crowe *et al.* (1973) proposed a correlation for the drag coefficients of particles in all flow regimes. This correlation has been successfully applied in simulations of burning particles in solid propellant rocket nozzles and was later simplified by Hermsen (1979) to the following expression

$$C_D = 2 + (C_{D,0} - 2)\exp\left[-\frac{3.07\sqrt{\gamma}\,g(Re_r)Ma}{Re_r}\right] + \frac{h(Ma)}{Ma\sqrt{\gamma}}\exp\left(-\frac{Re_r}{2Ma}\right), \quad (4.33)$$

where $C_{D,0}$ denotes the drag coefficient at $Ma = 0$, given by the standard drag curve, γ is the ratio of the specific heats of the carrier gas, and $g(Re)$ and $h(M)$ are empirical functions defined as follows

$$g(Re_r) = \frac{1 + Re_r(12.278 + 0.548Re_r)}{1 + 11.278Re_r} \quad and \quad h(Ma) = \frac{5.6}{1 + Ma} + 1.7\sqrt{\frac{T_d}{T_c}}. \quad (4.34)$$

In the last expressions T_c is the mean temperature of the carrier gas far from the particles, and T_d is the temperature of the particles. The drag coefficient given by

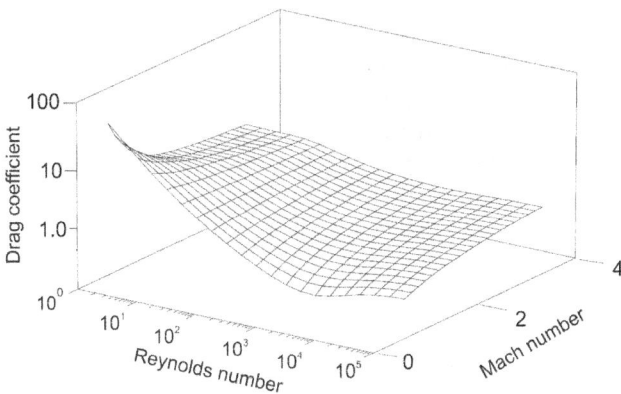

FIGURE 4.8 Variation of drag coefficient of solid spheres with the Mach number and Reynolds number.

Eqs. (4.33) and (4.34) approaches the drag coefficient of the standard drag curve as Ma approaches zero. Also, for large Knudsen numbers (large Ma/Re_r) that characterize the free molecular flow regime, the drag coefficient approaches the value 2.

4.3 NON-SPHERICAL PARTICLES

4.3.1 PARTICLES OF REGULAR SHAPES

Experiments have shown that the drag coefficients of non-spherical particles significantly differ from those of spheres. Oseen (1915) derived an analytical expression for the drag coefficient of disks that move at creeping flow conditions and with relative velocity perpendicular to their faces

$$C_{Ddisk} = \frac{64}{\pi \, Re_r} + \frac{32}{\pi^2} = \frac{20.37}{Re_r} + 3.24. \tag{4.35}$$

Oseen's expression has been used as the basis of experimental correlations for higher Reynolds numbers and for other shapes. For example, Pitter et al. (1973) derived the following correlations for disks and thin oblate spheroids

$$C_D = C_{Ddisk}\left(1 + 0.138\,Re_r^{0.792}\right) \quad for \quad 1.5 \le Re_r \le 100, \tag{4.36}$$

and

$$C_D = C_{Ddisk}\left(1 + 0.00871Re_r^{1.393}\right) \quad for \quad 100 < Re_r \le 300. \tag{4.37}$$

Other authors expressed their experimental results in graphical representations: Masliyah and Epstein (1970) developed a graph for the drag coefficients of spheroids and disks, which is shown in Figure 4.9. The aspect ratio, E, is the ratio of the equatorial (perpendicular to the flow) radius of the spheroid, b, to the axial (in the direction of the flow) radius, a. Thus $E = b/a$, with the case $E = 1$ corresponding to a sphere and $E \ll 1$ corresponding to a thin disk. Similarly, Zukauskas and Ziugzda (1985) developed a graph for the drag coefficients of long cylinders, which includes the transition to turbulent flow, and is depicted in Figure 4.10. This figure may also be used for long fibers. More recently, Zastawny et al. (2012) used numerical data to develop a correlation using parameters for the shape and directionality of particles with generalized spheroidal shapes that span the range from fibers to disks. They expressed the drag coefficients in terms of two parameters related to the drag in two perpendicular directions, denoted in the following equation by the angle φ, as $\varphi = 0°$ and $\varphi = 90°$

$$C_D(\phi) = C_D(\phi = 0) + \left(C_D(\phi = 90) - C_D(\phi = 0)\right)\sin^{a_0}(\phi)$$

$$C_D(\phi = 0) = \frac{a_1}{Re_r^{a_2}} + \frac{a_3}{Re_r^{a_4}} \tag{4.38}$$

$$C_D(\phi = 90) = \frac{a_5}{Re_r^{a_6}} + \frac{a_7}{Re_r^{a_8}}$$

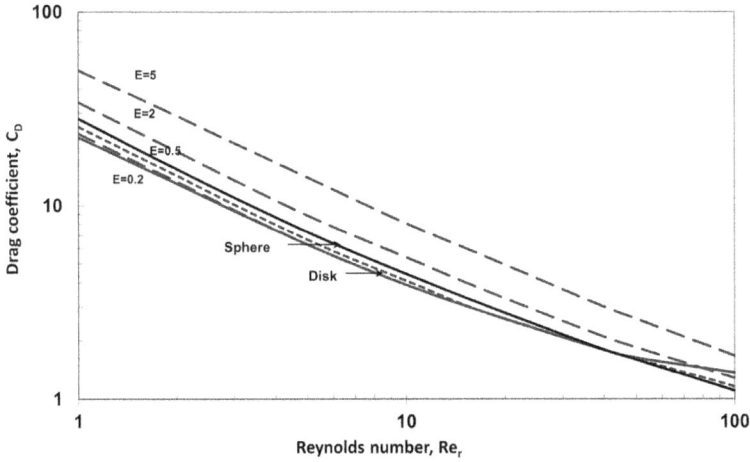

FIGURE 4.9 Drag coefficients for disks and spheroids.

Source: Data from Masliyah and Epstein, 1970.

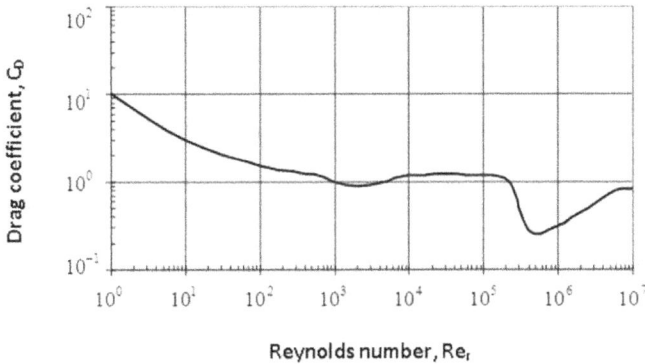

FIGURE 4.10 The drag coefficient of long cylinders.

Source: Data from Zukauskas and Ziugzda, 1985.

where the nine coefficients $a_0 - a_8$ represent empirical parameters that depend on the particle shape.

A great deal of information on the drag coefficients of particles that have specific shapes may be found in Clift *et al.* (1978) and Michaelides (2006) as well as in several journal publications.

4.3.2 Particles with Irregular Shapes

For particles that have irregular shapes, a variety of other factors—emanating from experiments or numerical computations—have been used to determine their drag coefficients. Among these factors are the *equivalent diameters*, which were defined in Eq. (3.1) as

$$D_V = \sqrt[3]{\frac{6V}{\pi}}, \quad D_A = \sqrt{\frac{A}{\pi}}, \quad D_{PA} = \sqrt{\frac{4A_p}{\pi}} \quad \text{and} \quad D_P = \frac{P}{\pi}.$$

A variable that has been extensively used to characterize the shape of irregular particles is the *sphericity*, Ψ, which was defined by Wadell (1933) and is used as a correction factor for the drag coefficients

$$\Psi = \frac{\pi^{1/3}(6V)^{2/3}}{A} = \frac{D_V^2}{D_A^2} \quad \text{with} \quad C_D = C_{Ds}\frac{\Psi}{K^2}, \tag{4.39}$$

where C_{Ds} is the drag coefficient of a sphere with diameter, D_V, and K is an experimentally determined factor given graphically in terms of Ψ and Re_r, $K = K(\Psi, Re_r)$. Eq. (4.38) may be used as an approximation for steady drag coefficients of near-spherical particles. However, Haider and Levenspiel (1989), who collected averaged drag data for shapes of particles like fibers and disks and proposed a correlation that depends on the sphericity, determined that (4.39) introduces significant errors with elongated particles, when the correction factors become very large.

It must be recognized that the cross-sectional area of significance for the drag coefficients of particles is the area, which is perpendicular to the flow direction, sometimes called the *projected area A_P*. The crosswise sphericity of the particle, Ψ_T, is based on this area and is defined in a manner similar to Eq. (4.39)

$$\Psi_T = \frac{\pi^{1/3}(6V)^{2/3}}{A_P} = \frac{D_V^2}{D_{PA}^2}. \tag{4.40}$$

An additional variable that also characterizes the shape of particles in motion is the *circularity*—sometimes also called *surface sphericity*—which is defined in terms of the equivalent diameter, D_{APr}, and the projected perimeter in the direction of the flow, P_{Pr}

$$c = \frac{\pi D_{PA}}{P_{Pr}}. \tag{4.41}$$

A correlation that emanates from experimental data and is accurate and simple enough to use and applies to several types of non-spherical particles was derived by Tran-Cong *et al.* (2004), who performed an experimental study with spheroids, prisms, star-shaped, H-shaped, elongated bars and cylinders, X-shaped, and cross-shaped particles with a wide range of sizes. They proposed the following expression for the drag coefficient of irregularly shaped particles

$$C_D = \frac{24}{Re_r\dfrac{D_{PA}}{D_V}}\left[1+\frac{0.15}{\sqrt{c}}\left(Re_r\frac{D_{PA}}{D_V}\right)^{0.687}\right]+\frac{0.42}{\sqrt{c}\left[1+4.25\times10^4\left(Re_r\dfrac{D_{PA}}{D_V}\right)^{-1.16}\right]}. \tag{4.42}$$

The Reynolds number in Eq. (4.42) is defined in terms of the volume-equivalent diameter, D_V, and hence, the two terms in the parentheses represent the Reynolds number based on the equivalent diameter of the projected area, D_{Apr}. Eq. (4.42) is valid in the ranges $0.15 < Re_r < 1,500$, $0.80 < D_{PA}/D_V < 1.50$, and $0.4 < c < 1.0$. These ranges cover most of the irregularly shaped particles in engineering applications.

A different type of correlation function for the drag on irregularly shaped particles requires the knowledge of an additional parameter related to the orientation of the particles relative to the direction of motion. This is the lengthwise sphericity of the particle, Ψ_{ll}, which is the ratio between the cross-sectional area of the volume equivalent sphere and the difference between half the surface area and the mean longitudinal (i.e. parallel to the direction of relative flow) projected cross-sectional area of the particle. Hölzer and Sommerfeld (2008) used this approach to correlate their numerical data and derived the following expression

$$C_D = \frac{8}{Re_r\sqrt{\Psi_{II}}} + \frac{16}{Re_r\sqrt{\Psi}} + \frac{3}{\sqrt{Re_r}\,\Psi^{3/4}} + 0.4210^{0.4(-\log\Psi)^{0.2}}\frac{1}{\Psi_T}. \quad (4.43)$$

There is a large number of more recent numerical studies (e.g. Ouchene et al., 2016; Sommerfeld and Lain, 2018; Sanjeevi et al., 2018) that examined the forces and the movement of non-spherical particles in carrier fluids. Most of these studies propose correlations for the drag and lift coefficients in terms of angles that indicate the orientation of the particle with respect to the main flow. However, these angles are only known when individual particles are considered. In the case of numerical modelling of heterogeneous mixtures using a Eulerian method (see Chapter 8), where the effect of groups of particles on the momentum of the carrier fluid is needed, correlations that include information on the particle orientation cannot be used.

It must be noted that experimental studies, which examined the flow of non-spherical particles (e.g. Tran-Cong et al., 2004; Dioguardi and Mele, 2015), also observed that such particles also "pitch" and "wobble" as they move within the carrier fluid. Several numerical studies (e.g. Sanjeevi et al., 2018) also observed significant (up to 50%) velocity fluctuations. Such observations are indications of transient effects that need to be accounted in the numerical simulations. In general, the transient effects may not be important at higher Re_r but significantly affect the flow and the hydrodynamic force on particles at lower values of Re_r, as it is explained in Section 4.11. Since the experimental data implicitly include the associated transient effects, correlations emanating from experiments may be preferable to be used in practice.

Another approach to the determination of the drag coefficients of irregularly shaped particles is to derive correlations for specific types of particles and applications. For example, Connolly et al. (2020) derived a correlation for the drag coefficients of Arizona road dust particles as a function of the Corey shape factor. Such correlations are strictly applicable to the materials and applications the experimental or computational results were originally derived.

4.3.3 THE STOKES OR HYDRODYNAMIC DIAMETER

A method that is often used to apply results derived for spheres to non-spherical and especially irregularly shaped particles is to define a diameter, often referred to as the

Stokes diameter, or *hydrodynamic diameter, D_{St}*, and the corresponding radius, a_{St}. The concept of the Stokes diameter stems from an equivalency of the drag force: A spherical particle with diameter, D_{St}, has the same Stokesian drag as the of irregular-shaped particle, whose dimensions may be difficult to measure. The Stokes diameter is often used to define a length scale in geophysical applications—for example, sediment movement or settling—and the value of this length scale is typically calculated from the terminal velocities of particles.

It must be noted that all the methods for the determination of the drag force on non-spherical particles are based on empirical correlations or asymptotic analytical methods. As such, the derived drag coefficients presented previously are approximate and laden with significant uncertainty. A more accurate determination of the drag force on particles may be achieved numerically for the specific types and shapes of particles in any application. Given the recent advances in computational methods (Chapter 8) and the availability of computational resources, this determination is feasible for individual particles as well as groups of tens (and possibly hundreds) of particles.

4.4   EFFECTS OF FLOW TURBULENCE

A number of experimental studies in the 1960s and 1970s suggested that the steady flow drag coefficients of particles is affected by the free-stream turbulence in the carrier fluid flow, and several correlations were proposed to account for this effect. However, a glance at the data and the derived correlations proves that there are many discrepancies in the data that were not resolved. Turbulence is inherently time-dependent, and the movement of all types of particles in turbulent flows is transient and not steady. As a consequence, some of the methods used in several of these studies to derive the steady drag coefficients of particles are flawed. A more recent experimental study by Warnica *et al.* (1994) with drops concluded that there is no experimental evidence to suggest that the free-stream turbulence has a significant effect on the drag coefficients. A numerical study by Bagchi and Balachandar (2003) with free-stream turbulence between 10% and 25% determined that turbulence has little effect on the average drag coefficient. Such recent studies suggest that the standard drag curve yields an adequate prediction of the drag force. It must be noted that while the free-stream turbulence has negligible effect on the drag of particles, the presence of particles significantly influences the turbulence intensity of the carrier fluid, a phenomenon that is known as *turbulence modulation*, which is examined in Section 5.7.

4.5   BLOWING EFFECTS

Several spray applications involve high carrier fluid temperatures, where drops evaporate and particles sublimate or burn. Mass transfer from the surface of particles generates an outward radial flow and the following three significant effects:

1. The temperature and concentration gradients between the particle surface and the bulk of the carrier fluid, in general, cause a reduction in the absolute viscosity of the carrier fluid.

2. Vaporization affects the boundary layer around the particle. In general, this reduces the friction drag and increases the form drag.
3. Regression of the surface of the particle, which is a consequence of the mass transfer to the carrier fluid and is referred to as *Stefan convection*. The particle shrinks.

Two dimensionless numbers, the heat transfer number, B_T, and the mass transfer number, B_M, are used in correlations for the blowing effects on particles

$$B_T = \frac{c_{pc}\left(T_s - T_\infty\right)}{h_{fg}} \quad and \quad B_M = \frac{\omega_s - \omega_\infty}{1 - \omega_s}, \tag{4.44}$$

where c_{pc} is the specific heat capacity of the carrier fluid and ω the mass fraction of the species of the particle. The effective latent heat of vaporization, h_{fg}^{eff}, which is the sum of the latent heat of the vapor at the drop surface and the sensible heat conducted to the interior of the particle, is sometimes used in the calculation of B_T. In the vast majority of vaporization and combustion processes, the heat conducted to the interior of the particle is much smaller than the latent heat and, therefore, $h_{fg} \approx h_{fg}^{eff}$.

Regarding the properties of the carrier fluid, Yuen and Chen (1976) recommend using the *film viscosity* of the carrier fluid, which is usually a gas

$$\mu_m = \mu_s + \frac{1}{3}(\mu_\infty - \mu_s). \tag{4.45}$$

This equation is sometimes referred to as the *1/3 rule* for the film properties. Renksizbulut and Yuen (1983) examined the effects of the radial flow and recommend that the following correction on the drag coefficient of drops be used in calculations

$$C_D = \frac{C_{D,0}}{1 + B_M}, \tag{4.46}$$

where $C_{D,0}$ represents the value of the drag coefficient in the absence of mass transfer, which is calculated from the correlations presented in section 4.1.2. The numerical study by Chiang et al. (1992) derived a different correlation for the drag coefficients of drops, valid in the range $30 < Re_{rm} < 200$

$$C_D = \frac{24.432}{(1 + B_T)^{0.27} (Re_{rm}/2)^{0.721}}. \tag{4.47}$$

The relative Reynolds number, Re_{rm}, in Eq. (4.47) is defined in terms of the film viscosity of the carrier gas—as given by Eq. (4.45)—and the free-stream carrier fluid density, $\rho_{c\infty}$.

Most of the experimental and numerical studies show that the drag of burning or evaporating droplets and particles is only slightly modified from the non-burning case unless the droplets are burning very fast in a pure oxidizing environment. A comprehensive review of droplet dynamics and evaporation in sprays can be found in Sirignano (1993, 1999).

4.6   TRANSVERSE (LIFT) FORCES DUE TO PARTICLE ROTATION AND FLOW SHEAR

Particle rotation combined with finite relative velocity, or fluid velocity gradients (shear), induces a transverse component of the hydrodynamic force on particles, which is often called *the lift force*. This transverse component of the hydrodynamic force alters the direction of movement of the particles in mid-flight and affects significantly industrial processes such as particle deposition and collection, particle resuspension, equipment erosion, channel blockage, and heat exchanger fouling. The transverse force of rotating balls also plays an important role in sports: The curve ball and breaking ball in baseball, the curve ball in golf, and the bending free kicks and corner kicks in soccer are all manifestations of the transverse hydrodynamic force on the balls. While examining the physics of the movement or spinning particles in fluids, it becomes apparent that two dimensionless parameters affect the transverse forces of spheres in viscous fluids, the Reynold number based on the rate of rotation, Re_R

$$Re_R = \frac{\rho_c D^2 |\vec{\Omega}|}{\mu_c},$$
(4.48)

and the *rotation parameter*, or *spin factor*

$$N_R = \frac{|\vec{\Omega}| D}{|\vec{u} - \vec{v}|},$$
(4.49)

where $\vec{\Omega} = 0.5 \nabla \times \vec{u} - \vec{\omega}$ represents the vector of the relative rate of rotation of the sphere in the fluid.

4.6.1   THE MAGNUS FORCE

When a spherical particle moves with relative velocity in an inviscid fluid and also rotates with respect to the far-field of the flow, the velocity asymmetry around the particle surface induces a transverse pressure difference, which generates the *Magnus force* (Magnus, 1861)

$$\vec{F}_{LM} = \frac{\pi}{8} D^3 \rho_c \vec{\Omega} \times (\vec{u} - \vec{v}),$$
(4.50)

It is apparent that the direction of the Magnus force is perpendicular to the plane of the relative velocity and the axis of rotation, as shown in part (a) of Figure 4.11.[2]

The magnitude of the lift force in viscous fluids is expressed as a function of a dimensionless *lift coefficient, C_{LM}*

$$C_{LM} = \frac{8 |\vec{F}_{LM}|}{\pi D^3 \rho_c |\vec{\Omega}| |\vec{u} - \vec{v}|}.$$
(4.51)

---

[2] This force is largely responsible for the lateral movement of the ball in the "curveballs" of soccer and baseball.

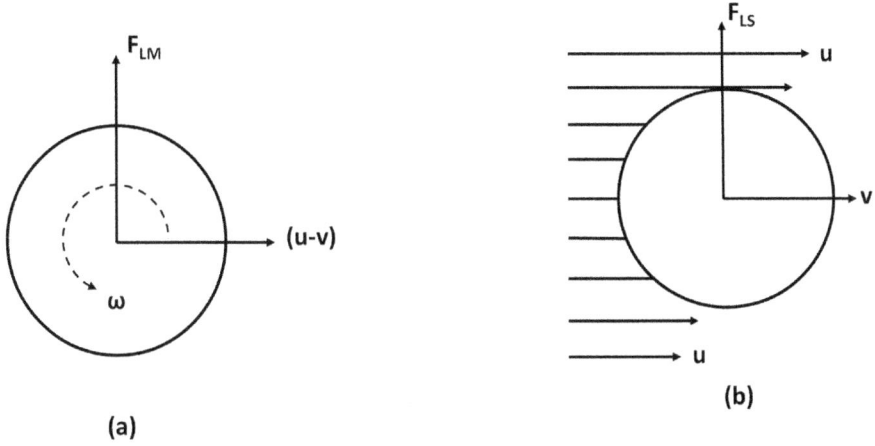

FIGURE 4.11 The Magnus force (a) and the Saffman force (b) on a spherical particle.

Experiments with solid spheres in viscous fluids have shown that the lift force is a function of the Reynolds numbers, Re_R and Re_r. The experimental study by Oesterle and Bui-Dihn (1998) recommends the following correlation

$$C_{LM} = 0.45 + (\frac{Re_R}{Re_r} - 0.45)exp(-0.05684Re_R^{0.4} Re_r^{0.3}) \quad for \ Re_R < 140, \quad (4.52)$$

The lift coefficient is, in general, a monotonically decreasing function of Re_R and increases with the rotation parameter, N_R. Experiments at relatively high Re_r and low values of N_R suggest that C_{LM} may become negative (Tanaka et al., 1990). This is most likely due to a higher relative velocity on one side of the sphere and a premature boundary layer transition to turbulent on that side, which reverses the direction of the lift force.

4.6.2 THE SAFFMAN FORCE

When a solid sphere is in shear flow, as shown in part (b) of Figure 4.11, the part of the sphere where the velocity is higher—the upper part in the figure—experiences lesser pressure. As a result of the pressure differential, a lateral force is generated on the sphere. Saffman (1965, 1968) calculated this force for a non-rotating sphere at the limit of vanishing Reynolds numbers (creeping flow) and derived the following expression

$$\vec{F}_{LS} = \frac{1.615D^2 \sqrt{\rho_c \mu_c}}{\sqrt{|\vec{\gamma}|}} (\vec{u} - \vec{v}) \times \vec{\gamma}, \quad (4.53)$$

where $\vec{\gamma}$ is the fluid velocity shear, evaluated at the center of the sphere. The direction of the Saffman force is the perpendicular direction to the plane defined by the relative velocity vector and the shear vector.

If the sphere is free to rotate within the shear flow field, the magnitude of this force becomes

$$
F_{LS} = \left[\frac{1.615D^2 \sqrt{\rho_c \mu_c}}{\sqrt{|\dot{\gamma}|}} - \frac{11}{64} \pi \rho_c D^3 \right] |\vec{u} - \vec{v}| |\dot{\gamma}| + \frac{\pi}{8} \rho_c |\vec{\Omega}| D^3 |\vec{u} - \vec{v}|. \tag{4.54}
$$

Eqs. (4.52) and (4.53) have been derived under these conditions

$$
Re_r \ll 1, \quad |\vec{v} - \vec{u}| \ll (\gamma v_c)^{1/2}, \quad \gamma D^2 / v \ll 1 \quad and \quad D^2 |\vec{\Omega}| / v_c \ll 1. \tag{4.55}
$$

These conditions imply one of the following alternatives:

1. Very low particle velocity, shear, and angular rotation for the particle
2. High kinematic viscosity for the carrier fluid if the particle has finite velocity, shear, or angular rotation

McLaughlin (1991) extended the theoretical analysis by Saffman to higher Re_r and derived a correction, part of which is given in tabular form. Dandy and Dwyer (1990) conducted a numerical study, also at higher Re_r, and their computational results were reduced to a useful correlation by Mei (1992)

$$
\vec{F}_{LS} = \left[\frac{1.615D^2 \sqrt{\rho_c \mu_c}}{\sqrt{|\dot{\gamma}|}} (\vec{u} - \vec{v}) \times \vec{\gamma} \right] \left[(1 - 0.3314\sqrt{\beta}) \exp(-\frac{Re_r}{10}) + 0.3314\sqrt{\beta} \right] \, for \, Re_r \leq 40
$$

$$
\vec{F}_{LS} = \left[\frac{1.615D^2 \sqrt{\rho_c \mu_c}}{\sqrt{|\dot{\gamma}|}} (\vec{u} - \vec{v}) \times \vec{\gamma} \right] 0.0524\sqrt{\beta Re_r} \, for \, Re_r > 40 \tag{4.56}
$$

$$
where \quad \beta = \frac{D |\nabla \times \vec{u}|}{2 |\vec{u} - \vec{v}|}
$$

The parameter β is a dimensionless measure of the flow shear. The recommended range of the correlation in Eq. (4.56) is $0.005 < \beta < 0.4$.

The transverse force on stationary particles in shear flow is significantly enhanced by the movement of other particles in the flow field. The numerical study by Feng and Michaelides (2002a) showed that the shear-induced lift on solid spheres that are deposited on a flat surface increases by a factor 2–4 when other particles in motion pass by in their vicinity.

Very little experimental information is available on the effect of rotation on the drag coefficient of spinning particles, primarily because of the difficulties in measuring drag coefficients and simultaneously quantifying the spin rate. Most of the available experimental studies and more recent numerical studies indicate that rotation does not significantly influence the drag coefficient of spinning particles (Barkla and Auchterlonie, 1971; Bagchi and Balachandar, 2002). An analytical study by Rubinow and Keller (1961) supports these observations. Similarly, Patnaik et al. (1992), who measured the drag coefficients of spheres suspended in turbulent boundary layers (shear flow), concluded that the drag coefficient, based on the velocity of the undisturbed flow at the center of the sphere, correlated well with the drag coefficients of the standard drag curve, which is depicted in Figure 4.3.

A glance at Eqs. (4.50) and (4.53) proves that the lift forces on small spheres are very much weaker than the longitudinal drag force. The ratio of the Magnus force to that of Stokes drag force is $D^2\omega/v_c$, and the ratio of the Safman force to the Stokes drag force is $D(\gamma/v_c)^{1/2}$. However, the transverse forces and the lateral migration of particles play a dominant role in the processes of boiling and evaporation from solid surfaces, diffusion and dispersion, wall deposition, channel erosion, mixing, and separation of particles.

4.7 EFFECTS OF SOLID BOUNDARIES

4.7.1 EFFECT OF ENCLOSURES

When the flow of the carrier fluid is constrained by an enclosure, the movement of particles is retarded. Faxen (1922) was the first to consider the rectilinear movement of a particle at the middle plane of two parallel walls. He used an asymptotic method and derived the wall effect on the hydrodynamic force. This effect is usually expressed in terms of a *wall drag multiplier*, K_{wall}, which is the ratio of the drag in the vicinity of the wall to the Stokesian drag in an unbounded flow. Bohlin (1960) extended Faxen's asymptotic method and obtained a higher order approximation for the wall drag multiplier in the case of a particle moving at the centerline of a cylinder at creeping flow conditions ($Re_r \ll 1$)

$$K_{wall} = \frac{|\vec{F}_D|}{3\pi D\mu_c|\vec{u}-\vec{v}|} = \begin{bmatrix} 1 - 2.01443\Xi + 2.088777\Xi^3 - 6.94813\Xi^5 \\ -1.372\Xi^6 + 3.87\Xi^8 - 4.19\Xi^{10} \end{bmatrix}^{-1}. \quad (4.57)$$

The variable, Ξ, is the dimensionless ratio of the particle diameter, D, to the diameter of the cylinder, and Eq. (4.57) is valid for $0 < \Xi < 0.5$. Paine and Scherr (1975) computed analytically the drag multiplier in the range $0 < \Xi < 0.9$ and $Re_r \ll 1$. Their results are correlated by the expression

$$K_{wall} = 4.9606\Xi - 9.8133\Xi^2 + 13.736\Xi^3, \quad (4.58)$$

and are depicted graphically in Figure 4.12, where it is apparent that the drag on the sphere increases significantly with Ξ. The dramatic increase of the drag coefficient at the higher values of Ξ is due to the significant updraft movement of the fluid through the small gap between the surface of the sphere and the wall of the enclosure. Effectively the movement of the particle in the enclosure creates a significant opposing movement of the fluid. It should be noted that the results of the three theoretical predictions by Faxen (1922), Bohlin (1960), and Paine and Scherr (1975) have been verified in the experimental studies by Iwaoka and Ishii (1979) and Miyamura et al. (1981).

Feng and Michaelides (2002b) conducted a more general numerical study in the ranges $0.1 < \Xi < 0.8$ and $0 < Re_r < 35$ for cylindrical and prismatic enclosures. They

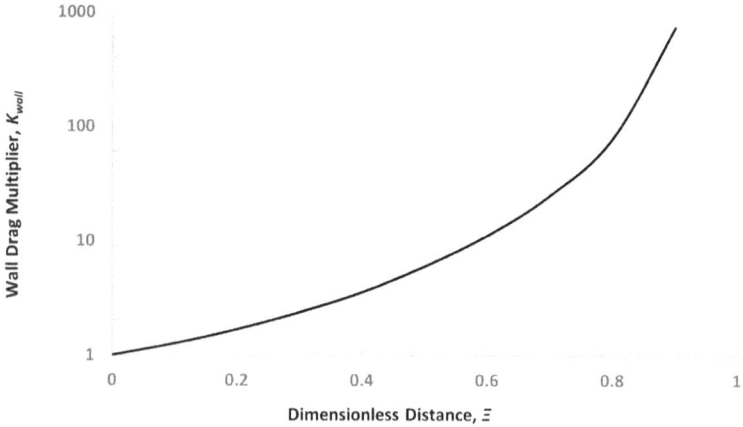

FIGURE 4.12 The wall drag multiplier as a function of the distance, Ξ.

determined that when $\Xi > 0.5$, the flow inertia contributes less than 10% of the value of K_{wall}, and surface friction is the dominant cause for drag enhancement. Their expression for the wall drag multiplier is

$$K_{wall} = \left(4.9606\Xi - 9.8133\Xi^2 + 13.736\Xi^3\right)$$
$$\left(1 + 0.0032\,Re_r + 0.0003\,Re_r^2\right), \quad Re_r < 35. \tag{4.59}$$

The study by Feng and Michaelides (2002b) also showed that the effect of boundaries on settling spheres is different in prismatic enclosures than in circular cylinders because secondary flows are induced in the corners of the prisms. Even though the secondary flows are relatively weak, they affect considerably the drag on the spheres.

4.7.2 EFFECT OF SOLID BOUNDARIES

When a particle is in motion close to a single solid wall, the carrier fluid velocity field is constrained by the no-slip condition at the side of the wall, but it is unconstrained at the side opposite the wall. This generates an asymmetry on the velocity and pressure fields around the particle, as shown in Figure 4.13 (from Feng et al., 2021). The result of the asymmetry is increased drag on the particle and a small lift component. The analytical studies by Cox and Hsu (1977) and Vasseur and Cox (1976) determined that spherical particles, which settle parallel to a vertical boundary with terminal velocity, V_T, also experience a lateral *migration velocity*, V_L, that drives the particles away from the wall

$$V_L = \frac{3}{16}\frac{DV_\infty^2}{v_c}\left[\left(\frac{v_c}{L_w V_T}\right)^2 + 2.219\left(\frac{v_c}{L_w V_T}\right)^{5/2}\right], \tag{4.60}$$

where V_∞ is the terminal velocity of the particle in the absence of the constraining wall (in an infinite flow field), and L_w is the distance of the center of the particle from

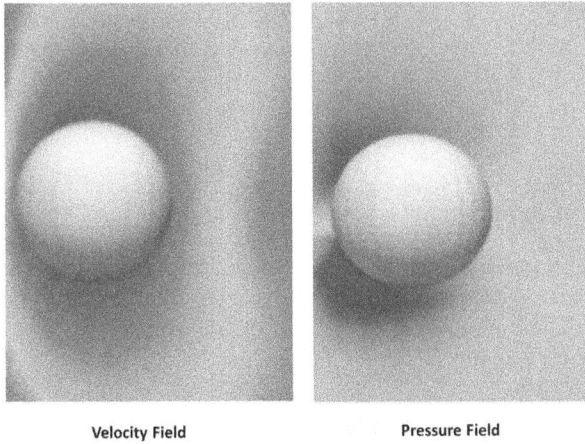

Velocity Field   Pressure Field

FIGURE 4.13   Asymmetry in the velocity and pressure fields generated in the vicinity of a solid boundary.

Source: From Feng *et al.*, 2021.

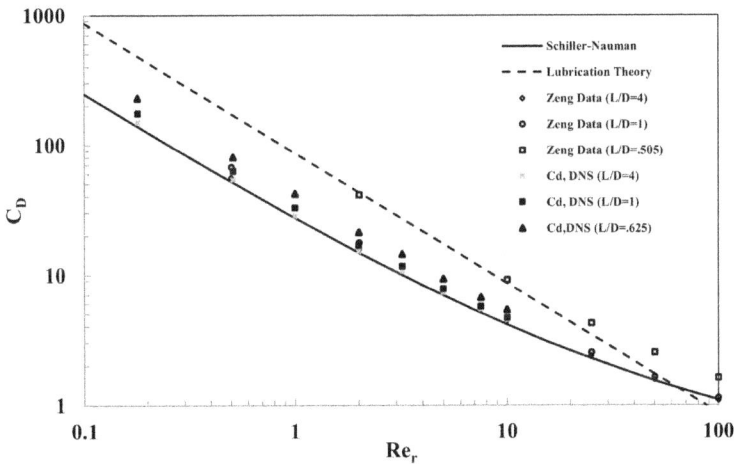

FIGURE 4.14   The effect of a solid wall on the drag coefficients of spherical particles. The two limits of lubrication flow and unconstrained flow are also depicted.

the wall. The terms in the two parentheses are the inverse of a Reynolds number with L_w as the characteristic length.

The drag on spherical particles that settle close to a vertical wall increases significantly as a result of the wall constraint and varies between two limits: At the lower limit is the drag in an infinite flow field and at the upper limit, the drag derived from the lubrication theory, which applies at distances extremely close to the wall ($L_w/D \approx 0.5$). Figure 4.14 shows the two limits as well as numerical data by Zeng *et al.* (2009) and Feng *et al.* (2021). Feng *et al.* (2021) also derived a simple correlation for the effect of the solid boundary on the drag coefficient of a spherical particle

$$C_D = \left(\frac{24}{Re_r} + \frac{4}{Re_r^{0.3}} \right) \left[1 - \frac{9}{32} \frac{D}{L_w} + \left(0.24 - 0.072 \frac{L_w}{D} \right) Re_r^{0.125} \right]^{-1}. \quad (4.61)$$

4.8 ELECTRICAL FORCES

Electric forces on particles are the result of electric charges on the surface of particles and imposed external electric fields. Hence, the electric forces scale as D^2 and are of far greater importance in systems of micro- and nanoparticles, for which the ratio of the surface to the volume forces scales as D^{-1}. For this reason, electric forces are extensively used in the characterization, separation, or fractionation[3] of polydisperse particulate systems. Electric effects are also used for the stabilization of several colloidal systems and the formation of gels.

Let us consider a negatively charged spherical particle in an electrolytic solution, as shown in Figure 4.15. Because of the negative charge on the surface of the particle, positive ions from the solution migrate in the vicinity of the particle, carry molecules of the solute, and form a fluid layer with predominantly positive charge around it. Given the equality of positive and negative ions in the electrolyte solution, this positive layer is surrounded by another layer of predominantly negative charges. The array of predominantly positive and negative fluid layers forms a loose fluid

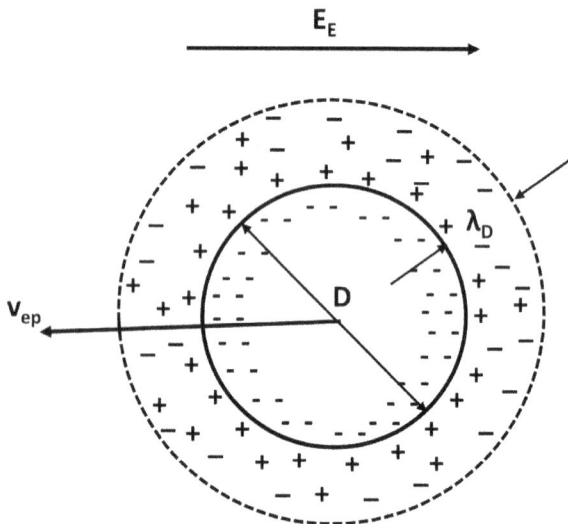

FIGURE 4.15 A negatively charged particle in an electrolyte solution with the double layer depicted. The external electric field causes the electrophoresis of the particle and the accompanying fluid layer.

---

[3] Fractionation is the separation of particles within the suspension in fractions of different sizes. Fractionation is typically caused by gravitational settling, centrifuging, or electric forces.

layer around the particle, which is referred to as *the Debye sheath, the Debye layer*, or *the double layer*. The system of the negatively charged particle and the double layer generate an electric potential within the carrier fluid that decays with the distance from the particle. The dimension of the double layer is characterized by the *Debye length*, λ_D, which is defined as the distance from the surface of the charged particle to an outer surface where the total electric potential is reduced by a factor $e^{-1} = 0.368$. A measure of the Debye length is (Probstein, 1994)

$$\lambda_D = \sqrt{\frac{\varepsilon\varepsilon_0 k_B T}{2000\, e^2 z_E^2 C N_{av}}}, \tag{4.62}$$

where ε is the dielectric constant of the solution; ε_0 is the electric permittivity of vacuum, $8.85*10^{-12}$ F/m; e is the electric charge of an electron, $e = 1.6*10^{-19}$ C; z_E is the charge number (valence) of the ions in the electrolytic solution; N_{av} is the Avogadro number $6.023*10^{-23}$ ions/mole; and C is the concentration of the solution in mol/L (moles per liter). For a water solution at 25°C the Debye length is (in m)

$$\lambda_D = \frac{3.041*10^{-10}}{z_E \sqrt{C}}. \tag{4.63}$$

The Debye length in aqueous solutions is on the order of 0.1–1 nm; hence, its effects are significant with small nanoparticles only. Non-aqueous fluids and especially organic fluids engender double layers that are significantly bulkier, and the mass of the fluid carried with the particle is significantly higher.

4.8.1 THE ZETA POTENTIAL

While the ions immediately close to the surface of the negatively charged sphere in Figure 4.15 are positive, the ions close to the end of the double layer are predominantly negative. The system of the electric charges around the negatively charged spherical particle may be modeled by considering two concentric spheres with opposite charges, q and $-q$, and with two diameters D and $D + 2\lambda_D$. The electric potential created by the two charged spheres is called the *zeta potential* and is given by the expression (Probstein, 1994)

$$\zeta = \frac{q}{2\pi\varepsilon\varepsilon_0 D} - \frac{q}{2\pi\varepsilon\varepsilon_0 \left(D + 2\lambda_D\right)}. \tag{4.64}$$

In the limit $D \gg 2\lambda_D$, which is called the *small Debye length limit*, the relationship between the surface charge per unit area, q_s, and the zeta potential simplifies to

$$\zeta = q_s \frac{\lambda_D}{\varepsilon\varepsilon_0} \quad with \quad q_s = \frac{q}{\pi D^2} \Rightarrow \zeta = \frac{q\lambda_D}{\pi D^2 \varepsilon\varepsilon_0}. \tag{4.65}$$

The zeta potential of particles offers a qualitative knowledge of the tendency of the particles to form aggregates. High values of the zeta potential in a suspension—either positive or negative—indicates that there are strong repulsion forces between the particles. The random Brownian movement of the particles is insufficient to overcome these repulsion forces, and particles do not aggregate. On the contrary, low values of the zeta potential signify weak inter-particle repulsion, which are easy to overcome by the Brownian movement, and the particles may aggregate and settle out of the suspension (Zhou *et al.*, 2007).

4.8.2 ELECTROPHORESIS

When an electric field of intensity E_E is externally imposed on the particulate mixture, as shown in Figure 4.15, the particle and the double layer move in the direction opposite to the field. This motion is referred to as *electrophoresis* and is resisted by the hydrodynamic drag on the combination of particle and the associated double layer. Hence, the particles attain steady motion with a constant velocity, v_{ep}, the *electrophoretic velocity*. In the case of very small particles that experience Stokesian drag, the electrophoretic velocity is

$$qE_E = 3\pi\mu_c D v_{ep} \Rightarrow v_{ep} = \frac{qE_E}{3\pi\mu_c D}. \tag{4.66}$$

The charge q around the particle is determined by the distribution of the electrolyte ions within the double layer. In the first limit, $D >> 2\lambda_D$, where the double layer is very thin, the negative and positive charges in the double layer are very close and interact. Accounting for the influence of the fluid double layer on the motion of the particle and the generated movement of counter-ions in the carrier fluid, one derives the following expression for the electrophoretic velocity in terms of the zeta potential (Probstein, 1994)

$$v_{ep} = \frac{\zeta\varepsilon\varepsilon_0 E_E}{\mu_c}, \quad 2\lambda_D << D. \tag{4.67}$$

This equation is referred to as the *Helmholtz-Smoluchowski equation*.

At the opposite limit, $2\lambda_D >> D$, the net charge, q, of the sphere is approximately equal to $4\pi\varepsilon\varepsilon_0\zeta a$, from Eq. (4.64). Assuming that the part of the fluid in the double layer is a spherical shell and does not deform by the movement of the particle, one obtains the following expression for the electrophoretic velocity

$$v_{ep} = \frac{2\zeta\varepsilon\varepsilon_0 E_E}{3\mu_c}, \quad 2\lambda_D >> D. \tag{4.68}$$

This equation is referred to as *the Hückel equation*.

It is apparent that the electrophoretic velocity in the two limits of the thickness of the double layer differs only by the factor 1 or 2/3. In the intermediate range, a shape factor C, which is a function of the dimensionless parameter, $D/2\lambda_D$, is often used in calculations

$$v_{ep} = \frac{2}{3}C\frac{\zeta\varepsilon\varepsilon_0 E_E}{\mu_c}. \tag{4.69}$$

The function $C(D/2\lambda_D)$ may be obtained analytically or experimentally. Oshima (1994) derived the following analytical form for this function

$$C = 1 + \frac{1}{2\left[1 + \dfrac{5\lambda_D}{D\left(1 + 2e^{-D/2\lambda_D}\right)}\right]^3}. \tag{4.70}$$

4.9 BODY FORCES

Gravity is the most common body force acting on particles. When the particles are within the mass of a carrier fluid, the buoyancy—also caused by the gravitational field—acts on the particle, and the net force on the particle becomes

$$F_G = V\left(\rho_d - \rho_c\right)g = \frac{\pi D_V^3}{6}\left(\rho_d - \rho_c\right)g, \tag{4.71}$$

where V is the volume of the particle, and D_V is the volume-equivalent diameter—the actual diameter for a spherical particle. When the density of the particle is much larger than that of the carrier fluid, as it happens in several gas-particle flows where $(\rho_d \gg \rho_c)$, the second term in the parenthesis may be neglected and $F_G \approx \pi D_V^3 \rho_d g/6$.

4.9.1 TERMINAL VELOCITY

Particles within the terrestrial environment are subjected to the gravity force. For particles rising or settling in stagnant fluids at steady conditions, the drag and the gravity force are opposite and equal in magnitude, and the prevailing mechanical equilibrium implies that the particles move at constant velocity, called the *terminal (or settling) velocity, v_T*

$$F_G = F_D \Rightarrow V\left(\rho_d - \rho_c\right)g = \frac{1}{2}C_D\rho_c A_{cs}v_T^2 \Rightarrow v_T = \sqrt{\frac{2Vg}{C_D A_{cs}}\frac{\left(\rho_d - \rho_c\right)}{\rho_c}}, \tag{4.72}$$

where V is the volume, and A_{cs} is the cross-sectional area of the particle. For a spherical particle, this equation yields

$$v_T = \sqrt{\frac{4Dg}{3C_D}\frac{\left(\rho_d - \rho_c\right)}{\rho_c}}. \tag{4.73}$$

Since C_D is a function of several variables, including the Reynolds number, $Re_r = Dv_T\rho_c/\mu_c$, this equation is typically solved by iteration. Exact solutions may be derived for the terminal velocity in the case of creeping flow of viscous and solid spheres. For example, in the case of a viscous sphere under creeping flow conditions, the drag force is modeled by Hadamard and Rybczynski (Eq. 4.9), and the expression for the terminal velocity is

$$v_T = \frac{gD^2\left(\rho_d - \rho_c\right)}{6\mu_c}\frac{\lambda + 1}{3\lambda + 2}. \tag{4.74}$$

For a solid spherical particle, $\lambda \to \infty$ and the terminal velocity is

$$v_T = \frac{gD^2(\rho_d - \rho_c)}{18\mu_c}. \tag{4.75}$$

In the more general case, when the flow around the solid particle is not creeping flow (Stokes flow), the drag coefficient may be written in terms of the drag factor, f

$$v_T = \frac{gD^2(\rho_d - \rho_c)}{18 f \mu_c}. \tag{4.76}$$

Since $f = f(Re_r)$, the friction factor is also a function of v_T ($Re_r = v_T D\rho_c/\mu_c$). Eq. (4.76) is implicit and may be solved by iteration.

A useful dimensionless number in calculations of the terminal velocity is the *Archimedes number*

$$Ar = \frac{gD^3 \rho_c (\rho_d - \rho_c)}{\mu_c^2} = \frac{gD^3 (\rho_d / \rho_c - 1)}{v_c^2}, \tag{4.77}$$

where $v_c = \mu_c/\rho_c$ is the kinematic viscosity of the carrier fluid. Hence, the terminal velocity may be expressed in terms of Ar as follows

$$v_T = \frac{Ar}{18f}\frac{\mu_c}{\rho_c D} = \frac{Ar}{18f}\frac{v_c}{D} \quad or \quad \frac{v_T D}{v_c} = \frac{Ar}{18f}, \tag{4.78}$$

An empirical relationship—originally developed by Camenen (2007) and modified to fit the data for an extended range of Re_r—yields the following expressions for the terminal velocity for a wide range of particle movements, extending from creeping flow up to the turbulent transition

$$v_T = \frac{v_c}{D}\left(\sqrt{22 + \sqrt{4.89 Ar}} - \sqrt{22}\right)^2 \quad for \quad Ar < 3 * 10^5$$
$$v_T = 1.74\frac{v_c}{D}\sqrt{Ar} \quad for \quad 3 * 10^5 \le Ar < 3 * 10^{10} \tag{4.79}$$

The value $Ar = 3 \times 10^{10}$ corresponds to the critical Reynolds number, $v_T D/v_c = 3 \times 10^5$, where transition to turbulence occurs. The terminal velocity obtained from Eq. (4.79) is within ±4% of the value that would be obtained using drag coefficient values from the standard drag curve.

Example: Calculate the terminal velocity of a 500 micron glass bead ($\rho_d = 2,500$ kg/m³) in air at standard conditions ($\rho_c = 1.2$ kg/m³ and $v_c = 1.51 \times 10^{-5}$ m²/s).

Solution: The Archimedes number for this particle may be written as

$$Ar = \frac{gD^3(\rho_d/\rho_c - 1)}{v_c^2} = \frac{9.81 \times (5 \times 10^{-4})^3 (2500/1.2 - 1)}{(1.51 \times 10^{-5})^2} = 1.12 * 10^4$$

Substituting this value into the first part of Eq. (4.79), we obtain the following

$$v_T = \frac{v_c}{D}\left(\sqrt{22 + \sqrt{4.89\,Ar}} - \sqrt{22}\right)^2 \Rightarrow v_T = \frac{1.51 \times 10^{-5}}{5 \times 10^{-4}} \times 128 = 3.86 \ m/s$$

4.9.2 CENTRIFUGING

Centrifuging or centrifugation is the process of applying a strong centrifugal force to a suspension of particles in a liquid. The suspension is placed in a tube, which is forced to rotate fast around a plane perpendicular to the tube axis. The generated centrifugal force may be on the order of thousands of times higher than gravity; it overcomes the Brownian dispersion and drives the particles toward the bottom of the tube. Effectively, centrifuging is a sedimentation process with a force much higher than gravity. If the angular velocity of the centrifuging process is denoted by ω, for a particle at a radius R from the center of rotation, the radial acceleration is $\omega^2 R$. At steady rotation, the centrifugal force is balanced by the drag force, as in the case of sedimentation, and generates an equivalent terminal velocity, which may be written in terms of the hydrodynamic (Stokes) diameter as

$$m_p \omega^2 R = 3\pi \mu_c D_{St} v_T \Rightarrow v_T = \frac{m_p \omega^2 R}{3\pi \mu_c D_{St}} = \frac{D_V^3 \rho_d \omega^2 R}{18\mu_c D_{St}}. \tag{4.80}$$

Since the terminal velocity of centrifuging is the rate of change of the radial position of the particle, one obtains an equation of motion for the particle

$$\frac{v_T}{\omega^2 R} = \frac{dR}{dt}\frac{1}{\omega^2 R} \Rightarrow \frac{1}{\omega^2 R}\frac{dR}{dt} = \frac{D_V^3 \rho_d}{18\mu_c D_{St}} = \frac{m_p}{3\pi \mu_c D_{St}}. \tag{4.81}$$

The right-hand side of the last equation is constant for a single particle or a group of identical particles; it is often called the sedimentation coefficient, S, and has the units of time (s). Integration of the last part of Eq. (4.81), between the limits of times t_1 and t_2, yields the following expression for the sedimentation coefficient

$$S = \frac{m_p}{3\pi D_{St}\mu_c} = \frac{\ln\left(R_2/R_1\right)}{\omega^2 \left(t_2 - t_1\right)}. \tag{4.82}$$

The sedimentation coefficient, S, is measured by recording the positions, R_1 and R_2, of a trace particle or of a layer of identical trace particles at any two times, t_1 and t_2, during the centrifuging process. Then Eq. (4.82) provides an expression for the determination of either the mass of the particles, m_p, or of their hydrodynamic diameter D_{St}. In the case of spheres, where $D = D_{St}$, this equation would determine the actual diameter of the spheres.

4.10 BROWNIAN MOVEMENT

The Brownian movement of fine particles was first observed in a microscope by Robert Brown in 1837 and was analytically described by Einstein (1905) early in the twentieth century. His scientific work on this subject was cited as one of Einstein's major achievements in awarding him the Nobel Prize for Physics in 1921.[4] The Brownian movement of particles is the result of the impulses of fluid molecules that move with very high velocities, which are on the order of 1,000 m/s and depend on the temperature of the fluid (Tien and Lienhard, 1979)

$$T = \frac{m}{3k_B} \overline{C^2}, \tag{4.83}$$

where m is the mass of a molecule; C is the magnitude of the velocity of the molecules with the overbar denoting average over all the molecules; and k_B is the Boltzman constant, $k_B = 1.38*10^{-23}$ J/K. A consequence of Eq. (4.83) is that molecules of fluids at higher temperatures have higher velocities, and therefore, the Brownian movement of particles is more intense in fluid domains of higher temperatures.

4.10.1 BROWNIAN DIFFUSION

Since the mass of the fluid molecules is by far smaller than the mass of the particles, the impacts of the individual molecular collisions on the particles are of very small magnitude. However, the number of molecular impacts per unit time is extremely large, and their aggregate effect is noticeable on the movement of smaller particles that have sizes less than 10 μm. Figure 4.16 depicts the Brownian movement of

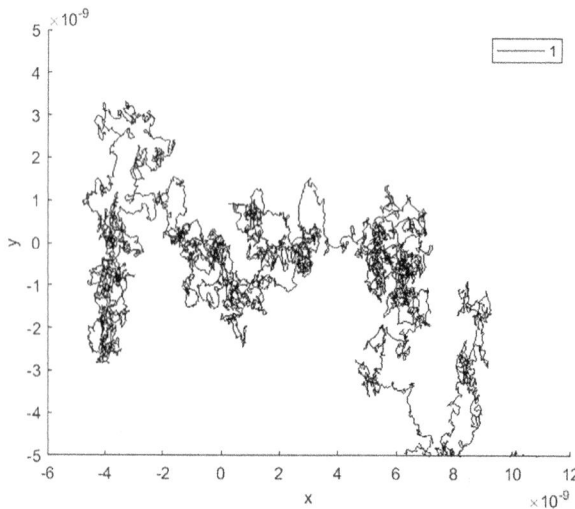

FIGURE 4.16 Brownian movement of a 40 nm alumina particle, over a time period 4,000τ_v (approx. 360 ns).

---

[4] Two relevant Nobel Prize awards were given in 1926 to J. B. Perrin (physics) and T. Svedberg (chemistry) for their work with heterogeneous dispersions.

an alumina particle, with D = 40 nm, over a period equal to 4,000 characteristic times of the particle ($t_v = D^2\rho_c/18\mu_c$), approximately 360 ns. It is observed that, even though this time period is very short, there is appreciable movement from the initial position of the particle. The aggregate movement of a group of particles because of the molecular impacts generates particle dispersion and, in the case of temperature gradients, thermophoresis.

During time periods on the order of τ_v, the effects of the Brownian movement are equivalent to the action of a random force, $\vec{F}_{Br}$, which acts on the particle continuously. This random Brownian force is resisted by the drag of the fluid. The magnitude of this force is such that it causes the same particle dispersion as the Brownian movement. Hence, one may write the equation of motion of a small particle as

$$m_p \frac{d\vec{v}}{dt} = 3\pi D\mu_c f_{Kn}(\vec{u} - \vec{v}) + \vec{F}_{Br}. \tag{4.84}$$

The variable f_{Kn} represents the Cunningham correction factor, $1/(1 + 2AKn)$, of Eq. (4.32). When the fluid is stagnant, we obtain the following equation for the distance covered by the particle

$$\frac{d}{dt}\left(\frac{d\vec{x}}{dt}\right) = -\frac{18\mu_c f_{Kn}}{D^2 \rho_d} \frac{d\vec{x}}{dt} + \frac{6}{D^3 \rho_d} \vec{F}_{Br}, \tag{4.85}$$

After taking the scalar product of the terms of the last equation with the position vector, $\vec{x}$, and obtaining the ensemble average of the motions of a large number of particles, we obtain the following expression

$$\frac{d}{dt}\left\langle \vec{x} \cdot \frac{d\vec{x}}{dt} \right\rangle - \left\langle \left(\frac{d\vec{x}}{dt} \cdot \frac{d\vec{x}}{dt}\right) \right\rangle = -\frac{18\mu_f f_{Kn}}{D^2 \rho_d} \left\langle \vec{x} \cdot \frac{d\vec{x}}{dt} \right\rangle + \frac{6}{D^3 \rho_d} \left\langle \vec{x} \cdot \vec{F}_{Br} \right\rangle, \tag{4.86}$$

where the angular brackets, < >, denote ensemble averages. The second term in Eq. (4.86) is equal to twice the ensemble average of the kinetic energy of the group of particles, divided by the mass, m_p. When the molecules of the carrier fluid and the particles are in equilibrium, the principle of equipartition of energy dictates that this term is equal to $3k_BT/2$. The last term in Eq. (4.86) vanishes because the force is random, and the movement of the ensemble does not have a preferred direction (Russel et al., 1989). Hence, the ensemble-average displacement of the particles becomes

$$\frac{d}{dt}\left\langle \vec{x} \cdot \frac{d\vec{x}}{dt} \right\rangle = \frac{9k_BT}{D^3 \rho_d} - \frac{18\mu_f f_{Kn}}{D^2 \rho_d}\left\langle \vec{x} \cdot \frac{d\vec{x}}{dt} \right\rangle. \tag{4.87}$$

Eq. (4.86) is integrated to yield the dispersion of the particles after a period of time, t

$$\langle \vec{x} \cdot \vec{x} \rangle = \frac{2k_BT}{3\pi D\mu_c f_{Kn}} t. \tag{4.88}$$

It is noted that the Brownian dispersion is independent of the density and the other characteristics of the particle and only depends on the size of the particle and the hydrodynamic drag. The dispersion coefficient, D_0, is defined as half the derivative of the Brownian dispersion

$$D_0 \equiv \frac{d}{2dt}\langle \vec{x} \cdot \vec{x} \rangle = \frac{k_B T}{3\pi D \mu_c f_{Kn}}. \tag{4.89}$$

When the interfacial slip effects on the drag are negligible and the Cunningham correction factor is 1 ($f_{Kn} = 1$), the last equation becomes the so-called *Stokes-Einstein diffusivity*.

Oftentimes, it is desirable to perform Lagrangian simulations for the motion of ensembles of particles, including the effects of the Brownian movement. In such cases, one must have an explicit form of the Brownian force, $\vec{F}_{Br}$, that acts on each particle. The three components of this force have random Gaussian distributions, and the resulting dispersion of particles as given by Eq. (4.89). A time interval, Δt, which is typically much higher than the characteristic time of molecular collisions, is chosen *a priori* for the numerical integration of the equation of motion. The magnitude of the time interval is typically in the range $0.1\tau_v < \Delta t < 0.2\tau_v$. The expression of the random Brownian force that causes dispersion, D_0, in the numerical scheme is (Li and Ahmadi, 1992)

$$\vec{F}_{Br} = \vec{R}\sqrt{\frac{6\pi k_B T D \mu_c f_{Kn}}{(\Delta t)}}, \tag{4.90}$$

where $\vec{R}$ is a random vector, whose components are Gaussian random numbers with zero mean and unit variance. This expression for the random Brownian force has been successfully used to calculate the isothermal particle dispersion in filtration systems (Kim and Zydney, 2004) and the hindered diffusion of nanoparticles in pores (Michaelides, 2017).

4.10.2 THERMOPHORESIS

When a temperature gradient is applied in the domain of the carrier fluid, particles in the side of higher temperatures experience stronger molecular impulses and a stronger Brownian force that induces faster particle movement in the hotter regions. As a result, particles migrate to the colder regions of the flow domain, where their concentration becomes higher. This phenomenon is called *thermophoresis* and is expressed quantitatively in terms of a phenomenological velocity, the *thermophoretic velocity*, v_{tp} (Brock, 1962; Talbot *et al.*, 1980). A *thermophoretic force*, F_{tp}, is also defined as the force that would induce this velocity in a Stokesian flow field

$$v_{tp} = -K_{tp}\frac{\mu_c}{\rho_c}\frac{\nabla T}{T_\infty} \quad and \quad F_{tp} = -3\pi\mu_c^2 DK_{tp}\frac{\nabla T}{\rho_c T_\infty}. \tag{4.91}$$

The thermophoretic coefficient, K_{tp}, depends on the Knudsen number and the properties of the carrier fluid and the solid particles. For spherical particles in gases, Brock (1962) derived analytically the expression

$$K_{tp} = \frac{2C_s \left(k_c + 2k_d Kn\right)}{(1 + 6C_m Kn)(2k_c + k_d + 4k_c C_t Kn)}, \tag{4.92}$$

where k_c and k_d are the thermal conductivities of the carrier fluid and of the particles, respectively, and the parameters C_s, C_m, and C_t are determined empirically from the flow field around the particles.

Talbot *et al.* (1980) used the velocity slip expression, recommended by Millikan (1923), and derived empirical correlations for these parameters. The experimental data for particles in gases have shown that reasonable results are obtained by treating these parameters as constants with the numerical values $C_s = 1.17$, $C_m = 1.14$ and $C_t = 2.18$ for $Kn < 0.1$.

McNabb and Meisen (1973) showed that the magnitude of the thermophoretic force is much lower in liquid suspensions—approximately 17% of the corresponding values for gaseous suspensions—primarily because of the significantly lower values of Kn in liquids.

Since thermophoresis is a consequence of the Brownian movement, its effects are significant in systems of very small particles, typically of less than 10 μm equivalent diameters. Michaelides (2015) incorporated the Brownian force, F_{Br}, of Eq. (4.90) in a stochastic Lagrangian numerical method to derive from first principles the functions for the thermophoretic coefficients, K_{tp}, for nanofluids

$$K_{tp} = A\left(\frac{D}{D_0}\right)^{-B} \Rightarrow v_{tp} = -A\left(\frac{D}{D_0}\right)^{-B} \frac{\mu_c}{\rho_c T_\infty} \nabla T, \tag{4.93}$$

where $D_0 = 2$ nm, and the constants A and B depend on the properties of the carrier fluid and the nanoparticles. The pairs of coefficients (A, B) for 20 common combinations of particles and fluids used in nanofluids are given in Table 4.2.

TABLE 4.2
Pairs of Coefficients (A, B) for the Correlation of Eq. (4.92) for Several Types of Spherical Particles in Common Carrier Fluids (from Michaelides, 2015)

	Water	Engine Oil	Ethyl Glycol	R-134a
Aluminum	(1264, 1.417)	(3.0920, 1.242)	(14.615, 1.869)	(4401, 1.774)
Alum. Oxide	(1227, 1.434)	(7.1026, 1.579)	(5.1095, 1.621)	(6270, 1.819)
CNT	(945.5, 1.263)	(5.8044, 1.445)	(3.6765, 1.406)	(8580, 1.894)
Copper	(2039, 1.870)	(7.1391, 1.724)	(2.3558, 1.587)	(4191, 1.659)
Gold	(3155, 1.799)	(6.6483, 1.917)	(4.2431, 1.672)	(2721, 1.603)

It must be emphasized that thermophoresis is the consequence of the molecular impulses—that also cause the Brownian movement of particles—in fluid domains with temperature gradients. The thermophoretic force, F_{tp}, and the Brownian force, F_{Br}, are not independent forces to be accounted separately in simulations of particle motion. The simple inclusion of the Brownian force in the Lagrangian simulations of particle motion in fluids with temperature gradients will reproduce the effects of the thermophoretic force and will generate the thermophoretic velocity of the particles toward the colder regions of the carrier fluid (Michaelides, 2015).

4.11   TRANSIENT DRAG-ADDED MASS AND HISTORY (BASSET) FORCE

When the carrier flow field is time-dependent and the transients are on the same order of magnitude as the response time of the particles, the hydrodynamic force is significantly modified. Exact analytical expressions for the hydrodynamic force in unsteady flows have been derived for creeping flow (Stokes flow) conditions ($Re_r \ll 1$). Asymptotic expressions have been derived for finite but small Reynolds numbers ($Re_r < 1$). Semi-empirical expressions that originate from a combination of experimental and numerical data have also been developed for higher Re_r.

4.11.1   CREEPING (STOKES) FLOW (RER << 1)

Late in the nineteenth century, Boussinesq (1885) and Basset (1888a) independently derived an analytical solution for the transient hydrodynamic force on a solid sphere at creeping flow conditions ($Re_r \ll 1$), where the inertia in the momentum equation of the fluid are neglected in comparison to the viscous effects. Maxey and Riley (1983) performed a more detailed analysis for a rigid sphere in an arbitrary non-uniform flow field and obtained the following expression for the equation of motion of the solid sphere

$$
\begin{aligned}
m_p \frac{dv_i}{dt} &= -3\pi D \mu_c \left(v_i - u_i - \frac{D^2}{24} u_{i,jj}\right) - \frac{1}{2} m_c \frac{d}{dt}\left(v_i - u_i - \frac{D^2}{40} u_{i,jj}\right) \\
&- \frac{3\pi D^2 \mu_c}{2\sqrt{\pi \nu_c}} \int_0^t \frac{\frac{d}{d\tau}\left(v_i - u_i - \frac{D^2}{24} u_{i,jj}\right)}{\sqrt{t-\tau}} d\tau + (m_p - m_c) g_i + m_c \frac{Du_i}{Dt}
\end{aligned}
\tag{4.94}
$$

where m_p is the mass of the sphere, m_c is the mass of the carrier fluid that occupies the same volume as that of the sphere, τ is a dummy variable used in the integration, the repeated index with comma ($,jj$) denotes the Laplacian operator, and the derivative D/Dt is the total (Lagrangian) derivative following the center of the sphere. All the spatial derivatives are evaluated at the center of the sphere. The velocity function of the fluid, $u_i(x_i, t)$, is a solution of the Navier-Stokes equations and not an arbitrary function.

The left-hand side in Eq. (4.94) represents the acceleration of the sphere. Of the terms on the right-hand side, the first is the steady drag; the second term is referred to as the *added mass* (or virtual mass) of the fluid and represents the mass of the fluid that is accelerated with the sphere; and the third term is referred to as *history term*, or the *Basset term*, and represents the effects of the fluid viscous boundary layer around the sphere. The remaining terms on the right-hand side are the buoyancy force and the Lagrangian acceleration term of the fluid. The Laplacian terms in the three parentheses account for the non-uniformity of the fluid velocity field; they are referred to as the *Faxen terms* and vanish when the fluid velocity field is uniform. The Faxen terms scale as D^2/L^2, where L is the macroscopic characteristic length of the fluid velocity field. In most applications of dispersed multiphase flows, where $D/L \ll 1$, the Faxen terms are negligible.

If the relative velocity of the particle at time $t = 0$, $v_i(0) - u_i(0)$, is finite, the history term is modified as follows (Michaelides, 2003)

$$\frac{3\pi D^2 \mu_c}{2\sqrt{\pi v_c}} \int_0^t \frac{\frac{d}{d\tau}[v_i - u_i - \frac{a^2}{6} u_{i,jj}]}{\sqrt{t-\tau}} d\tau + \frac{3\pi D^2 \mu_c [v_i(0) - u_i(0)]}{2\sqrt{\pi v_c t}}. \qquad (4.95)$$

Eq. (4.94) is an integrodifferential equation where the dependent variable, v_i, is implicit in the integral. Such equations must be solved numerically by iteration methods that are computationally intensive. A method has been devised to convert this integrodifferential equation to a second-order differential equation that is explicit in v_i and may be numerically solved without iteration (Michaelides, 1992; Vojir and Michaelides, 1994).

Eq. (4.94) applies to solid spheres alone, where the relevant timescale is the timescale of the carrier fluid. For viscous spheres in a carrier fluid, there are two pertinent timescales: One for the motion of the viscous fluid in the interior, and the second for the motion of the external fluid. For this reason, an analytical expression for their equation of motion is impossible to derive in the time domain. An expression for the hydrodynamic force on a viscous sphere under transient creeping flow conditions in the Laplace domain was derived by Galindo and Gerbeth (1993). Michaelides and Feng (1995) derived an expression for the hydrodynamic force on viscous spheres with slip at the interface, also in the Laplace domain. The effects of flow compressibility on the transient hydrodynamic force at transient creeping flow conditions were examined numerically by Parmar *et al.* (2011), while Ling *et al.* (2011a, 2011b) determined the effects of the unsteady terms on the dispersion of small particles in shock waves.

Of the two transient drag terms in Eq. (4.94), the added mass term may not present any difficulties in the computations. The effect of this term is, effectively, an increase of the mass of the sphere. The history term renders the equation implicit in the dependent variable, v_i; suggests a solution by iteration; and significantly adds to the computational time of the solution. In an extensive numerical study Vojir and Michaelides (1994) determined that the history term enhances the steady drag by up to 30% in liquid-solid flows, depending on the frequency of the transients. The effects of the history term are negligible: (a) in gas solid flows where the density ratio, ρ_d/ρ_c, is less than 500; (b) for time-averaged results with random fluid velocity fluctuations; and (c) in transients where the dimensionless frequency of variation (defined as the product of the characteristic frequency of the transient and the characteristic time, τ_v, of the sphere) is less than 0.5.

4.11.2 FLOW AT FINITE REYNOLDS NUMBERS

When the Reynolds number is finite, the fluid advection around the sphere must be considered. For finite but small Reynolds numbers ($Re_r < 1$), asymptotic analyses may be used to derive expressions for the transient hydrodynamic force. Sano (1981) and Lovalenti and Brady (1993) conducted such studies and derived asymptotic expressions for the movement of spheres. Such studies show that the decay of the history term is faster than in the case of creeping flow because the finite relative velocity of the sphere induces faster advection and evolution of the vorticity field around the sphere. While in the creeping flow case the vorticity field around the sphere is transported by diffusion alone, in the case of advection, the finite relative velocity of the fluid "carries" the vorticity field far from the sphere and reduces its effect on the transient drag. Hinch (1993) presented several physical arguments and quick asymptotic methods to explain the physics of the movement of a sphere at $Re_r < 1$. He reduced the effects of flow advection to the action of sources and sinks associated with the presence of the wake behind the sphere and deduced that any memory of the initial velocity of the sphere fades at long times. Also, because of differences of the velocity field developed in the carrier fluid, the acceleration and deceleration processes of a sphere are not achieved with the same force, even though the initial and final velocities of the sphere may be the same. The asymmetry in the acceleration and deceleration processes at finite values of Re_r is in contrast to the case of the creeping flow solution in Eq. (4.94), which is invariant with respect to time.

Mei and Adrian (1992) obtained a solution for the motion of a solid sphere, valid at very low frequencies ($Sl \ll Re_r < 1$). Their results revealed a different history term for the hydrodynamic force acting on the particle, which may be written as follows

$$\int_0^t \frac{\dfrac{dv_i}{dt} - \dfrac{du_i}{dt}}{\left[\left(\dfrac{4\pi v_c}{D^2}(t-\tau) \right)^{1/4} + \left(\dfrac{\pi}{Dv_c} \left(\dfrac{|u_i(\tau)|}{0.75 + 0.105\,Re_r} \right)^3 (t-\tau)^2 \right)^{1/2} \right]^2} d\tau, \quad (4.96)$$

For particulate flows at higher Reynolds numbers, Odar and Hamilton (1964) proposed an empirical modification to extend the validity of the three terms of the Boussinesq-Basset expression in Eq. (4.94) to higher Reynolds numbers

$$F_i = 3\pi D\mu_c (v_i - u_i)\left(1 + 0.15\,Re_r^{0.687}\right) + \Delta_A \frac{1}{2} m_c \frac{d(v_i - u_i)}{dt}$$

$$+ \Delta_H \frac{D^2}{4} \sqrt{\pi \rho_c \mu_c} \int_0^t \frac{\dfrac{d(v_i - u_i)}{d\tau}}{\sqrt{t - \tau}} d\tau. \qquad (4.97)$$

The first term in this equation is the Schiller and Nauman term for the steady drag. The variables Δ_A and Δ_H, multiply the added mass and the history term respectively, are given by empirical correlations.

It must be noted that the added mass term—the second term in Eq. (4.97)—emanates from the far field contribution to the hydrodynamic force. It is independent of the viscous effects around the sphere and, therefore, must be independent of Re_r. As a consequence, the value of Δ_A is constant and equal to 1. The experimental data by Bataille *et al.* (1990) and the analysis by Auton *et al.* (1998) confirm that $\Delta_A = 1$. Based on these observations, Michaelides and Roig (2011) re-interpreted and re-correlated the original data by Odar and Hamilton (1964) and recommended the following expressions for the functions Δ_A and Δ_H in Eq. (4.97)

$$\Delta_A = 1 \quad and \quad \Delta_H = 6.00 - 3.16\left[1 - \exp\left(-0.14\,Re_r\,Sl^{0.82}\right)^{2.5}\right], \qquad (4.98)$$

where *Sl* is the Strouhal number that characterizes the transient flow.

4.12   SUMMARY

A large number of experimental, numerical, and analytical studies have been performed for the determination of the hydrodynamic force (drag) for single particles and drops. The force largely depends on the Reynolds number of the flow. Velocity and temperature discontinuities (slip) on the surface of the particle, the viscosity of drops, and the shape of the particles also affect the drag force, while the free flow turbulence does not seem to have a significant effect. Mass transfer from the surface of the particles—for example, because of evaporation, sublimation, or condensation—and the presence of boundaries in the vicinity of the particles is associated with momentum transfer and significantly influences the drag force. The lateral forces on the particles, generated by particle spin and the flow shear, are of lower magnitude but contribute significantly in particle dispersion. The Brownian movement also contributes to the dispersion of very small particles and causes thermophoresis when temperature gradients exist in the flow field. Thermophoresis, electrical forces, settling under gravity, and centrifuging contribute to practical methods for the separation and collection of particles. During the transient movement, the history term contributes significantly to the drag at creeping flow conditions, but its effects diminish when the flow inertia become substantial.

REFERENCES

Achenbach, E., 1974, Vortex shedding from spheres, *J. Fluid Mech.*, **62**, 209–221.
Allen, M.D. and Raabe, O.G., 1982, Re-evaluation of Millikan's oil drop data for the motion of small particles in air, *J. Aerosol Sci.*, **13**, 537–546.
Allen, M.D. and Raabe, O.G., 1985, Slip correction measurements of spherical solid aerosol particles in an improved Millikan apparatus, *J. Aerosol Sci. Tech.*, **4**, 269–282.

Auton, T.R., Hunt, J.R.C. and Prud'homme, M., 1998, The force exerted on a body in inviscid unsteady non-uniform rotational flow, *J. Fluid Mech.*, **197**, 241–257.

Bagchi, P. and Balachandar, S., 2002, Shear versus vortex-induced lift force on a rigid sphere at moderate, *Re. J. Fluid Mech.*, **473**, 379.

Bagchi, P. and Balachandar, S., 2003, Effect of turbulence on the drag and lift of a particle, *Phys. Fluids*, **15**, 3496.

Barkla, H.M. and Auchterlonie, L.J., 1971, The Magnus or Robins effect on rotating spheres, *J. Fluid Mech.*, **47** (3), 437.

Basset, A.B., 1888a, *Treatise on Hydrodynamics*, Bell, London.

Basset, A.B., 1888b, On the motion of a sphere in a viscous liquid, *Philos. Trans. Roy. Soc. London*, **179**, 43–63.

Bataille, J., Lance, M. and Marie, J.L., 1990, Bubbly turbulent shear flows, in *ASME-FED, 99* (Eds. J. Kim, U. Rohatgi and M. Hashemi), ASME, New York, pp. 1–7.

Bohlin, T., 1960, Terminal velocities of solid spheres in cylindrical enclosures, *Transactions of the Royal Institute of Technology*, Stockholm, Report # 155.

Boussinesq, V.J., 1885, Sur la Resistance qu' Oppose un Liquide Indéfini en Repos . . ., *Comptes Rendu, Acad. Sci. Paris*, **100**, 935–937.

Brock, J.R., 1962, On the theory of thermal forces acting on aerosol particles, *J. Colloid and Interface Science*, **17**, 768–780.

Camenen, B.S., 2007, Simple and general formula for the settling velocity of particles, *J. Hydraulic Engr.*, **133**, 229.

Chiang, C.H., Raju, M.S. and Sirignano, W.A., 1992, Numerical analysis of a convecting, vaporizing fuel droplet with variable properties, *Int. J. Heat Mass Transfer.*, **35**, 1307–1327.

Clift, R., Grace, J.R. and Weber, M.E., 1978, *Bubbles, Drops and Particles*, Academic Press, New York.

Connolly, B.J., Loth, E. and Smith, C.F., 2020, Shape and drag of irregular angular particles and test dust, *Powder Technol.*, **363**, 275–285.

Cox, R.G. and Hsu, S.K., 1977, The lateral migration of solid spheres in a laminar flow near a plane, *Int. J. Multiphase Flow*, **3**, 201–222.

Crowe, C.T., Babcock, W.R. and Willoughby, P.G., 1973, Drag coefficient for particles in rarefied, low Mach number flows, *Prog. Heat and Mass Trans.*, **6**, 419.

Crowe, C.T., Babcock, W.R., Willoughby, P.G. and Carlson, R.L., 1969, Measurement of particle drag coefficients in flow regimes encountered by particles in a rocket nozzle, United Technology Report 2269-FR.

Cunningham, E., 1910, On the velocity of steady fall of spherical particles through a fluid medium, *Proc. Roy. Soc. Series A*, **83**, 357–364.

Dandy, D.S. and Dwyer, H.A., 1990, A sphere in shear flow at finite Reynolds number: Effect of particle lift, drag and heat transfer, *J. Fluid Mech.*, **226**, 381–398.

Dioguardi, F. and Mele, D., 2015, A new shape dependent drag correlation formula for non-spherical rough particles: Experiments and results, *Powder Technology*, **277**, 222–230.

Einstein, A., 1905, Über die von der molekularkinetischen Theorie der Wärme geforderte Bewegung von in ruhenden Flüssigkeiten suspendierten Teilchen, *Ann. Phys.*, **17**, 549–560.

Faxen, H., 1922, Der Widerstand gegen die Bewegung einer starren Kugel in einer zum den Flussigkeit, die zwischen zwei parallelen Ebenen Winden eingeschlossen ist, *Ann. Phys.*, **68**, 89–119.

Feng, Z.G., 2010, A correlation of the drag force coefficient on a sphere with interface slip at low and intermediate Reynolds numbers, *J. of Dispersion Science and Technology*, **31**, 968–974.

Feng, Z.G., Gatewood, J., Michaelides, E.E., 2021, Wall effects on the flow dynamics of a rigid sphere in motion, *J. Fluids Engin.*, **143** (8), 081106.

Feng, Z.-G. and Michaelides, E.E., 2001, Drag coefficients of viscous spheres at intermediate and high Reynolds numbers, *J. Fluids Eng.*, **123**, 841–849.

Feng, Z.-G. and Michaelides, E.E., 2002a, Inter-particle forces and lift on a particle attached to a solid boundary in suspension flow, *Phys. Fluids*, **14**, 49–60.

Feng, Z.-G. and Michaelides, E.E., 2002b, Hydrodynamic force on spheres in cylindrical and prismatic enclosures, *Int. J. Multiphase Flow*, **28**, 479–496.

Feng, Z.-G., Michaelides, E.E. and Mao, S.-L., 2012, On the drag force of a viscous sphere with interfacial slip at small but finite Reynolds numbers, *Fluid Dynamics Research*, **44** 025502, 1–16.

Galindo, V. and Gerbeth, G., 1993, A note on the force on an accelerating spherical drop at low Reynolds numbers, *Phys. Fluids*, **5**, 3290–3292.

Hadamard, J.S., 1911, Mouvement Permanent Lent d' une Sphere Liquide et Visqueuse dans un Liquide Visqueux, *Compte-Rendus de' l' Acad. des Sci.*, Paris, **152**, 1735–1738.

Haider, A.M. and Levenspiel, O., 1989, Drag coefficient and terminal velocity of spherical and nonspherical particles, *Powder Technol.*, **58**, 63–70.

Hermsen, R.W., 1979, Review of particle drag models, *JANAF Performance Standardization Subcommittee 12th Meeting Minutes*, CPIA, 113.

Hinch, E.J., 1993, The approach to steady state in oseen flows, *J. of Fluid Mechanics*, **256**, 601–603.

Hölzer, A. and Sommerfeld, M., 2008, New simple correlation formula for the drag coefficient of non-spherical particles, *Powder Technol.*, **184**, 361–365.

Hutchins, D.K., Harper, M.H. and Felder, R.L., 1995, Slip correction measurements for solid spherical particles by modulated dynamic light scattering, *Aerosol Science and Technol.*, **22**, 202–218.

Iwaoka, M. and Ishii, T., 1979, Experimental wall correction factors of single solid spheres in circular cylinders, *J. Chem. Eng. Jpn.*, **12**, 239–242.

Kim, J.H., Mulholland, G.W., Pui, D.Y.H. and Kukuck, S.R., 2005, Slip correction measurements of certified psl nanoparticles using a nanometer Differential Mobility Analyzer (nano-DMA) for Knudsen number from 0.5 to 83, *J. Res. Natl. Inst. Stand. Technol.*, **110**, 31–54.

Kim, M.M. and Zydney, A.L., 2004, Effect of electrostatic, hydrodynamic, and Brownian forces on particle trajectories and sieving in normal flow filtration, *J. of Colloid and Interface Science*, **269**, 425–431.

Knudsen, M. and Weber, S., 1911, Resistance to motion of small particles, *Ann. D. Phys.*, **36**, 981–985.

Levich, V.G., 1962, *Physicochemical Hydrodynamics*, Prentice-Hall, Englewood Cliffs, NJ.

Li, A. and Ahmadi, G., 1992, Dispersion and deposition of spherical particles from point sources in a turbulent channel flow, *Aerosol Science and Techn.*, **16**, 209–226.

Ling, Y., Haselbacher, A. and Balachandar, S., 2011a, Importance of unsteady contributions to force and heating for particles in compressible flows: Part 1: Modeling and analysis for shock-particle interaction, *Int. J. Multiphase Flow*, **37**, 1026–1044.

Ling, Y., Haselbacher, A. and Balachandar, S., 2011b, Importance of unsteady contributions to force and heating for particles in compressible flows: Part 2: Application to particle dispersal by blast waves, *Int. J. Multiphase Flow*, **37**, 1013–1025.

Lovalenti, P.M. and Brady, J.F., 1993, The hydrodynamic force on a rigid particle undergoing arbitrary time-dependent motion at small reynolds numbers, *J. Fluid Mechanics*, **256**, 561–601.

Magnus, G., 1861, A note on the rotary motion of the liquid jet, *Ann. Phys. Chem.*, **63**, 363–365.

Masliyah, J.H. and Epstein, N., 1970, Numerical study of steady flow past spheroids, *J. Fluid Mech.*, **44**, 493–512.

Maxey, M.R. and Riley, J.J., 1983, Equation of motion of a small rigid sphere in a non-uniform flow, *Phys. Fluids*, **26**, 883–889.

McLaughlin, J.B., 1991, Inertial migration of a small sphere in linear shear flows, *J. Fluid Mech.*, **224**, 261–274.

McNabb, G.S. and Meisen, A., 1973, Thermophoresis in liquids, *J. Colloidal and Interphase Sci.*, **44**, 339–346.

Mei, R., 1992, An approximate expression of the shear lift on a spherical particle at finite Reynolds numbers, *Int. J. Multiphase Flow*, **18**, 145–160.

Mei, R. and Adrian, R.J., 1992, Flow past a sphere with an oscillation in the free-stream and unsteady drag at finite reynolds number, *J. of Fluid Mechanics*, **237**, 323–341.

Michaelides, E.E., 1992, A novel way of computing the Basset term in unsteady multiphase flow computations, *Phys. Fluids A*, **4**, 1579–1582.

Michaelides, E.E., 2003, Hydrodynamic force and heat/mass transfer from particles, bubbles and drops: The freeman scholar lecture, *J. Fluids Eng.*, **125**, 209–238.

Michaelides, E.E., 2006, *Particles, Bubbles and Drops: Their Motion, Heat and Mass Transfer*, World Scientific Publishing, Singapore.

Michaelides, E.E., 2014, *Nanofluidics-Thermodynamic and Transport Properties*, Springer, New York.

Michaelides, E.E., 2015, Brownian movement and thermophoresis of nanoparticles in liquids, *Int. J. Heat and Mass Transf.*, **81**, 179–187.

Michaelides, E.E., 2017, Nanoparticle diffusivity in narrow cylindrical pores, *Int. J. of Heat and Mass Transfer*, **114**, 607–612.

Michaelides, E.E. and Feng, Z.-G., 1995, The equation of motion of a small viscous sphere in an unsteady flow with interface slip, *Int. J. Multiphase Flow*, **21**, 315–321.

Michaelides, E.E. and Roig, A., 2011, A reinterpretation of the odar and hamilton data on the unsteady equation of motion of particles, *A.I. Ch. E. J.*, **57** (11), 2997–3002.

Millikan, R.A., 1923, The general law of fall of a small spherical body through a gas and its bearing upon the nature of molecular reflection from surfaces, *Phys. Review*, **22**, 1–23.

Miyamura, A., Iwasaki, S. and Ishii, T., 1981, Experimental wall correction factors of single solid spheres in triangular and square cylinders, and parallel plates, *Int. J. Multiphase Flow*, **7**, 41–46.

Odar, F. and Hamilton, W.S., 1964, Forces on a sphere accelerating in a viscous fluid, *J. Fluid Mech.*, **18**, 302–303.

Oesterle, B. and Bui-Dinh, 1998, Experiments on the lift of a spinning sphere in the range of intermediatte Reynolds numbers, *Exper. Fluids*, **25**, 16–22.

Oliver, D.L. and Chung, J.N., 1987, Flow about a fluid sphere at low to moderate Reynolds numbers, *J. Fluid Mech.*, **177**, 1–18.

Oseen, C.W., 1913, Uber den Goltigkeitsbereich der Stokesschen Widerstandsformel, *Ark. Mat. Astron. Fysik*, **9** (19), 1–15.

Oseen, C.W., 1915, Uber den Wiederstand gegen die gleichmassige Translation eines Ellipsoides in einer reibenden Flussigkeit, *Arch. Math. Phys.*, **24**, 108–114.

Oshima, H., 1994, Simple expression for Henry's function for the retardation effect in electrophoresis of spherical colloidal particles, *J. Colloid and Interface Sci.*, **168**, 269–271.

Ouchene, R., Khalij, M., Arcen, B. and Tanière, A., 2016, A new set of correlations of drag, lift and torque coefficients for non-spherical particles and large Reynolds numbers, *Powder Technol.*, **303**, 33–43.

Paine, P.L. and Scherr, P., 1975, Drag coefficients for the movement of rigid spheres through liquid-filled cylindrical pores, *Biophys. J.*, **15**, 1087–1091.

Parmar, M., Haselbacher, A. and Balachandar, S., 2011, Generalized Basset-Boussinesq-Oseen equation for unsteady forces on a sphere in a compressible flow, *Phys. Rev. Lett.*, **106**, 084501.

Patnaik, P.C., Vittal, N. and Pande, P.K., 1992, Drag coefficient of a stationary sphere in gradient flow, *J. Hydraulic Research*, **30**, 389.

Pitter, R.L., Pruppacher, H.R. and Hamielec, A.E., 1973, A numerical study of viscous flow past a thin oblate spheroid at low and intermediate Reynolds numbers, *J. Atmosph. Sci.*, **30**, 125–134.

Probstein, R. F., 1994, *Physicochemical Hydrodynamics*, 2nd edition, Elsevier, New York.

Proudman, I. and Pearson, J.R.A., 1956, Expansions at small Reynolds numbers for the flow past a sphere and a circular cylinder, *J. Fluid Mech.*, **2**, 237–262.

Rader, D.J., 1990, Momentum slip correction factor for small particles in nine common gases, *J. Aerosol Sci.*, **21**, 161–168.

Renksizbulut, M. and Yuen, M.C., 1983, Experimental study of droplet evaporation in high temperature air stream, *J. Heat Transfer*, **105**, 364–388.

Rubinow, S.I. and Keller, J.B., 1961, The transverse force on spinning sphere moving in a viscous fluid, *J. Fluid Mech.*, **11**, 447.

Russel, W.R., Saville, D.A. and Schowalter, W.R., 1989, *Colloisal Dispersions*, Cambridge University Press, Cambridge.

Rybczynski, W., 1911, On the translatory motion of a fluid sphere in a viscous medium, *Bull. Acad. Sci.*, *Krakow*, Series A, **40**, 40–46.

Saffman, P.G., 1965, The lift on a small sphere in a slow shear flow, *J. Fluid Mech.*, **22**, 385–398.

Saffman, P.G., 1968, The lift on a small sphere in a slow shear flow-corrigendum, *J. Fluid Mech.*, **31**, 624–625.

Sanjeevi, S.K.P., Kuipers, J.A.M. and Padding, J.T., 2018, Drag, lift and torque correlations for non-spherical particles from Stokes limit to high Reynolds numbers, *Int. J. Multiph. Flow*, **106**, 325–337.

Sano, T., 1981, Unsteady flow past a sphere at low reynolds number, *J. of Fluid Mechanics*, **112**, 433–441.

Schaaf, S.A. and Chambré, P.L., 1958, Fundamentals of gas dynamics, in *High Speed Aerodynamics and Jet Propulsion* (Ed. H.W. Emmons), Vol. 3, pp. 687–739, Princeton University Press, Princeton, NJ.

Schiller, L. and Nauman, A., 1933, Uber die grundlegende Berechnung bei der Schwekraftaufbereitung, *Ver. Deutch Ing.*, **44**, 318–320.

Seeley, L.E., Hummel, R.L. and Smith, J.W., 1975, Experimental velocity profiles in laminar flow around spheres at intermediate Reynolds numbers, *J. Fluid Mech.*, **68**, 591–608.

Sirignano, W.A., 1993, Fluid dynamics of sprays, *J. Fluids Eng.*, **115**, 345–378.

Sirignano, W.A., 1999, *Fluid Dynamics and Transport of Droplets and Sprays*, Cambridge University Press, Cambridge.

Sommerfeld, M. and Laín, S., 2018, Stochastic modelling for capturing the behaviour of irregular-shaped non-spherical particles in confined turbulent flows, *Powder Technology*, **332**, 253–264.

Stokes, G.G., 1851, On the effect of the internal friction of fluids on the motion of a pendulum, *Trans. Cambridge Philos. Soc.*, **9**, 8–106.

Talbot, L., Cheng, R.K., Schefer, R.W. and Willis, D.R., 1980, Thermophoresis of particles in a heated boundary layer, *J. Fluid Mech.*, **101**, 737–758.

Tanaka, T., Yamagata, K. and Tsuji, Y., 1990, Experiment on fluid forces on a rotating sphere and a spheroid, *Proceedings of the 2nd KSME-JSME Fluids Engineering Conference*, 1, 266–378, Seoul, Korea.

Taneda, S., 1956, Experimental investigation of the wake behind a sphere at low Reynolds numbers, *J. Phys. Soc. Japan*, **11**(10), 1104–1108.

Tien, C.L. and Lienhard, J.H., 1979, *Statistical Thermodynamics, Revised Printing*, Hemisphere, New York.

Tran-Cong, S., Gay, M. and Michaelides, E.E., 2004, Drag coefficients of irregularly shaped particles, *Powder Technology*, **139**, 21–32.

Vasseur, P. and Cox, R.G., 1976, The lateral migration of spherical particles in two-dimensional shear flow, *J. Fluid Mech.*, **78**, 385–413.

Vojir, D.J. and Michaelides, E.E., 1994, The effect of the history term on the motion of rigid spheres in a viscous fluid, *Int. J. Multiphase Flow*, **20**, 547–556.

Wadell, H., 1933, Sphericity and roundness of rock particles, *J. Geol.*, **41**, 310–331.

Warnica, W.D., Renksizbulut, M. and Strong, A.B., 1994, Drag coefficient of spherical liquid droplets. Part 2: Turbulent gaseous fields, *Exp. Fluids*, **18**, 265.

Yuen, M.C. and Chen, L.W., 1976, On drag of evaporating droplets, *Combust. Sci. Technol.*, **14**, 147–154.

Zastawny, M., Mallouppas, G., Zhao, F. and van Wachem, B., 2012, Derivation of drag and lift force and torque coefficients for non-spherical particles in flows. *Int. J. Multiphase Flow*, **39**, 227–239.

Zeng, L., Najjar, F., Balachandar, S. and Fischer, P., 2009, Forces on a finite-sized particle located close to a wall in a linear shear flow, *Physics of Fluids*, **21**, 033302–1–17.

Zhou, H., Zhang, C., Tang, Y., Wang, J., Ren, B. and Yin, Y., 2007, Preparation and thermal conductivity of suspensions of graphite nanoparticles, *Carbon*, **45**, 226–228.

Zukauskas, A. and Ziugzda, J., 1985, *Heat Transfer of a Cylinder in Crossflow*, Hemisphere, Washington.

PROBLEMS

4.1 A particle is in a harmonically oscillating gas flow field. The flow velocity is given by

$$u = u_0 e^{i\omega t}$$

Assume Stokes drag is applicable so the equation of motion is

$$\frac{dv}{dt} = \frac{u - v}{\tau_V}$$

1. Determine the amplitude and phase lag between the gas and the particle in terms of ω and τ_A. Provide a physical interpretation of the result.
2. Evaluate the amplitude ratio and phase lag of a 10 micron glass ($\rho_p = 2{,}500$ kg/m³) in air at standard conditions oscillating with a frequency of 100 Hz.

4.2   A charged particle is injected into a channel with quiescent fluid and a uniform electric field strength. The initial velocity of the particle is U_0, and the distance from the injection point to the wall (collecting surface) is L. Find the axial distance the particle will travel before it impacts the wall in terms of electric field strength, E, the aerodynamic response time of the particle, the charge to mass ratio on the particle (q/m), the initial velocity, and the distance to the wall. Neglect gravitational effects.

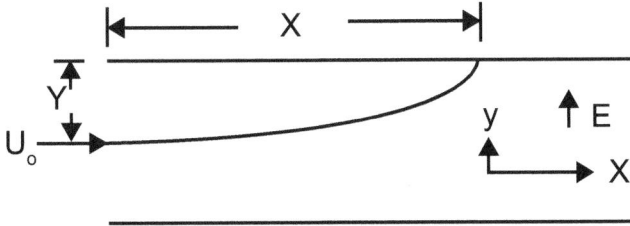

PROBLEM 4.2

4.3   A cubical coal particle, with a side equal to 1 mm and a material density of 1,400 kg/m³ drops in air at standard conditions. Find the terminal velocity.

4.4   Assume a 100 micron glass bead (density = 2,500 kg/m³) is injected into stagnant water at 5 m/s at 20°C. Determine numerically the velocity and distance as a function of time by including the added mass and the history (Basset) term. Compare your results with those obtained if the history term were neglected.

4.5   An expression sometimes used for the drag coefficient of a particle is

$$C_D = \frac{1}{2} + \frac{24}{Re}$$

A 100 micron particle with a material density of 2,000 kg/m³ is fired into still air at a velocity of 10 m/s. The air is at standard conditions. Using the preceding drag law, calculate how far it will travel before stopping. What will the velocity be after one aerodynamic response time? How does this velocity compare with the velocity calculated using Stokes drag? Explain the difference. Neglect the gravitational effects.

4.6   A 100 nm particle with a material density of 800 kg/m³ falls in air at standard conditions (P = 101 kPa, T = 20°C). Find the terminal velocity, assuming the Cunningham correction is valid. What is the terminal velocity based on Stokes drag?

4.7   How important do you think the history (Basset) term is for a prismatic particle (with angular edges)? Explain.

4.8   Particles are injected into a cross-flow as shown. The initial particle velocity is v_0, and the gas velocity has a linear profile from the wall in the form $u = ky$. The particles have diameter D, density ρ_p, and the fluid viscosity is μ_c. Find an expression for the particle trajectory. Assume Stokes drag is valid. Upon what parameters does the maximum penetration from the wall depend?

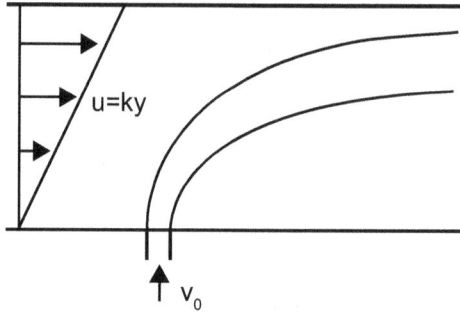

PROBLEM 4.8

4.9   A glass particle with a material density of 2,500 kg/m³ is located in the laminar sublayer of a turbulent boundary layer in airflow in a smooth (no roughness) 5 cm duct. The mean air velocity is 10 m/s, and the temperature is 20°C. Assuming the particle has zero velocity, for what particle diameter will the Saffman lift force just balance the particle weight?

4.10   The equation of motion for a particle in a gas flow field is

$$\frac{dv}{dt} = \frac{f}{\tau_V}(u - v) + g$$

It is difficult to solve this equation for small τ_v because $u - v$ also becomes small. By using $(v - u)/\tau_v$ as the dependent variable, show that

$$v = u + \frac{\tau_V g}{f} - \frac{\tau_V}{f}\frac{du}{dt} + \left(\frac{\tau_V}{f}\right)^2 \frac{d^2 u}{dt^2} + O\left(\tau_V^3\right)$$

5 Particle-Fluid Interactions

Particle-fluid interaction refers to the exchange of mass, momentum and energy between phases and is responsible for coupling in dispersed phase flows. Based on the force mechanisms for single particles and droplets, which were introduced in Chapter 4, the phenomena responsible for the mass, momentum, and energy transfer between phases are presented. The purpose of this chapter is not to present an extensive review of fluid-particle[1] interactions, but to present the basic ideas, methods, and observations for the complex interactions of multiple particles and drops with the carrier fluids. Reference to books by Clift *et al.* (1978), Michaelides (2006) and the *Multiphase Flow Handbook* (Michaelides, *et al.* [eds], 2017) as well as review papers in the literature provide more details.

5.1 FUNDAMENTAL MULTIPHASE FLOW EQUATIONS

The fundamental equations—sometimes referred to as *governing equations* and *conservation laws*—are mathematical expressions of scientific principles, usually given in terms of differential or integral equations. The fundamental equations that govern the majority of multiphase mixtures emanate from the following general physical principles:

1. Conservation of mass
2. Conservation of energy[2]
3. Conservation of linear momentum
4. Conservation of angular momentum
5. Principle of entropy increase

The principle of entropy increase, which is a consequence of the second law of thermodynamics, is expressed as an inequality that determines the directionality of mass flow and heat transfer as well as constitutive equations for the properties of multiphase mixtures (Kestin, 1978; Michaelides, 2021).

The fundamental equations of a multiphase and multi-species mixture are given here in their continuum formulation and are expressed for a general case and for an arbitrary phase, *i*. Following the discussion in section 2.1, this implies that the volume of interest, ΔV_c, is large enough to contain a large number of particles or drops. A second implicit assumption is that all the phases present in the mixture are continua.

---

[1] For brevity, the word "particle" will be used in this chapter to denote both solid particles and liquid drops. When this generalization does not apply, clarifications will be given as "solid particles" and "liquid drops" or just "drops."

[2] When nuclear reactions take place, the conservation of mass and energy are combined to a single mass-energy conservation principle, using the equation $E = mc^2$.

DOI: 10.1201/9781003089278-5

5.1.1 MASS CONSERVATION EQUATION

$$\frac{\partial \rho_i}{\partial t} + \nabla_k \bullet (\rho_i \mathbf{v}_k) = \sum_{j=1}^{N} J_{ij}. \tag{5.1}$$

The density, ρ_i, is the density of the *i-th* phase and the parameter J_{ji} represents the rate of mass transfer from the *j-th* to the *i-th* phase. Its units are mass per unit volume of the mixture and it is often called a *source term*. The number N in Eq. (5.1) is equal to the number of phases in the multiphase mixture. From the conservation of total mass of all the components we derive the following condition for this parameter: $J_{ij} = -J_{ji}$ and $J_{ii} = 0$. Eq. (5.1) stipulates that the rate of change of the mass of each component is equal to the flux of this component through the surrounding area, plus the mass generated—for example, because of combustion, evaporation, etc.

5.1.2 LINEAR MOMENTUM EQUATION FOR THE *I-TH* PHASE

$$\frac{\partial \rho_i \mathbf{v}_j}{\partial t} + \nabla_k \bullet (\rho_i \mathbf{v}_j \mathbf{v}_k) = \nabla_k \bullet \sigma_{jk} + \rho_i \mathbf{g}_j + \sum_{i=1}^{N} F_{ij}, \tag{5.2}$$

where ρ_i is the density of the *i-th* phase, σ_{jk} is the shear stress tensor applied to this phase, and the vector F_{ij} represents the forces interactions between the phase *i* and the other phases of the heterogeneous mixture. The forces are determined from fluid-particle interactions that are described in Chapter 4.

5.1.3 ANGULAR MOMENTUM EQUATION

The application of the angular momentum conservation principle results in the simple result that the tensor σ_{jk} is symmetric

$$\sigma_{kj} = \sigma_{jk}. \tag{5.3}$$

This relationship may be used to determine three of the nine components of the stress tensor.

5.1.4 ENERGY EQUATION

The energy conservation equation emanates from the first law of thermodynamics for open systems and may be written as follows for the total energy of the *i-th* phase, e_i

$$\frac{\partial \rho_i e_i}{\partial t} + \nabla_k \bullet (\rho_i e_i \mathbf{v}_k) = \nabla_k \bullet (q_{ki} - w_{ki}) + \rho_i (\mathbf{g}_k \bullet \mathbf{v}_k) + (\nabla_k \bullet \tilde{A}_{kj}) \bullet \mathbf{v}_j$$
$$+ \sum_{j=1}^{N} (E_{ij} - J_{ij} e_i), \tag{5.4}$$

where q_{ki} and w_{ki} represent the heat flux entering and the work flux leaving the *i-th* phase, and the term E_{ij} denotes the sources term for the energy per unit volume exchanged between the phases. As in the previous cases with these source terms: $E_{ij} = -E_{ji}$.

5.1.5 THE ENTROPY INEQUALITY

The entropy inequality is a consequence of the second law of thermodynamics and may be expressed as follows

$$\frac{\partial(\rho_i s_i)}{\partial t} + \nabla_k \bullet (\rho_i s_i \mathbf{v}_k) - \nabla_k \bullet \frac{\mathbf{q}_{ki}}{T} = \Phi_i + \nabla_k \bullet \Theta_{ik} \geq 0. \tag{5.5}$$

The terms Φ_i and Θ_{ik} denote the entropy production inside the material volume in the *i-th* phase and the entropy production at the interface of the *i-th* phase. These entropy sources cannot be measured, are difficult to quantify in practice, and must be computed indirectly. Because of this, the entropy inequality is not used for the determination of any parameters but is used to derive constitutive equations and constraints between material properties and flow parameters.

5.1.6 GENERALIZED FORM OF THE FUNDAMENTAL EQUATIONS

It is computationally efficient to use a single form of a generalized expression that generates each one of the expressions (5.1) through (5.5). This generalized equation is

$$\frac{\partial \rho_i \psi_i}{\partial t} + \nabla_k \bullet (\rho_i \mathbf{v}_k \psi_i) = \nabla_k \bullet \mathbf{H}_k + \rho_i \phi_i + \sum_{j=1}^{N} S_{ij}. \tag{5.6}$$

The generalized variables ψ_i, H_k, ϕ_i, and S_{ij} are given in the following Table 5.1.

TABLE 5.1
Variables in the General Form of the Conservation Equations

Balance Equation	ψ_i	H_k	ϕ_i	S_{ij}
Mass	1	0	0	J_{ij}
Linear Momentum	v_i	σ_{jk}	g_i	F_{ij}
Energy	e_i	$-w_{ki} + qi_k$	$g_k v_k$	$E_{ij} - J_{ij} e_i$
Entropy	s_i	$(q_i)/T$	Φ_i	Θ_i

5.2 APPLICATIONS IN EVAPORATION AND COMBUSTION—MASS COUPLING

Mass coupling can occur through a variety of mechanisms such as evaporation, condensation, or chemical reactions.

5.2.1 Evaporation or Condensation

Evaporation or condensation entails a phase change at the interface and the transfer of mass from one phase to another. The driving force for both processes is the difference in concentration of the droplet vapor between the droplet surface and the free stream. It is assumed, in general, that the carrier fluid is a binary mixture consisting of the carrier gas and the vapor of the liquid droplets. For example, water droplets evaporating in a nitrogen stream would be a binary mixture of water vapor and nitrogen. Because the ambient air has a constant composition, it is common practice to consider air as a single species and to regard water or any other type of vapor in air as a binary mixture.

Figure 5.1 shows a schematic diagram for the evaporation or condensation of a droplet translating in a carrier fluid with velocity U. The velocity of the vapor entering or leaving the surface of the droplet is v_s, and the outer normal direction is denoted by the vector n. The mass flux at the surface of evaporating or condensing droplets in a binary mixture is given by *Fick's law* as follows

$$\rho_s v_s = -\mathcal{D}_{AB} \frac{\partial \rho_A}{\partial n}, \tag{5.7}$$

where the left-hand side represents the mass flux to/from the particulate phase, $\mathcal{D}_{AB}$ is the diffusion coefficient of the droplet vapor in the carrier fluid, and ρ_A represents the partial density of the droplet vapor within the carrier phase.

The mass flux may be written in terms of the mass fraction $\omega_A \, (= \rho_A/\rho_s)$ as follows

$$\rho_s v_s = -\rho_s \mathcal{D}_{AB} \frac{\partial \omega_A}{\partial n}. \tag{5.8}$$

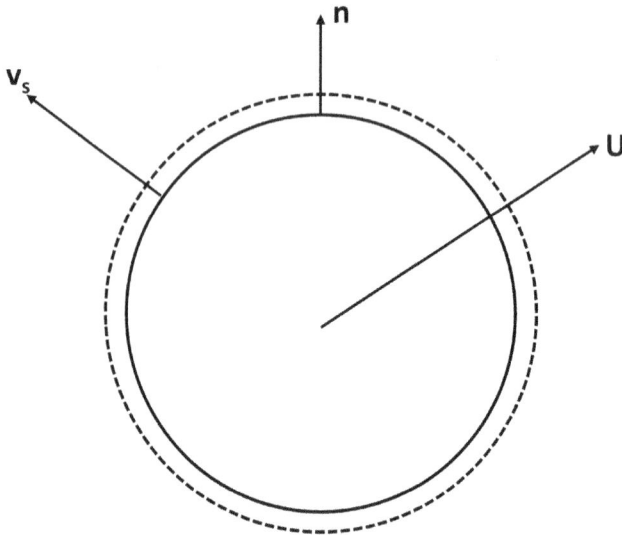

FIGURE 5.1 Schematic diagram of evaporation or condensation from the surface of a droplet.

In most practical applications, the mass fraction of the vapor of the particulate species within the carrier fluid, ω_A, is much less than 1 ($\omega_A \ll 1$). At thermodynamic equilibrium, the mass fraction of the vapor at the droplet surface is a function of the saturation temperature of the droplet, T, which also determines the partial pressure of the vapor at the surface of the droplet, $P_A = P_{sat}(T)$

$$\omega_{As} = \frac{M_A}{M_M} \frac{P_A}{P}, \tag{5.9}$$

where M_A is the molecular weight of the species A, M_M is the effective molecular weight of the carrier fluid, and P is the total pressure of the heterogeneous mixture.

For a droplet of diameter, D, the mass fraction gradient must be proportional to the mass fraction difference between the surface and the freestream and inversely proportional to the droplet diameter. Therefore, we may write the mass from the surface of droplets

$$\rho_s v_s = -\rho_s \mathcal{D}_{AB} \frac{\partial \omega_A}{\partial n} \sim \rho_c \mathcal{D}_{AB} \frac{\omega_{As} - \omega_{A\infty}}{D}, \tag{5.10}$$

where ω_{As} is the mass fraction of the vapor corresponding to the droplet/particulate phase at the droplet surface and $\omega_{A\infty}$ in the freestream. The density ρ_c is a representative density and is typically taken as the average density between the surface and the freestream. Such average properties between the surface and the freestream are called the *film conditions*. Except for very high evaporation rates, the density change through the boundary layer is small, and ρ_c is considered the density of the continuous phase. The sign on the difference in mass fraction between the surface and freestream indicates evaporation or condensation—for example, for evaporation, $\omega_{As} > \omega_{A\infty}$, and the mass flux is positive, $\rho_s w > 0$.

Therefore, the rate of change of the mass of droplets may be written as

$$\frac{dm}{dt} = -\rho_s v_s S \sim \pi D^2 \rho_c \mathcal{D}_{AB} \frac{\omega_{A\infty} - \omega_{As}}{D}, \tag{5.11}$$

The constant of proportionality is the *Sherwood number*, and the last equation becomes

$$\frac{dm}{dt} = \pi Sh D \rho_c \mathcal{D}_{AB} \left(\omega_{A\infty} - \omega_{As} \right). \tag{5.12}$$

The Sherwood number for the mass transfer is analogous to the Nusselt number for the heat transfer. Analytical and experimental expressions for the Sherwood number are typically obtained from the corresponding expressions for the Nusselt number according to the rules of the heat/mass transfer analogy that are presented in section 5.6.1. For example, the expression for the Sherwood number corresponding to Eq. (5.43) is (Feng and Michaelides, 2001)

$$Sh\left(Pe_m, Re_r\right) = 0.852 Pe_m^{1/3} \left(1 + 0.233 Re_r^{0.287}\right) + 1.3 - 0.182 Re_r^{0.355}. \tag{5.13}$$

Other correlations for the Sherwood number are found in Michaelides (2006).

5.2.2 THE D-SQUARE LAW

The rate of mass transfer from the droplet may also be expressed in terms of the diameter shrinkage or expansion as

$$\frac{dm}{dt} = \frac{d}{dt}\left(\rho_d \frac{\pi D^3}{6}\right) = \frac{\pi}{2}\rho_d D^2 \frac{dD}{dt}. \tag{5.14}$$

Therefore,

$$\frac{1}{2}\rho_d D^2 \frac{dD}{dt} = Sh D \rho_c \mathcal{D}_{AB}\left(\omega_{A\infty} - \omega_{As}\right), \tag{5.15}$$

This expression may be rewritten for the droplet diameter as

$$\frac{d\left(D^2\right)}{dt} = \frac{4 Sh \rho_c \mathcal{D}_{AB}\left(\omega_{A\infty} - \omega_{As}\right)}{\rho_d}. \tag{5.16}$$

The right-hand side of Eq. (5.16) may be considered as constant—the evaporation or condensation or combustion constant—and this yields the solution

$$D^2 = D_0^2 + \frac{4 Sh \rho_c \mathcal{D}_{AB}\left(\omega_{A\infty} - \omega_{As}\right)}{\rho_d} t, \tag{5.17}$$

where D_0 is the diameter of the droplet at the inception of the process, t = 0. This expression is often referred to as the *D-Square Law* for evaporation, condensation, or combustion.

This form of the mass transfer equation has been proven to be sufficiently accurate and is extensively used for industrial applications, where data for droplet evaporation, condensation, and combustion have been frequently reported as values for the constant in Eq. (5.16). It must be noted that the latter is not a constant in a flow with changing freestream conditions, but for many situations, the approximation may be adequate.

A related variable is the *lifetime or evaporation time* (depending on the process or condensation or combustion), τ_m, of a droplet, which is obtained by setting D = 0

$$\tau_m = \frac{-D_0^2}{\dfrac{Sh \rho_c \mathcal{D}_{AB}\left(\omega_{A\infty} - \omega_{As}\right)}{\rho_d}} = \frac{-D_0^2 \rho_d}{Sh \rho_c \mathcal{D}_{AB}\left(\omega_{A\infty} - \omega_{As}\right)}, \tag{5.18}$$

5.2.3 MASS TRANSFER FROM SLURRY DROPLETS

The mass transfer from a slurry droplet represents an important technological problem because it is extensively used for drying in the food industry—for example, the production of powdered milk. Food slurries are atomized and sprayed into a hot gas stream where the water is driven off and the dried products are collected.

The material in the atomized droplets is a porous medium formed by the solids in the food product. As the drying proceeds, the mass of the droplets decreases as the water/moisture is removed.

The drying process is generally regarded as happening in two stages: the *constant rate* and the *falling rate periods*. As the droplet dries and water is removed from the surface, more liquid is conveyed to the surface by capillary forces (surface tension). During the constant rate stage, a liquid layer covers the surface of the droplet, and the drying rate proceeds as if the slurry droplet was a liquid droplet. In this case, the rate of mass transfer is given by Eq. (5.12). Because the solids inside the droplet form a porous medium and the droplet diameter does not change significantly with time, the mass removal rate is almost constant, and this stage is called the *constant rate* stage.

The moisture in the droplet is quantified by the moisture ratio, x, or *wetness* defined as the ratio of the mass of water/moisture that is contained in the droplet to the mass of the solids, $x = m_w/m_s$. When the wetness reaches a *critical moisture content*, which is a characteristic of the drying materials, the liquid film on the surface of the droplet disappears and the drying process enters the *falling rate* stage. In this stage the surface of the droplet is dry, moisture is transferred through the pores, the resistance to the mass transfer increases, and the rate of mass transfer rate is considerably slower, hence the name *falling rate* period.

Figure 5.2 shows the typical variation of the drying rate, the moisture content, and the droplet temperature as functions of time. The two stages of drying, constant rate, and falling rate—following the attainment of the critical moisture content—are evident in the figure. After the critical moisture content is reached, the drying rate decreases, and the droplet temperature rises.

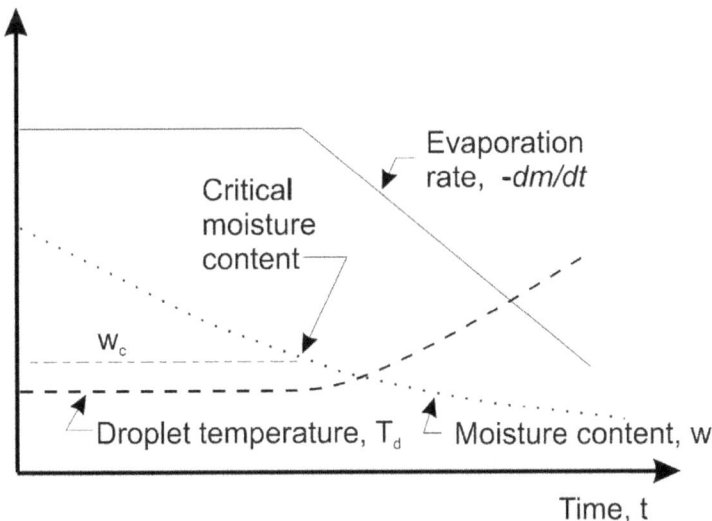

FIGURE 5.2 Slurry droplet property variation during a drying process.

A common approach to modelling this drying process is to assume that the rate of moisture removal is proportional to the moisture content, x. This approach results in an exponentially decreasing moisture content

$$\frac{dm}{dt} = m_s \frac{dx}{dt} \sim m_s x \Rightarrow x \sim \exp(-kt). \tag{5.19}$$

The correct modelling of the drying process and the determination of the model constants depend on the nature of the solids that constitute the porous material, the shape of the porous medium, and the thermal conductivity of the materials in the droplet.

5.2.4 COMBUSTION

The combustion of a single droplet is modeled as a liquid fuel droplet surrounded by a flame, which is composed of hot gases, and is shown in Figure 5.3. The flame front is defined at the surface where the fuel vapor and oxidizer meet and react. The position of the flame front is determined by the heat transfer process, which evaporates the droplet and supplies the fuel vapor to support the flame. As the liquid in the droplet evaporates, the surface of the droplet recedes.

A simple model, formulated by Hedley *et al.* (1971), is based on the spherically symmetric flow of vapor from the droplet and the oxidizer from the surroundings. From the energy balance at the surface of the burning droplet, one obtains the following equation for the rate of mass of the vapor that enters the flame region

$$\dot{m} = \frac{4\pi k_c \ln\left(1 - \frac{c_p \Delta T}{h_{fg}}\right)}{c_p\left(\frac{1}{r_d} - \frac{1}{r_f}\right)}, \tag{5.20}$$

where $\dot{m}$ is the evaporation rate of the droplet, k_c is the thermal conductivity of the carrier gas, c_p is the specific heat of the carrier gas at constant pressure, and ΔT is the

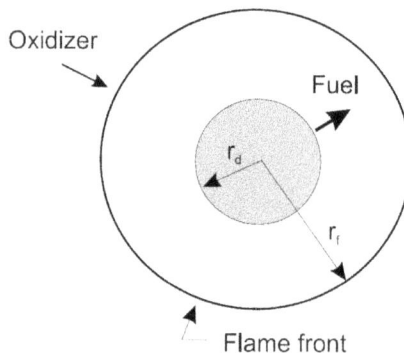

FIGURE 5.3 The fuel droplet combustion process.

temperature difference between the flame and droplet surface. The distance between the flame and droplet surface is called the *standoff distance, δ*

$$\delta = r_f - r_d. \tag{5.21}$$

Hence, one may write Eq. (5.20) as follows

$$\dot{m} = \frac{dm}{dt} = \frac{2\pi k_c D \ln\left(1 - \dfrac{c_p \Delta T}{h_{fg}}\right)}{c_p \dfrac{\delta}{r_f}}. \tag{5.22}$$

It is observed that Eqs. (5.22) and (5.12) show the same dependence with respect to the diameter of the droplet, that is, both the burning rate and the rate of evaporation/ condensation are proportional to the diameter, D, of the droplet. This leads to the D^2-law for the rate of burning

$$D^2 = D_0^2 - \left[\frac{8k_c \ln\left(1 - \dfrac{c_p \Delta T}{h_{fg}}\right)}{\rho_d c_p \dfrac{\delta}{r_f}} \right] t. \tag{5.23}$$

The value for the standoff distance δ depends on the heat released by the flame. When the droplets are very small, δ≈r_f. The parameter inside the square brackets is called the *burning constant*, λ, and its units are L²/T. The burning time of a droplet may be calculated from Eq. (5.23) by setting $D^2 = 0$

$$\tau_m \approx \frac{D_0^2}{\lambda}. \tag{5.24}$$

The approximate sign in the last equation is inserted because continuum theory does not apply in the last stage of a burning droplet (the theory of molecular dynamics must be applied).

The following Table 5.2 gives typical values for the burning constant of four common fuels.

TABLE 5.2
Burning Rate Constants, λ, for Different Fuels Burning in Air at 20°C and 100 kPa

Fuel	Burning Constant λ, m ²/s
Benzene	9.9×10^{-7}
Kerosene	9.6×10^{-7}
Diesel oil	7.9×10^{-7}
Iso-octanes	11.4×10^{-7}

Example: Calculate the burning time of a 200 micron diesel oil droplet.

Answer: A good approximation of the burning time is given by Eq. (5.24). From Table 5.2 the burning rate constant is $7.9 \times 10^{-7} m^2/s$, so the burning time is $(200 \times 10^{-6})^2/7.9 \times 10^{-7} = 0.051s$.

It must be noted that, when there is relative velocity between the droplet and the carrier fluid (the oxidizer), the burning rate is increased. An estimate for this enhancing effect is obtained by applying an expression for the convective heat/mass transfer to the burning constant. For example, Williams (1990) used the Ranz-Marshall correlation for convection to obtain

$$\lambda = \lambda_0 \left(1 + 0.24 \, Re_r^{0.5} \, Sc^{0.33}\right), \tag{5.25}$$

where λ_0 is the burning rate of a stationary droplet (with no convection effects). Models for droplet combustion in sprays (Chigier, 1995) suggest that combustion does not occur as flames around individual droplets but rather as flames around groups of droplets.

It must be noted that the combustion of coal particles does not follow the droplet model, described in this section. Coal particles consist of four components: volatiles, moisture, char, and ash. Some volatiles are combustible gases, such as methane and hydrogen. Coal combustion takes place in two steps: First, the volatiles and moisture are driven off. The coal particles start the burning process and contribute to a gas phase flame. This happens quickly and depends on the heating rate of the particles. After the volatiles are driven off and burned, the solid char (carbon) burns very slowly. Char burning is not well-quantified and depends strongly on the coal type and grade.

5.3   LINEAR MOMENTUM INTERACTIONS

Linear momentum coupling between phases occurs as the result of mass transfer, body forces (e.g. gravity), magnetic, electric, and hydrodynamic forces, such as drag, lift, and Brownian movement. The movement of single particles and drops is expressed by the equation

$$m_p \frac{d\vec{v}}{dt} = \sum_i \vec{F}_i, \tag{5.26}$$

where the right-hand side represents the sum of all the forces, $\vec{F}_i$, acting on the particle. A detailed exposition of all the forces that may act on a single particle or droplet is given in Sections 4.1–4.10, including transient forces. For the determination of the movement and trajectory of single particles, one must add all the applicable forces and then solve Eq. (5.26), usually by a numerical method. It must be noted that momentum transfer from mass efflux due to evaporation, condensation, or burning vanishes if the mass transfer process is uniform around the surface of the particle. If the mass transfer is not uniform, detailed numerical computations should be conducted to determine the additional force/impact from the momentum associated with the mass efflux.

5.3.1   MOMENTUM INTERACTIONS WITH GROUPS OF PARTICLES

While Eq. (5.26) and the expressions for the drag coefficients of particles in Chapter 4 may be used in all flow where the particles are far from each other and the wakes may develop unimpeded (dilute flows), understanding and modelling the momentum interactions at higher volume fractions is a formidable problem. Several early studies (e.g. Ergun, 1952) attempted to model the particle-fluid momentum interactions at higher volume fractions, assuming that the group of particles maintain a certain configuration in the flow. However, in all suspensions of finite concentration, individual particles interact with their neighboring particles as well as with the interstitial fluid carrier, and their configuration continuously changes. One may realize the difficulty of analytically modelling the problem of finite particle concentration by considering only two particles where one is immersed in the wake of the preceding particle: Due to the sheltering effect of the preceding particle, the following particle experiences lower relative velocity. In addition, the following particle does not experience a spatially and temporally uniform velocity. As a result, its Reynolds number is not known, and the hydrodynamic force on it cannot be analytically determined. When more particles are added in the flow field, the complex interactions between individual particles and between particles and the interstitial fluid make the accurate determination of hydrodynamic forces impossible to determine analytically.

An examination of the several correlations that attempted to quantify the drag coefficients of particles at finite concentrations (Ergun, 1952; Richardson and Zaki, 1954; Wen and Wu, 1966; Di Felice, 1994) proves that the drag coefficients of clusters of particles are significantly lower than the corresponding coefficients of single particles. This has been corroborated by numerical computations, especially those using methods that are extensively described in Chapter 8. For example, an early numerical study with 48 particles and initial concentration 34% (Xu and Michaelides, 2003) determined the changes of the particle configuration and determined the coefficient 2.5 for the Richardson-Zaki correlation. All the numerical studies with groups of particles point to the complexities of the flows and interactions and corroborate the fact that the drag on clusters is significantly lower than that on individual particles (e.g. Feng and Michaelides, 2005; Dietzel and Sommerfeld, 2013). A more recent numerical study that followed a group of 140 particles allowed to freely interact at constant concentration at 30% (Hardy *et al.*, 2022) determined that the averaged drag force exerted on the particles during the time of the simulation (a) exhibits a large variation and (b) is lower than the drag force experienced by individual particles in undisturbed flows. One easily concludes from all these studies that only accurate numerical calculations may model quantitatively the complex interactions of fluids with groups of particles.

5.4   ANGULAR MOMENTUM INTERACTIONS

The rotation of solid particles is induced by wall impacts and shear flow. Wall and inter-particle collisions convert part of the linear momentum of the particles to angular momentum and may induce extremely high angular velocities for the particles. Following a collision process, the particles have a distinct angular velocity, which is different from the surrounding fluid. As a consequence, a viscous torque acts on the particles and reduces this angular velocity. Particle rotation influences the subsequent wall collisions and, additionally, imposes a transverse (lift) force, the Magnus force (see Section 4.6.1).

The angular velocity of the particle, $\vec{\omega}$, is determined by the following equation

$$I_p \frac{d\vec{\omega}}{dt} = -\vec{T}, \tag{5.27}$$

where I_p is the moment of inertia of the particle, and T is the torque. For a spherical particle, the moment of inertia is $m_p D^2/10$. The torque exerted by a fluid on a rotating sphere at $Re_R \ll 1$ is (Rubinow and Keller, 1961)

$$\vec{T} = \pi \mu_c D^3 \left(\frac{1}{2} \nabla \times \vec{u} - \vec{\omega} \right). \tag{5.28}$$

The quantity in the parenthesis is the relative rotation between particle and fluid, $\vec{\Omega}$. Under steady-state conditions, a particle with no externally applied torque moving in a flow with shear at the same longitudinal velocity will reach a steady-state angular velocity

$$\vec{\omega} = \frac{1}{2} \nabla \times \vec{u}. \tag{5.29}$$

This is, simply, the local rotational rate of the fluid.

Dennis et al. (1980) performed an analytical study on the torque required to rotate a rigid sphere in a viscous fluid, which is at rest far from the sphere in the range. These results apply in the range $80 < Re_R < 8,000$, and a good representation of the torque required to rotate the sphere is

$$\vec{T} = -2.01 \mu_c D^3 \vec{\omega} \left(1 + 0.1005 \sqrt{Re_R} \right). \tag{5.30}$$

As with the linear momentum drag coefficient, a rotational drag coefficient, C_R, is introduced to model the torque at higher rotational Reynolds numbers

$$\vec{T} = \frac{1}{2} C_R \rho_c \left(\frac{D}{2} \right)^5 |\vec{\Omega}| \vec{\Omega}. \tag{5.31}$$

A glance at the last two equations proves that the rotational drag coefficient in Stokesian flow conditions ($Re_R \ll 1$) is

$$C_R = \frac{64\pi}{Re_R}. \tag{5.32}$$

Figure 5.4 depicts numerical data of C_R by Dennis et al. (1980), experimental data by Sawatzki (1970), and the following correlation that emanates from the two data sets

$$C_R = \frac{12.9}{\sqrt{Re_R}} + \frac{128}{Re_R} \qquad 32 < Re_R < 1000. \tag{5.33}$$

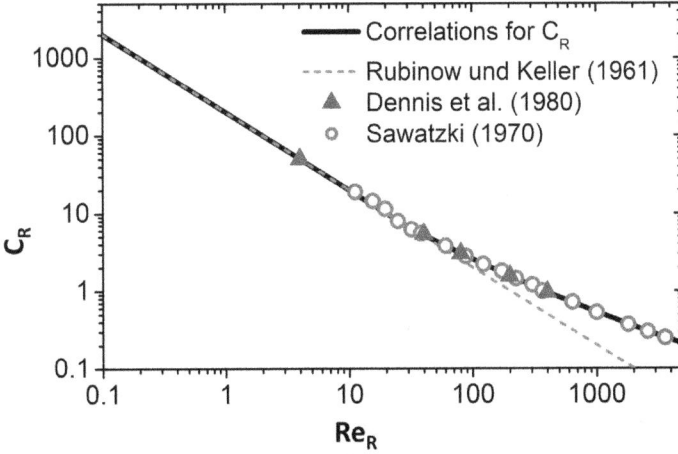

FIGURE 5.4   Rotational drag coefficients for spherical particles.

The results of the Rubinow and Keller (1961) are also shown by the dashed line in Figure 5.4.

The rotational response time has been defined in Eq. (2.29). The ratio of the rotational to the translational response time in the Stokesian regime is 3/10.

5.4.1   TRANSIENT ROTATION

Feuillebois and Lasek (1978) examined analytically the subject of transient rotation of a rigid sphere at creeping rotational flow conditions ($Re_R \ll 1$). They derived the following expression for the angular velocity of the rigid sphere when the applied torque undergoes stepwise change from 0 to $\vec{T}$, commencing at time, $t = 0$, in a viscous and initially quiescent fluid

$$\left(\frac{1}{60}\rho_d D^5\right)\frac{d\vec{\omega}}{dt} = \vec{T} - \pi\mu_c D^3\vec{\omega} - \pi\mu_c D^3\left(\frac{D}{6\sqrt{\pi v_c}}\int_0^t \frac{\frac{d\vec{\omega}}{d\sigma}}{\sqrt{t-\sigma}}d\sigma\right) +$$

$$+\frac{1}{3}\pi\mu_c D^3\left[\int_0^t \frac{d\vec{\omega}}{d\sigma}\exp\left(\frac{4v_c(t-\sigma)}{D^2}\right)erfc\sqrt{\frac{4v_c(t-\sigma)}{D^2}}d\sigma\right]$$

(5.34)

Eq. (5.34) is analogous to Eq. (4.94) for the transient linear velocity of the sphere. The last two terms are history terms that result from the step-change of the torque. As in the case of the transient heat transfer, Eq. (5.71), the transient expression for the angular velocity does not include an added mass term. This is due to the absence of the pressure gradient term in the angular momentum equation. The result is also supported by the analogy between the energy diffusion (heat transfer) and the vorticity diffusion processes. At the end of the transient process, the rotating sphere will reach asymptotically a terminal angular velocity, which is equal to

$$\vec{\omega} = \frac{\vec{T}}{\pi \mu_c D^3}.$$
(5.35)

5.5 ENERGY INTERACTIONS—HEAT TRANSFER

5.5.1 HEAT-MASS TRANSFER SIMILARITY

The processes of heat and mass transfer from particles are governed by similar governing equations. As a consequence, with the same boundary and initial conditions, all the solutions of the heat transfer equations and all the experimental correlations derived from heat transfer experiments apply to the similar systems for mass transfer as well. Table 5.3 lists important variables for the transport of heat and the corresponding similarity variables for the transport of mass (Michaelides, 2017).

The rates of heat and mass transfer from particles are, in general, given by the following equations

$$\dot{Q} = hA_S \left(T_S - T_\infty\right) \quad and \quad \dot{m}_i = h_m A_S \rho_c \left(\omega_{iS} - \omega_{i\infty}\right),$$
(5.36)

where h is the *heat transfer coefficient*; A_S is the surface area of the particle; h_m is the *mass transfer coefficient*; ω_i represents the mass fraction of the species i, which is generated by the particle or condenses on the particle; and the subscripts S and ∞ denote local conditions at the surface of the particle and far from the particle respectively. In general, the heat and mass transfer coefficients are given in terms of two dimensionless numbers, the Nusselt and the Sherwood number

$$Nu = \frac{hD}{k_c} \quad and \quad Sh = \frac{h_m D}{\mathcal{D}_c}.$$
(5.37)

For particles carried by a fluid, the dimensionless Nusselt number is usually expressed as a function of the Peclet number—based on the relative velocity of the particle, Pe_r—and Prandtl number, Pr.

$$Pe_r = Re_r Pr = \frac{D|\vec{u} - \vec{v}|\rho_c c_{pc}}{k_c} \quad and \quad Pr = \frac{c_{pc} \mu_c}{k_c}.$$
(5.38)

TABLE 5.3
Corresponding Variables for the Heat and Mass Transfer Processes

Heat Transfer Variables	Mass Transfer Variables
Temperature, T	Mass Concentration, C
Thermal Diffusivity, $k_c/\rho_c c_{pc}$	Mass Diffusivity, $\mathcal{D}_c$
Nusselt Number, Nu	Sherwood Number, Sh
Peclet Number, Pe	Peclet number for Mass, Pe_m
Prandtl Number, Pr	Schmidt Number, Sc

Because two independent variables, Pe_r and Pr—or equivalently, Re_r and Pr—come into the functions for the heat transfer coefficient for solid particles, these functions are more complex than those of momentum transfer, where Re_r is the single pertinent variable, and the drag is simply expressed in terms of one function, given by the standard drag curve of Figure 4.3.

The corresponding Sherwood number for particles carried by fluids is expressed in terms of the Peclet number for mass—based on the relative velocity, Pr_m—and the Schmidt number, Sc.

$$Pe_m = Re_r Sc = \frac{D|\vec{u} - \vec{v}|}{\mathcal{D}_c} \quad and \quad Sc = \frac{\mu_c}{\rho_c \mathcal{D}_c}. \tag{5.39}$$

For brevity, in the following sections, only results for the heat transfer process will be presented. It must be emphasized that all the results apply to the mass transfer processes when the similarity conditions are satisfied.

5.5.2 STEADY HEAT TRANSFER FROM SPHERES

In general, the steady heat interactions of solid particles with fluids affect the temperature of the particles. In the case of droplets, heat transfer will affect the temperature as well as the rate of evaporation. For particles and droplets that do not undergo mass transfer (due to evaporation, condensation, or combustion), the temperature change is described by the expression

$$\frac{dT_d}{dt} = \dot{Q} = \oint_A -\vec{q}\vec{n}dA, \tag{5.40}$$

where $\dot{Q}$ is the total heat transfer to the particle, the vector q denotes the heat flux (rate of heat per unit area), and n is the outward normal unit vector at the surface of the particle.

When liquid drops in thermodynamic equilibrium with the carrier fluid evaporate or vapor condenses, the heat transfer causes the evaporation or condensation of a mass rate $\dot{m}$

$$\dot{Q} = \oint_A -\vec{q}\vec{n}dA = \dot{m}h_{fg}, \tag{5.41}$$

where h_{fg} is the latent heat of the liquid in the drops.

The total rate of heat transfer, $\dot{Q}$, is given in general by a heat transfer coefficient, h, and the Nusselt number, Nu, as in Eqs. (5.36) and (5.37). In the case of pure conduction heat transfer $Nu = 2$ and $h = 2k_c/D$. The following subsections describe the heat transfer process of particles and drops under different flow conditions in terms of the pertinent Nusselt numbers.

5.5.2.1 Solid Spheres

At creeping (Stokes flow) conditions and Prandtl numbers on the order or less than 1, which imply $Pe_r \ll 1$, heat conduction around solid spheres dominates the process, and the Nusselt number is equal to 2. At higher Peclet numbers, Whitaker (1972) developed an empirical correlation for solid spheres, which is valid up to $Re_r = 10^4$

$$Nu = 2 + \left(0.4\,\mathrm{Re}_r^{1/2} + 0.06\,\mathrm{Re}_r^{2/3}\right)\mathrm{Pr}^{0.4}. \tag{5.42}$$

A more recent and more accurate expression for Nu that includes the two independent variables Pe_r and Re_r was derived by Feng and Michaelides (2001)

$$Nu\left(Pe_r, \mathrm{Re}_r\right) = 0.852 Pe_r^{1/3}\left(1 + 0.233\,\mathrm{Re}_r^{0.287}\right) + 1.3 - 0.182\,\mathrm{Re}_r^{0.355}. \tag{5.43}$$

5.5.2.2 Viscous Spheres

Feng and Michaelides (2000, 2001) conducted numerical studies on the heat transfer from viscous spheres without mass transfer (blowing effects) at the interface and derived accurate and useful correlations for Nu, which are summarized as follows:

A. At small but finite values of Re_r ($0 < Re_r < 1$) and $Pe_r > 10$

$$Nu = \left(\frac{0.651}{1 + 0.95\lambda}Pe_r^{1/2} + \frac{0.991\lambda}{1+\lambda}Pe_r^{1/3}\right)\left[1 + f\left(\mathrm{Re}_r\right)\right]$$
$$+ \left(\frac{1.65\left[1 - f\left(\mathrm{Re}_r\right)\right]}{1 + 0.95\lambda} + \frac{\lambda}{1+\lambda}\right), \tag{5.44}$$

where the function $f(Re_r)$ is defined as

$$f\left(\mathrm{Re}_r\right) = \frac{0.61\mathrm{Re}_r}{\mathrm{Re}_r + 21} + 0.032. \tag{5.45}$$

B. At higher Re_r and Pe_r, the analysis of numerical data revealed that the most accurate functional relationship, $Nu = f(Re_p, Pe_r)$, is generated—as with the drag coefficient of Eqs. (4.19), (4.20), and (4.21)—in terms of three functions that pertain to specific values of the viscosity ratio, λ

B1. For inviscid spheres ($\lambda = 0$):

$$Nu\left(0, Pe_r, \mathrm{Re}_r\right) = 0.651 Pe_r^{1/2}\left(1.032 + \frac{0.61\mathrm{Re}_r}{\mathrm{Re}_r + 21}\right)$$
$$+ \left(1.60 - \frac{0.61\mathrm{Re}_r}{\mathrm{Re}_r + 21}\right). \tag{5.46}$$

B2. For solid spheres ($\lambda = \infty$), the expression of Eq. (5.43):

$$Nu\left(\infty, Pe_r, \mathrm{Re}_r\right) = 0.852 Pe_r^{1/3}\left(1 + 0.233\mathrm{Re}_r^{0.287}\right)$$
$$+ 1.3 - 0.182\mathrm{Re}_r^{0.355}. \tag{5.47}$$

B3.   For viscous spheres with $\lambda = 2$:

$$Nu(2, Pe_r, Re_r) = 0.64 Pe_r^{0.43} \left(1 + 0.233 Re_r^{0.287}\right) + 1.41 - 0.15 Re_r^{0.287}. \quad (5.48)$$

Using these functions, the overall correlations for the Nusselt numbers are given by the following expressions in two ranges of the viscosity ratio, $0 \le \lambda < 2$ and $2 < \lambda \le \infty$

$$Nu(Pe_r, Re_r, \lambda) = \frac{2 - \lambda}{2} Nu(Pe_r, Re_r, 0) + \frac{4\lambda}{6 + \lambda} Nu(Pe_r, Re_r, 2) \quad (5.49)$$

$$for \quad 0 \le \lambda \le 2, \quad and \quad 10 \le Pe_r \le 1000$$

and

$$Nu(Pe_r, Re_r, \lambda) = \frac{4}{\lambda + 2} Nu(Pe_r, Re_r, 2) + \frac{\lambda - 2}{\lambda + 2} Nu(Pe_r, Re_r, \infty). \quad (5.50)$$

$$for \quad 2 \le \lambda \le \infty, \quad and \quad 10 \le Pe_r \le 1000$$

For smaller values of $Pe_r < 10$, it is not possible to obtain a simple correlation of the numerical results, $Nu(Pe_r, Re_r, \lambda)$, with any satisfactory degree of accuracy. For applications in the range $0 < Pe_r < 10$, the use of the original numerical data in Feng and Michaelides (2001) is recommended.

As in the case of the hydrodynamic force on a viscous sphere, it was determined that, for fixed values of Re_r and λ, variations of the density ratio, ρ_d/ρ_c, have only a minimal effect on the external flow field, and its effect on the heat transfer coefficients is less than 0.1%.

5.5.2.3   Mixed Convection

When the external velocity field is moderate and forced convection is not strong, the free (natural) convection from a sphere must be considered. Since the free convection takes place in the direction of gravity and the forced convection is in the direction of the external flow, which is arbitrary, the heat flux directions of the two types of convection are not necessarily the same. Musong and Feng (2014) examined numerically the mixed convection from heated spheres at arbitrary incident flow angles and determined that the overall heat transfer is significantly influenced by the direction of the forced flow. Based on numerical simulation results, they developed the following correlation for mixed convection

$$Nu = a Re_r^b Ri^{0.5b} \theta^2 + c Re_r^d Ri^{0.5d} \theta + n Re_r^f Ri^{0.5f}$$
$$+ g Re_r^{1+h} Ri^{0.5} + k Re_r^m, \quad (5.51)$$

where the constants are $a = -0.0208$; $b = 0.4851$; $c = 0.0250$; $d = 0.05$; $n = 1.436$; $f = 0.3642$; $g = 0.2534$; $h = -0.722$; $k = -0.2602$; $m = 0.2412$. The units of the incident angle θ are radians. This expression is valid for laminar flows in the range $1 \le Re_r \le 100$ and $1 \le Ri \le 5$. The Richardson number, Ri, is defined as follows

$$Ri = \frac{gD|T_S - T_\infty|\beta_c}{|\vec{u} - \vec{v}|^2}, \tag{5.52}$$

where β_c is the thermal expansion coefficient of the carrier fluid.

5.5.2.4 Velocity Slip and Temperature Difference (Temperature Slip)

The expressions for Nu in the previous subsections are based on two stipulations: (a) No-slip velocity condition at the interface and (b) equal temperatures of the sphere and the carrier fluid at the interface. However, it has been experimentally observed that, when the size of particles is comparable to the mean free path of the base fluid, there is a significant discontinuity of both velocity and temperature at the interface (Michaelides, 2014). As with the drag coefficients, the net effect of the two discontinuities is to decrease the magnitude of the convective heat transfer. The temperature discontinuity is modeled in terms of an *accommodation coefficient*, $\zeta = f(T)$, defined in terms of the molecular collisions on the surface of the sphere. Accordingly, the temperature discontinuity at the solid-fluid interface is expressed by the boundary condition

$$T_f - T_s\big|_{r=a} = \frac{2\gamma(2-\zeta)L_{mf}}{(\gamma+1)\zeta\,\mathrm{Pr}}\frac{\partial T_f}{\partial r} = \lambda_t\frac{D}{2}\frac{\partial T_f}{\partial r}, \tag{5.53}$$

where γ is the ratio of the specific heats of the carrier fluid, c_{pc}/c_{vc}. For liquids, $\gamma \approx 1$, and for gases, $1 < \gamma < 1.67$. λ_t, is a dimensionless thermal slip parameter defined as

$$\lambda_t = \frac{4(2-\zeta)\gamma\,Kn}{(\gamma-1)\zeta\,\mathrm{Pr}}. \tag{5.54}$$

Feng and Michaelides (2012) derived an asymptotic solution to the general problem of heat transfer from viscous and solid spheres with velocity slip and thermal slip

$$Nu = \frac{2}{1+\lambda_t} + \frac{1}{2(1+\lambda_t)^2}Pe_r + \frac{f_s}{4(1+\lambda_t)^2}Pe_r^2\ln(Pe_r) + \frac{1}{2(1+\lambda_t)^2}\times$$

$$\left[\frac{-156+148f_s-152\lambda_t+341\lambda_t f_s+129f_s^2+528\lambda_t f_s^2}{960(1+2\lambda_t)} - (0.62+0.55\lambda_t)f_s\right], \tag{5.55}$$

$$Pe_r^2 + \frac{1}{16(1+\lambda_t)^3}Pe_r^3\ln(Pe_r) + O(Pe_r^3),$$

where f_s is a dimensionless parameter related to the interfacial velocity slip. For solid spheres, f_s is defined as

$$f_s = \frac{\beta a + 2\mu_c}{\beta a + 3\mu_c} = \frac{1+2Sp}{1+3Sp}. \tag{5.56}$$

And for viscous spheres

$$f_s = \frac{3\mu_d + 2\mu_c}{3\mu_d + 3\mu_c} = \frac{3\lambda+2}{3\lambda+3}. \tag{5.57}$$

Since λ_t is always positive, the effect of the interfacial thermal slip is to reduce Nu and by extent the convective heat transfer coefficient of the solid and fluid spheres.

5.5.2.5 Blowing Effects

During evaporation, sublimation, and combustion processes, changes in the properties of the gaseous boundary layer and phase change on the surface of particles significantly influence the heat and mass transfer coefficients. The blowing dimensionless numbers defined in Eq. (4.44), B_T and B_M, also determine the heat and mass transfer effects on the surface of the sphere. Several empirical correlations emanating from experimental and numerical data have been widely used to quantify the effects of mass transfer from the surface of particles and the shrinking of particles. Among these, Chiang *et al.* (1992) derived the following correlations for the heat and mass transfer coefficients

$$Nu = 1.275(1 + B_T)^{-0.678} Re_{rm}^{0.438} Pr_m^{0.619}, \tag{5.58}$$

and

$$Sh = 1.224(1 + B_M)^{-0.568} Re_{rm}^{0.385} Sc_m^{0.492}. \tag{5.59}$$

The subscript m in the Reynolds, Prandtl, and Schmidt numbers, Re_{rm}, Pr_m, and Sc_m, signifies that the carrier fluid properties are calculated using the film transport coefficients, as defined in Eq. (4.45), and in the case of Re_{rm}, the free-stream gas density, $\rho_{f\infty}$, far from the droplet. The correlations lead to the following expressions for the rates of heat and mass transfer from spheres

$$
\begin{aligned}
\dot{Q} &= 1.275\pi k_c D(T_s - T_\infty)(1 + B_T)^{-0.678} Re_{rm}^{0.438} Pr_m^{0.619} \\
\dot{m} &= 1.224\pi D\rho_c \mathcal{D}(Y_s - Y_\infty)B_M(1 + B_M)^{-0.568} Re_m^{0.385} Sc_m^{0.492}.
\end{aligned} \tag{5.60}
$$

Renksizbulut and Yuen (1983) have devised alternative correlations to calculate the heat and mass transfer from drops. Sirignano (1999) confirmed that the results for the heat and mass transfer from these two methods do not differ substantially.

It must be noted that the expressions for the Nusselt and Sherwood numbers for evaporating and burning droplets differ. This happens because the boundary conditions of the governing equations are different, and the heat and mass transfer processes are not similar.

5.5.2.6 Effects of Rotation

Several studies have focused on the problem of heat transfer from rotating spheres, typically at high rotation Reynolds numbers, $Re_R > 1,000$, where the fluid boundary layer is laminar and well formed. Among these, Kreith *et al.* (1963) studied the cooling and heating of a solid sphere in several fluids, including oil, water, and air. Dorfman and Serazetdinov (1973) used the boundary layer theory and derived analytically an expression for the Nusselt number. Tieng and Yan (1993) studied the mixed convection (forced and free) of rotating spheres in air. Feng (2014) employed a direct numerical simulation and obtained numerical data for the heat transfer from a rotating sphere in the range $0 < Re_R < 1,000$. Figure 5.5 shows the relationship between the Nusselt numbers and the Reynolds number, using the data and correlations from Feng (2014) and Kreith *et al.* (1963). The first two are applicable in the range $0 \leq Re_R \leq 1,000$, and the last correlation is applicable for $1,000 < Re_R$.

Considering all the data and correlations, the following expressions for the effects of the rotation on the heat transfer from spheres are recommended:
In the range $0 < Re_R \leq 500$, the expression in Feng (2014)

$$Nu = 2 + 0.00051 Re_R^{1.5}. \tag{5.61}$$

In the range $500 < Re_R \leq 1000$, a correlation obtained from Feng's (2014) data

$$Nu = 7.167 \ln(Re_R) - 36.712. \tag{5.62}$$

In the range $1,000 < Re_R$, the Kreith *et al.* (1963) correlation, which is also supported by the results of Tieng and Yan (1993) and the theory for developed laminar boundary layers

$$Nu = 0.43 Re_R^{0.5} Pr^{0.4}. \tag{5.63}$$

5.5.2.7   Effects of Flow Turbulence

As in the case of the drag coefficients, a number of experimental studies in the past suggested that the steady flow heat transfer coefficients of particles are affected by the free-stream turbulence. However, a glance at the data proves that there is high experimental uncertainty and several discrepancies that were not resolved. One

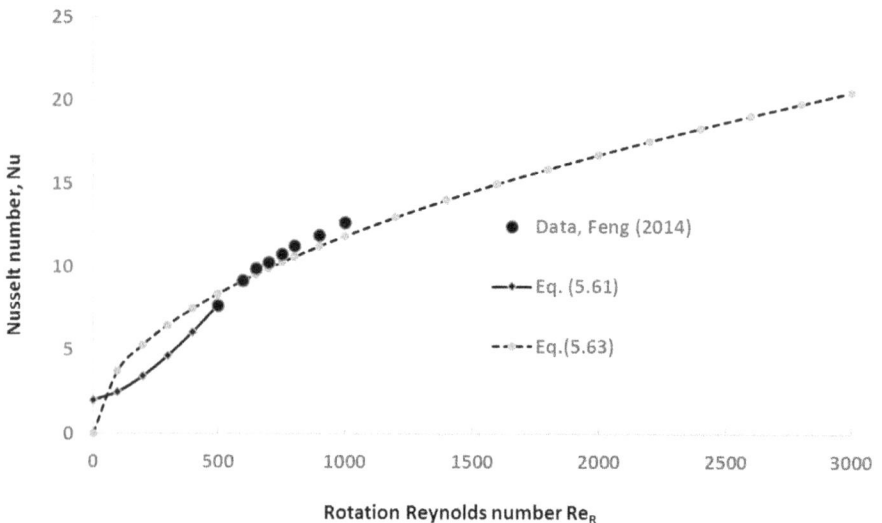

FIGURE 5.5   The effect of rotation on the heat transfer of a sphere, *Nu* vs. *Re_R*.

must consider that turbulence is inherently a transient process, and the movement of all types of particles in turbulent flows is unsteady. As a consequence, some of the methods used in several of these studies to derive the steady drag coefficients of particles are flawed. For this reason, it is recommended to use the results for transient convection, which are given in Section 5.6.5.

5.5.3 RADIATION

Radiation is the mode of heat transfer via electromagnetic waves (or photons) and is intercepted by the cross-sectional area of particles. A sphere, which has a cross-sectional area equal to πD^2, with surface temperature T_s, emits a rate of heat by radiation

$$\dot{Q}_{rad}^{em} = \sigma \varepsilon \pi D^2 T_s^{\,4}, \tag{5.64}$$

where σ is the Stefan-Boltzmann constant, $5.669*10^{-8}$ W/m²K⁴, and ε is the emissivity of the sphere ($0 \leq \varepsilon \leq 1$) with the emissivity of a black body being equal to 1. The sphere also absorbs heat from all the objects in its surroundings. In the simple case where the sphere is enclosed by a single medium at temperature T_w, the sphere absorbs thermal radiation equal to

$$\dot{Q}_{rad}^{ab} = \sigma \alpha \pi D^2 T_w^{\,4}, \tag{5.65}$$

where α is the absorptivity of the sphere ($0 \leq a \leq 1$). In general, the absorptivity and the emissivity of a material are functions of its temperature and are equal in magnitude: $\alpha(T_s) = \varepsilon(T_s)$.

The net rate of energy that enters the sphere as a result of thermal radiation is equal to the difference of the previous two equations. In analogy with convection heat transfer, it is given in terms of a radiation coefficient, h_{rad}

$$\dot{Q}_{net} = \dot{Q}_{rad}^{ab} - \dot{Q}_{rad}^{em} = \sigma \varepsilon \pi D^2 (T_w^{\,4} - T_s^{\,4}) = \pi D^2 h_{rad} (T_w - T_s). \tag{5.66}$$

The radiation coefficient is a strong function of temperature, which must be in absolute units (K or °R)

$$h_{rad} = \sigma \varepsilon \left(T_w^3 + T_w^2 T_s - T_w T_s^2 + T_s^3 \right). \tag{5.67}$$

The spherical particle exchanges radiation with any other object or surface that it "sees," that is, objects connected to the sphere by a straight line without interception by another object. Hence, one has to account not only for the carrier fluid but also for all the boundaries and other objects in the vicinity of the sphere. In a general case, when particles are surrounded by several other particles and solid boundaries within a carrier fluid, the determination of the net thermal radiation must be carried out by carefully evaluating the effects of all the other particles and surfaces, a challenging task. Extensive descriptions of the processes and methods for the determination of radiative heat transfer may be found in specialized treatises, such as the one by Siegel and Howel (1981).

When the temperature difference of the sphere and its surroundings is sufficiently high, the net energy exchanged by radiation is significant and may actually surpass the energy exchanged due to conduction and convection. In the general case of convection and radiation, the total rate of heat entering the sphere is the sum of the convection and radiation

$$\dot{Q}_t = 4\pi D^2 h(T_\infty - T_s) + \pi D^2 h_{rad}(T_w - T_s). \tag{5.68}$$

Radiation may cause a significant deviation of the temperature of a particle from that of the surrounding fluid, while the two are in thermal equilibrium. In extreme cases, it may also cause significant free convection from particles. Let us consider a spherical particle in a fluid stream of temperature, T_f, and velocity, u, inside an enclosure at a different temperature, T_w. The particle reaches an equilibrium temperature, which is defined by zero net heat gain at its surface. This implies equal and opposite magnitudes of convection and radiation. In the simplest case of black-body radiation and a non-absorbing carrier fluid, the equilibrium yields the following expression for the temperature of the particle

$$\dot{Q}_t = 4\pi D^2 h(T_\infty - T_s) + \pi D^2 h_{rad}(T_w - T_s) = 0 \Rightarrow T_s = \frac{4hT_\infty + h_{rad}T_w}{4h + h_{rad}}. \tag{5.69}$$

This analysis implies that the particle temperature will be close to the temperature of the gas, T_∞, only when $h \gg h_{rad}$. The particle's temperature will be close to that of the enclosure, T_w, when $h \ll h_{rad}$. In all cases, the particle will reach an equilibrium temperature that is not equal to the temperature of the carrier fluid, T_∞. An implication of this phenomenon is that a thermometer exposed to the sun does not necessarily measure the temperature of the air that surrounds it.

5.5.4 DIELECTRIC HEATING

Dielectric, or microwave, heating is used in industrial processes such as plywood manufacturing, drying of textiles, food drying, and rubber curing. Dielectric materials do not have free electrons and do not conduct electric currents. When placed in an electric field, the positive and negative electric charges within atoms and molecules form dipoles that are displaced and aligned in a direction opposite to that of the imposed field. When the direction of the external field reverses, the dipoles move and attempt to rotate, a movement that dissipates energy, which increases the temperature of the dielectric medium. The power dissipated by the alternating electric field is given by the expression

$$\dot{Q} = 0.556 * 10^{-10} V f \varepsilon_{eff} E_{rms}^2, \tag{5.70}$$

where V is the volume of the dielectric material, E_{rms} is the electric field intensity, f is the frequency of the electric field in Hz, and ε_{eff} is the effective dielectric constant of the material. If a material experiences magnetic loss as well, Eq. (5.70) must be extended to include the energy dissipation due to the magnetic effects.

5.5.5 TRANSIENT HEAT TRANSFER

The analytical study by Michaelides and Feng (1994) determined that, for a rigid, isothermal sphere in a time-variable, non-uniform carrier fluid temperature field $T_c(x_i, t)$, and at creeping flow conditions ($Pe_r \ll 1$), the transient energy equation is

$$m_d c_d \frac{dT_d}{dt} = -m_c c_c \frac{DT_c}{Dt} - 2\pi D k_c \left[T_d - T_c - \frac{1}{24} D^2 T_{c,jj} \right]$$

$$-\pi D^2 k_c \int_0^t \frac{\frac{d}{d\tau} \left[T_d - T_c - \frac{1}{24} D^2 T_{c,jj} \right]}{\left[\pi \alpha_c (t - \tau) \right]^{1/2}} d\tau . \qquad (5.71)$$

where c_c and c_d are the specific heat capacities of the carrier fluid and the sphere, respectively; α_c in the denominator of the last term is the thermal diffusivity of the fluid, which is equal to $k_c / \rho_c c_d$; the differential operator D/Dt represents the Lagrangian derivative with a reference system of coordinates at the center of the sphere; and the temperature T_c is the temperature function of the fluid far from the sphere. The Laplacian derivatives in the square brackets account for the non-uniformities of the temperature field. They are analogous to the Faxen terms of the equation of motion and scale as D^2/L^2. In most practical applications, the particles are significantly smaller than the characteristic dimension of the fluid, $D^2/L^2 \ll 1$, these corrections are insignificant and may be neglected. The left-hand side of Eq. (5.71) denotes the change of the temperature of the sphere due to heat transfer. The first term on the right-hand side is analogous to the inertia term in the momentum equation and accounts for the change of temperature of the mass of displaced fluid. The second term is the steady conduction term from the sphere to the fluid and is analogous to the steady drag term of the equation of motion and represents the rate of heat transfer due to the bulk temperature difference between the carrier fluid and the sphere. The last term in the equation is a history integral, which is generated by the diffusion of the temperature gradients in the fluid temperature field. The history term depends on the temporal as well as the spatial variation of the temperature field and is analogous to the history term of the transient equation of motion.

If at the commencement of the transient process the temperatures of the carrier fluid and the sphere are not equal, the history term is substituted by the expression

$$\pi D^2 k_c \int_0^t \frac{\frac{d}{d\tau} \left[T_d - T_c - \frac{1}{24} D^2 T_{c,jj} \right]}{\left[\pi \alpha_c (t - \tau) \right]^{1/2}} d\tau + \pi D^2 k_c \frac{T_d(0) - T_c(0)}{\sqrt{\pi \alpha_c t}}. \qquad (5.72)$$

Gay and Michaelides (2002) conducted a numerical study to determine the effect of the history term on the heat transferred to a sphere by considering three types of variation of the carrier fluid temperature: (a) a step change, (b) a ramp change, and (c) a sinusoidal variation. The main parameters for this study are the volumetric heat capacity ratio $\beta = \rho_c c_c / \rho_d c_d$ and a thermal Stokes number, St_T, defined as the ratio of the thermal timescale of the particle, $\tau_T = D^2\rho_c c_c/4k_c$ to the characteristic timescale of the fluid, τ_c

$$St_T = \frac{D^2 \rho_c c_c}{4k_c} \frac{|\vec{u} - \vec{v}|}{L_c}. \tag{5.73}$$

This study concluded that the history term is important in computations when the ratio of the volumetric heat capacities, β, is between 0.002 and 0.5, the range of liquid-solid flows and particle flows in dense gases. In this range, the history term may contribute up to 35% of the instantaneous heat transfer from a particle. There is almost no effect of the history term on the heat transfer for bubbles, which are in the range $\beta > 10$ and of solid particles in ambient air, where $\beta < 0.001$. Results of these computations for a sinusoidal variation of the fluid temperature field with frequency f, which is equal to the last ratio of Eq. (5.73), are shown in Figure 5.6. The figure depicts the amplitude of the particle to fluid temperatures when the history term is neglected and when it is taken into account in the calculations for two values of β.

Feng and Michaelides (1998) performed a study on the transient heat transfer from a particle with arbitrary motion, in an arbitrary temperature field, with finite but small inertia ($Pe_r < 1$). They determined that the effect of flow inertia is the faster advection and evolution of the temperature gradients around the sphere, which results in the faster decay of the history term. Details on the transient hydrodynamic force and heat transfer for particles at finite Re_r and Pe_r as well as particles of non-spherical shapes may be found in Michaelides (2003, 2006).

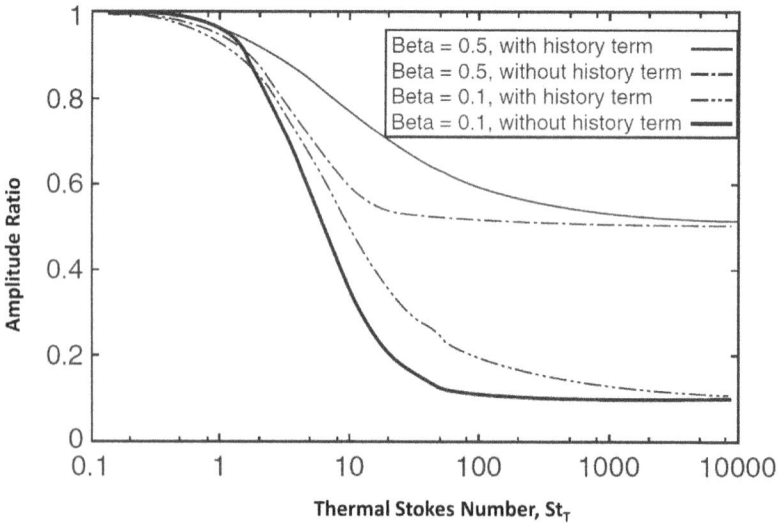

FIGURE 5.6 The effect of the history term on the amplitude of particle-to-fluid temperature ratio when the fluid temperature varies sinusoidally.

5.5.6 ENERGY INTERACTIONS WITH GROUPS OF PARTICLES

When one considers the problem of heat transfer from two neighboring particles, of which the second is in the wake of the first, one realizes that the heat transfer to or from the second particle is lower because the temperature of the fluid that encounters the second particle is modified by the heat transfer from the first. This neighboring effect on heat transfer in a multiphase flow can be viewed at the level of particle pairs and also at the level of an ensemble of particles. In all cases, one realizes that in multiphase flows consisting of groups of particles, even when the Reynolds and Prandtl number of the particles are about the same, their individual heat transfer rates will be substantially different due to the different interactions of each particle with its surrounding fluid and its neighbors. A deeper understanding of such complex interactions is currently emerging with more detailed numerical simulations (Balachandar and Michaelides, 2022).

Gunn (1978) was among the first to combine experimental and analytical results and derive a semi-empirical expression for the heat transfer from fixed and fluidized beds of particles with random particle distributions as a function of the volume fraction, α_d, of the solids in the suspension and a superficial Reynolds number $Re_{su} = Re_r(1 - \alpha_d)$

$$
Nu = \left[7 + 10(1-\alpha_d) + 5(1-\alpha_d)^2\right]\left(1 + 0.7 Re_{su}^{0.2} Pr^{0.33}\right) +
$$
$$
\left[1.33 - 2.4(1-\alpha_d) + 1.2(1-\alpha_d)^2\right] Re_{su}^{0.7} Pr^{0.33}.
$$

(5.74)

With the improvement of computational methods and computer power to model large ensembles of particles, several more recent studies have considered the problem of heat transfer from particle groups and clusters. In particular, computations using an extension of the immersed boundary method (Feng and Michaelides, 2008a, 2008b) allow for the variation of the fluid as well as the particles' temperature. Recent simulations have typically considered idealized models where the particles maintain a constant temperature, for example, because of combustion. Two reviews of the subject (Deen et al., 2014; Balachandar and Michaelides, 2022) describe several of these studies and their main results.

Let us consider a volume, V, where fluid-particle heat transfer takes place with the particles maintained at a constant temperature, T_d. The carrier fluid enters the volume with an average temperature of $\langle T \rangle_{c,in}$ and exits at temperature $\langle T \rangle_{c,out}$, where the angled brackets indicate an ensemble average. Because of the heat transfer between the particles and the fluid, the temperature field is inhomogeneous along the flow direction, while the flow field may be assumed statistically homogeneous if the heat transfer does not significantly affect the local density of the carrier fluid. The total heat transfer between the fluid and particles may be written as

$$
\dot{Q} = A_{cs}\rho_c c_{Pc}\langle u_c \rangle \left[\langle T_{c,out} \rangle - \langle T_{c,in} \rangle\right],
$$

(5.75)

where A_{cs} represents the cross-sectional area perpendicular to the flow direction. Eq. (5.75) may be rendered dimensionless to define the average Nusselt number for the interactions of the fluids with the ensemble of particles

$$Nu = \frac{2D^2}{3V\alpha_d k_c} \frac{\dot{Q}}{\left[\langle T_{c,out} \rangle - \langle T_{c,in} \rangle\right]} = \frac{2D^2 A_{cs} \rho_c c_{Pc} \langle u_c \rangle}{3V\alpha_d k_c}, \tag{5.76}$$

Sun *et al.* (2015) offers the following correlation for *Nu* based on numerical data

$$Nu = \frac{1}{\left(1-\alpha_d\right)^3}\left[-0.46 + 1.77\left(1-\alpha_d\right) + 0.69\left(1-\alpha_d\right)^2\right] + \\ \left[1.37 - 2.4\left(1-\alpha_d\right) + 1.2\left(1-\alpha_d\right)^2\right]\text{Re}_{su}^{0.7}\,\text{Pr}^{0.33}. \tag{5.77}$$

It is noted that Eqs. (5.74) and (5.77) do not asymptotically reduce to the heat transfer correlations for single particles—Eq. (5.42) or Eq. (5.43) in the limit $\alpha_d \to 0$. Also, that Eq. (5.74) does not approach asymptotically the conduction limit ($Nu = 2$) when $Re_{su} \to 0$, while Eq. (5.77) does approach this limit. This is an indication of the complexity of the particle-particle and particle-fluid interactions. As with the case of momentum interactions with groups of particles, for accurate determinations of the heat transfer in non-dilute particulate flows, the models must take into account these complex interactions.

5.6 TURBULENCE MODULATION BY PARTICLES

Turbulence modulation is the effect of inertial particles and droplets on the turbulence of the carrier phase. The presence of the disperse phase and its interactions with the carrier fluid phase may either augment or attenuate turbulence relative to single-phase turbulence. Turbulence modulation can have a significant impact on industrial flows, energy conversion efficiency, etc.

5.6.1 EXPERIMENTAL STUDIES

Research on the physics of how particles affect the fluid turbulence has gained some traction since the late 1980s. While prior to that time some practitioners were aware that the presence of particles may modify the rates of heat transfer and chemical reaction—a fact that could only be explained through the effect of the particles on the fluid turbulence—the direct measurements of turbulence in the presence of particles with *Laser Doppler Velocimetry* (LDV) made possible the quantitative and direct measurement of fluid turbulence in the presence of particles. Detailed studies of turbulence profiles in particle-laden flows showed that both attenuation and augmentation of turbulence may occur, depending on particle size and concentration. Gore and Crowe (1989) and Hetsroni (1989) summarized the then available data on turbulence modulation. The two studies suggest that suggest the length scale ratio D/L_e, where L_e is the length scale of the most energetic turbulent eddies, and the particle Reynolds number, Re_p, play important roles in the modulation of single-phase turbulence. A summary of the then available data is shown in Figure 5.7, where it is

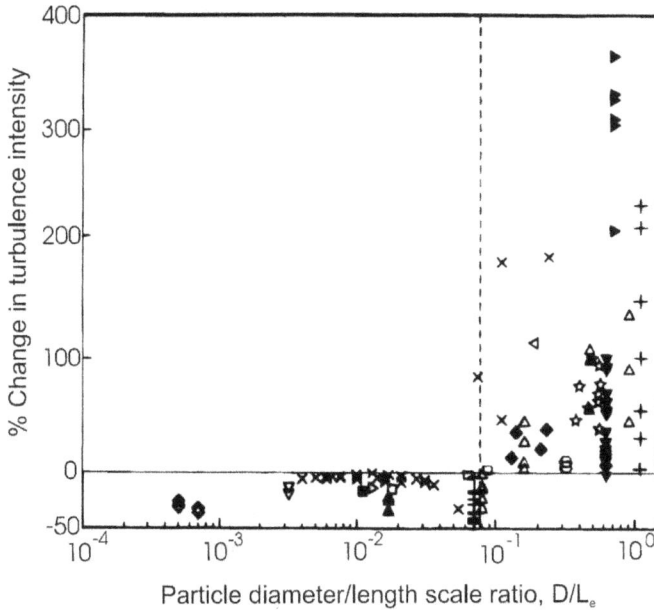

FIGURE 5.7   A summary of the data for the change in turbulent intensity due to the presence of the dispersed phase.

observed that turbulence intensity is attenuated when D/L_e is less than ~ 0.1, while the turbulence level is enhanced for larger ratios (Gore and Crowe, 1989). The effects of particles on fluid turbulence are similar to those of a stationary grid or screen in a turbulent flow: Turbulence is generated or attenuated depending on the size of the grid and turbulence intensity.

Additional experimental studies on particulate channel flows (Tsuji *et al.*, 1984; Kussin and Sommerfeld, 2002) showed that augmentation occurred mostly at the center of the channel when the particle Reynolds number was high. Varaksin *et al.* (1998) showed attenuation of turbulence kinetic energy (TKE) in pipe flows for a low particle Reynolds number and varying concentration. Ferrante and Elghobashi (2003) also showed a decrease in TKE with certain types of particle properties, but also observed an increase in TKE and no change in TKE all relative to the unladen flow. It appears from the experimental studies that the following factors contribute to turbulence modulation by particles and droplets (Crowe, 1993; Balachandar and Eaton, 2010):

1. Size effects: particle size and shape normalized by a length scale
2. Loading effects: particle concentration or mass loading
3. Inertial effects: particle Reynolds number
4. Response effects: particle response time, or Stokes number
5. Interaction effects: particle-particle and particle-wall interactions, formation of particle wakes

5.6.2   TURBULENCE MODULATION MODELS

One of the earlier models for the modulation of turbulence by particles was developed by Yuan and Michaelides (1992). They modeled the interaction of turbulence eddies with particles and considered the time of interaction between eddies and particles, τ. The particles absorb energy from the turbulent eddies when their velocity approaches that of the fluid and produce energy in their wakes. The total energy absorbed and produced is expressed by the equation

$$\Delta E_k = -\frac{\pi}{12} D^3 \rho_d (u-v)^2 \left[1 - \exp\left(-\frac{2f\tau}{\tau_v} \right) \right] + \frac{\pi}{12} D^3 \rho_c L_w (u^2 - v^2) \qquad (5.78)$$

where f is the drag factor in Eq. (4.26), τ_v is the timescale of the particle, and L_w is a measure of the wake behind the particle. An extension to this model was developed by Kenning and Crowe (1997), who considered separately the turbulence production and dissipation mechanisms by the fluid alone and by the particles. They defined a *hybrid length scale*, l_h, as the dissipation length scale and concluded that the fractional change in turbulence intensity for particles transported by air in a vertical duct is

$$\frac{\Delta E_k}{E_0} = \left[\frac{l_h}{l_\varepsilon^{sp}} + \frac{l_h}{\left(k^{sp} \right)^{3/2}} \frac{f}{\tau_v} \frac{\bar{\rho}_d}{\bar{\rho}_c} (u-v)^2 \right]^{1/3} - 1, \qquad (5.79)$$

where l_ε^{sp} and k^{sp} are the dissipation length scale and turbulence energy of the single-phase carrier fluid in the absence of particles. The two models capture qualitatively the mechanisms of turbulence modulation by particles and show good agreement with the experimental data for channel flows and jets (Levy and Lockwood, 1981; Tsuji *et al.*, 1984; Lee and Durst, 1982). A review on the modelling of particle-laden turbulence, Eaton and Longmire (2017) state the need for a more general model that accounts for the all factors listed at the end of Section 5.6.1

The experimental data show that when particles are introduced, the statistics of the continuous phase turbulence are altered. Depending on particle characteristics, the level of turbulent kinetic energy and dissipation changes relative to the corresponding single-phase flow. Because of the complex fluid-particle interactions, no general turbulence model exists for particle-laden flows. Crowe and Gillandt (1998) derived a transport equation for the volume averaged turbulent kinetic energy, within the framework of the k-ε turbulence model. They added terms to represent the production and energy redistribution mechanisms that were due to the effect of particle surfaces. Several researchers adopted the additional modulation terms and proposed a dissipation equation with a similar effect. Among them, Zhang and Reese (2003) used the additional terms proposed by Crowe and Gillandt (1998) to form a modulation term within the dissipation equation of the form

$$\alpha_c \rho_c \frac{D\varepsilon}{Dt} = P_\varepsilon + D_\varepsilon - C_{\varepsilon 2} \alpha_c \rho_c \frac{\varepsilon^2}{k} + C_{\varepsilon 3} \frac{\varepsilon}{k} \Delta k, \qquad (5.80)$$

where the Δk term is obtained from Eq. (5.79). Such methods have been successfully used by several researchers, including Lain *et al.* (1999) and Kartushinski *et al.* (2010).

REFERENCES

Balachandar, S. and Eaton, J.K., 2010, Turbulent dispersed multiphase flow, *Annual Rev. Fluid Mech.*, **42**, 111.

Balachandar, S. and Michaelides, E.E., 2022, Dispersed multiphase heat and mass transfer, *Annual Rev. of Heat Transfer*, ARHT-42092, DOI: 10.1615/AnnualRevHeatTransfer. 2022042092.

Chiang, C.H., Raju, M.S. and Sirignano, W.A., 1992, Numerical analysis of a convecting, vaporizing fuel droplet with variable properties, *Int. J. Heat Mass Transfer.*, **35**, 1307–1327.

Chigier, N.A., 1995, Spray combustion, *Proc. ASME Fluids Engr. Div.*, FED, **223**, 3.

Clift, R., Grace, J.R. and Weber, M.E., 1978, *Bubbles, Drops and Particles*, Academic Press, New York.

Crowe, C.T., 1993, Modeling turbulence in multiphase flows, *Proc. 2nd Int. Symp. on Engr. Turbulence Modeling and Measurements*, Elsevier, Amsterdam, 899.

Crowe, C.T. and Gillandt, I., 1998, Turbulence modulation of fluid-particle flows: A basic approach, *3rd Int. Conf. on Multiphase Flows*, Lyon, France, 8–12 June.

Deen, N.G., Peters, E., Padding, J.T. and Kuipers, J., 2014, Review of direct numerical simulation of fluid: Particle mass, momentum and heat transfer in dense gas-solid flows, *Chem. Eng. Science*, **116**, 710–724.

Dennis, S.C.R., Singh, S.N. and Ingham, D.B., 1980, The steady flow due to a rotating sphere at low and moderate reynolds numbers, *J. Fluid Mech.*, **101**, 257–279.

Dietzel, M. and Sommerfeld, M., 2013, Numerical calculation of flow resistance for agglomerates with different morphology by the Lattice-Boltzmann Method, *Powder Technology*, **250**, 122–137.

Di Felice, R., 1994, The voidage function for fluid: Particle interaction systems, *Int. J. Multiphase Flow*, **20**, 153–162.

Dorfman, L.A. and Serazetdinov, A.Z., 1973, Laminar flow and heat transfer near rotating axisymmetric surface, *Int. J. Heat Mass Transfer*, **8**, 317–327.

Eaton, J.K. and Longmire, E.K., 2017, Turbulence interactions, Ch. 12. in *Multiphase Flow Handbook*, 2nd edition (Eds. E.E. Michaelides, et al.), CRC, CRC Press, Boca Raton.

Ergun, S., 1952, Fluid flow through packed columns, *Chem. Eng. Prog.*, **48**, 89–94.

Feng, Z.G., 2014, Direct numerical simulation of forced convective heat transfer from a heated rotating sphere in laminar flows, *J. Heat Transf.*, **136**, 041707.

Feng, Z.-G. and Michaelides, E.E., 1998, Transient heat transfer from a particle with arbitrary shape and motion, *J. Heat Transfer*, **120**, 674–681.

Feng, Z.-G. and Michaelides, E.E., 2000, A numerical study on the transient heat transfer from a sphere at high Reynolds and Peclet numbers, *Int. J. Heat Mass Transfer*, **43**, 219–229.

Feng, Z.-G. and Michaelides, E.E., 2001, Heat and mass transfer coefficients of viscous spheres, *Int. J. Heat Mass Transfer*, **44**, 4445–4454.

Feng, Z.G. and Michaelides, E.E., 2005, *Proteus*-A direct forcing method in the simulation of particulate flows, *J. Computational Physics*, **202**, 20–51.

Feng, Z.-G. and Michaelides, E.E., 2008a, Heat transfer in particulate flows with Direct Numerical Simulation (DNS), *Int. J. Heat Mass Transf.*, **52**, 777–786.

Feng, Z.-G. and Michaelides, E.E., 2008b, Inclusion of heat transfer computations for particle laden flows, *Phys. Fluids*, **20**, #040604.

Feng, Z.G. and Michaelides, E.E., 2012, Heat transfer from a nano-sphere with temperature and velocity discontinuities at the interface, *Int. J. Heat Mass Transf.*, **55**, 6491–6498.

Ferrante, A. and Elghobashi, S., 2003, On the physical mechanisms of two-way coupling in particle-laden isotropic turbulence, *Phys. Fluids*, **15**, 315.

Feuillebois, F. and Lasek, A., 1978, On the rotational historic term in non-stationary Stokes flow, *J. Mech. & Appl. Math.*, **31**, 435.

Gay, M. and Michaelides, E.E., 2002, Effect of the history term on the transient energy equation of a sphere, *Int. J. Heat Mass Transfer*, **46**, 1575–1586.

Gore, R.A. and Crowe, C.T., 1989, The effect of particle size on modulating turbulent intensity, *Intl. J. Multiphase Flow*, **15**, 279.

Gunn, D., 1978, Transfer of heat or mass to particles in fixed and fluidized beds, *Int. J. of Heat and Mass Transfer*, **21**(4), 467–476.

Hardy, B., Simonin, O., de Wilde, J. and Winckelmans, G., 2022, Simulation of the flow past random arrays of spherical particles: Microstructure-based tensor quantities as a tool to predict fluid-particle forces, *Int. J. Multiphase Flow*, **149**, 103970.

Hedley, A.B., Nuruzzaman, A.S.M. and Martin, G.F., 1971, Progress review no. 62: Combustion of single droplets and simplified spray systems, *J. Inst. Fuel*, **44**, 38–45.

Hetsroni, G., 1989, Particles-turbulence interaction, *Int. J. Multiphase Flow*, **15**, 735–748.

Kartushinski, A., Michaelides, E.E., Rudi, Y. and Graham, N., 2010, RANS Modeling of a particulate turbulent round jet, *J. of Chemical Engineering Science*, **65**, 3384–3393.

Kenning, V.M. and Crowe, C.T., 1997, On the effect of particles on the carrier phase turbulence in gas-particle flows, *Int. J. Multiphase Flows*, **23**, 403.

Kestin, J., 1978, *A Course in Thermodynamics*, Vol. 2, Hemisphere, Washington, DC.

Kreith, F., Roberts, L.G., Sullivan, J.A. and Sinha, S.N., 1963, Convection heat transfer flow phenomena of rotating sphere, *Int. J. Heat Mass Transfer*, **6**, 881–895.

Kussin, J. and Sommerfeld, M., 2002, Experimental studies on particle behavior and turbulence modification in horizontal channel flow with different wall roughness, *Exp. in Fluids*, **33**, 143.

Lain, S., Bröder, D. and Sommerfeld, M., 1999, Experimental and numerical studies of the hydrodynamics in a bubble column, *Chem. Eng. Sci.*, **54**, 4913.

Lee, S.L. and Durst, F., 1982, On the motion of particles in turbulent duct flows, *Int. J. Multiphase Flow*, **8**, 125.

Levy, Y. and Lockwood, F.C., 1981, Velocity measurements in a particle laden turbulent free jet, *Combustion and Flame*, **40**, 333–346.

Michaelides, E.E., 2003, Hydrodynamic force and heat/mass transfer from particles, bubbles and drops: The freeman scholar lecture, *J. Fluids Eng.*, **125**, 209–238.

Michaelides, E.E., 2006, *Particles, Bubbles and Drops: Their Motion, Heat and Mass Transfer*, World Scientific Publishers, Singapore.

Michaelides, E.E., 2014, *Nanofluidics: Thermodynamic and Transport Properties*, Springer, New York.

Michaelides, E.E., 2017, Nanoparticle diffusivity in narrow cylindrical pores, *Int. J. of Heat and Mass Transfer*, **114**, 607–612.

Michaelides, E.E., 2021, *Exergy and the Conversion of Energy*, Cambridge University Press, Cambridge.

Michaelides, E.E., Crowe, C.T. and Schwarzkopf, J.D. (eds.), 2017, *Multiphase Flow Handbook*, 2nd edition, CRC Press, Boca Raton.

Michaelides, E.E. and Feng, Z.-G., 1994, Heat transfer from a rigid sphere in a non-uniform flow and temperature field, *Int. J. Heat Mass Transfer*, **37**, 2069–2076.

Musong, S. and Feng, Z.G., 2014, Mixed convective heat transfer from a heated sphere at an arbitrary incident flow angle in laminar flows, *Int. J. Heat and Mass Transf.*, **78**, 34–44.

Renksizbulut, M. and Yuen, M.C., 1983, Experimental study of droplet evaporation in high temperature air stream, *J. Heat Transfer*, **105**, 364–388.

Richardson, J.F. and Zaki, W.N., 1954, The fall velocities of spheres in viscous fluids, *Trans. Inst. Chem. Eng.*, **32**, 35–41.

Rubinow, S.I. and Keller, J. B., 1961, The transverse force on a spinning sphere moving in a viscous fluid, *J. Fluid Mech.*, **11**, 447–459.

Sawatzki, O., 1970, Das Strömungsfeld um eine Rotierende Kugel, *Acta Mech.*, **9**, 159–214.

Siegel, R. and Howel, J.R., 1981, *Thermal Radiation Heat Transfer*, McGraw-Hill, New York.

Sirignano, W.A., 1999, *Fluid Dynamics and Transport of Droplets and Sprays*, Cambridge University Press, Cambridge.

Sun, B., Tenneti, S. and Subramaniam, S., 2015, Modeling average gas: Solid heat transfer using particle-resolved direct numerical simulation, *Int. J. of Heat and Mass Transfer*, **86**, 898–913.

Tieng, S.M. and Yan, A.C., 1993, Experimental investigation on convective heat transfer of heated rotating sphere, *Int. J. Heat Mass Transfer*, **36**, 599–610.

Tsuji, Y., Morikawa, Y. and Shiomi, H., 1984, LDV measurements of an air-solid two-phase flow in a vertical pipe, *J. Fluid Mech.*, **139**, 417–429.

Varaksin, A.Y., Kurosaki, Y., Satch, I., Polezhaev, Y.V. and Polyahov, A.F., 1998, Experimental study of the direct influence of small particles on carrier air turbulence intensity for pipe flow, *3rd Int. Conf. Multiphase Flow*, Lyon, France, 8–12 June.

Wen, C.Y. and Wu, Y.H., 19666, Mechanics of fluidization, *Chem. Eng. Prog. Symp. Ser.*, **62**, 100–125.

Whitaker, S., 1972, Forced convection heat transfer correlations for flow in pipes past flat plates, single cylinders, single spheres, and for flow in packed beds and tubes bundles, *A.I.Ch.E. J.*, **18**, 361–371.

Williams, A., 1990, *Combustion of Liquid Fuel Sprays*, Butteworths, Sevenoaks, Kent.

Xu, Z.-J. and Michaelides, E.E., 2003, The effect of particle interactions on the sedimentation process of non-cohesive particles, *Int. J. Multiphase Flow*, **29**, 959–982.

Yuan, Z. and Michaelides, E.E., 1992, Turbulence modulation in particulate flows: A theoretical approach, *Int. J. Multiphase Flow*, **18**, 779–782.

Zhang, Y. and Reese, J.M., 2003, Gas turbulence modulation in a two-fluid model for gas-solid flows, *AIChE J.*, **49**, 3048–3054.

PROBLEMS

5.1   Measurements indicate that the mass transfer rate on the windward side of an evaporating droplet is 10% higher than that on the leeward side. Assume that the local mass efflux rate can be subdivided into a uniform flow over the windward hemisphere and a uniform flow over the leeward hemisphere. Determine the non-dimensional thrust on the droplet in the form

$$\frac{T}{\rho_f \left(\frac{\dot{m}}{\rho_c S} \right)^2 A_d} = f\left(\frac{\rho_d}{\rho_c} \right)$$

where $\dot{m}$ is the evaporation rate of the droplet, A_d is the projected area of the droplet, ρ_c is the density of the gas adjacent to the droplet surface, S is the surface area of the droplet, and ρ_d is the density of the droplet material.

5.2 A spherical droplet is rotating with an angular velocity ω_i as it moves with a translational velocity of U_i at its center of mass. Assume that the droplet is evaporating uniformly. Is the expression

$$m\frac{dU_i}{dt} = F_i$$

where F_i is the force acting on the particle still applicable? Determine by applying Reynolds transport theorem to the rotating, evaporating droplet.

5.3 A rigid spherical droplet is spinning with an angular velocity about its z-axis, as shown. The magnitude of the velocity due to spin is

$$|v_i| = \omega r \sin\phi$$

The velocity with respect to an inertial reference frame is $U_i + v_i$, where U_i is the velocity of the droplet center with respect to an inertial reference frame. The specific kinetic energy is

$$e_k = \frac{|U_i + v_i|^2}{2}$$

1. Show that the kinetic energy of the droplet is

$$\int_{cv} \rho_d e_k \, dV = m\frac{U^2}{2} + AI\omega^2$$

where A is a constant and I is the moment of inertia of the droplet. Evaluate the constant A.

2. The velocity at the surface of a non-evaporating spherical particle is given by

$$v_{i,s} = v_i + v_t t_i$$

where v_t is the magnitude of the tangential velocity, and t_i is the unit vector in the tangent plane of the particle surface. Show that the work term in the energy equation due to surface forces becomes

$$W = -v_i F_{i,s} - \int_{cs} v_t t_i \tau_{ij} n_j \, dA$$

5.4   A coal particle is modeled as a cube, as shown. Mass is ejected from the surface due to burning, but the size does not change with time. The efflux is uniform and equal to w over the surface except for the leeward face where it is $2w$. Find the equation of motion for the particle in the x-direction in the form

$$m\frac{dv}{dt} + A\dot{m}w = F_{d,x}$$

where $F_{d,x}$ is the drag force in the x-direction, m is the mass of the particle, and $\dot{m}$ is the burning rate.

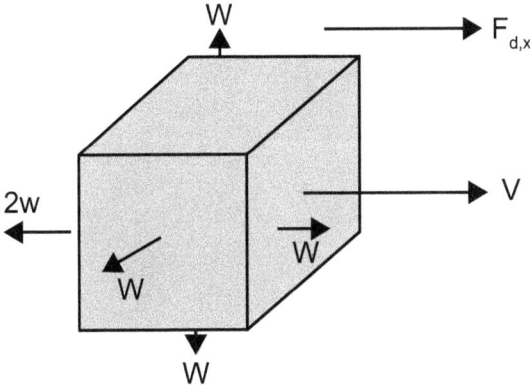

PROBLEM 5.4

5.5   The following questions refer to the energy equation for a droplet (or particle).

1. How would the energy equation be altered if the dependence of surface tension on temperature was to be included?
2. What changes would occur to the energy equation if internal circulation of the fluid was included? Do not work the problem out, only indicate what changes you might expect and explain why.
3. How would the equation change if the particle was porous and the surface was a solid material with pores?

5.6   A droplet is injected with a velocity v_0 into a quiescent medium (no motion). The droplet evaporates according to the D^2-law, and Stokes drag is applicable. The drag force is the only force acting on the droplet (no gravity), and the mass flux is uniform over the surface.

1. Derive an equation for the droplet velocity and distance as a function of v_0, t, and τ_V/τ_m.
2. Setting $t = \tau_m$, evaluate the distance traveled for τ_V/τ_m much less than unity and much greater than unity.

5.7   The energy equation for an evaporating droplet is

$$mc\frac{dT_d}{dt} = Nu\pi D_d k\left(T_g - T_d\right) + Sk\pi\rho_c D_\upsilon D_d\left(\omega_\infty - \omega_s\right)h_{fg}$$

1. Discuss the relative interaction of the terms ensuing as a cold droplet is injected into a hot stream.
2. Determine the wet-bulb temperature for a water droplet in air at 20°C and 40% relative humidity. Assume that Nu = Sh and Sc = 0.65. You will need properties from a thermodynamics table for water to do this problem.
3. Write out an equation which represents a perturbation about the wet-bulb temperature; that is,

$$T_g = T_{g,0} + T_g'$$

and

$$T_d = T_{wb} + T_d'$$

yielding

$$\frac{dT_p'}{dt} = f\left(T_p', T_g'\right)$$

5.8   During the falling rate period, the wetness of a spherical slurry droplet is assumed to vary as

$$\frac{dW}{dt} = -kW$$

where k is a rate constant. The wetness is defined as the ratio of the moisture to the dry solids in the droplet. Derive an equation for the velocity variation with time during this time, assuming Stokes law is valid, the gas is quiescent, and the droplet size remains constant. The result should be in the form

$$\upsilon = \upsilon_0 f\left(W_0, D, \mu, k, m_s, t\right)$$

where υ_0 is the initial velocity, W_0 is the initial wetness, D is the droplet diameter, μ is the gas viscosity, and m_s is the mass of dry solids.

5.9   An evaporating droplet is accelerated from rest in a uniform flow with a velocity of υ_0. The droplet evaporates according to the D²-law and Stokes drag is applicable. Derive an expression for the droplet velocity as a function of aerodynamic response time based on initial droplet diameter, the fluid velocity, and droplet evaporation time. How far does the droplet travel before completely evaporating? Neglect gravity.

5.10   An evaporating droplet is released from rest in a quiescent medium and drops due to gravity. Buoyancy effects are negligible. The D²-law and Stokes law apply. Find an expression for the maximum velocity in terms of aerodynamic response time based on initial diameter, acceleration due to gravity, and evaporation time.

5.11   The rate of mass decrease of a porous particle is modeled as

$$\frac{dm_w}{dt} = -k\frac{m_w}{m_s}$$

where m_w is the mass of the water in the particle, and m_s is the mass of the solids. As the particle dries, its diameter remains constant, but the mass decreases. Assuming Stokes law is valid, derive an expression for the penetration distance of a wet particle in terms of the initial velocity, initial wetness, particle diameter, gas viscosity, and mass of the solid component. Neglect gravity.

5.12   Neglecting the transient terms, the equation for the rotational motion of a spherical particle in a viscous fluid is

$$I\frac{d\omega_{d,i}}{dt} = T_i - \pi\mu_c D^3 \omega_{d,i}$$

Determine an expression for the rotational response time and compare with the velocity response time, τ_V.

6 Particle-Particle Interactions

Particle-particle collisions may be only neglected in very dilute systems, where the mean free paths between subsequent binary collisions of particles are very long (Sommerfeld *et al.*, 2008). A classification of different dispersed particle-laden systems based on the particle volume fraction was introduced by Crowe *et al.* (2012) considering the dominant transport mechanisms, namely, the fluid-dynamic transport ($\alpha_p < 0.0005$), inter-particle collisions ($0.0005 < \alpha_p < 0.1$) and particle-contact dominance ($\alpha_p > 0.1$). As the particle number density becomes higher, particles collide with each other, and the loss of particle kinetic energy due to particle deformation during a collision cannot be neglected. It must be noted that particle interactions may occur through the interstitial fluid and do not necessarily require particle contact (Feng and Michaelides, 2002, 2003). Hence, the interaction/collision between particles moving in a flow involves primarily three stages: (a) fluid dynamic interaction that occurs before the particles come into contact, (b) contact and collision between the particles, and (c) possible long-term contact between the particles depending on the adhesion forces.

The fluid dynamic interactions between particles are a very complex process, especially if systems with a higher particle volume fraction are considered. This effect depends on the relative position and velocity of the interacting particles (Feng and Michaelides, 2003, Prahl *et al.*, 2007). Such a fluid dynamic interaction may arise from the hindered settling effect of particles, which is mostly described only in an integral (Eulerian) manner through correlations that enhance the drag dependence on the particle volume fraction (Deen *et al.*, 2007; Zhu *et al.*, 2007; Michaelides, 2017). The collision process itself may be described by the two most common methods, namely, the hard-sphere and the soft-sphere model, which will be described later. Modern numerical methods, which are described in Chapter 8, are adequate to describe the fluid dynamics interactions between particles. This chapter primarily describes the contact/collision stage and the possible aggregation stage.

The *hard-sphere collision model* assumes that the collisions are binary and quasi-instantaneous—the temporal evolution of the collision process is not resolved. Such a collision model is easy to apply numerically, and the velocity change of the particles is determined by a set of algebraic equations. The relation between the pre- and post-collision velocities is given explicitly using the coefficient of restitution and the Coulomb's friction coefficient. However, if the particle number density becomes too high, the tracking time step has to be reduced accordingly in order to ensure the occurrence of binary collisions only and exclude multi-particle contacts.

The *soft-sphere collision model* resolves the collision time and, therefore, is also suitable for contact-dominated flows with multi-particle contacts. In the soft-sphere model, the whole process of collision or contact is solved by the numerical

DOI: 10.1201/9781003089278-6

integration of the equations of motion, and this makes the computational time to be significantly higher. This model is primarily used in the discrete-element method (DEM) computations by introducing mechanical elements between the particles, such as springs, sliders, and dash-pot elements.

6.1   BINARY HARD-SPHERE PARTICLE COLLISIONS

When applying a Lagrangian approach for simulating the particle phase, there are several elementary processes that should be considered in describing a binary hard-sphere collision process. The first step is the detection of the possibility of a collision between two particles among several hundred thousand of simultaneously tracked numerical particles. Such a collision detection also needs to consider the impact efficiency, which is relevant for collisions between small and large particles (Ho and Sommerfeld, 2002). In such cases the smaller particle might move around the larger particle, driven by the fluid flow. This becomes important in technological processes, where the particles have a size distribution or the droplets within a spray have a wide spectrum of droplet sizes. This step includes the determination of the collision point on both particle surfaces, which is the basis for the computation of the particle velocity change. The most computationally expensive step in modelling hard-sphere collisions is the search for possible collision partners in the system.

6.1.1   BINARY COLLISION DETECTION

The detection of a possible particle collision pair may be done in three ways, as shown in Figure 6.1. In the *fully deterministic approach*, the collision between any pair of particles (or parcels) in the computational domain (or a certain meaningful

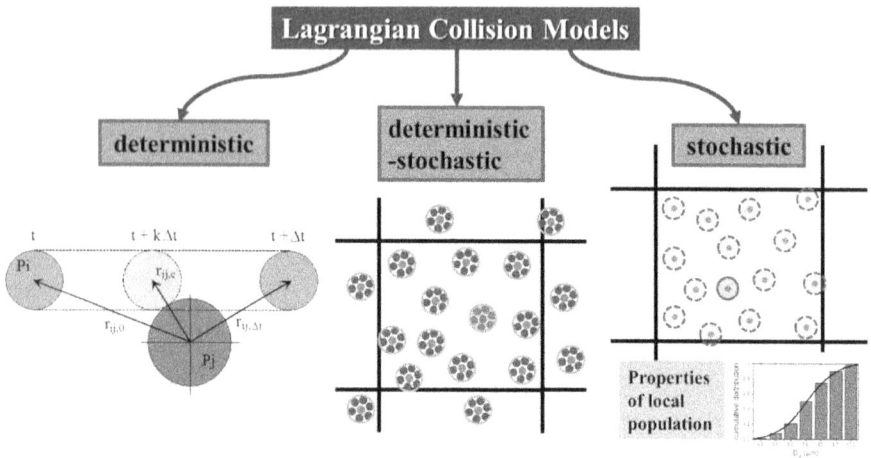

FIGURE 6.1   Summary of Lagrangian particle collision models.

Source: From Sommerfeld, 2017.

subdomain) is determined by checking the crossing of their trajectories during a time step, and considering their finite particle size (e.g. Tanaka and Tsuji, 1991; Sundaram and Collins, 1997; Ernst and Sommerfeld, 2012). This collision detection model may be also termed geometrical collision detection. Figure 6.1 (left) shows that the point of impact on the surface of the particles is known. Such an approach is time-consuming and requires $N_P(N_P - 1)/2$ examinations of particle pairs for detecting a possible collision.

The collision model originally developed for spray computations is the *deterministic-stochastic parcel collision model* (Figure 6.1 middle), which does not examine whether parcel trajectories are crossing during a time step, that is, geometrical determination of collisions. Instead, any pair of parcels residing in one computational cell (or any other meaningful volume) are examined for the occurrence of a collision, regardless of their real spacing and relative motion. A possible collision is determined from the collision probability of two parcels (O'Rourke, 1981). If a uniformly distributed random number is larger than the probability of no collision, a collision is taking place. The number of collisions between particles within the parcel is obtained from the collision probability times the number of particles in the other parcel. This collision detection algorithm is numerically more efficient, but the result strongly depends on the grid size, because close parcels located in neighboring cells are not allowed to collide (see Figure 6.1 middle). This effect may be reduced by using an adapted collision grid (Zhang *et al.*, 2012). Several hybrid models were suggested in order to improve the performance of the classical O'Rourke (1981) droplet parcel collision model. A detailed analysis and validation of such hybrid models was conducted by Pischke *et al.* (2015) followed by guidelines for a proper selection of hybrid droplet collision models. After the collision detection, the geometry of collisions and the lateral displacement of the particles are also generated by random processes.

The *stochastic collision model* (Figure 6.1 right) does not require the knowledge of the neighboring parcels. For each computational cell, the averaged statistical properties of the particle phase are needed (Sommerfeld and Zivkovic, 1992; Oesterle and Petitjean, 1993; Rüger *et al.*, 2000; Sommerfeld, 2001; Sgrott Junior and Sommerfeld, 2019). In unsteady flows, this requires a sufficient number of parcels to be tracked for providing the necessary statistical information. An important requirement for the stochastic model is that during the Lagrangian tracking, the relevant local particle properties (i.e. within a computational cell) and distributions have to be collected a priori and stored. This collection includes particle concentration, particle size distribution, and particle size-velocity correlations with mean- and rms-values. However, this step is only necessary if particle phase properties are determined from a two-phase flow computation. Then, during the tracking of the considered parcels (full particle shown in Figure 6.1 right) at each time step, a fictitious collision partner is drawn from the local population (Sommerfeld, 2001), with size and velocity determined by stochastic processes. The fictitious particle velocity fluctuations need, however, to be obtained through a pair-correlation function with respect to the real particle, which depends on the particle Stokes number (i.e. the ratio of particle response time to the integral timescale of turbulence). This accounts for the velocity correlations of the responsive particles through turbulence, whereas inertial particle velocities are completely uncorrelated (Sommerfeld, 2001; Sommerfeld *et al.*, 2008). It should be noted that only a deterministic collision modelling approach provides the real instantaneous velocities of both colliding particles. Having all the information about the fictitious collision partners, the collision probability can be calculated as

$$P = \frac{\pi}{6}(d_S + d_L)^2 \left| \overrightarrow{u_S} - \overrightarrow{u_L} \right| n_d \, \Delta t_L \qquad (6.1)$$

Here d_i are the diameters of small (S) and large (L) particles, n_d the particle number concentration (i.e. particles/m³), Δt_L is the current Lagrangian tracking time step, and $u_{rel} = \left| \overrightarrow{u_S} - \overrightarrow{u_L} \right|$ is the magnitude of the relative velocity between both particles. If an equally distributed random number ($0 < RN < 1$) is smaller than this probability, then a collision occurs. It should be mentioned that the mean inter-particle collision time is an additional criterion for limiting the Lagrangian tracking time step Δt_L in order to ensure only binary collisions, that is, the collision probability is actually much smaller than one. As a result of the implied linearization of the collision probability, the selected Lagrangian time step should be close to 0.05 τ_P, with the averaged inter-particle collision timescale, τ_P, for allowing a maximum relative error of 2.5% (Sommerfeld and Zivkovic, 1992).

The collision itself is then calculated in a coordinate system where the fictitious particle is stationary, as shown in Figure 6.2, where the collision point on the particle surfaces can be randomly determined (Sommerfeld, 2001) and the lateral displacement b of the collision partners is known. Such a stochastic model is numerically

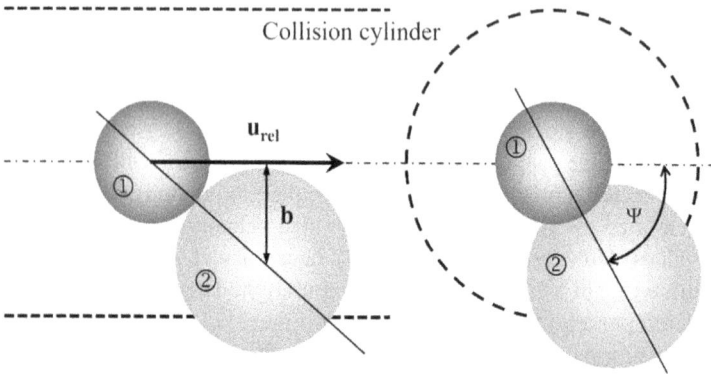

FIGURE 6.2 Particle-particle collision configuration in a coordinate system, where the fictitious particle (2) is stationary, the real particle (1) moves, and the axis of the collision cylinder is aligned with the relative velocity vector.

very efficient, and the algorithm can be easily parallelized since it is not necessary to know the location of the possible collision partners explicitly.

6.1.2   IMPACT EFFICIENCY

An essential element that must be integrated in the collision detection is the impact efficiency, which is relevant for collisions between small and large particles (Ho and Sommerfeld, 2002; Sgrott Junior and Sommerfeld, 2019). In such a situation, the smaller particles might move around the larger ones with the relative velocity field, and this frequently occurs in particle-laden flows, if a wide particle size spectrum exists. A good example for such a situation is a spray scrubber used for fine particle separation from gases, where the larger spray droplets capture the fine solid particles. This application was the motivation of the developments conducted by Schuch and Löffler (1978). Effectively, the impact efficiency yields a reduction of collision cross-section and, hence, of the collision rate, which is depicted in Figure 6.3. This phenomenon may be also described in a coordinate system where the larger collector particle is stationary. Considering only inertial effects, the impact efficiency is then defined as the ratio of the circular cross-section, from which the small particles are coming in contact and just hit the larger particle in relation to the effective collector cross-section. Accounting for the so-called *blocking effect*, this cross-section diameter is the sum of the large and small particle diameter, as shown in Figure 6.3. The impact efficiency for inertial effects in the work of Schuch and Löffler (1978) was calculated in terms of the relative Stokes number. This is the ratio of the Stokesian relaxation time of the small particle to the

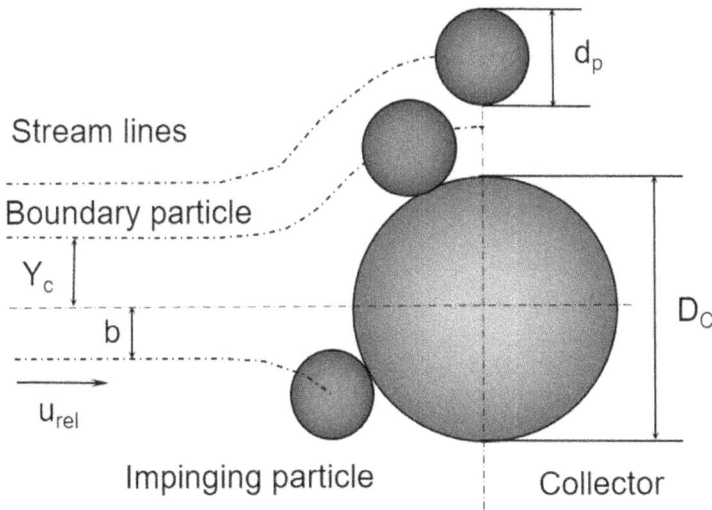

FIGURE 6.3   Inertial impact efficiency for the collision of a small particle with a larger collector particle.

time it needs to pass the collector particle with the applicable relative velocity (Ho and Sommerfeld, 2002)

$$St_{rel} = \frac{\tau_P}{T_{Pass}} = \frac{\rho_S \, d_P^2 \, u_{rel}}{18 \, \mu \, D_C} \tag{6.2}$$

Here the d_p and D_C refer to the diameters of the small and large (collector) particle, respectively, ρ_S is the small particle density (normally identical for all particles), and u_{rel} is the instantaneous relative velocity.

The impact efficiency may be related to the relative Stokes number using the two parameters a and b, which depend on the collector particle Reynolds number (Schuch and Löffler, 1978)

$$\eta = \left(\frac{2 \, Y_C}{d_S + d_L} \right)^2 = \left(\frac{St_{rel}}{St_{rel} + a} \right)^b \tag{6.3}$$

It was shown that the impact efficiency for inertial impaction depends on the large particle Reynolds-number Re_c, determined with the relative velocity. This is motivated by the fact that, with increasing collector Reynolds number, the flow deflection in front of the collector particle is increasingly compressing toward the collector and so are the streamlines. The numerically obtained values for the constants a and b are given in Table 6.1, and the results of the different correlations are shown in Figure 6.4, together with simulations, which agree fairly well with the data. The impact efficiency increases with increasing Stokes number from zero (i.e. no collisions) to one (100%) for highly inertial particles. For a given relative Stokes-number the increase in Reynolds number yields higher impact efficiency (Sommerfeld and Lain, 2009) due to the increasing small particle inertia as a result of the stronger flow deflection. A collision only occurs if a randomly drawn lateral displacement b (see Sommerfeld, 2001) is smaller than the value of Y_C plus the radius of the small particle

$$b \le Y_C + \frac{d_p}{2} \tag{6.4}$$

TABLE 6.1
Constants for the Correlation Proposed by Schuch and Löffler (1978) for the Impact Efficiency (Eq. 6.3)

Re_c	a	b
>> 1	0.25	2.0
60–80	0.506	1.84
40	1.03	2.07
10-, 20	1.24	1.95
< 1	0.65	3.7

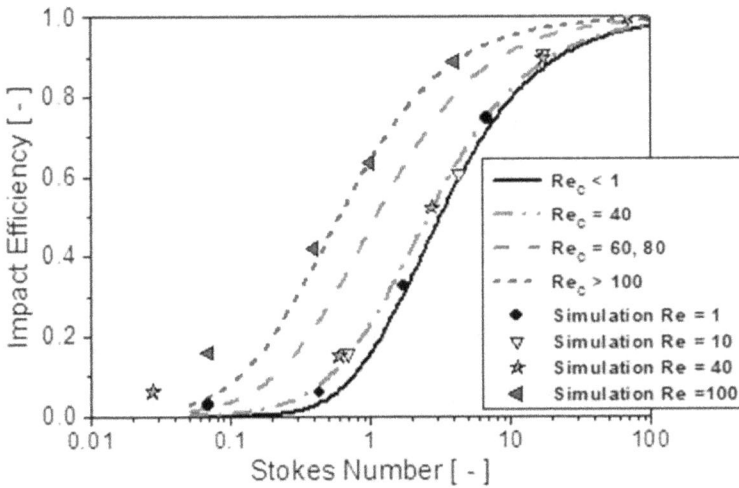

FIGURE 6.4 Impact efficiency as a function of relative Stokes number. Lines: correlations proposed by Schuch and Löffler (1978); symbols: laminar numerical computations by Sommerfeld and Lain (2009).

This instantaneous lateral displacement of the fine particle from the axis of the collision cylinder (see Figure 6.2) as well as the angular position in the cross-section is determined by independent random processes (Sommerfeld, 2001).

The preceding model of the impact efficiency may be only applied to turbulent flows if the collision pair is smaller than the smallest turbulent eddies, namely, the Kolmogorov length scale. Otherwise, the impact efficiency is still very important, but the collision geometry is much more complex. In such a situation, the trajectories of the small colliding particles are affected by turbulent transport so that even small droplets moving initially outside the collision cross-section might hit the collector. This regime is called turbulent diffusional impact. The impact efficiency will increase again when reducing the Stokes number below 0.1 (Figure 6.4). This will happen in a similar way if the colliding particles are very small so that their motion is affected by Brownian diffusion.

The importance of the impact efficiency on the collision and agglomeration rate was clearly demonstrated by Ho and Sommerfeld (2002) and Sommerfeld and Lain (2009) through Euler/Lagrange calculation of particle separation in a gas cyclone. Disregarding the impact efficiency increases the effective collision rate and possible agglomeration. This leads to a remarkable overprediction of separation efficiency compared to experiments for particles up to about 2.5 μm, as it was obtained by Sommerfeld and Lain (2009). Therefore, the consideration of the impact efficiency is essential for correctly predicting the collision and agglomeration rates. An example of the high importance of the impact efficiency is a spray with a rather wide size distribution, e.g. between 10 and 100 μm for a pressure atomizer. Recent Euler/Lagrange calculations of Lain and Sommerfeld and (2020) for a hollow cone water spray showed that impact efficiency is essential to be accounted for and reduces the collision rate by 20% for all types of collision outcomes—bouncing, coalescence, and separation.

6.1.3 PARTICLE VELOCITY CHANGE

The calculation of the particle velocity change due to hard-sphere inter-particle collisions is based on solving the impulse equations for particle translation and rotation. For solving these equations, the following assumptions are made:

- The collision process occurs instantly so that the forces and moments acting on the particles appear only as integral values (integrated over the very short collision time).
- The deformation of the particles during the collision process is not explicitly considered, but the dissipation of energy due to the deformation is described by a restitution ratio.
- A sliding collision is described by *Coulomb's law of friction*, which relates to the wall normal force and friction force.

The final equations for obtaining the velocities after the collision include two necessary model parameters: the normal restitution ratio and a friction coefficient, which are dependent on particle material properties as well as the impact angle and velocities. Hence, the change of linear and angular velocity components can be calculated by solving the momentum equations so that relations are obtained for sliding and non-sliding collisions. The impulse equations for the change of the translational and angular velocities of two particles denoted by the indices 1 and 2, are

$$m_{P1}\left(\vec{u}_{P1}^{*}-\vec{u}_{P1}\right)=-m_{P2}\left(\vec{u}_{P2}^{*}-\vec{u}_{P2}\right)=\vec{J}, \qquad (6.5)$$

$$\frac{I_{P1}}{R_{P1}}\left(\vec{\omega}_{P1}^{*}-\vec{\omega}_{P1}\right)=\frac{I_{P2}}{R_{P2}}\left(\vec{\omega}_{P2}^{*}-\vec{\omega}_{P2}\right)=-\vec{n}\times\vec{J}, \qquad (6.6)$$

where m_p is the particle mass, I_p the moment of inertia (*i.e.* $I_p = 0.1 m_p D_P^2$ for a sphere), R_{P1} and R_{P2} are the radii of both particles, $\vec{n}$ is the unit vector, and $\vec{J}$ the vector of the impulse force. By solving these equations, one can calculate the change of linear and angular particle velocity components during a collision. Hence, one obtains the following set of equations to calculate the post-collision linear and angular velocity components (superscript *) of both particles in terms of the relative velocity components before collision

$$u_{P1}^{*}=u_{P1}+\frac{J_{x}}{m_{P1}},$$

$$v_{P1}^{*}=v_{P1}+\frac{J_{y}}{m_{P1}},$$

$$w_{P1}^{*}=w_{P1}+\frac{J_{z}}{m_{P1}},$$

$$u^*_{P2} = -\frac{J_x}{m_{P2}},$$

$$v^*_{P2} = -\frac{J_y}{m_{P2}},$$

$$w^*_{P2} = -\frac{J_z}{m_{P2}} \qquad (6.7)$$

$$\omega^{*x}_{P1} = \omega^x_{P1},$$

$$\omega^{*y}_{P1} = \omega^y_{P1} - \frac{5\ J_z}{m_{P1}\ D_{P1}},$$

$$\omega^{*z}_{P1} = \omega^z_{P1} + \frac{5\ J_y}{m_{P1}\ D_{P1}},$$

$$\omega^{*x}_{P2} = \omega^x_{P2},$$

$$\omega^{*y}_{P2} = \omega^y_{P2} + \frac{5\ J_z}{m_{P2}\ D_{P2}},$$

$$\omega^{*z}_{P2} = \omega^z_{P2} - \frac{5\ J_y}{m_{P2}\ D_{P2}} \qquad (6.8)$$

Here, m_{P1} and m_{P2} are the masses of the two particles, and J_x, J_y and J_z are the components of the impulsive force. With the definition of the normal restitution ratio

$$e = -\frac{u^*_{P1} - u^*_{P2}}{u_{P1}}, \qquad (6.9)$$

and the conservation of the x-component of the momentum for particle 2

$$J_x = -m_{P2}\ u^*_{P2}, \qquad (6.10)$$

one finally obtains the following expression for J_x

$$J_x = -(1+e)\ u_{P1}\frac{m_{P1}\ m_{P2}}{m_{P1} + m_{P2}}. \qquad (6.11)$$

By applying Coulomb's law of friction, one obtains furthermore the condition for a non-sliding or sticking collision as a function of the static coefficient of friction μ_s

$$\sqrt{J_y^2 + J_z^2} < \mu_s\ |J_x|. \qquad (6.12)$$

The components of the impulse force are introduced into Eq. (6.12), and the condition for a non-sliding collision is obtained as a function of the velocities before collision

$$|u_R| < \frac{7}{2} \mu_s (1+e) |u_{P1}|. \tag{6.13}$$

The relative velocity at the point of contact is determined with the linear and angular velocity components of both particles.

$$u_R = \sqrt{u_{Ry}^2 + u_{Rz}^2},$$

$$u_{Ry} = v_{P1} + \frac{D_{P1}}{2} \omega_{P1}^z + \frac{D_{P2}}{2} \omega_{P2}^z,$$

$$u_{Rz} = w_{P1} - \frac{D_{P1}}{2} \omega_{P1}^y - \frac{D_{P2}}{2} \omega_{P2}^y. \tag{6.14}$$

The components of the impulsive force J_y and J_z depend on the type of collision. For a *non-sliding collision*, one obtains the following

$$J_y = -\frac{2}{7} u_{Ry} \frac{m_{P1} m_{P2}}{m_{P1} + m_{P2}},$$

$$J_z = -\frac{2}{7} u_{Rz} \frac{m_{P1} m_{P2}}{m_{P1} + m_{P2}}. \tag{6.15}$$

For a *sliding collision*, the components of the impulsive force depend on the dynamic coefficient of friction μ_d

$$J_y = -\mu_d \frac{u_{Ry}}{u_R} |J_x|,$$

$$J_z = -\mu_d \frac{u_{Rz}}{u_R} |J_x|. \tag{6.16}$$

The preceding equations show that the parameters involved in the collision model are the restitution coefficient e and the static μ_s and dynamic μ_d coefficient of friction. Generally, these properties depend on the particle material, collision velocity, and impact angle and mostly require experimental studies for evaluating proper values. In most applications, however, constant values of these parameters are employed.

The equations may be further simplified in the fully stochastic collision model by transforming the particle velocities into a coordinate system, where one of the particles is stationary, for example, the fictitious particle ② in Figure 6.2. For such a collision geometry, where the relative velocity vector coincides with the axis of the collision cylinder, the relations for the calculation of the post-collision properties of both particles are reduced to those for an oblique central collision (Sommerfeld and Zivkovic, 1992; Oesterle and Petitjean, 1993; Sommerfeld, 1996; Sommerfeld, 2001). Once the new velocities (translational and rotational) for the considered or tracked particle (i.e. particle ① in Figure 6.2) are obtained, they are re-transformed into the original coordinate system. The velocities of the fictitious particles, which are neglected after this collision calculation, are not of interest.

The stochastic inter-particle collision model, which also accounts for the velocity correlation of colliding particles, and is numerically efficient (Sommerfeld, 2001), involves the following steps:

1. At each time step of the trajectory calculation, a fictitious collision partner is generated from the local particle population with previously sampled properties, that is, with size and velocity sampled from local distribution functions.
2. In randomly creating the fictitious particle velocity, a possible correlation due to turbulence, which depends on the particle Stokes number, is considered.
3. The collision probability (i.e. the product of collision frequency and time step size) is calculated on the basis of Eq. (6.1). If a uniform random number in the interval between zero and one becomes smaller than this probability, a collision is occurring.
4. By transforming the particle velocities into a coordinate system where the fictitious particle is stationary, it is possible to sample the point of impact, which can only be located on the hemisphere facing the fictitious particle.
5. The new velocities of the considered particle are calculated based on the previous equations, Eq. (6.5) through Eq. (6.16).
6. Finally, the particle velocities are re-transformed into the original coordinate system. The fictitious particle is not of further interest.

Fluid dynamic effects during the collision process may be neglected if the duration of the collision process is negligibly small compared to the time of the collision-less motion, the sizes of the colliding particles are not too different, and the ratio of solid particle density to the fluid density is much larger than one. Hence, under such conditions, the collision efficiency may be assumed to be 100 %, that is, neglecting impact efficiency.

6.1.4 Physical Effects of Inter-Particle Collisions

From Eq. (6.1) for the collision probability, it is clear that the collision frequency increases with growing collision cross-section and, hence, particle size, augmented relative velocity between particles and local particle number concentration. The relative velocity is largely influenced by particle inertia and the response to the mean flow as well as turbulence. For small particle Stokes numbers, particles tend to follow the turbulence, and the collision frequency is quite low approaching the Saffman and Turner (1956) limit, derived for the collision rate due to turbulent shear (see also Sommerfeld et al., 2008). On the other side, for very large Stokes numbers, colliding particles will have completely uncorrelated velocities, and the collision frequency is determined by kinetic theory (Abrahamson, 1975). In the intermediate regime of

Stokes numbers (i.e. St ≈ 1), the collision frequency will be highest (Sommerfeld, 2001; Sommerfeld *et al.*, 2008). Besides particle inertia and their response, also shear flows and particle Brownian motion (Smoluchowski, 1916) will induce particle relative motion and inter-particle collisions. In technical equipment, inter-particle collisions may lead to agglomeration or breakage of particles but also a modification of the particle concentration distribution and associated secondary effects. A very obvious effect is the dispersion of particles out of developing denser dust regions due to inter-particle collisions. These may be produced by gravitational or centrifugal settling of particles in confined flows, such as pneumatic conveying pipelines (Sommerfeld, 2003), or by inertial particle clustering in turbulence or vortical structures (Ernst *et al.*, 2019). The re-dispersion effect by inter-particle collisions in a horizontal channel airflow is illustrated in Figure 6.5 for 30 μm and 110 mm spherical and mono-sized glass beads, obtained by numerical computations using the stochastic inter-particle collision model. The upper graphs of Figure 6.5 a and b are obtained without inter-particle collisions and show the gravitational settling of both small and large particles with the gravitational effect being more pronounced for the larger 110 μm particles. The fluctuations observed on the trajectories of the small particles (Figure 6.5a) are due to their response to turbulence. When considering inter-particle collisions with the stochastic modelling approach, particles are strongly re-dispersed (lower graphs of Figures 6.5 a and b), and gravitational settling is reduced. Actually, the kinks that are clearly observed on the trajectories of the larger particles result from collisions of the real tracked particles with

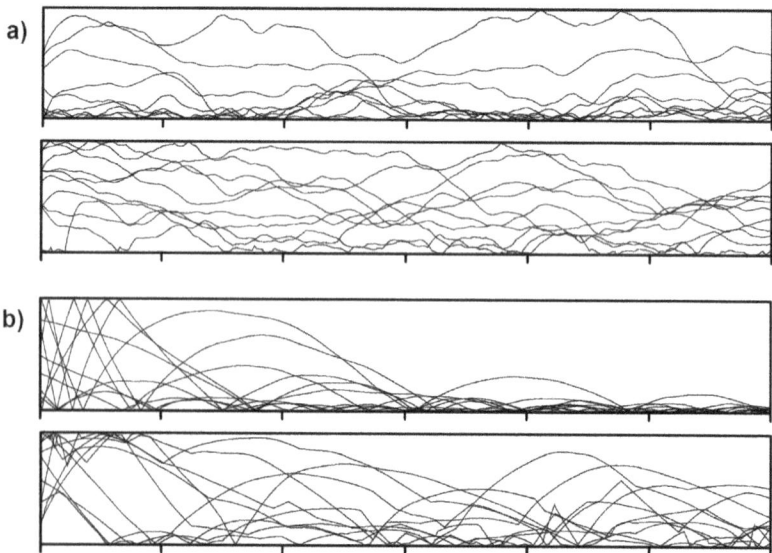

FIGURE 6.5 Horizontal channel flow (H = 35 mm, L = 6 m) with length and height scaled appropriately. Calculated particle trajectories for (a) 30 μm, (b) 110 mm. The upper graphs are without and the lower graphs with inter-particle collisions (average air velocity U_{av} = 18 m/s, particle mass loading η = 1.0; from Sommerfeld, 2003).

fictitious particles. In this special situation, particles approaching the high concentration region near the bottom are mostly colliding with already rebounding particles and establish a high relative velocity. Even in this case with the moderate particle mass loading of 1.0, a strong effect of inter-particle collisions is observed.

However, there are also other observations where particles are trapped in regions of higher particle concentration since multiple inter-particle collisions hinder the particles to be kicked out of this dense region; this was observed in a stirred vessel with particle accumulation behind the baffles (Sommerfeld, 2017). In addition, a dense particle layer formed in a pneumatic conveying bend protects the wall from particle impacts, as they will collide already before they reach the wall with other rebounding particles. By this shielding effect, particle erosion in a pipe bend will be reduced at higher mass loading (Lain and Sommerfeld, 2019).

6.2 SOFT-SPHERE PARTICLE COLLISION/CONTACT

Soft-sphere models, frequently also referred to as discrete element models (DEM), were first proposed by Cundall and Strack (1979). They approximate the actual deformation in the contact areas of two colliding particles and predict the resulting forces from this. These forces are then used to compute the subsequent velocities and positions of the particles. In DEM, each actual collision is resolved by a model, by using a so-called force-displacement model. This model describes the relationship between the displacement (a measure of the deformation of the contact plane between to particles) and the resulting force from the deformation. This deformation can be clearly seen observing a tennis ball as it collides with a hard surface. An example of the stress patterns in a sphere impacting on a flat surface is shown in Figure 6.6.

Most often, the deformation in the normal direction is expressed in terms of a single parameter, the displacement or the overlap, which is indicated as δ in Figure 6.7.

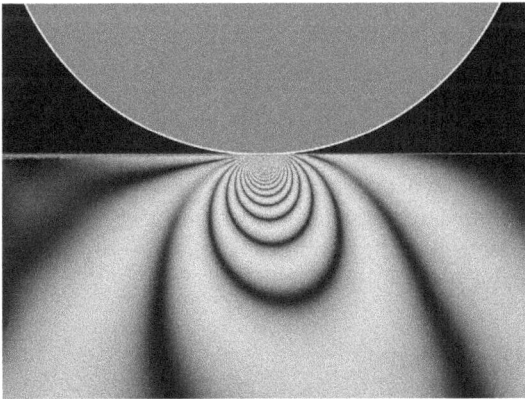

FIGURE 6.6 Stress resulting from the impact between a sphere and a flat surface. Although no visible deformation can be seen, there is a deformation, and the resulting stresses in the contact point are clearly visible.

Source: https://commons.wikimedia.org/wiki

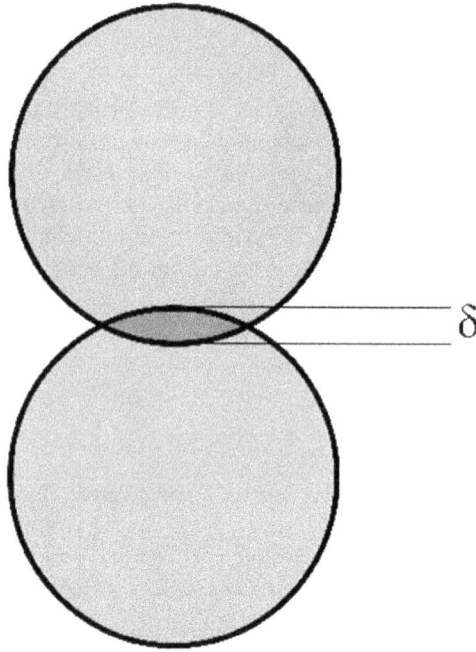

FIGURE 6.7   The definition of the normal displacement, δ, also commonly referred to as the "overlap." The solid material of two colliding particles does not really overlap but deforms. The displacement is merely a measure of the amount of deformation.

The simplest relationship between the force and the displacement is Hooke's law, which merely states that the force in the normal direction between two colliding spheres is proportional to the displacement, or overlap

$$\vec{F}_n = -k\delta\vec{n} \qquad\qquad (6.17)$$

where k is the spring constant, δ is the displacement, and $\vec{n}$ is the normal vector in the direction connecting the two sphere centers. This force is in the direction perpendicular to the direction of overlap, in the normal direction of the particle-particle contact. Although simple to program, and therefore used in many simulations, using a linear spring model to describe the interaction force between two colliding spheres may physically not be very realistic.

6.2.1   Elastic Deformation

The original work in contact mechanics dates back to 1881 with the publication of the paper "On the Contact of Elastic Solids" (Über die Berührung fester elastischer Körper) by Heinrich Hertz (Hertz, 1882). Hertz was attempting to understand how the optical properties of multiple, stacked lenses might change with the force holding them together. Hertzian contact stress refers to the localized stresses that develop as two curved surfaces come in contact and deform slightly under the imposed loads. The amount of deformation under a given load depends on the modulus of elasticity of the material in contact. It provides an expression for the contact stress as a function of the normal contact force, the radii of curvature of both bodies, and the modulus of elasticity of both bodies. Hertzian contact stress forms the foundation for the equations for load bearing capabilities and fatigue life in bearings, gears, and any other bodies where two surfaces are in contact. Hertz predicted the stress and the strain to be linearly related.

Stress is the physical quantity that expresses the relative internal forces that neighboring particles in a continuum exert on each other, while strain is the measure of the relative deformation of the material. The stress is defined as the applied force over the area where the force is applied,

$$\sigma = \frac{F}{A},$$ (6.18)

where F is the applied force, and A is the corresponding area. The strain, ϵ, is the relative deformation, and the stress can be related to the strain as

$$\sigma = E\varepsilon,$$ (6.19)

where σ is the applied stress, and E is a material constant called Young's modulus, or elastic modulus. When the particle is colliding and deforms, Hertz showed that the contact area radius for a given displacement is

$$a = \sqrt{\frac{D}{2}\delta},$$ (6.20)

where a is the contact area radius, D is the diameter of the particle, and δ is the displacement. This is also illustrated in Figure 6.8.

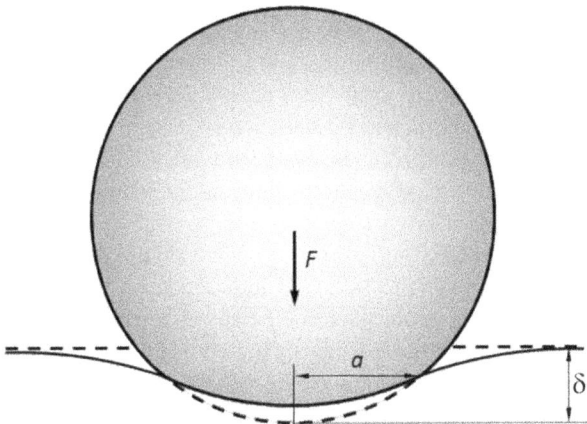

FIGURE 6.8   Contact of an elastic sphere with an elastic half-space, where the displacement δ is related to the contact area radius a.

Using the definitions of the stress and strain

$$\sigma = \frac{F}{\pi a^2} \tag{6.21}$$

$$\varepsilon = \frac{\delta}{a} \tag{6.22}$$

the following relation between force and displacement is derived

$$F_n = \frac{2\sqrt{2}}{3} E^* D^{\frac{1}{2}} \delta^{\frac{3}{2}} \tag{6.23}$$

where the reduced elastic modulus is

$$\frac{1}{E^*} = \frac{1-v_1^2}{E_1} + \frac{1-v_2^2}{E_2} \tag{6.24}$$

and E_1 and E_2 are the elastic moduli and v_1 and v_2 the Poisson's ratios associated with each of the colliding particles. It is clear that with an elastic material, the relationship between the force and the displacement is not linear but to the power of $\frac{3}{2}$. Eq. (6.23) is a so-called force-displacement law, which is required for a DEM simulation and can be used to explain the behavior of elastic particles. However, fully elastic interactions are not realistic in practice because they describe the interactions of particles with a very low relative velocity. To realistically describe most particle interactions, a more complicated force-displacement law is required, or an extension, which accounts for energy dissipation during particle-particle interactions.

6.2.2 Dissipation in the Normal Direction

The equations in the last subsection on the effect for normal impact describe the same behavior in the case of loading (the particles approaching each other) and as unloading (the particles moving away from each other). Thus, during the collision, the kinetic energy is conserved, and the collision is perfectly elastic. As in the hard-sphere model, the kinetic energy can be dissipated due to the visco-elastic or plastic nature of the materials during collision. In addition, the vibrations of the solid material caused by the collision often lead to non-ideal behavior during the collision. This implies that the force during unloading is less than during the loading stage. There are a number of models to take this into account. Among these, the hard-sphere approach introduces a constant coefficient of restitution, e, by relating the unloading spring constant to the loading spring constant, as it is done in the linear force displacement model (Walton and Braun, 1986)

$$e = \sqrt{\frac{k_n^{unloading}}{k_n^{loading}}} \tag{6.25}$$

where the *loading* or *unloading* of the collision depends on the relative normal velocity of the two spheres, loading

$$\vec{u}_{1,2} \cdot \vec{n} < 0$$

unloading

$$\vec{u}_{1,2} \cdot \vec{n} > 0$$

In a more general framework, the coefficient of restitution is not constant but depends on the relative velocity before the impact and possibly on other factors as well. The damping can then be modelled as

$$\vec{F}_n^{diss} = -\eta_n \dot{\delta}_n^\alpha \delta_n^\beta \tag{6.26}$$

where η_n, α and β are constants and define the model. If $\beta = 0$, which is a popular choice in the literature, it means the dissipation scales with $\dot{\delta}_n^\alpha$, which is equal to v_n, the relative velocity between the particles at each point in time. Although this is counterintuitive, because the dissipation will be maximal as the displacement itself is maximal, the overall behavior of a collision seems to be accurately captured. The two most popular models, the Tsuji model (Tsuji, 1993) and the Kuang model (Kuang *et al.*, 2008), use the expression

$$\vec{F}_n^{diss} = -\eta_n \dot{\delta}_n \tag{6.27}$$

The parameter η_n is given as follows in the two studies

$$\eta_n = \begin{cases} \alpha_n \left(M_{12} k_n \right)^{\frac{1}{2}} & \text{Tsuji (Tsuji, 1993)} \\ \alpha_n \left(6 M_{12} E^* \sqrt{R_{12}} \right)^{\frac{1}{2}} & \text{Kuang et al (Kuang et al., 2008)} \end{cases} \tag{6.28}$$

where α_n is the damping coefficient and depends on the material and for both models $\beta = \frac{1}{4}$. In both models, the coefficient of restitution is not constant but depends on the material properties as well as the relative velocities of the particles.

6.2.3 ROTATION

When the particles rotate, not only a normal force, but also a tangential force should be considered during the particle-particle or particle-wall interactions. To account for the tangential force between the particles, the slip velocity in the tangential direction needs to be determined. The slip velocity in the contact point is given by the expression

$$\vec{q} = \left(\vec{u}_1 - \vec{u}_2 \right) - \left(\frac{D_1}{2} \vec{\omega}_1 + \frac{D_2}{2} \vec{\omega}_2 \right) \times \vec{n} \tag{6.29}$$

From this equation, the normal and tangential slip velocity can be determined as follows

$$\vec{u}_n = \vec{q} \cdot \vec{n} \tag{6.30}$$
$$\vec{u}_t = \vec{q} - \vec{u}_n \tag{6.31}$$

There is some confusion in the literature on how to formulate a realistic model for the tangential force as there is no direct simple Hertzian contact model for the tangential force. It is important to realize that the *tangential* force or overlap is a result of the *history* of the collision, as shown in Figure 6.9. This means that the previous time step of the tangential force needs to be stored because it is required to construct the tangential force at the new time level. Also that the orientation of the previous tangential force needs to be projected into the reference frame at the new time step. The reference frame at the new time step can differ from the reference frame of the old time step because the relative positions and orientation of the two colliding particles changes due to translational and rotational movement.

Determining the tangential force at the new time step required a number of calculations.

First, the tangential force, which was determined at the previous time step and denoted by $\vec{F}_t^o$, is mapped onto the reference frame of the new collision, as the framework of the two colliding particles have moved and/or rotated

$$\vec{F}_t^* = \vec{n} \times \vec{F}_t^o \times \vec{n}$$
$$= \vec{F}_t^o - \vec{n}\left(\vec{n} \cdot \vec{F}_t^o\right) \tag{6.32}$$

Here, $\vec{F}_t^*$ is the tangential force defined at the previous time step mapped into the reference frame of the new time step. However, this projection may have changed the magnitude of the tangential force, and this needs to be corrected.

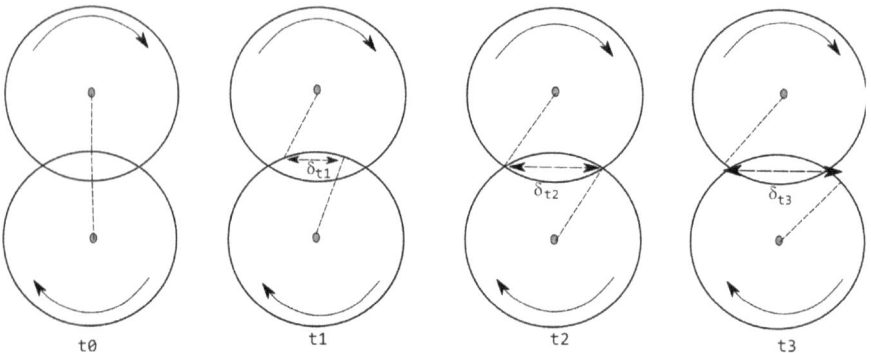

FIGURE 6.9 Schematic diagrams of a sticking collision as time progresses from t_0, assuming no-slip in the point of contact. The tangential force is a result of the history of the collision.

To ensure that the *length* of the tangential force at the previous time step in the new reference frame is the same as in the previous frame, a scaling is introduced

$$\vec{F}_t^{**} = \left| \frac{F_t^o}{F_t^*} \right| \vec{F}_t^* \tag{6.33}$$

So $\vec{F}_t^{**}$ is the *old* tangential force with the correct magnitude in the new frame of reference, that is, the new orientation of the normal vector. From this, the tangential direction normal of the tangential force is determined by the expression

$$\vec{t} = \frac{\vec{F}_t^{**}}{\left| \vec{F}_t^{**} \right|} \tag{6.34}$$

Second, the new contribution to the new tangential force is determined. For this, the relative surface displacement that occurred during the current time step is calculated as follows

$$\Delta \vec{s} = \vec{u}_t \Delta t \tag{6.35}$$

This displacement is then split into a displacement continuing along the direction of the old tangential force and a new direction

$$\Delta \vec{s}_\parallel = \left(\Delta \vec{s} \cdot \vec{t} \right) \vec{t} \tag{6.36}$$

$$\Delta \vec{s}_\perp = \Delta \vec{s} - \Delta \vec{s}_\parallel \tag{6.37}$$

The additional displacement in the parallel direction is treated as *additional* overlap,

$$\vec{F}_{t,\parallel} = \vec{F}_t^{**} + K_T \Delta \vec{s}_\parallel \tag{6.38}$$

the additional displacement in the perpendicular direction is treated as *new* overlap,

$$\Delta \vec{F}_{t,\perp} = K_0 \Delta \vec{s}_\perp \tag{6.39}$$

and the new tangential force is found by adding the old overlap and the new

$$\vec{F}_t^n = \vec{F}_{t,\parallel} + \Delta \vec{F}_{t,\perp} \tag{6.40}$$

$$= \vec{F}_t^{**} + K_T \Delta \vec{s}_\parallel + K_0 \Delta \vec{s}_\perp \tag{6.41}$$

Hence, the tangential force is determined in terms of a change per time step. The actual model for the incremental tangential force for the sticking regime has been derived in Mindlin and Deresiewicz (1953) and is given as

$$\Delta \vec{F}_t = -k_t \Delta \vec{s} = -k_t \Delta \vec{u}_t \Delta t \tag{6.42}$$

where the tangential spring constant is given as

$$k_t = 8\sqrt{a^* \delta} G^* \qquad (6.43)$$

with a^* being the reduced radius and G^* is the reduced shear modulus of the particles. The shear modulus for each particle is defined as

$$G = \frac{E}{2(1+v)} \qquad (6.44)$$

The magnitude of the tangential force as analyzed by the previous approach is for a sticking regime and should never exceed the maximum allowable contact force given by the Coulomb friction criterion. If this happens, slip of the contact point occurs, and the maximum tangential force is kept at the value of the Coulomb sliding force

$$\left| \vec{F}_t^{max} \right| = \mu \left| \vec{F}_n \right| \qquad (6.45)$$

in the same tangential direction. Therefore, the DEM model can correctly predict the combination of sticking and sliding of the contact point, whereas in the hard-sphere model, the collision is treated as sticking or sliding.

6.2.4 ADHESION

The soft-sphere model is particularly suitable to study the behavior of particles with adhesion. Adhesive forces between particles have been a subject of study since the initial investigations by Bradley (1932). Adhesion between elastic spheres is well understood and usually described using the DMT (Derjaguin et al., 1975) or JKR (Johnson et al., 1971) models. The discrepancy between the behavior predicted by these two models was explained by Maugis (1992), who recognized that the DMT model is applicable in the case where adhesive forces are weak and particles are hard, whereas the JKR approach accurately describes the behavior of soft, sticky particles.

In the DMT model, the adhesive force is formulated in terms of a *pull-off* force. This force is defined as the maximum force required to pull two particles, which are in contact, out of contact. This pull-off force is given as

$$\vec{F}_{adh} = -2\pi a^* \Delta \gamma \, \vec{n}, \qquad (6.46)$$

where $\Delta \gamma$ is the work of adhesion, which can also be expressed in terms of the Hamaker constant. In the DMT model, this adhesive force is added to each pair of particles in contact.

6.2.5 DISSIPATION IN THE TANGENTIAL DIRECTION

Several models also use a dissipation scheme in the tangential direction, although it is not precisely clear where this originates from. According to these models, similarly as the normal component of the force, the kinetic energy losses are associated with the tangential velocity

$$\vec{F}_t^{diss} = -\eta_t \vec{u}_t \qquad (6.47)$$

where $\bar{u}_t$ is the slip velocity between the particles in the tangential direction and η_t represents the tangential damping coefficient. For the two models described earlier, these are given as follows

$$
\eta_t = \begin{cases}
\eta_n & \text{Tsuji (Tsuji, 1993)} \\[2em]
\alpha_t \left(6\mu M_{12} \left| \vec{F}_n \right| \dfrac{\sqrt{1 - \dfrac{\delta_t}{\delta_{t,max}}}}{\delta_{t,max}} \right)^{\frac{1}{2}} & \text{Kuang et al (Kuang et al., 2008)}
\end{cases}
\tag{6.48}
$$

6.2.6 PARTICLE COORDINATE REFERENCE FRAME

When the particles move and rotate relatively to each other, the *frame of reference* of particles will also move, and the normal and tangential vectors of the collisions will change. For the normal force, this is not a problem because the normal force vector does not, generally, depend on time. However, for the tangential force, this is different, and the previous contributions to the tangential force must be mapped into the new frame of reference.

There are two coordinate transformations required to project the tangential force from the absolute Cartesian coordinates to the two-particle frame of reference. The first transformation arises from the *change of orientation* of the reference frame over time, as indicated in Figures 6.10 and 6.11. This change can be expressed as

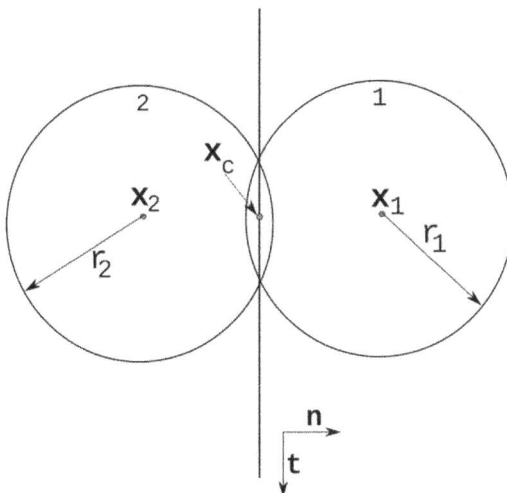

FIGURE 6.10 Two particles colliding in the reference frame of the previous time step.

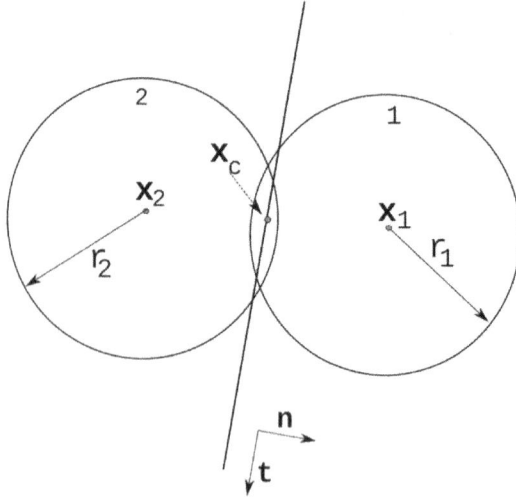

FIGURE 6.11   The same two particles colliding in the new reference frame, at the next time step. The reference frame change can be expressed by the evolution of the collision normal, n, during the time step, Δt.

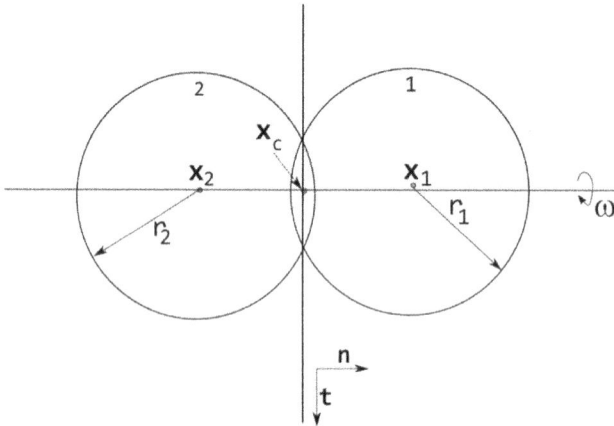

FIGURE 6.12   The rotation of the reference frame along the normal of the collision plane, indicated by ω, does not change the orientation of the collision plane and the orientation of the collision normal, but it does change the direction of the tangent t.

the change in the normal of the collision plane. The second transformation arises from a frame rotation around the normal direction of the collision plane, as shown in Figure 6.12. The first coordinate transformation may be expressed as

$$F_i^1 = F_j \left(\delta_{ij} - e_{ijk} e_{kmn} n_m^{old} n_n \right) \tag{6.49}$$

where δ_{ij} is the Kronecker delta, and e_{ijk} is the Levi-Civita symbol. This yields the three components of the force as follows

$$F_1^1 = F_1 + F_2 \left(n_2^{old} n_1 - n_1^{old} n_2 \right) + F_3 \left(n_3^{old} n_1 - n_1^{old} n_3 \right) \tag{6.50}$$

$$F_2^1 = F_1 \left(n_1^{old} n_2 - n_2^{old} n_1 \right) + F_2 + F_3 \left(n_3^{old} n_2 - n_2^{old} n_3 \right) \tag{6.51}$$

$$F_3^1 = F_1 \left(n_1^{old} n_3 - n_3^{old} n_1 \right) + F_2 \left(n_2^{old} n_3 - n_3^{old} n_2 \right) + F_3 \tag{6.52}$$

The second transform, to be applied upon the preceding one, gives

$$F_i^2 = F_j^1 \left(\delta_{ij} - e_{ijk} \bar{\omega}_k \Delta t \right) \tag{6.53}$$

where $\bar{\omega}_k$ is the average rotation of the frame of reference around the normal

$$\bar{\omega}_i = \frac{1}{2} \left(\omega_j^1 + \omega_j^2 \right) n_i n_j \tag{6.54}$$

Therefore, the force components may be written as

$$F_1^2 = F_1^1 + F_2^1 \left(-\bar{\omega}_3 \Delta t \right) + F_3^1 \left(\bar{\omega}_2 \Delta t \right) \tag{6.55}$$

$$F_2^2 = F_1^1 \left(\bar{\omega}_3 \Delta t \right) + F_2^1 + F_3^1 \left(-\bar{\omega}_1 \Delta t \right) \tag{6.56}$$

$$F_3^2 = F_1^1 \left(-\bar{\omega}_2 \Delta t \right) + F_2^1 \left(\bar{\omega}_3 \Delta t \right) + F_3^1 \tag{6.57}$$

This completes the mapping of the force in the framework at a previous time to the force in the current framework and is crucial to perform when considering time-dependent force models, such as the tangential forces.

6.2.7 INTEGRATION OF THE EQUATIONS OF MOTION

The equations of motion for each sphere in a DEM framework are typically given as

$$m_p \vec{a}_p = \sum_{k=contacts} \left(\vec{F}_{n,k} + \vec{F}_{t,k} \right) + \vec{b}_p \tag{6.58}$$

where $\vec{b}_p$ represents all other body forces, arising from the gravity, drag, pressure gradient, etc.

To solve this ordinary differential equation (ODE), a suitable integration scheme and time step must be chosen. Since evaluating the collision forces is computationally expensive, this is done as infrequently as possible. Therefore, numerical schemes such as the Leapfrog, Verlet, or Beeman integration scheme are adopted (Allen and Tildesley, 1989).

The time step must be chosen sufficiently small to accurately capture the dynamics of each collision. In practice, a minimum of 30 time steps are required to accurately capture the physical mechanism of a collision, from the first time when a particle contacts, to the last time when a particle contacts. It can be shown that, for an elastic model, the time of the complete collision scales as (Hemph *et al.*, 2006)

$$\tau \propto R\left(\frac{\rho_s}{E^* \sqrt{v}}\right). \tag{6.59}$$

Hence, the smaller the particles or the larger the elasticity modulus (indicating high hardness for the particle), the smaller the collision time and, therefore, the required time step. Many researchers performed simulations with artificially soft particles to enable the application of a larger time step, but this significantly affects the physics of the modelled phenomena (Hemph *et al.*, 2006).

Most often, the time step will vary during a simulation (Hamaker, 1937). An example of the Verlet scheme with a varying time step is given by the expression

$$\vec{x}_p^{n+1} = \vec{x}_p^n + \left(\vec{x}_p^n - \vec{x}_p^{n-1}\right)\frac{\Delta t^n}{\Delta t^{n-1}} + \left(\frac{\sum_{k=contacts}\left(\vec{F}_{n,k}^n + \vec{F}_{t,k}^n\right) + \vec{b}_p^n}{m_p}\right)\left(\Delta t^n\right)^2 + O\left(\Delta t^4\right) \tag{6.60}$$

where the superscripts on the variables denote the time-level. The velocity of the particle is approximated by the expression

$$\vec{u}_p^{n+\frac{1}{2}} = \frac{\vec{x}_p^{n+1} - \vec{x}_p^n}{\Delta t} + O\left(\Delta t^2\right) \tag{6.61}$$

Although the approximation for the new velocity of the particle is less accurate, this is not crucial as determining the new position does not depend on the velocity itself. The torque on the particle is given by

$$\vec{T}_p = \frac{1}{2}D\sum_{k=contacts}\vec{n}^k \times \vec{F}_t^k \tag{6.62}$$

and the rotation of the particle is approximated as

$$\vec{\omega}^{n+\frac{1}{2}} = \vec{\omega}^{n-\frac{1}{2}} + \vec{T}_p^n I^{-1}\Delta t + O\left(\Delta t^2\right) \tag{6.63}$$

where *I* is the tensor of inertia of the particle. The velocity and rotation are computed at the mid-intervals of $n\pm\frac{1}{2}$, wheras the position, forces, and torque are determined at the primary intervals, *n*, as is mostly the case for Verlet algorithms.

6.3 AGGLOMERATION AND FLOCCULATION MODELLING

In many technical and industrial processes, particles may come into close contact, through collisions or external forces, so that short-range forces become active. Oftentimes particles stick together and form a more or less stable cluster of particles. There are several such short-range forces, which are relevant for different technical situations such as liquid bridges and adsorption layers on the particle surface, van der Waals forces, and electrostatic forces (Marshall and Li, 2014). The magnitude of the different short-range forces depends on the carrier fluid (i.e. particles dispersed in a gas or liquid continuum), the particle size, and the contact distance between the particles. The inter-molecular van der Waals force has a relatively small magnitude and is only of importance at very small contact distances between about 4 Å and 400 Å (i.e. 0.4 nm to 40 nm). Therefore, in press agglomeration or tabletting, van der Waals forces are most important since the contact area between neighboring particles is increased by external forces. In most situations, the adhesion forces, due to liquid bridges, are the strongest short-range forces, and therefore they are used in several processes, including spray granulation processes where fine particles are "glued" together by liquid bridges to form granulates. For particles moving in a gaseous environment, there is normally no deterministic defined electrostatic force between contacting particles as charging mostly occurs through collisions with the walls. This process is called triboelectric charging and may be regarded as a random process. A deterministic charging of particles is done in an electrostatic separator where the bombardment of particles by ions results eventually in particle charging. The strong field force then directs the particles to a collection electrode. However, for particles suspended in a liquid, repulsive electrostatic forces are in competition with the attractive van der Waals force yielding the so-called interaction potential. Depending on the shape of the interaction potential, agglomeration is favored or avoided, and the suspension is called unstable or stable (non-agglomerating).

Different terms are being used for describing clusters of attached or bonded primary particles, namely, agglomeration, aggregation, coagulation, and flocculation. The terms *agglomerate* and *aggregate* are very common in the field of powder technology to describe clusters of fine particles that are formed by dry powders in a gas-solid system, but also for powders in a liquid suspension. There are many definitions of these two terms in the available standards, and they often are used interchangeably, which makes their discussion and calculations difficult (Nichols *et al.*, 2002). The standard ISO14887 gives the following definitions: the term aggregation implies that clusters are formed having strong chemical bounds between primary particles, whereas agglomerates are clusters having weak physical interactions between primary particles. The latter are sometimes referred to as *loose agglomerates*.

Very often also the term *coagulation* is met, defined differently depending on the application. Originating in the medical discipline, coagulation is the process in which blood changes from a liquid to a kind of gel, forming blood clots through the clustering of blood cells, actually for humans a dangerous incident. Also, in colloidal science, the binding together of primary particles is called coagulation. Commonly, coagulation is also used in sewage water treatment, meaning the formation of effectively larger particles from a very fine solid suspension through the competition of attractive van der Waals and repulsive electrostatic forces (the so-called interaction potential, see Marshall and Li, 2014). The relative importance of these forces may be modified by the addition of coagulants. Naturally, the formed larger particles have higher terminal velocity and are therefore more easily separated by sedimentation. The competitive interaction between van der Waals and electrostatic force is mostly described by the DLVO-theory, which was established by Derjaguin, Landau, Verwey, and Overbeek in the 1940s (Derjaguin and Landau, 1941; Verwey and Overbeek, 1948). This theory assumes that the electrostatic double layer forces and the van der Waals forces are independent and therefore can be superimposed or added at each interacting distance between two particles. Variations in the ionic strength of the suspension control the distance of the double-layer interaction, while the van der Waals force is insensitive to the ionic strength. The van der Waals attraction increases with a power-law dependence with the reduction of the separation distance of the particles. On the other hand, the double layer force grows exponentially with reducing particle separation and depends on the ionic strength of the suspension. The resulting curve yields the interaction potential, which often exhibits a maximum of the repulsive force, the energy barrier (Trefalt *et al.*, 2014). When the interaction potential is strongly repulsive, the particles will repel each other and form a so-called stable suspension. When the interaction potential is attractive, the particles will approach each other and eventually stick together, creating agglomerates, in which case the suspension is called unstable (Trefalt *et al.*, 2014).

Flocculation is a very similar process also occurring in water-based systems or aqueous suspensions, frequently called colloidal systems. The term flocculation is also widely used in industrial applications, such as biotechnology, mineral processing, papermaking, water and wastewater treatment, where filtration of particles is important. Very often the terms stable and unstable suspensions are found in technological practice as mentioned previously. These states can be reached by additives or specific agents, the flocculants, which simply change the pH value in an aqueous system. A suspension is called stable when the electrostatic repulsion high enough for the flocculation not to take place. This implies that the so-called energy barrier (the maximum in the repulsion force) is not overcome by the dynamics of the particle motion (due to Brownian movement, turbulence, or some other energy input) so that the primary particles are far enough not to be attracted and remain separated. When flocculation is desired, for example, for particle filtration, the suspension is destabilized by an additive flocculant so that the repulsive electrostatic forces are reduced and the attractive van der Waals force becomes dominant, yielding primary particle attachment and a strong flocculation. On the other hand, when particle aggregation and flocculation is undesirable, for example, in the production of nanofluids, a dispersant is added, which typically increases the *zeta-potential* of the primary particles, prevents flocculation, and maintains the nano-size particles separate (Michaelides, 2014).

6.3.1 Characteristics of Agglomerates

To analyze the temporal evolution of agglomeration and to compare agglomerates obtained under different conditions, information on the agglomerate morphology is required. Fundamental agglomerate properties are the number of primary particles included, as well as volume and surface area, which provide the basis for calculating further quantities such as volume-specific surface area, porosity, and sphericity. The sphericity, ψ, of an agglomerate is defined as the ratio of the surface area of the compact volume equivalent sphere, A_{VES}, considering all agglomerated primary particles, n_{pp}, and the total surface of the given agglomerate, A_{Agg}, accounting also for a possible penetration of primary particles

$$\psi = \frac{A_{VES}}{A_{Agg}} \qquad (6.64)$$

The sphericity is a measure of particle shape and becomes less meaningful if the surface of the particle is more rough or structured. The disadvantage of the sphericity is that it does not carry any information about the spatial dimensions and the agglomerate structure.

Another property characterizing the compactness of agglomerates is the porosity, ε. This value is quantified by the ratio of the void-space V_V between the primary particles to the volume occupied by the agglomerate, that is, primary particle volume plus void volume $(V_S + V_V)$

$$\varepsilon = \frac{V_V}{V_S + V_V} = \frac{V_V}{V_H} = 1 - \frac{V_S}{V_H} \qquad (6.65)$$

where the agglomerate volume corresponds to the sum of void volume V_V plus the solids volume V_S, which is very often assumed to be given by the volume of the enwrapping sphere. It is more appropriate to use the volume occupied by the agglomerate as the volume of the convex hull wrapped around the agglomerate V_H, as shown in Figure 6.13a (Dietzel and Sommerfeld, 2013; Ernst et al., 2013). The sphere enwrapping the entire agglomerate will give very high porosity values in many cases as it is apparent in Figure 6.13b. The diameter of the enwrapping sphere D_o (sometimes called the *interception diameter*) is a measure of the absolute agglomerate size, which corresponds to the largest extension of the agglomerate with respect to its center of mass. With the convex hull volume, V_H, one can also define an equivalent

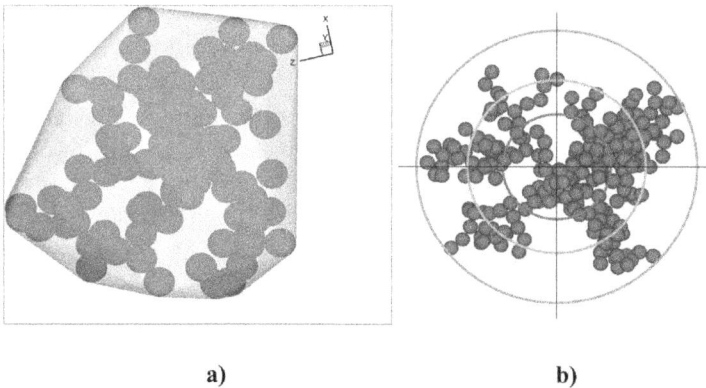

a)   b)

FIGURE 6.13 Agglomerates generated by a model simulation from mono-sized primary particles with characteristic properties: (a) Convex hull wrapped around an agglomerate; (b) agglomerate with the indication of the possible equivalent diameters (number of primary particles n_{pp} = 266).

diameter, which is smaller than the diameter of the enwrapping sphere, as shown in Figure 6.13b. Indicators of the mass distribution inside agglomerates are the radius of gyration R_g or the fractal dimension D_f. Typically, R_g is determined by calculating the root mean square of the distance of each primary particle $\overline{x_{P,i}}$ to the center of mass of the agglomerate $\overline{x_{A,C}}$

$$R_g = \sqrt{\frac{1}{n_{PP}} \sum_{i=1}^{n_{PP}} \left| \overrightarrow{x_{A,C}} - \overrightarrow{x_{P,i}} \right|^2} \tag{6.66}$$

As a result, more dendritic structured agglomerates will give larger gyration radii as compared to compact agglomerates with the same volume.

Another parameter used with agglomerates is the fractal dimension, which is explained in detail in Section 3.1.1 and is often indicative of the spatial mass distribution of the primary particles (Mandelbrot, 1967). A relation between agglomerate mass, m_A, and its size, R_o (e.g. interception radius), can be formulated incorporating the fractal dimension, D_f, as the exponent

$$m_A \sim R_o^{D_f} \tag{6.67}$$

Based on Eq. (6.67), the number of mono-sized primary particles (without overlap) in agglomerates may be approximated by the expression

$$n_{pp} = k_f \left(\frac{R_g}{r_p} \right)^{D_f} \tag{6.68}$$

where r_p is the radius of the primary particles, and k_f is the fractal pre-factor, which also depends on the fractal dimension (Gmachowski, 1995; Sorensen and Roberts, 1997). The linear part of this correlation may be approximated by the expression derived by Vanni (2000)

$$k_f = 0.414 \cdot D^f - 0.211 \quad \left(1.5 < D_f < 2.75 \right) \tag{6.69}$$

Several approaches have been developed for calculating the fractal dimension of three-dimensional fractal objects, which, however, often lead to quite different results (Nelson et al., 1990). The iterative box counting method, proposed by Martinez-Lopez et al. (2001), is suggested to determine the fractal dimension of complex polydisperse agglomerates. The number of boxes that are necessary to completely include the agglomerate structure is determined as a function of the box size. With these data a double-logarithmic plot of the box number n_b over the reciprocal value of the box size δ is generated (Dietzel and Sommerfeld, 2013; Brasil et al., 2001). The slope of the linear segment of the resulting curve corresponds to the fractal dimension, $D_{F,B}$, of the agglomerate, as shown in Figure 6.14a

$$D_{F,B} = -\frac{\log(n_b)}{\log(1/\delta)} \tag{6.70}$$

a)

b)

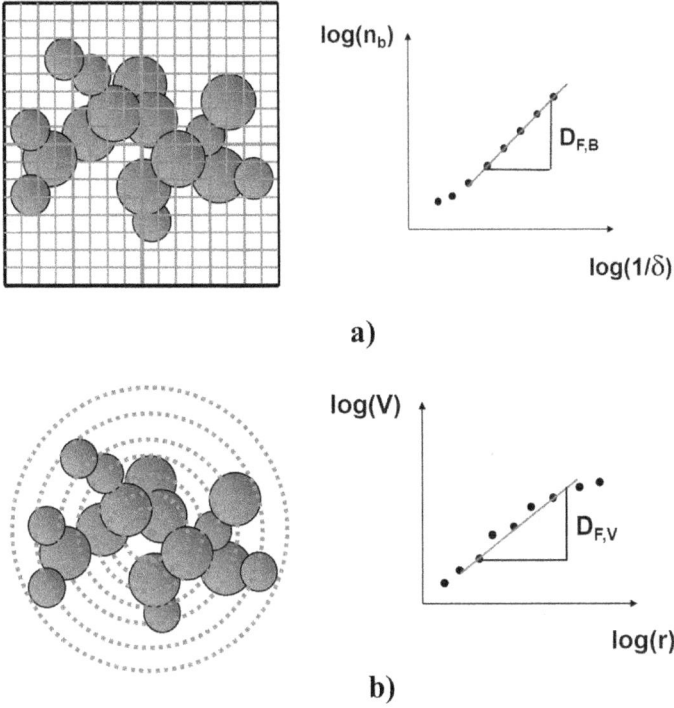

FIGURE 6.14   Methods for the determination of the fractal dimensions of simulated small fractal aggregates: (a) Box-counting method and (b) volume-radius method.

A less common method for the determination of a fractal dimension is the mass-radius (or volume-radius) method, where the entire agglomerate structure is discretized by a series of concentric spheres with growing radius, r, as shown in Figure 6.14b. The volume occupied by the agglomerate structure is determined by using an appropriate discretization, such as boxes (bins). Again, a double logarithmic plot is produced, and the fractal dimension is obtained as the slope of the resulting curve:

$$D_{F,V} = -\frac{\log(V)}{\log(r)} \qquad (6.71)$$

The estimation of the fractal dimension in both ways is only reliable when the structure of the growing agglomerates may be considered as scale-invariant (Gmachowski, 1995), an assumption that is not realistic with actual agglomerates composed of a finite number of particles and bounded by a finite volume (see Section 3.1.1). The volume-radius method is less accurate since it is more difficult to determine the volume occupied by the agglomerate within a spherical object accurately. It should be noted that the fractal dimensions obtained by the two methods are not necessarily identical.

6.3.2 Models of the Agglomeration Process

Consider fine particles moving in a turbulent gas or liquid flow so that for the development of the agglomerate, first, the particles must collide through the effects of inertia, turbulence, or Brownian motion, and then they must stick together by adhesion forces to form agglomerates consisting eventually of many primary particles that may have different structures, for example, more compact or dendritic configuration in soot agglomerates (Eggersdorfer and Pratsinis, 2012; Dietzel and Sommerfeld, 2013; Michaelides, 2014). Consequently, the agglomerate that is formed should not be considered as a volume equivalent sphere for the correct calculation of the fluid forces acting on such particles or to determine the collision cross-section. As shown in Figure 6.13b, the outer diameter of the agglomerate (the longest of the characteristic dimensions) is much larger than the volume equivalent sphere and naturally has much higher porosity.

The modelling of fine particle agglomeration in numerical calculations of powder handling processes (e.g. gas cleaning by cyclone separators or sedimentation of suspensions) is possible in a very mechanistic and descriptive way when applying the Euler/Lagrange approach. This method allows to combine the fluid dynamic transport of fine particles, inter-particle collisions, as well as short-range interaction forces to judge a possible agglomeration. Finally, the Lagrangian approach also offers the opportunity to describe the possible breakup of agglomerate structures. As a result, a Lagrangian agglomeration model should consist of three fundamental steps, each including the relevant physical mechanisms:

1. Searching for a possible collision between two particles based on either deterministic (Sundaram and Collins, 1997; Ernst and Sommerfeld, 2012) or stochastic (Sommerfeld, 2001) collision detection approaches. The latter is numerically much more efficient and is largely independent of the grid size (Lain and Sommerfeld, 2020). The collision detection should include the consideration of impact efficiency, especially for cases with wide particle size distributions and if agglomerates are present in the particle spectrum. This effect implies that smaller particles might move with the relative flow around larger ones, also called collectors without a collision. In this case, the collision frequency is effectively reduced (Ho and Sommerfeld, 2002; Sommerfeld and Lain, 2009). It should be emphasized that the model describing this effect was developed for inertial particle impact. With the increasing importance of diffusional particle transport, the impact efficiency will increase again, a situation typically observed in filtration. Such a scenario, however, requires further model extensions.

2. Assessment whether the collision will result either in a rebound of particles, or they might stick together forming a new particle or agglomerate with velocity and size (equivalent diameter) that must be determined. In a gas-particle system, which is considered here, the most relevant adhesion force is the van der Waals molecular interaction force. This force is only relevant, however, if the particle separation is very small. Electrostatic forces are not important unless a charging of particles is employed, or tribo-charging is likely to occur. Contact forces through liquid adsorption layers around

the particle will also become of relevance if the considered process takes place in a humid environment and vapor may adsorb on the particle surface. Criteria on how to decide on sticking in dry systems will be discussed later.
3. Description of the collision outcome and on how the newly formed particle pair and later on particle cluster is treated with regard to size and velocity. The velocity of the new agglomerate may be determined by a momentum balance. However, which particle size has to be used in further tracking needs also to be determined. Assuming a new volume equivalent sphere is surely a very crude assumption. Hence, it is more realistic to use other equivalent diameters that far better represent the real shape of the agglomerate. This issue is important for the determination of the fluid dynamic forces acting on the agglomerate and the collision rate with new primary particles because of the effectively larger collision cross-section.

One of the first numerical calculations using the Euler/Lagrange approach for analyzing particle agglomeration in particle-laden flows, in case of homogeneous isotropic turbulence, was introduced by Ho and Sommerfeld (2002). The stochastic collision model in combination with the impact efficiency was used to decide about the occurrence of a collision. The probability of particle sticking and agglomerate formation was based on an energy balance accounting for dissipated and van der Waals energy. After agglomeration occurred, the new particles were considered as compact volume equivalent spheres (illustrated in Figure 6.15 left) and the new velocity components followed from a momentum balance. Simulations for a gas cyclone (Ho and Sommerfeld, 2005) revealed that only small particles can stick to other particles so that the grade efficiency was improved up to particle sizes of about 2.5 μm. For larger particles, inertia is too high for the relative velocity to fall below the critical velocity and agglomeration to occur.

Initiated by the need to numerically predict the properties of agglomerates generated by spray drying (i.e. particle property design), models were developed in the past allowing the penetration of mushy particles (Blei and Sommerfeld, 2007) and finally yielding also the structure of agglomerates, including porosity and fractal dimension (Sommerfeld and Stübing, 2012; Stübing et al., 2011). These different modelling approaches for particle agglomeration are summarized in Figure 6.15.

The most straightforward approach, the simple agglomeration model, assumes that agglomerating particles form a new volume equivalent compact sphere with a velocity resulting from a momentum balance. This model does not provide any information on the realistic structure and size of agglomerates. Only statistics on particle contacts and the penetration of mushy particles may be obtained by such a model, and naturally, the number of primary particles collected in an agglomerate can be counted. The real agglomerate size is not considered in the determination of the fluid dynamic forces nor the calculation of the collision cross-section.

The second model, called sequential (or history) model, assumes that for every two-particle agglomeration, a new spherical particle is formed with total volume consisting of the volume of the involved two spherical particles plus the void volume resulting from the convex hull wrapped around the two particles, as may be seen in the middle section of Figure 6.15. The new particle is now tracked with a size being

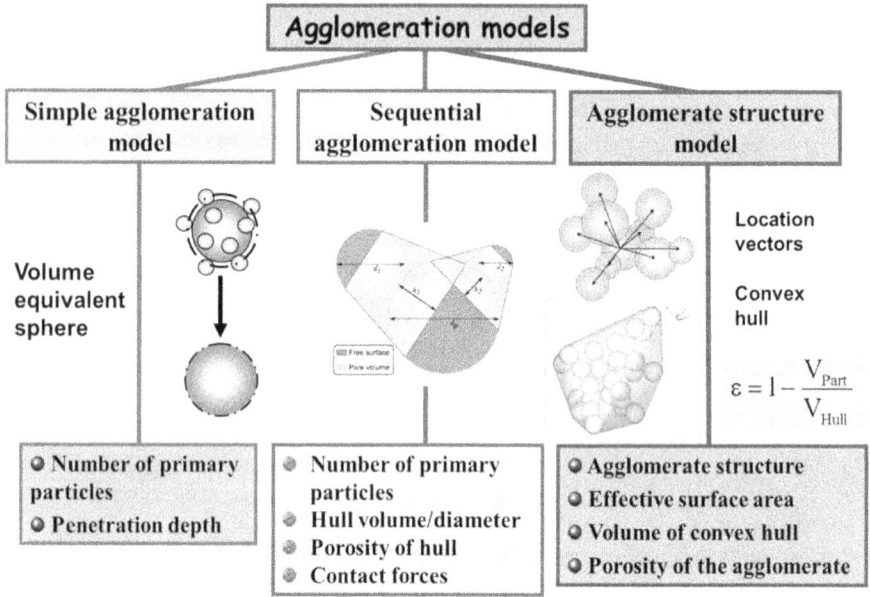

FIGURE 6.15 The different approaches for modelling particle agglomeration outcomes: Simple (volume equivalent sphere) agglomeration model (Ho and Sommerfeld, 2002), sequential agglomeration model (Lipowsky and Sommerfeld, 2008; Sgrott Junior and Sommerfeld, 2019), and agglomerate structure model (Sommerfeld and Stübing, 2017).

larger than that resulting from the volume of the two primary particles since it also includes the porosity resulting from the void of a two-particle configuration. This procedure is successively repeated for each collision of a primary particle with an agglomerate of equivalent size, considering that this spherical agglomerate has porosity. Hence, fluid dynamic forces and the inter-particle collision cross-section are calculated with this equivalent diameter. Using the free surface area, which is continuously updated (dark gray area in Figure 6.15 middle), it is decided by a random process, for which primary particle within the structure the new primary particle adheres. This sequential porous sphere model was applied for calculating particle separation in a gas cyclone (Lipowsky and Sommerfeld, 2008; Sgrott Junior and Sommerfeld, 2019). The agglomerate history is stored in a tree structured way (linked list) with the agglomerate diameter, porosity, and van der Waals adhesion forces, which may be further on used to decide about possible agglomerate breakage.

A comparison of the volume equivalent sphere and the sequential porous sphere agglomeration model is presented by Sgrott Junior (2019) together with the improvement of fine particle separation in cyclones through agglomeration. The comparison of the porous sphere agglomeration model with the sophisticated structure model, being described in the next section, showed an excellent agreement (see Figure 22 of Sgrott Junior, 2019) with respect to the predicted porosity and the size of the agglomerates. Hence, the porous sphere agglomeration model is numerically very efficient and suitable to be applied to simulations of industrial-scale processes.

A similar porous-sphere approach was used in the work of Breuer and Almohammed (2015), where the agglomerate is considered to have a closed-packed sphere diameter with fixed porosity (i.e. the porosity is not changing during agglomerate growth). It is, however, not made clear whether these diameters were also used in the calculation of the fluid dynamic forces on the agglomerates and the determination of the collision cross-section. This more advanced model does not enable the determination and prediction of agglomerate structure parameters or porosity, which will be essential for obtaining information on the product quality, but also the correct fluid forces and collision cross-section.

The most complex and advanced agglomeration model, which is depicted in the right part of Figure 6.15, allows for the estimation of the complete structure of agglomerates through statistical approaches. From an initial primary particle position, location vectors are stored for each new primary particle being collected by the agglomerate. This information is carried throughout the flow field for all produced Lagrangian agglomerates. Here, it is possible to obtain a realistic agglomerate size from the convex hull volume, agglomerate porosity, effective particle density, and other properties for agglomerate characterization, such as the fractal dimension or the gyration diameter, as used by Sommerfeld and Stübing (2017). It should be emphasized that the agglomerate is treated as a point-particle. Such an agglomerate structure model may be combined with different short-range particle interaction forces and also accounts for a possible penetration of highly viscous droplets or mushy particles (Blei and Sommerfeld, 2007).

The most important step in modelling the agglomeration of two primary particles or a primary particle with primary particles in an already existing agglomerate is to decide under which conditions sticking or rebound will occur. Recently, modelling the occurrence of agglomeration was classified in energy-based and momentum-based approaches (Almohammed and Breuer, 2016a). One version of the energy-based model stems from studies related to filtration where the single-fiber separation is calculated as the product of impact efficiency and adhesion probability. The adhesion probability was estimated applying an energy balance considering kinetic energy of impact, dissipated energy (i.e. due to plastic deformation), and van der Waals energy (Hiller, 1981; Löffler and Muhr, 1972). From this, a limiting upper impact velocity for sticking was derived; higher velocities result in rebound. The model was first integrated into Euler/Lagrange calculations by Ho and Sommerfeld (2002) for calculating the collision and agglomeration processes of dispersed solid particles in a turbulent flow. The collision detection was based on the stochastic collision model of Sommerfeld (2001), in which the collision process was eventually calculated in a transformed coordinate system where the larger particle or collector is stationary. In this configuration, the impact point was stochastically determined by considering the influence of impact efficiency. Hence, the normal component of the relative velocity vector was compared with the critical maximum impact velocity for agglomeration derived from the energy balance (Sommerfeld and Lain, 2009). In this model, particle rotation was neglected since turbulent flows with very small particles were considered (e.g. particle separation in cyclones), where such low inertia particles immediately rotate with the flow after some induction of rotation. Hence, relative rotation between colliding particles should be negligibly small.

Considering a primary particle hitting another primary particle within an existing agglomerate in a dry gas, it is possible to derive a critical velocity, below which sticking occurs from an energy balance. It is assumed that the agglomerate is stationary, and the primary particle moves toward the agglomerate (i.e. velocities must be transformed accordingly). Hence, the kinetic energy of particle 1 before collision, E_{ki}, is set equal to the possible rebound kinetic energy, E_{kr}; the dissipated energy, E_d; and the energy due to van der Waals attraction, ΔE_{vdW}, yielding (Hiller, 1981)

$$E_{ki} = E_{kr} + E_d + \Delta E_{vdW} \tag{6.72}$$

Only if zero kinetic energy is available for the rebound (i.e. $E_{kr} = 0$), the two particles are sticking together. The relative kinetic energy, E_{ki}, for the moving primary particle is derived by using the relative velocity from

$$E_{ki} = \frac{\pi}{12} \rho_p \, d_{p1}^3 \, u_{rel}^2 \tag{6.73}$$

The energy dissipated during the collision process may be expressed in terms of an energy restitution ratio, which is a measure of the irreversible deformation of the two involved spherical particles:

$$k^2 = \frac{E_{ki} - E_d}{E_{ki}} \tag{6.74}$$

Assuming that the deformation of both particles during collision is very small compared to the particle sizes (diameters $d_{p,1}$ and $d_{p,2}$), the van der Waals energy for the separation of two planar surfaces, which is shown in Figure 6.16, follows from the

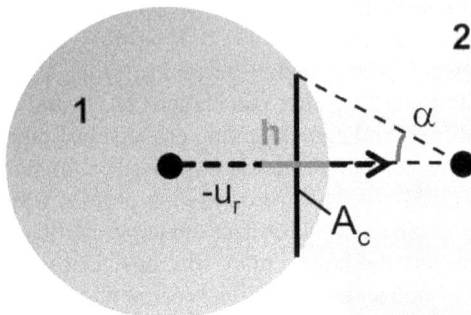

FIGURE 6.16 Collision configuration for a smaller particle (1) with diameter $d_{p,1}$ hitting a larger particle (2) with diameter $d_{p,2}$ and deformation of both particles with contact area A_c. In this geometry, the particle (2) is assumed to be stationary so that particle (1) hits with the relative velocity u_r.

adhesion force per unit area (the adhesion pressure) and the deformation area, A_c, with the radius a:

$$\Delta E_{vdW} = \frac{H \, A_c}{12 \, \pi \, z_0^2} \tag{6.75}$$

where H is the Hamaker constant and z_0 the minimum contact distance (values are provided in the next section). Under the assumption of small deformations, the deformation area may be approximated as $A_c = \pi \, d_{p,1} \, h_{d,1} = \pi \, d_{p,2} \, h_{d,2}$, where the flattening of both particles $h_{d,1}$ and $h_{d,2}$ is determined from the definition of plastic deformation energy (Eq. 6.74). The energy of the plastic deformation for ductile materials is derived from the integration of the yield strength p_y and the contact area over the deformation depth and eventually yields for $h_{d,1}$:

$$E_d = \frac{\pi}{2} p_y \left(d_{p,1} \, h_{d,1}^2 + d_{p,2} \, h_{d,2}^2 \right) = \left(1 - k^2 \right) E_{ki} \tag{6.76}$$

$$h_{d,1} = d_{p,1} \, u_{rel} \left(\frac{\left(1 - k^2 \right) \rho_p}{6 \, p_y \left(1 + d_{p,1} / d_{p,2} \right)} \right)^{\!\! \frac{1}{2}} \tag{6.77}$$

Based on Eq. (6.72), which yields $k^2 \, E_{ki} = \Delta E_{vdW}$ for the case of $E_{kr} = 0$, and introducing Eq. (6.75) with $A_c = \pi \, d_{p,1} \, h_{d,1}$ and Eq. (6.77), the following critical velocity for agglomeration is obtained:

$$u_{crit} = \frac{\left(1 - k^2 \right)^{\frac{1}{2}}}{k^2} \cdot \frac{H}{\pi \, d_{p1} \, z_0^2 \sqrt{6 \, p_c \, \rho_p \left(1 + d_{p1} / d_{p2} \right)}} \tag{6.78}$$

Since the collision configuration also allows for oblique collisions between primary particles (see Figure 6.2), the normal component of the relative velocity obtained with the collision angle φ should be smaller than the critical velocity to yield agglomeration (Ho and Sommerfeld, 2002):

$$\left| \overrightarrow{u_{rel}} \right| \cos \varphi \leq u_{krit} \tag{6.79}$$

Several additional properties are needed for the calculation of the critical velocity. The minimum contact distance is normally set to the value 0.4 nm ($0.4 \text{\AA} 10^{-9}$ m) and k, the restitution ratio, depends on the degree of plastic deformation. The Hamacker constant, H, depends on the materials of the particles and the carrier fluid (Bergström, 1997). The yield pressure depends on the particle material and may be specified with 10^7 to 10^8 Pa for plastic materials and 5×10^8 Pa for steel (Hiller, 1981). For small glass beads, the yield pressure is 5×10^9 Pa.

Similar formulations of energy-based models for describing fine particle sticking on surfaces were derived in the frame of particle deposition on solid walls (e.g. Brach and Dunn, 1992; Dahneke, 1995; Thornton and Ning, 1998), which will be described in Chapter 7 as they relate to wall deposition.

The momentum-based agglomeration model is incorporated into the impulse equations of the standard inter-particle collision calculation to obtain the particle velocities after the rebound (Kosinski and Hoffmann, 2009, 2010). It is assumed that the normal component of the impulse consists of two contributions: (a) The part due to the mechanical deformation and (b) the part due to cohesive forces as proposed by Kosinski and Hoffmann (2009, 2010). The cohesive contribution is evaluated by introducing the duration of collision (Kosinski and Hoffmann, 2010) for fully elastic collisions. This assumption was relaxed by Breuer and Almohammed (2015), introducing the collision time for an inelastic collision. Both models (energy- and momentum-based) were compared by Almohammed and Breuer (2016a), who studied agglomeration of small mono-sized particles in a particle-laden vertical turbulent channel flow. The agglomeration rates were found to be relatively low so that the produced agglomerates only consisted of up to about five primary particles. Also, both models provided the same answers, and it was concluded that the momentum-based model is preferable because it includes fewer parameters.

An essential step in Lagrangian agglomeration modelling is the selection of an appropriate equivalent diameter to be used in the further tracking of the agglomerates through the flow field. This is, first, the agglomerate linear dimension or cross-section to be used in the calculation of the fluid dynamic forces, that is, drag force and/or transverse lift force. The drag force is normally calculated from a drag coefficient, which may depend on the particle shape, the outer surface structure, and the inner porosity. However, information on these dependencies are rather limited and appropriate correlations are not available.

The evaluation of the drag coefficient for different agglomerate structures (i.e. compact or dendritic agglomerates) with between, 20 and 50 primary particles using the Lattice-Boltzmann method (LBM) revealed that the volume equivalent diameter does not provide the correct drag coefficients (Dietzel and Sommerfeld, 2013). Instead, it was shown that the cross-section of the convex hull perpendicular to flow direction and the resulting diameter provides drag coefficients that are only slightly below the standard drag curve for the same diameter. In addition, it was found that the drag coefficient determined in this way is almost independent of agglomerate structure and orientation.

The effect of porosity on the drag coefficient of compact nearly spherical agglomerates consisting of between 400 and 1,400 primary particles was also determined by DNS computations using the LBM (Dietzel et al., 2016). For these agglomerate structures, the projected area of the agglomerate perpendicular to flow direction was found to be more suitable for the calculation of the drag coefficient. With increasing agglomerate porosity (from 30% to 80%), the drag coefficient showed a continuous decrease, and the values were lower than those for compact spherical particles with the same diameter, as to be expected (i.e. drag coefficient 7% smaller for e = 80%). The drag coefficient calculated from the simulated force using the volume equivalent sphere diameter was much larger than that predicted by the standard correlation for rigid spheres. For a Lagrangian tracking of agglomerates this implies that the use of the volume equivalent sphere diameter for drag force determination will generate lower values. For rather compact agglomerates with porosities up to about 80%, the error when using the drag coefficient of a compact sphere with the diameter of a sphere wrapping the agglomerate is relatively small, as shown by Dietzel *et al.* (2016).

The dispersion of agglomerates in decaying grid turbulence was analyzed by Stübing *et al.* (2011) by assuming different equivalent diameters of the agglomerate. The results revealed that the volume equivalent diameter causes the highest dispersion, whereas the volume equivalent diameter of the convex hull around the agglomerate gives much lower dispersion. The difference was found to be about 20%, which highlights the importance of selecting the correct equivalent diameter for agglomerate tracking.

In addition, the presumed equivalent diameter of the agglomerate is important for collision detection, since the inter-particle collision probability depends on the collision cross-section of the two colliding particles, that is, $\pi/4 \, (D_C + D_P)^2$, where D_C is the collector or agglomerate diameter, and D_P is the primary particle diameter. When tracking the agglomerate with the volume equivalent sphere diameter, the collision rate will be very low and will impede the growth rate of the agglomerates.

REFERENCES

Abrahamson, J., 1975, Collision rates of small particles in a vigorously turbulent fluid, *Chem. Eng. Sci.*, **30**, 1371–1379.

Allen, M.P. and Tildesley, D.J., 1989, *The Computer Simulation of Liquids*, Vol. 42, Oxford University Press, Oxford.

Almohammed, N. and Breuer, M., 2016a, Modeling and simulation of agglomeration in turbulent particle-laden flows: A comparison between energy-based and momentum-based agglomeration models, *Powder Technology*, **294**, 373–402.

Bergström, L., 1997, Hamaker constants of inorganic materials, *Advances in Colloid and Interface Science*, **70**, 125–169.

Blei, S. and Sommerfeld, M., 2007, CFD in drying technology-spray drying simulation, in *Modern Drying Technology: Volume 1 Computational Tools at Different Scales* (Eds. E. Tsotsas and A.S. Majumdar), WILEY-VCH, Weinheim, pp. 155–208.

Brach, R.M. and Dunn, P.F., 1992, A mathematical model of the impact and adhesion of microspheres, *Aerosol Science and Technology*, **16**, 51–64.

Bradley, R.S., 1932, The cohesive force between solid surfaces and the surface energy of solids, *The London, Edinburgh, and Dublin Philosophical Magazine and J. of Science*, **13**, 853–862.

Brasil, A.M., Farias, T.L., Carvalho, M.G. and Koyu, U.O., 2001, Numerical characterization of morphology of aggregated particles, *Aerosol Science*, **32**, 489–508.

Breuer, M. and Almohammed, N., 2015, Modeling and simulation of particle agglomeration in turbulent flows using a hard-sphere model with deterministic collision detection and enhanced structure models, *Int. J. of Multiphase Flow*, **73**, 171–206.

Crowe, C.T., Schwarzkopf, J.D., Sommerfeld, M. and Tsuji, Y., 2012, *Multiphase Flows with Droplets and Particles*, 2nd edition, CRC Press, Taylor & Francis Group, Boca Raton, U.S.A.

Cundall, P.A. and Strack, O.D.L., 1979, A discrete numerical model for granular assemblies, *Geotechnique*, **29**, 47–65.

Dahneke, B., 1995, Particle bounce or capture: Search for an adequate theory: I. Conservation-of-energy model for a simple collision process, *Aerosol Science and Technology*, **23**, 25–39.

Deen, N.G., van Sint Annaland, M., van der Hoef, M.A. and Kuipers, J.A.M., 2007, Review of discrete particle modeling of fluidized beds, *Chemical Engineering Science*, **62**, 28–44.

Derjaguin, B. and Landau, L. D., 1941, Theory of the stability of strongly charged lyophobic sols and of the adhesion of strongly charged particles in solutions of electrolytes, *Acta Phys. Chim.*, **14**, 633–662.

Derjaguin, B., Muller, V.M. and Toporov, Y.P., 1975, Effect of contact deformations on the adhesion of particles, *J. of Colloid and Interface Science*, **53**, 314–326.

Dietzel, M., Ernst, M. and Sommerfeld, M., 2016, Application of the Lattice-Boltzmann method for particle-laden flows: Point-particles and fully resolved particles, *Flow, Turbulence and Combustion*, **97**, 539–570.

Dietzel, M. and Sommerfeld, M., 2013, Numerical calculation of flow resistance for agglomerates with different morphology by the Lattice-Boltzmann Method. *Powder Technology*, **250**, 122–137.

Eggersdorfer, M.L and Pratsinis, S.E., 2012, The structure of agglomerates consisting of polydisperse particles, *Aerosol Science and Technology*, **46**, 347–353.

Ernst, M., Dietzel, M. and Sommerfeld, M., 2013, A lattice Boltzmann method for simulating transport and agglomeration of resolved particles, *Acta Mechanica*, **224**, 2425–2449.

Ernst, M. and Sommerfeld, M., 2012, On the volume fraction effects on inertial colliding particles in homogeneous isotropic turbulence, *J. of Fluids Engineering, Transactions of the ASME*, **134** (031302).

Ernst, M., Sommerfeld, M. and Lain, S., 2019, Quantification of preferential concentration of colliding particles in a homogeneous isotropic turbulent flow, *Int. J. Multiphase Flow*, **117**, 163–181.

Feng, Z.-G. and Michaelides, E.E., 2002, Inter-particle forces and lift on a particle attached to a solid boundary in suspension flow, *Physics of Fluids*, **14**, 49–60.

Feng, Z.-G. and Michaelides, E.E., 2003, Equilibrium position for a particle in a horizontal shear flow, *Int. J. Multiphase Flow*, **29**, 943–957.

Gmachowski, L., 1995, Mechanism of shear aggregation, *Water Research*, **29**, 1815–1820.

Hamaker, H., 1937, The London: van der Waals attraction between spherical particles, *Physica*, **4**, 1058–1072.

Hemph, R.M., van Wachem, B.G.M. and Almstedt, A.E., 2006, DEM modeling of hopper flows: Comparison and validation of models and parameters, *World Conference on Particle Technology*, Orlando, Florida.

Hertz, H., 1882, Über die Berührung fester elastischer Körper, *J. Reine Angew. Math.*, **92**, 156–171.

Hiller, R.B., 1981, *Der Einfluss von Partikelstoß und Partikelhaftung auf die Abscheidung in Faserfiltern*. Dissertation, Universität Karlsruhe, VDI-Verlag GmbH Düsseldorf.

Ho, C.A. and Sommerfeld, M., 2002, Modelling of micro-particle agglomeration in turbulent flow, *Chem. Eng. Sci.*, **57**, 3073–3084.

Ho, C.A. and Sommerfeld, M., 2005, Numerische Berechnung der Staubabscheidung im Gaszyklon unter Berücksichtigung der Partikelagglomeration, *Chemie Ingenieur Technik, Jg.*, **77**, 282–290.

Johnson, K.L., Kendall, K. and Roberts, A.D., 1971, Surface energy and the contact of elastic solids, *Proc. R. Soc. Lond.*, **A324**, 301–313.

Kosinski, P. and Hoffmann, A.C., 2009, Extension of the hard-sphere particle-wall collision model to account for particle deposition, *Physical Review E*, **79**, 061302.

Kosinski, P. and Hoffmann, A.C., 2010, An extension of the hard-sphere particle-particle collision model to study agglomeration, *Chemical Engineering Science*, **65**, 3231–3239.

Kuang, S.B., Chu, K.W., Yu, A.B., Zou, Z.S. and Feng, Y.Q., 2008, Computational investigation of horizontal slug flow in pneumatic conveying, *Industrial and Engineering Chemistry Research*, **47**, 470–480.

Lain, S. and Sommerfeld, M., 2019, Numerical prediction of particle erosion of pipe bends, *Advanced Powder Technology*, **30**, 366–383.

Lain, S. and Sommerfeld, M., 2020, Influence of droplet collision modelling in Euler/Lagrange calculations of spray evolution, *Int. J. Multiphase Flow*, **132**, 103392.

Lipowsky, J. and Sommerfeld, M., 2008, Influence of particle agglomeration and agglomerate porosity on the simulation of a gas cyclone, *Proceedings 6th Int. Conf. on CFD in Oil & Gas, Metallurgical and Process Industries*, Trondheim Norway, Paper No. CFD08–043, June.

Löffler, F. und Muhr, W., 1972, Die Abscheidung von Feststoffteilchen und Tropfen an Kreiszylindern infolge von Trägheitskräften, *Chem. Ing. Techn.*, **44**, (8).

Mandelbrot, B., 1967, How long is the coast of Britain? Statistical self-similarity and fractional dimension, *Science*, **156**, 636–638.

Marshall, J.S. and Li, S., 2014, *Adhesive Particle Flow: A Discrete-Element Approach*, Cambridge University Press, Cambridge.

Martinez-Lopez, F., Cabrerizo-Vilches, M.A. and Hidalgo-Alvarez, R., 2001, An improved method to estimate the fractal dimension of physical fractals based on the Hausdorff definition, *Physica A*, **298**, 387–399.

Maugis, D., 1992, Adhesion of spheres: The JKR-DMT transition using a Dugdale model, *J. of Colloid and Interface Science*, **150**, 243–269.

Michaelides, E.E., 2014, *Nanofluidics: Thermodynamic and Transport Properties*, Springer, Heidelberg.

Michaelides, E.E., 2017, Nanoparticle diffusivity in narrow cylindrical pores, *Int. J. of Heat and Mass Transfer*, **114**, 607–612.

Mindlin, R.D. and Deresiewicz, H., 1953, Elastic spheres in contact under varying oblique forces, *J. of Appl. Mech.*, **20**, 327–344.

Nelson, J.A., Crooks, R.J. and Simmons, S., 1990, On obtaining the fractal dimension of a 3D cluster from its projection on a plane-application to smoke agglomerates, *J. Phys. D: Appl. Phys.*, **23**, 465–468.

Nichols, G., Byard, S.J., Bloxham, M.J., Botterill, J., Dawson, N., Dennis, A., Diart, V., North, N.C. and Sherwood, J.D., 2002, A review of the terms agglomerate and aggregate with a recommendation for nomenclature used in powder and particle characterization, *J. of Pharmaceutical Sciences*, **91**, 2103–2109.

Oesterle, B. and Petitjean, A., 1993, Simulation of particle-to-particle interactions in gas-solid flows, *Int. J. Multiphase Flow*, **19**, 199–211.

O'Rourke, P.J., 1981, *Collective Drop Effects on Vaporizing Liquid Sprays*. Dissertation, Los Alamos National Laboratory, New Mexico.

Pischke, P., Kneer, R. and Schmidt, D.P., 2015, A comparative validation of concepts for collision algorithms for stochastic particle tracking, *Computers & Fluids*, **113**, 77–86.

Prahl, L., Hölzer, A., Arlov, D., Revstedt, J., Sommerfeld, M. and Fuchs, L., 2007, On the interaction between two fixed spherical particles, *Int. J. of Multiphase Flow*, **33**, 707–725.

Rüger, M., Hohmann, S., Sommerfeld, M. and Kohnen, G., 2000, Euler/Lagrange calculations of turbulent sprays: The effect of droplet collisions and coalescence, *Atomization and Sprays*, **10**, 47–81.

Saffman, P.G. and Turner, J.S., 1956, On the collision of drops in turbulent clouds, *J. Fluid Mech.*, **1**, 16–30.

Schuch, G. and Löffler, F., 1978, Über die Abscheidewahrscheinlichkeit von Feststoffpartikeln an Tropfen in einer Gasströmung durch Trägheitseffekte, *Verfahrenstechnik*, **12**, 302–306.

Sgrott Junior, O.L., 2019, *Influence of Inter-Particle Interactions on the Performance of Cyclone Separators*. Ph.D. Thesis, Fakultät für Verfahrens- und Systemtechnik, Otto-von-Guericke University Magdeburg.

Sgrott Junior, O.L. and Sommerfeld, M., 2019, Influence of inter-particle collisions and agglomeration on cyclone performance and collection efficiency, *Canadian J. Chemical Engineering*, **97**, 511–522.

Smoluchowski, M., 1916, Drei Vorträge über Diffusion, Brownsche Bewegung und Koagulation von Kolloidteilchen, *Physik. Zeitschr.*, **17**, 557–585.

Sommerfeld, M., 1996, *Modellierung und numerische Berechnung von partikelbeladenen turbulenten Strömungen mit Hilfe des Euler/Lagrange-Verfahrens. Habilitationsschrift.* Habilitation thesis, Universität Erlangen-Nürnberg, Shaker Verlag, Aachen.

Sommerfeld, M., 2001, Validation of a stochastic Lagrangian modelling approach for inter-particle collisions in homogeneous isotropic turbulence, *Int. J. of Multiphase Flows*, **27**, 1828–1858.

Sommerfeld, M., 2003, Analysis of collision effects for turbulent gas-particle flow in a horizontal channel: Part I. Particle transport, *Int. J. Multiphase Flow*, **29**, 675–699.

Sommerfeld, M., 2017, Numerical methods for dispersed multiphase flows, in *Particles in Flows* (Eds. T. Bodnár, G.P. Galdi and Š. Nečasová), pp. 327–396, Series Advances in Mathematical Fluid Mechanics, Springer Int. Publishing, New York, NY.

Sommerfeld, M. and Lain, S., 2009, From elementary processes to the numerical prediction of industrial particle-laden flows, *Multiphase Science and Technology*, **21**, 123–140.

Sommerfeld, M. and Stübing, S., 2012, Lagrangian modeling of agglomeration for applications to spray drying, *9th Int. ERCOFTAC Symposium on Engineering Turbulence Modeling and Measurements*, Thessaloniki, Greece, 6–8 June.

Sommerfeld, M. and Stübing, S., 2017, A novel Lagrangian agglomerate structure model, *Powder Technology*, **319**, 34–52.

Sommerfeld, M., van Wachem, B. and Oliemans, R., 2008, *Best Practice Guidelines for Computational Fluid Dynamics of Dispersed Multiphase Flows*. ERCOFTAC, ISBN 978-91-633-3564-8.

Sommerfeld, M. and Zivkovic, G., 1992, Recent advances in the numerical simulation of pneumatic conveying through pipe systems, in *Computational Methods in Applied Science* (Eds. Hirsch, et al.), pp. 201–212, First European Computational Fluid Dynamics, Brussels.

Sorensen, C.M. and Roberts, G.C., 1997, The prefactor of fractal aggregates, *J. of Colloid and Interface Science*, **186**, 447–452.

Stübing, S., Dietzel, M. and Sommerfeld, M., 2011, Modelling agglomeration and the fluid dynamic behaviour of agglomerates, *Proceedings of ASME-JSME-KSME Joint Fluid Engineering Conference, 2011 (AJK2011-FED)*, Hamamatsu, Shizuoka, Japan, July, Paper No. AJK2011–12025.

Sundaram, S. and Collins, L.R., 1997, Collision statistics in an isotropic particle-laden turbulent suspension. Part 1. Direct numerical simulations, *J. Fluid Mech.*, **335**, 75–109.

Tanaka, T. and Tsuji, Y., 1991, Numerical simulation of gas-solid two-phase flow in a vertical pipe: On the effect of inter-particle collision, in *Gas-Solid Flows* (Eds. D.E. Stock, Y. Tsuji, J.T. Jurewicz, M.W. Reeks and M. Gautam), FED-Vol. 121, pp. 123–128, ASME, New York, NY.

Thornton, C. and Ning, Z., 1998, A theoretical model for the stick/bounce behaviour of adhesive, elastic-plastic spheres, *Powder Technology*, **99**, 154–162.

Trefalt, G., Montes Ruiz-Cabello, F.J. and Borkovec, M., 2014, Interaction forces, heteroaggregation and deposition involving charged colloidal particles, *J. Phys. Chem. B*, **118**, 6346–6355.

Tsuji, Y., 1993, Discrete particle simulation of gas-solid flows, *KONA Powder and Particle J.*, **11**, 57–68.

Vanni, M., 2000, Approximate population balance equations for aggregation-breakage process, *J. Colloid and Interface Sci.*, **221**, 143–160.

Verwey, E.J.W. and Overbeek, J.T.G., 1948, *Theory of Stability of Lyophobic Colloids*, Elsevier Amsterdam, Holland.

Walton, O.R. and Braun, R.L., 1986, Viscosity, granular-temperature, and stress calculations for shearing assemblies of inelastic, frictional disks, *J. of Rheology.*, **30**, 949–980.

Zhang, J., Mi, J. and Wang, H., 2012, A new mesh-independent model for droplet/particle collision, *Aerosol Science and Technology*, **46**, 622–630.

Zhu, H.P., Zhou, Z.Y., Yang, R.Y. and Yu, A.B., 2007, Discrete particle simulation of particulate systems: Theoretical developments, *Chemical Engineering Science*, **62**, 3378–3396.

7 Particle-Wall Interactions

Particle-wall interaction directly affects the particle transport behavior in confined systems, such as pneumatic conveying, fluidized beds, and cyclone separators. In pneumatic conveying systems, the momentum loss of particles caused by inelastic wall collisions results in a reacceleration of the particles by the fluid after the rebound. Hence, momentum is transferred from the fluid phase to accelerate the particles causing the additional pressure loss (Michaelides, 1987; Sommerfeld and Kussin, 2004). The pressure loss depends on the average wall collision frequency or mean free path between subsequent particle-wall collisions. Several parameters determine the wall collision frequency in pneumatic conveying (Siegel, 1991)

- Particle mass loading
- Dimensions of the confinement (e.g. pipe diameter in pneumatic conveying)
- Particle response time or response distance
- Conveying velocity and turbulence intensity
- Particle shape and wall roughness
- The materials of the particles and the wall

An approximate estimate of the importance of particle-wall collisions is based on the ratio of the particle response distance, λ_P, to the dimension of the confinement, for example, the diameter of the pipe D (Sommerfeld, 1992). The particle response distance can be estimated from the following equation with the particle response time, τ_P

$$\lambda_P = \frac{\rho_P D_P^2}{18 \mu_F f} w_t \tag{7.1}$$

where w_t is the terminal velocity of the particles resulting from the force balance in the direction of gravity (i.e. gravity = drag plus buoyancy), and f is the friction factor, defined in Eq. (4.25). When λ_P is larger than the dimension of the confinement, D, the particles are not able to respond to the flow before they collide with the wall again; hence their motion is dominated by wall collisions. When $\lambda_P < D$, the particles adapt to the flow field before they collide with the wall again.

In addition to the mentioned effects, the wall collision process may be affected by short-range fluid dynamic interactions just before impact, which eventually causes a deceleration of the particle as described in section 4.7. This effect, however, is only of importance for viscous fluids and small particle Reynolds numbers. Such effects are very difficult to be experimentally determined but may be discerned with fully resolved direct numerical simulations that determine the wall corrections to the drag force (Feng and Michaelides, 2002a, 2002b; Zeng et al., 2009).

DOI: 10.1201/9781003089278-7

Particle-wall collisions in a confined multiphase system will have several different effects on individual particles and, by extent, on the bulk behavior of the particle
phase. In addition to the additional particle-phase-induced pressure drop, one will
have the following:

- Transfer between linear and rotational momentum, particle rotation and the
  associated magnus effect causes lateral displacement
- Possible breakage/fracture of the particles
- Erosion/deformation of the wall material
- Additional heat transfer during the particle-wall contact period

7.1 MOMENTUM AND ENERGY EXCHANGES

The hard-sphere model for the wall collision will be described, which implies that
the wall collision process is not temporarily resolved but treated as an instantaneous
process to obtain the particle properties after rebound. Particle and wall deformations are not explicitly resolved, but the dissipated energy is expressed through a
restitution coefficient. Coulomb's law of friction, which relates the wall normal and
the friction force (parallel to the wall), is assumed to hold for the sliding period of a
collision. For such an inelastic collision process, one may identify a compression and
a recovery period. The change of the particle's translational and rotational velocities
during the bouncing process can be calculated from the momentum equations of
classical mechanics. Three types of collisions are distinguished:

Type 1: The particle stops sliding in the compression period.
Type 2: The particle stops sliding in the recovery period.
Type 3: The particle continues sliding along the wall during the whole colli-
sion process.

The type of collision is determined by the static coefficient of friction, μ_s; the restitution ratio of the normal velocity components, e; and the velocity of the particle
surface relative to the contact point, u_{R0}. Since the temporal evolution of the normal contact force $\vec{F}_n$ is unknown, the integrated version of Newton's second law
is used to determine the change of the linear velocities (the three components are:
x—streamwise, y—wall-normal, z—lateral or sideways) from

$$m_p\left(\vec{u}_{p2} - \vec{u}_{p0}\right) = \int_{t_0}^{t} \vec{F}_n dt = \vec{J} \tag{7.2}$$

Similarly, the equation for the change of the angular impulse momentum is

$$\frac{I_p}{R_p}\left(\vec{\omega}_{p2} - \vec{\omega}_{p0}\right) = -\vec{n} \times \vec{J} \tag{7.3}$$

In Eqs. (7.2) and (7.3), m_p is the particle mass, I_p the moment of inertia (i.e.
$I_p = 0.1 m_p D_p^2$ for a sphere), $\vec{n}$ is the unit vector, and $\vec{J}$ the vector of the impulse.

For clarity, the collision process is separated into the compression and recovery period, respectively; the index 0 refers to the velocities before impact; and the indices 1 and 2 indicate the velocities at the end of the compression period and after full rebound, respectively. The normal restitution ratio is defined as the ratio of the normal impulses in the y—direction (i.e. recovery impulse to compression impulse):

$$e = \frac{J_{y,2}}{J_{y,1}} = \frac{J_{y,r}}{J_{y,c}} \tag{7.4}$$

where the subscripts r and c denote the rebound and compression period, respectively. Only if the particle is spherical the impulse ratio may be replaced by the velocity ratio so that the normal restitution ratio is given by

$$e_n = \frac{v_{p2}}{v_{p1}} \tag{7.5}$$

Similarly, it is also useful to define a tangential (in the x-z plane) as well as a total restitution ratio:

$$e_t = \frac{u_{p2}}{u_{p1}} \quad e_{tot} = \frac{|U_{p2}|}{|U_{p1}|} \tag{7.6}$$

Evaluating Eqs. (7.2) and (7.3) in connection with the appropriate boundary conditions, two sets of equations are obtained for sliding and non-sliding collisions, which allow the determination of the change of the linear and angular velocity of the particle during the wall collision process. A *non-sliding (sticking) wall collision* (defined for type 1 and 2 collisions) results by introducing Coulombs law of friction and asserting that the friction impulse—defined as static friction coefficient times the wall-normal impulse, $\mu_s J_n$—becomes larger than the wall tangential impulse, J_t

$$|J_t| \le \mu_s J_n \tag{7.7}$$

with: $J_t = \sqrt{J_x^2 + J_z^2} \quad J_n = J_y$

$$|u_{R0}| \le \frac{7}{2} \mu_s (1+e) v_{P0} \tag{7.8}$$

$$u_{R0} = \sqrt{\left(u_{P0} + \frac{D_P}{2} \omega_{P0}^z\right)^2 + \left(w_{P0} - \frac{D_P}{2} \omega_{P0}^x\right)^2} \tag{7.9}$$

where, u_p, v_p, and w_p are the translational velocity components in the x-, y- and z-direction and ω_P^x, ω_P^y, and ω_P^z are the angular velocity components of the particle around the axis of the coordinate system shown in Figure 7.1. The subscripts 1 and 2 refer to the conditions before and after collision, respectively. For the

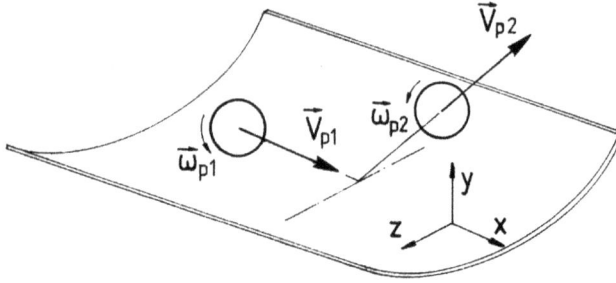

FIGURE 7.1 A spherical particle colliding with a pipe wall with initial translational and rotational velocity.

non-sliding collision, the change of particle velocities is obtained by the following set of equations

$$u_{P2} = \frac{5}{7}\left(u_{P0} - \frac{D_P}{5}\, \omega_{P0}^z\right)$$

$$v_{P2} = -e\, v_{P0}$$

$$w_{P2} = \frac{5}{7}\left(w_{P0} + \frac{D_P}{5}\, \omega_{P0}^x\right) \tag{7.10}$$

and

$$\omega_{P2}^x = \frac{2\, w_{P2}}{D_P}$$

$$\omega_{P2}^y = \omega_{P0}^y$$

$$\omega_{P2}^z = -\frac{2\, u_{P2}}{D_P} \tag{7.11}$$

The collision type 3 is the so-called full *sliding collision*, which occurs for

$$|u_{R0}| \geq \frac{7}{2}\, \mu_s\, (1+e)\, v_{P0} \tag{7.12}$$

The change of translational and rotational velocities throughout the *sliding collision* is obtained by the following sets of equations

$$u_{P2} = u_{P0} + \mu_d\ \varepsilon_x\, (1+e)\, v_{P0}$$

$$v_{P2} = -e\ v_{P0}$$

$$w_{P2} = w_{P0} + \mu_d\ \varepsilon_z\, (1+e)\, v_{P0} \tag{7.13}$$

and

$$\omega_{P2}^x = \omega_{P0}^x - 5\,\mu_d\,\,\varepsilon_z\,(1+e)\frac{v_{P0}}{D_P}$$

$$\omega_{P2}^y = \omega_{P0}^y$$

$$\omega_{P2}^z = \omega_{P0}^z + 5\,\mu_d\,\,\varepsilon_x\,(1+e)\frac{v_{P0}}{D_P} \tag{7.14}$$

In Eqs. (7.13) and (7.14), the terms ε_x and ε_z determine the direction of the motion of the particle surface with respect to the wall:

$$\varepsilon_x^2 + \varepsilon_z^2 = 1 \begin{cases} \varepsilon_x = \left(\dfrac{u_{P1} + \dfrac{D_P}{2}\,\omega_{P1}^z}{u_{R1}}\right) \\[4mm] \varepsilon_z = \left(\dfrac{w_{P1} - \dfrac{D_P}{2}\,\omega_{P1}^x}{u_{R1}}\right) \end{cases} \tag{7.15}$$

In the preceding equations, μ_s and μ_d are the static and dynamic coefficients of friction, and e_n is the normal restitution coefficient. These parameters are not only dependent on the material of particle and wall but also on impact velocity and angle (Sommerfeld and Huber, 1999). Thereby the deformation of particle and wall is affected and, hence, the degree of energy dissipation. Based on numerous experiments for different particle sizes and wall materials, bi-linear correlations were suggested for the dynamic coefficient of friction, μ_d, and the coefficient of restitution, e, as a function of the impact angle, α_1 (expressed in degrees), as shown in Figure 7.2. The experimental results show that the normal restitution coefficient decreases linearly from one to a constant value for a wide range of larger impact angles. The estimated transition angle is labelled as α_e, and the restitution ratio for larger particle impact angles, α_e, may be assumed to be roughly constant for most of the wall materials, as illustrated in Figure 7.3a. When the wall material becomes "softer," the normal

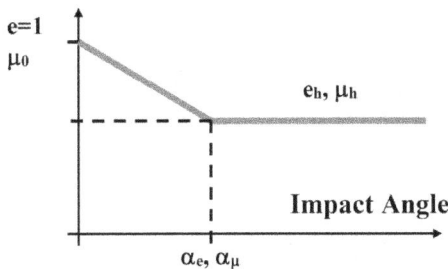

FIGURE 7.2 Functional relationship for a bi-linear dependence of the normal restitution ratio and the friction coefficient on the impact angle (Sommerfeld and Huber, 1999).

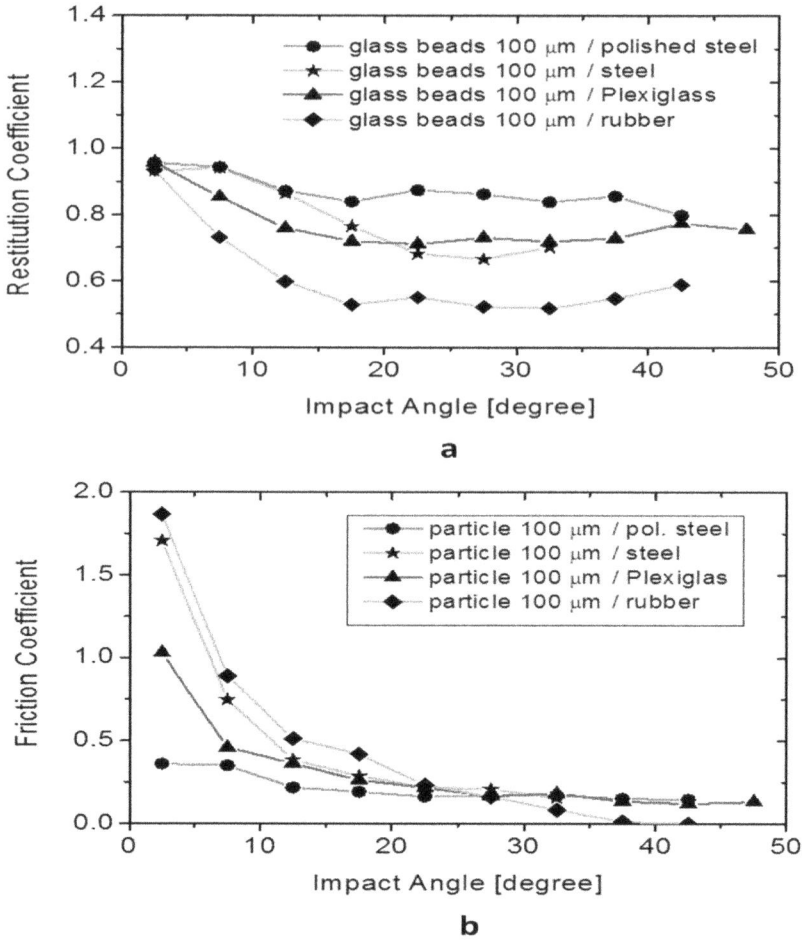

FIGURE 7.3 Measured dependence of the restitution coefficient (a) and the friction coefficient (b) on impact angle for a spherical glass bead of 100 μm hitting different wall materials in a horizontal channel flow; averaged velocity magnitude between 10–14 m/s.

Source: From Sommerfeld and Huber, 1999.

restitution coefficient decreases. Note that in the conducted channel flow experiments (Sommerfeld and Huber, 1999), wall impact angles larger than 40° could not be realized. A similar bi-linear dependence of the friction coefficient on the particle impact angle, α_1, may be identified from Figure 7.3b. Hence for both particle-wall collision parameters, the following two conditions may be applied:

$$e_n = max\left\{e_h,\, 1 - \frac{1-e_h}{\alpha_e}\alpha_1\right\} \qquad (7.16)$$

$$\mu_d = max\left\{\mu_h,\, \mu_0 - \frac{\mu_0 - \mu_h}{\alpha_\mu}\alpha_1\right\} \qquad (7.17)$$

TABLE 6.1
Summary of Experimental Conditions and Parameters for the Wall Collision Model, Including Definitions for the Bi-linear Distribution (Figure 7.2) Suggested for the Normal Restitution Ratio and the Friction Coefficient (- indicates that no data could be defined)

Particles	Wall Material	U_{p1} [m/s]	U_{p2} [m/s]	e_h	α_e	μ_0	μ_h	α_μ
Glass 100 μm	Steel polished	12.82	11.52	0.9	22°	0.4	0.15	20°
Glass 100 μm	Steel	13.32	11.34	0.7	22°	0.5	0.15	20°
Glass 100 μm	Plexiglas	9.75	8.33	0.73	18°	0.4	0.15	27°
Glass 100 μm	Rubber	11.31	9.16	0.5	18°	0.8	0.02	35°
Glass 500 μm	Steel polished	5.56	4.93	0.75	15°	0.35	0.1	25°
Glass 500 μm	Steel	5.91	5.15	0.7	22°	0.4	0.15	20°
Quartz 100 μm	Steel polished	16.04	13.37	0.55	27°	–	–	–
Quartz 100 μm	Steel	15.02	12.05	–	–	–	–	–
Quartz 100 μm	rubber	14.12	10.86	0.4	27°	–	–	–

A set of parameters included in Eq. (7.16) and (7.17) was suggested for spherical glass beads with sizes between 100 and 500 μm and non-spherical 100 μm quartz particles hitting different wall materials in Table 6.1 (Sommerfeld and Huber, 1999).

Exemplarily, measurement results for the restitution coefficient and the dynamic friction coefficient in dependence of impact angle are shown in Figure 7.3 for the 100 μm glass beads and different wall materials. The different wall materials considered have also a range of surface roughness, from very low for Plexiglass to around 25 μm for the stainless steel. The resulting effects were corrected for the distribution of the restitution coefficient by introducing the assumption that $e = 1$ for zero impact angle (Figure 7.2). The coefficient of friction was derived with the following equation:

$$\mu = \frac{\left|u_{P1} - u_{p2}\right|}{\left(1 + e\right) v_{p1}} \tag{7.18}$$

With this approach, no difference is made between the static and dynamic coefficient of friction. The velocity magnitude of 100 μm particles hitting the wall was between 10–14 m/s, for 500 μm glass beads around 6 m/s and for 100 μm quartz sand between 14–16 m/s.

7.2   WALL ROUGHNESS EFFECTS AND IRREGULAR BOUNCING

The wall collisions of particles in confined flows alone will introduce a momentum loss to the particle phase, and at higher particle mass loading also, an additional pressure loss in a pneumatic conveying system will result (Michaelides, 1987). However, the process of particle-wall impact will be remarkably different in cases where the particles are non-spherical (Sommerfeld, 2002; Quintero *et al.*, 2021) or when the wall surface has a certain roughness structure resulting from the material processing (Sommerfeld and Kussin, 2004). In industrial equipment, such as pneumatic conveying lines and cyclone separators, mostly stainless steel material is used, which has a mean roughness height between 10 and 50 μm, depending on the manufacturing process. In addition, the wall surface may be modified during the processes through particle erosion, resulting in a temporal variation of the wall roughness (Novelletto Ricardo and Sommerfeld, 2020).

7.2.1 Modelling Approaches for Irregular Bouncing

Experiments have shown that the wall roughness remarkably increases the wall collision frequency, and thereby the additional pressure drop due to the particle phase (Sommerfeld, 2003; Sommerfeld and Kussin, 2003; Sommerfeld and Kussin, 2004; Lain and Sommerfeld, 2008; Sommerfeld and Lain, 2015). In addition, in rather narrow confinements, the particle velocity fluctuation is enhanced considerably (Sommerfeld and Kussin, 2004), an effect which is not induced by turbulence. For understanding the particle behavior in confined flows and for an accurate numerical prediction, a model for particle-wall collisions should additionally consider the wall roughness and the resulting stochastic nature of the particle rebound process. The influence of surface roughness on the wall collision process depends on the roughness structure, which is a result of the manufacturing process of the equipment (Volk, 2005) and the range of particle sizes considered (Sommerfeld and Lain, 2009). The wall roughness may be characterized by a number of parameters, but the most important ones for model derivation are the mean roughness depth (H_r, which is twice the arithmetic mean roughness value R_a, DIN EN ISO 4287) and the mean cycle of roughness, L_r, as shown in Figure 7.4. When a small particle with a diameter less than the cycle of roughness ($D_P < L_r$) is considered, one may estimate the maximum change of the collision angle due to roughness by the expression:

$$\gamma_{max} = \frac{2H_r}{L_r} \tag{7.19}$$

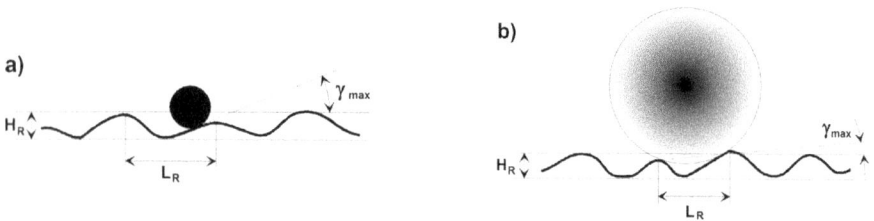

FIGURE 7.4   Wall roughness effect for (a) small and (b) large particles.

On the other hand, for particles larger than the cycle of roughness, the maximum roughness angle is much lower, as it is apparent in Figure 7.4. Assuming that the smallest roughness height is about $H_r/2$, the maximum roughness angle is given as

$$\gamma_{max} = \frac{H_r}{2 L_r} \tag{7.20}$$

The roughness height for a machined metal plate lies typically in the range of 5–20 μm, and the mean distance between the roughness peaks (i.e. roughness cycle) is about 10–20 times larger. This yields a maximum roughness angle of $\gamma_{max} = \pm 11°$ to $\pm 5.7°$ for small particles and $\gamma_{max} = \pm 3°$ to $\pm 1.4°$ for large particles. Consequently, the wall roughness will affect the wall collision process much more for smaller particles (i.e. smaller than about 100 μm in a gas flow) than for large particles where the effective roughness angle is much lower. Hence, the effect of wall roughness will depend on both the roughness structure and the particle diameter. Such a correlation needs to be known in order to model the wall roughness effect properly. In addition, small particles will, after departure from the wall, quickly follow the fluid flow, whereas large particles may cover several roughness structures and therefore feel "less" of the roughness since they have large inertia and maintain the velocity change due to roughness for much longer periods. This eventually causes the wall roughness to be more important for the bulk behavior of larger particles in a flow (Sommerfeld, 1992, 1996).

One of the first models that considers the influence of the wall roughness and the effect of particle shape on particle-wall bouncing was proposed by Matsumoto and Saito (1970a, 1970b) and is based on a stochastic treatment of the collision process. In these studies, the rough wall was represented by a sinusoidal shape, where A_r is the amplitude of the waves, and L_r is the cycle of roughness. The phase angle of the roughness (α_r) was randomly sampled from a uniformly distributed random number in the range $[0, 2\pi]$ to avoid a correlation between the particle collision point and the surface roughness. For the calculation of the collision process, the impulse equations given in the previous section were applied after a transformation of the particle velocities to a coordinate system, which is aligned with the statistically sampled local wall (roughness) inclination. The comparison of numerical simulations with experimental results, obtained in a horizontal two-phase channel flow using 500 μm glass beads showed reasonable agreement for the particle velocity and concentration profiles at $L_r/D_p = 10$ and $A_r/D_p = 1/40$. The dynamic coefficient of friction was assumed to be $\mu_d = 0.4$ and the coefficient of restitution to be $e_n = 0.97$. The experiments of Matsumoto and Saito (1970b) were, however, conducted in a channel made of glass plates that are perfectly smooth. Therefore, their model simulates the effect of slight non-sphericities in the particle shape rather than wall roughness, which also explains the choice of the model parameters. Tsuji et al. (1985, 1987) considered the particle-wall collision for a pipe flow in a different way. Their "abnormal bouncing" model states that the plane wall is replaced by a virtual wall when the particle collision angle is below a certain shallow value. This method was introduced to eliminate particle settling in horizontal conveying, so as to conform with experimental observations. The virtual wall increases the impact collision angle α_1 by a value γ, which is given by the following equations:

$$\gamma = \begin{cases} -\delta(\alpha_1 - \beta) \cdots \cdots (\alpha_1 \leq \beta) \\ 0 \qquad\qquad \cdots \cdots (\alpha_1 > \beta) \end{cases} \tag{7.21}$$

$$\delta = \frac{2.3}{Fr} - \frac{91}{Fr^2} + \frac{1231}{Fr^3} \tag{7.22}$$

$$Fr = \frac{u_P}{\sqrt{gh}}; \quad \beta = 7° \tag{7.23}$$

where Fr is the Froude number, u_P the particle velocity, h is the pipe diameter or channel height, and g is the gravitational constant. These empirical correlations were derived from two-phase pipe flow experiments. The randomness of the bouncing process is obtained by additionally introducing a randomly distributed yaw angle, which is confined within certain limits. This angle determines the angle of reflection in the pipe cross-section. Irregularities in the bouncing process for a horizontal channel flow (Tsuji et al., 1987) were introduced through a randomly distributed coefficient δ, according to $\delta = cR^k\delta_0$, where c = 5, k = 4, δ_0 is obtained from Eq. (7.20), and R is a random number in the range [0, 1]. In the numerical simulations of Tsuji et al. (1987), the model described by these equations yields reasonably good agreement with experimental data for the particle concentration and velocity, considering the lower gas bulk velocities (e.g. U_0 = 7 m/s). In the case of higher velocities (e.g. U_0 = 15 m/s), the particle concentration profile obtained by the numerical simulation showed a large peak near the bottom of the channel, which was not observed in the experiments. However, a major drawback of the model introduced by Tsuji et al. (1985, 1987) is that it does not account for roughness effects in the collision process for collision angles larger than α_1 = 7°. Hence, it does not reproduce the physical effects of such collisions since the roughness effect is always present regardless of the collision angle. Only the consequences for particle rebound are different.

7.2.2 WALL ROUGHNESS NORMAL PDF MODEL

This model, introduced by Sommerfeld (1992), differs from the previous, in that a random inclination of the local wall with respect to the particle trajectory is assumed for all collision angles. The model is based on the assumption that the instantaneous particle impact angle, α_1, is composed of the particle trajectory angle, α_0, plus a stochastic contribution due to the wall roughness, which is modelled as follows

$$\alpha_1 = \alpha_0 + \Delta\gamma\,\xi \tag{7.24}$$

where $\Delta\gamma$ is the standard deviation of the roughness angle distribution and depends on the roughness structure itself and on the considered particle size. For validation, the roughness induced stochastic contribution, $\Delta\gamma\,\xi$, was initially created from different distributions with certain constraints, namely, a uniformly distributed probability [-0.5, 0.5], a Gaussian distribution of the wall roughness, and a sinusoidal roughness distribution at the wall according to Matsumoto and Saito (1970b). The best agreement with the results of Grant and Tabakoff (1975) was achieved with the normal roughness distribution function (Sommerfeld, 1992). This was one of the first studies emphasizing the importance of wall roughness on inertial particle velocity fluctuations, which was remarkably enhanced for inertial particles. The proposed modelling approach initiated further experimental studies and a further refinement and validation of the wall roughness model, especially the important concept of the *shadow effect*, introduced by Sommerfeld and Huber (1999). The effect implies that, depending on the particle trajectory, not all roughness structures may be reached by the point-particle, for example, the lee-side of roughness hills as it is apparent in Figure 7.5. Only for particle impact angles larger than the magnitude of γ is this possible. Therefore, the PDF (probability density function) of the assumed roughness angle, being symmetric around zero (second term of Eq. 7.24), would effectively not be completely seen by the particle. As a result, the roughness angle distribution would show a cutoff if the statistically drawn roughness angle would be $\gamma < -\alpha_1$ (Sommerfeld and Huber, 1999). Naturally, this phenomenon generates a shift of the effective roughness angle distribution to larger values, and eventually a positive mean value of the roughness angle will result. This effect decreases with increasing particle-to-wall impact angle, and hence, the roughness mean angle is diminished and approaches zero for larger particle impact angles, α_0. The effective standard deviation of the roughness angle distribution is reduced for shallow impact angles and approaches (for increasing α_0) the originally assumed standard deviation, $\Delta\gamma$. For an assumed normal distribution, the shadow effect model was analyzed and validated through experiments (Sommerfeld and Huber, 1999). The shadow effect (at small wall impact angles) yields a noticeable transfer of horizontal to transverse particle momentum associated with a strong increase of particle fluctuation velocity and, therefore, wall collision frequency in channel flows.

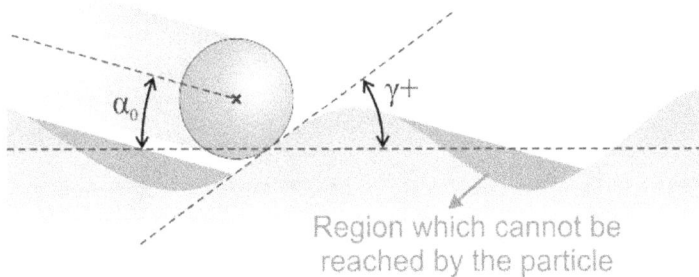

FIGURE 7.5 The *shadow effect* for small impact angles of particles showing that the lee side of a roughness structure cannot be reached by the particle.

Only if these details are incorporated in the wall collision model, it can be expected to correctly predict particle velocities and the pressure drop (Lain and Sommerfeld, 2008).

An experimental analysis of several surface roughness structures suggests that the wall roughness angle is almost normally distributed, and the standard deviation of this distribution depends on the scanning distances used for the surface profiler (Sommerfeld and Huber, 1999). This scanning distance is closely related in some way to the particle size (see Figure 7.4). As a result, the distribution of the roughness angle is given by

$$p(\gamma) = \frac{1}{\sqrt{2\pi \Delta\gamma^2}} \exp\left(\frac{\gamma^2}{2\Delta\gamma^2}\right) \tag{7.25}$$

where the random process ξ in Eq. (7.24) is based on a Gaussian distribution with zero mean and a standard deviation of unity. The value of the roughness angle, $\Delta\gamma$, depends mainly on the structure of wall roughness and the particle size. The remaining task is to build correlations between the roughness angle, $\Delta\gamma$, and the particle size. From experiments (Sommerfeld and Kussin, 2004) and several numerical simulations (Lain and Sommerfeld, 2008), comparing the simulated pressure drop with measurements, such correlations have been built (Sommerfeld and Lain, 2009), as shown in Figure 7.6. A particle-laden channel flow was considered in these experiments with interchangeable wall plates having three degrees of roughness (a polished steel wall has a mean roughness height H_r of 2.3 μm, whereas a rolled stainless steel sheet gave a roughness height of 17 μm). Different spherical glass beads with a range of diameters between 60–625 μm were considered, providing a large set of experimental data for the validation of the numerical simulations. The

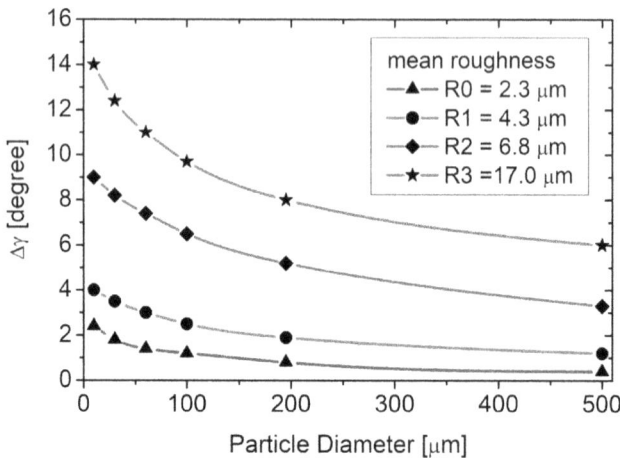

FIGURE 7.6 Correlations of roughness standard deviation, $\Delta\gamma$, with the particle diameter, developed from experiments and numerical simulations for different degrees of wall roughness.

dependence of the roughness angle, $\Delta\gamma$, on the particle size continuously decreased for each degree of roughness and approached a lower limiting value for very large particles.

The standard deviation of the roughness angle was only measured for the 100 and 500 μm glass beads (Sommerfeld and Huber, 1999). Therefore, the correlations that are shown in Figure 7.6 were further developed to allow for the proper consideration of the roughness angle as a function of particle size. Two degrees of roughness correspond to the measurements performed by Kussin and Sommerfeld (2002). The correlations for both degrees of roughness are given as follows (LR: low roughness R1 = 4.8 μm; HR: high roughness R2 = 6.8 μm):

$$\Delta\gamma_{HR} = 3.4963 + 5.797 \cdot \exp\left(-\frac{D_P}{154.12}\right) \tag{7.26}$$

$$\Delta\gamma_{LR} = 1.551 + 3.438 \cdot \exp\left(-\frac{D_P}{161.55}\right) \tag{7.27}$$

The other two curves were determined from experiments with very smooth and high roughness walls (Sommerfeld and Kussin, 2004). With these data the effect of wall roughness for the wall collisions of spherical particles may be described for a wide range of particle diameters and wall roughness. For non-spherical particles colliding with walls of the same roughness degree, the model needs to be extended for capturing the specific phenomena related to non-spherical particles (Quintero et al., 2021).

Consequently, the procedure to model the wall collisions of spherical particles with rough walls is as follows:

1. A wall collision occurs if the trajectory of a point-particle crosses the plane wall boundary. The finite particle size should be considered for obtaining the actual contact point.
2. The appropriate instantaneous roughness angles are obtained from the presumed normal distribution function with the standard deviation, $\Delta\gamma$. It should be noted that this roughness angle is located in the plane of the mainstream flow direction and the wall normal direction.
3. The three-dimensionality of the wall roughness is considered by turning the collision plane around the vertical axis by an independent stochastic process. Since no preferential direction of the roughness was assumed, this angle was sampled from a uniform distribution in the range [-π, π], as described in Sommerfeld and Zivkovic (1992) and Radenkovic and Simonin (2018).
4. The particle velocities are then transformed to a coordinate system, which is aligned with the new collision plane accounting for the roughness angle. The translational and rotational velocities of the particles after rebound are calculated, according to the equations provided in the previous section.
5. The particle rebound velocities are re-transformed to a Cartesian coordinate system aligned with the flow system or the pipe axis.

It should be noted that the particle collision angles in cylindrical or rectangular channels are usually very small, that is, below about 20° (Sommerfeld, 1992). Therefore, it is possible that, when a negative roughness angle is drawn from the normal distribution function, the resulting effective wall collision angle becomes negative as shown in Figure 7.5, that is, the particle hits the lee-side roughness structure from the wall side. This is an impossible situation, and a new roughness angle is drawn from the normal distribution function, allowing for positive roughness angles γ. Such situations especially occur during pneumatic conveying (Lain and Sommerfeld, 2008) and cyclone separators (Sgrott Junior and Sommerfeld, 2019).

The effective particle-wall impact angle is very small. After the rebound and including the momentum loss, the particles are not able to leave the near wall region and are not fully dispersed into the flow. In order to capture this effect, Konan *et al.* (2009) proposed a stochastic approach to account for multiple rebounds during the interaction of particles with rough walls. Particles may collide with a subsequent rough structure with a certain probability only when the rebound angle from the first collision is greater than zero. The higher the resultant angle from the first collision, the smaller is the probability of the particle to collide with another roughness structure. In order to describe this effect, Konan *et al.* (2009) proposed the following analytic correlation, which is the probability that only one rebound is occurs

$$P = \begin{cases} \left| tanh\left(1.5 \dfrac{\alpha_2}{\Delta\gamma} \right) \right| & \text{if } \alpha_2 \geq 0 \\ 0 & \text{if } \alpha_2 \leq 0 \end{cases} \tag{7.28}$$

where α_2 is the rebound angle of the particle, and the $\Delta\gamma$ is standard deviation of the wall roughness structure. Another particle-wall interaction takes place if the probability, P, becomes smaller than a uniform random number in the range of [0, 1]. In such a case, a new wall roughness angle is drawn from the distribution.

The effect of wall roughness on particle transport in a narrow channel is demonstrated in Figure 7.7 (Sommerfeld, 2003; Sommerfeld and Kussin, 2003), which is analogous to Figure 6.5 for inter-particle collisions. The upper graphs of Figures 7.7a and 7.7b are without considering wall roughness and the lower graphs with wall roughness. The height to length ratio of the channel has been considerably changed in the graphs for clarity—the channel length is 6 m, and the height is 35 mm. The results demonstrate that small particles (30 µm, Figure 7.7a) are slightly affected by turbulence (fluctuations on the particle trajectories), whereas the motion of larger particles is completely governed by inertia. Considering an integral timescale of turbulence in the central region of the channel ($T_L = 3.8$ ms), the Stokes numbers of small and large particles are 1.4 and 11.2, respectively. It is obvious in Figure 7.7 that the wall roughness has a strong effect on the particle motion, especially for larger particles, whereby they bounce from wall to wall. It is noted that the horizontal distance between subsequent wall collisions is approximately 0.5 m in the case of roughness (Figure 7.7 b). Also, that the small particles are better dispersed when the wall roughness effects are accounted.

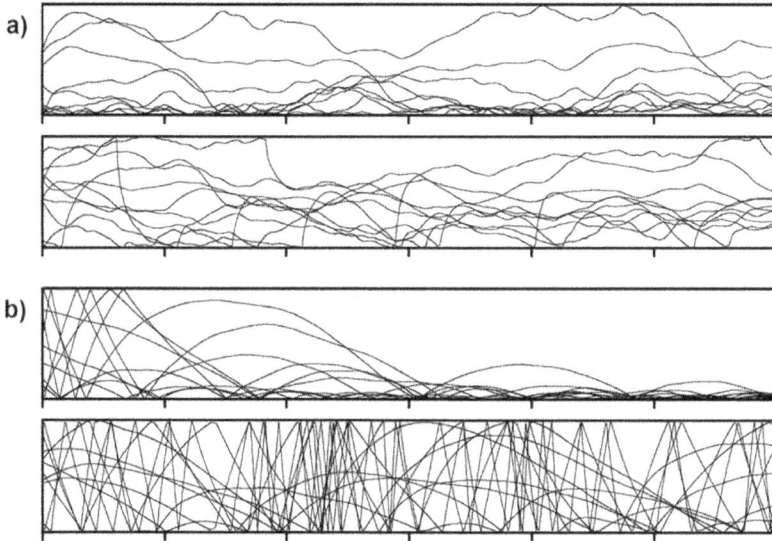

FIGURE 7.7 Calculated particle trajectories in a horizontal channel flow ($H = 35$ mm, $L = 6$ m) illustrating the effect of wall roughness: (a) 30 μm particles, (b) 110 μm particles (bulk velocity $U_{av} = 18$ m/s). The upper part calculations are in the absence of wall roughness, and the lower parts include wall roughness.

7.3 PARTICLE DEPOSITION AND WALL ADHESION

Particle adhesion and deposit formation on walls or other devices and inserts within engineering facilities is mostly an unwanted phenomenon, which may modify flow structures, increase pressure drop, reduce heat transfer, und eventually also lead to clogging. Depending on the kind of continuous phase (i.e. gas or liquid), different types of adhesion forces may be responsible for particle capturing at walls. In a gaseous environment, van der Waals and electrostatic forces are important, but at elevated humidity, liquid adsorption layers may also develop on the particles, and small liquid bridges can result in wall adhesion. Depending on the contact distance (i.e. the gap between particles in close contact with a wall), liquid bridges generally yield the strongest adhesion forces. On the other hand, electrostatic forces diminish in a humid environment. In a liquid environment, the interaction potential (DLVO theory in Section 6.3), describing the competition between van der Walls force and electrostatic force, is important. In powder technology, there are also numerous technical processes and applications where particle deposition on rigid walls is a wanted phenomenon, namely, in a number of particle separation devices, such as electrostatic precipitators, deep bed filtration, and surface filtration operated in both gaseous and liquid environments. A classical application example is the impact and deposition of fine particles on a fiber matrix or individual fibers.

In order to yield particle deposition on rigid walls, two criteria must be fulfilled: First, the particles need to be transported toward the wall through inertial effects, fluid dynamic forces, as well as turbulent and Brownian diffusion. Second, the particles need to stick to the wall by the adhesion forces that depend on particle and wall materials.

Common deposition models are introduced in this Section that consider the oblique elastic-plastic wall collisions of small spherical particles with plane walls and account for adhesion. Typically, larger inertial particles will not deposit since the adhesion forces are very small. The process of particle-wall impact, including all the details of material deformation, is rather complicated and only persists over a very short time period. Therefore, in the frame of a Lagrangian particle tracking, models are required for deciding on a possible deposition, and in case of rebound, the associated recoil velocities are needed. For very small, low inertia particles, the effect of particle rotation is negligible. A possible particle-wall adhesion may be decided based on two approaches, namely, an energy-balance-based derivation or a momentum-based model. Models using an energy balance yield a critical velocity below which deposition occurs (Dahneke, 1995). In the momentum or impulse-based approach, an adhesion impulse is integrated into the hard-sphere wall collision models (Almohammed and Breuer, 2016b; Kosinski and Hoffmann, 2009). It should be noted that the momentum-based approach does not consider the material properties of both involved partners, the particle, and the wall. Only a restitution coefficient and a friction coefficient are used in this model.

The relevant energies in the particle-wall impact are the kinetic energy of impact and rebound, a dissipated energy, mainly due to plastic deformation, and an adhesion energy caused in dry systems by the van der Waals force for the impact and rebound phase. Electrostatic effects are not considered because in most cases, the particle charge is not known, and it is assumed that the charge equalization with the carrier fluid occurs very fast. To derive the different energy contributions, the size of the contact area (i.e. radius and depth) is required, which may be derived by the Hertzian theory (Hertz, 1882) or the JKR model (Johnson et al., 1971). This deformation area depends on the material properties of particle and wall, such as Poisson ratio and Young's modulus. Other relevant material properties are the Hamaker constant, the yield pressure, and the interface energy. Several deposition models are presented based on different derivation concepts, also considering different effects during the particle impact on a plane wall. Here three deposition models are considered for delivering a critical impact velocity. Based on the energy balance, deposition will take place if the rebound kinetic energy becomes zero, yielding the condition

$$E_{K1} \leq E_D + \Delta E_A \tag{7.29}$$

where the impact kinetic energy is $E_{K1} = 1/2\, m_P\, V_{Pn1}^2$, the adhesion energy is ΔE_A, and E_D is the dissipated energy due to irreversible deformation. Note that for an oblique particle-wall impact, V_{Pn1} is the wall normal velocity component. Such an energy-based model was proposed by Hiller (1981) for fine particle deposition on single fibers as related to gas filtration. For solving the energy balance, the adhesion energy and the dissipated energy are specified as

$$\Delta E_A = -\int_{z_0}^{\infty} \frac{H}{6\pi z^3} \pi a^2 \, dz \tag{7.30}$$

$$E_D = \left(1 - k^2\right) E_A \tag{7.31}$$

where H is the Hamaker constant, $A = \pi a^2$ is the contact area of the deformation region, z is the wall-normal coordinate, and k is the energy restitution ratio $k^2 = (1 - E_D / E_{K1})$. In order to obtain a solution to this problem, models or assumptions are necessary for the maximum contact area radius, a_m, and/or the deformation depth, h_m. For the case of a small spherical particle hitting a rigid planar wall, the dissipated energy is given by Bitter (1963a, 1963b):

$$E_D = \frac{1}{2} \pi d_P h_m^2 p_y \qquad (7.32)$$

with the yield pressure of the softer material (the particle in this case) given by p_y. Thereby, the resulting critical velocity for deposition in the normal direction is obtained as

$$V_{crit} = \frac{\sqrt{1-k^2}}{\pi k^2 d_P z_0^2} \frac{H}{\sqrt{6 \rho_P p_y}} \qquad (7.33)$$

The minimum contact distance z_0 is normally set to 0.4 nm (i.e. 0.4×10^{-9} m).

The deposition model of Brach and Dunn (1992), in the following called B&D, was developed for an oblique elastic-plastic wall impact of small particles. The maximum contact area radius is determined by the Hertzian contact theory (Hertz, 1882). Setting the kinetic energy of the impact (i.e. normal component for oblique impact angles) equal to the adhesion or elastic strain energy, one obtains

$$\frac{1}{2} m_P V_{Pn1}^2 = \frac{8}{15} \frac{E^*}{R_P^2} a_m^5 \qquad (7.34)$$

with E^* being the reduced Young's modulus, defined as

$$E^* = \left(\frac{1-v_S^2}{E_S} + \frac{1-v_P^2}{E_P} \right)^{-1} \qquad (7.35)$$

Here, the properties of surface (S) and particle (P) are considered, namely, the Young's modulus E_i and the Poisson ratio v_i. This yields for the maximum contact area radius:

$$a_m = \left\{ \frac{15}{16} \frac{R_P^2}{E^*} m_P V_{Pn1}^2 \right\}^{1/5} \qquad (7.36)$$

In the analysis of Brach and Dunn (1992), the interface adhesion energy, a dissipated surface energy, is equal to the stored elastic energy, which is also the work needed to break the contact, and is denoted by E_A:

$$E_A = \pi a_m^2 \gamma^*$$

$$E_A = -\pi \left(\frac{5}{4} \pi \right)^{2/5} \left(\frac{\rho_P}{E^*} \right)^{2/5} R_P^2 V_{Pn1}^{4/5} \gamma^* \qquad (7.37)$$

The critical velocity is obtained from the impulse equations for the condition that $V_{Pn2}^2 + V_{Pt2}^2 = 0$, namely, that the particle velocity at the contact point is zero (Brach and Dunn, 1992), which eventually yields

$$V_{crit} = -\frac{1+\eta^2}{k^2}\frac{2E_A}{m_P}$$
(7.38)

Inserting the adhesion energy (Eq. 7.37) and replacing the normal impact velocity by the critical velocity, one obtains the critical velocity for deposition to happen

$$V_{crit} = 2.212\left(\frac{1+\eta^2}{k^2}\right)^{5/6}\left(\frac{1}{E^*}\right)^{1/3}\left(\frac{1}{\varrho_P}\right)^{1/2}\left(\frac{\gamma^*}{R_P}\right)^{5/6}$$
(7.39)

In the work of Brach and Dunn (1992), an experimental based correlation for the surface energy in dependence of impact velocity (at least between 1 and 10 m/s), obtained by Wall *et al.* (1990) for different wall materials and ammonium fluorescein (NH$_4$ Fl) spherical particles, is inserted

$$\gamma^* = 0.34\sqrt{|V_{Pn1}|}$$
(7.40)

This yields the final form of the critical normal impact velocity provided by Brach and Dunn (1992), in a slightly modified form

$$V_{crit} = 0.835\left(\frac{1+\eta^2}{k^2}\right)^{10/7}\left(\frac{1}{E^*}\right)^{4/7}\left(\frac{1}{\varrho_P}\right)^{6/7}\left(\frac{1}{R_P}\right)^{10/7}$$
(7.41)

In the last equation, E^* is defined in the most common way given by Eq. (7.34), and η is a friction coefficient, also known as the *impulse ratio*, and k is the energy restitution ratio (Brach and Dunn, 1992). Both values generally depend on the magnitude of the impact velocity and the impact angle. The influence of these two parameters was already shown by Brach and Dunn (1992) and will be demonstrated later. Beyond the critical normal impact velocity, that is, when particle rebounds, a three-dimensional hard-sphere wall collision model may be solved (see Chapter 7.1) for obtaining the rebound velocities. However, when the particles are small and have very low inertia, particle rotation and the change of rotation velocity, due to wall collisions, may be neglected without significant errors. It should be noted that only for very small particles, (below 10 µm in gas flows) wall adhesion is of major importance.

Another deposition model frequently used (see e.g. Venturini, 2010) is that of Thornton and Ning (1998), in the following called Th&N, where the normal wall impact of elastic-plastic adhesive spheres was considered. Again, based on an energy balance, the critical velocity for deposition was derived considering the JKR theory (Johnson *et al.*, 1971). For this situation, one obtains

$$\frac{1}{2} m_P V_{Pnl}^2 = E_A = 7.09 \left(\frac{\gamma^{*5} R_P^4}{E^{*2}} \right)^{\frac{1}{3}} \qquad (7.42)$$

where the work of separating the surfaces was derived by integrating over the total normal deformation. This yields the critical velocity as

$$V_{crit} = 1.84 \left\{ \left(\frac{1}{E^*} \right)^{\frac{1}{3}} \left(\frac{1}{\varrho_P} \right)^{\frac{1}{2}} \left(\frac{\gamma^*}{R_P} \right)^{\frac{5}{6}} \right\} \qquad (7.43)$$

where $\gamma^* = \sqrt{\gamma_P \gamma_S}$ is the interface energy composed of both the particle energy and the surface energy. Except for the constant and the term including the restitution coefficient, this equation is identical to the formulation of Brach and Dunn (1992). The impact velocity beyond which the material begins to creep resulting in an irreversible deformation is given by

$$V_Y = 1.56 \left\{ \left(\frac{1}{E^*} \right)^2 \left(\frac{1}{\varrho_P} \right)^{\frac{1}{2}} p_y^{\frac{5}{2}} \right\} \qquad (7.44)$$

For the calculation of the particle rebound velocity, correlations for the restitution coefficient in dependence of this yield velocity are presented by Thornton and Ning (1998).

A comparison of the critical deposition velocity using the Th&N-model in dependence of particle size with experimental data of Wall *et al.* (1990) was conducted by Xie *et al.* (2018), showing a remarkable underprediction of the critical velocity for the three wall materials (i.e. Silicon, Molybdenum, and Tedlar). Therefore, another correlation (like Eq. 7.42) was proposed by tuning the constant and the exponents as well as adding a term including the yield pressure. However, this correlation is dimensionally incorrect and not generally applicable.

To validate the different deposition models, the experimental data by Wall *et al.* (1990) were considered, where mono-size ammonium fluorescein ($NH_4 Cl$) spheres with diameters between 2.6 and 6.9 μm impinge on different surface materials. These materials were silicon, molybdenum, and Tedlar (polyvinyl fluoride), all carefully prepared to yield smooth surfaces. The results are shown in Figure 7.8. The model of Brach and Dunn (1992), referred to as B&D model, with k = 1.0 and η = 0, underpredicts the measured critical velocity. However, when including constant values for the restitution coefficient k = 0.8 and the friction η = 0.1, the agreement is almost perfect for the two wall materials, both with respect to the slope and magnitude. The model of Thornton and Ning (1998) underpredicts the critical velocity's dependence

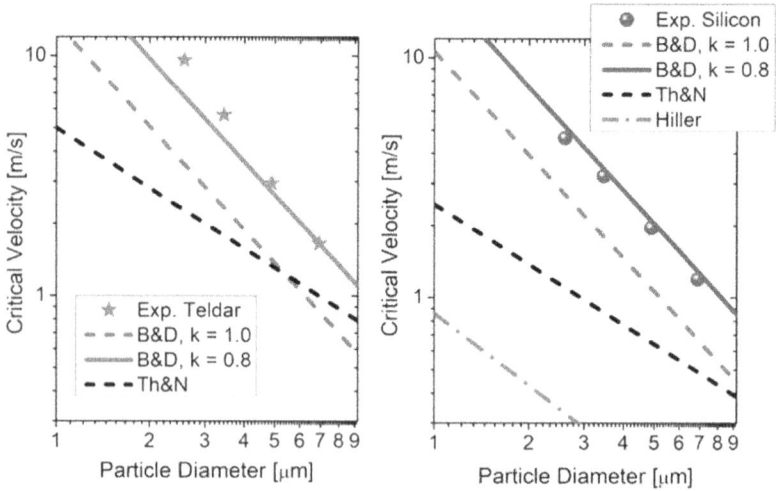

FIGURE 7.8 Critical velocity dependence on particle diameter (ammonium fluorescein) for different wall materials (left: Teldar (polyvinyl-fluoride), right: Silicone), comparison of model predictions (B&D: Brach and Dunn (1992) k = 1.0, η = 0 and k = 0.8, η = 0.1; Th&N: Thornton and Ning (1998); Hiller (1981)) with experiments (Wall *et al.*, 1990).

on particle size, and the slope of the curve is completely different. Even when a restitution ratio k = 0.8 is selected, the critical velocity obtained by the Hiller (1981) model is more than one order of magnitude lower than the experimental values. This model does not show strong sensitivity on the wall material, probably because of the selection of the Hamaker constant.

7.4 WALL EROSION BY PARTICLE IMPACT

Particle-wall collisions frequently occur in confined particle-laden flows depending on the dimensions of the confinement, the particle response behavior, the properties and shapes of particles, and the wall properties and surface structure or roughness. Such wall impacts may also be associated with different ways of material abrasion. The lifetime of powder transport systems (e.g. solid powder conveying lines or cyclone separators) may be considerably limited by local particle erosion and degradation (Verma *et al.*, 2018). Solid particle erosion may occur in gas as well as liquid flow systems, but it is more pronounced in gas-solid flows due to the much higher particle impact velocities and missing lubrication effects as the particles hit the wall. Following El Togby *et al.*, (2005), such erosion of materials caused by the impact of hard particles is one of the forms of material degradation classified as wear.

There are several definitions of erosion in the literature: According to Bitter (1963a), it is defined as "material damage caused by the attack of particles entrained in a fluid system impacting the surface at high speed," while Hutchings and Winter (1974) define it as "erosion is an abrasive wear process in which the repeated impact of small particles entrained in moving fluid against a surface result in the removal of material from that surface."

In oil and gas industry, the problems resulting from wear, caused by particle impacts, are especially severe as they adversely influence not only the transport process efficiency but also compromise safety (Pereira *et al.*, 2014). In oil extraction, sand particles may destroy the function of the transport process components, such as pipes, valves, and separation devices. Failure due to erosion in such elements might result in oil spill, which can generate an environmental damage as well as a considerable liability (Zhang *et al.*, 2007). The transport of chemicals in pneumatic conveying pipelines and the erosion in bends and fittings is equally catastrophic (Westman *et al.*, 1987). Therefore, it is of paramount importance to develop strategies aimed to predict erosion rates depending on operating conditions (Chen *et al.*, 2004). It would not only allow the service lifetime estimation but also to preview the geometrical locations where critical erosion is likely to occur. This is not an easy task as particulate erosion depends on several factors; some of which contribute in a synergistic manner to intensify wall material degradation.

The mathematical description or correlations for erosion rates have been subject of research at least since the work of Finnie (1960). Meng (1994) and Meng and Ludema (1995) conducted a literature survey of more than 5,000 papers and found hundreds of equations for the prediction of friction and wear. The great majority of the equations apply to specific conditions and are not applicable beyond the range of parameters under which they were developed. In the following sections, some of the main findings on the parameters affecting single particle erosion are summarized (see also Lain and Sommerfeld, 2019).

There are several different definitions related to the degree of particle erosion. Mostly, the erosion damage of equipment is defined as the mass or volume of removed wall material per mass of particles passing through the equipment. In order to better quantify the lifetime of any equipment subject to erosion also the penetration depth is often used, defined as the erosion depth per kilogram of impinging particles, for example, in [mm/kg]. Another definition of the erosion rate is the weight of eroded material (material mass loss), over the surface element area, and the erosion time (i.e. $kg/(m^2 s)$).

Numerically, the wear is typically predicted as the sum of the damage caused by each individual particle-wall collision for a certain wall element (i.e. mostly the area of a numerical mesh) divided by the total mass of impinging erodent [kg/kg]:

$$ER_S = \sum_{i=1}^{N_s} \frac{m_{pi} E_i}{m_T} \tag{7.45}$$

where ER_S is the total erosion at each relevant surface element area, E_i is the erosion caused by a single particle i, which must be computed by an erosion correlation, some of which are provided in the next section, m_{pi} is the mass of the particle i and m_T is the total mass of particles impinging the considered area. Quite often the erosion damage is also expressed in terms of the surface area related penetration ratio, $PR_S \left[\dfrac{m}{kg} \right]$, given by the expression

$$PR_S = \frac{ER_S}{\rho_w A_S} \tag{7.46}$$

where ρ_w is the density of the wall material. It is accepted that solid particle erosion acts through two main mechanisms depending on the ductility of the surface. For ductile materials, the main mechanisms are micro-cutting together with work hardening of the surface (Finnie, 1958; Levy, 1995). For brittle materials, it is accepted that erosion is due to crack formation (Kleis and Kulu, 2008). A more detailed description of both kinds of erosion mechanisms is found in Parsi *et al.* (2014). These authors include particle physical properties (shape, size, and material), fluid properties, target wall properties, particle impact speed and angle, and inter-particle collisions.

Angular *particle shapes* can result in erosion four times larger than that of round particles (Levy and Chik, 1983), and the impact angle for maximum erosion varies with particle angularity. Larger and heavier particles naturally cause higher erosion ratio (defined as mass of eroded material over the mass of impacting particles) than smaller and lighter ones because they have larger kinetic energy. However, the erosion ratio is found to be nearly independent of particle size for particles larger than 100 μm. Smaller sand particles are also more affected by turbulent transport affecting first the wall impact frequency. For particles with similar size and shape, higher density and hardness usually cause higher erosion rates (depending on hardness of the target material). Higher particle density increases the kinetic energy and impact force and increases the erosion rate. Larger hardness of the particles (up to about 700 Hv, Vickers hardness) also increases the erosion rate.

It was believed in the past that materials with lower hardness were more prone to erosion than those with higher hardness. However, Levy and Hickey (1982) found in some cases the opposite trend and suggest using toughness of the target material as a better indicator of erosion performance than hardness. As a result, and after numerous experiments, nowadays there is no definitive correlation between target wall material and particle erosion rate.

The *particle impact velocity* strongly determines the erosion rate. The relation between them follows an exponential functional relationship, with a constant exponent n, whose value can vary between 0.3 and 4.5 depending on the investigator. The prevalent theoretical value of this exponent is 2.0. However, it has been suggested (Oka et al., 2005; Oka and Yoshida, 2005) that such an exponent is not constant but depends on the hardness of the eroded material. More reasonable values for n are between 1.6 and 2.6.

Another important influence on erosion rate is *particle impact angle*, coupled however with the type of wall material used. For brittle materials, the main erosion mechanism is the formation of radial and lateral cracks when particles hit the surface, so the maximum erosion happens for near normal impacts corresponding to angles close to 90° (see Figure 7.9). In the case of ductile material surfaces, the cutting action and subsequent crater formation is predominant, so the maximum erosion values are obtained for more shallow angles as shown in Figure 7.9. However, most material surfaces have characteristics of both ductile and brittle, and consequently, a variety of angle functions have been proposed in the literature. Most of the proposed angle functions are empirical and are only valid for limited conditions. Figure 7.9 presents the impact angle dependence of erosion as obtained with the Oka-model (Oka et al., 2005; Oka and Yoshida, 2005) according to Eq. (7.57) for ductile and brittle material.

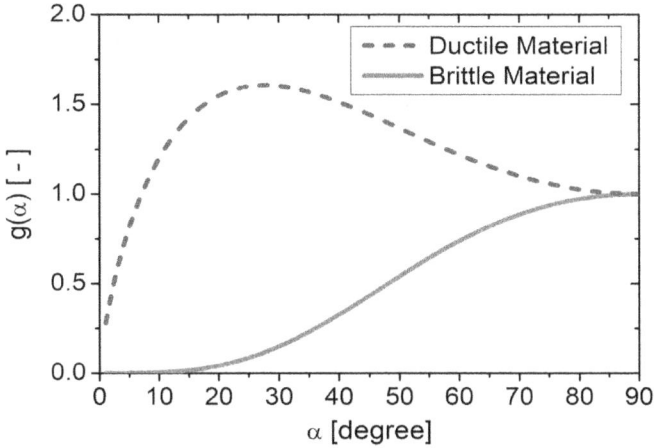

FIGURE 7.9 Dependence of eroded volume per mass of impinging particles, expressed as the function $g(\alpha)$ from Eq. (7.57).

In addition to erosion damage, there are a number of technical processes making use of erosion phenomena for surface finish and sandblasting. An interesting application is hydro-erosive grinding, where a slurry is forced through the individual holes of a fuel injection nozzles in order to achieve desired contours and surface smoothness (Rizkalla, 2007; Weickert et al., 2011). In both of these studies, the erosion process was simulated by an Euler/Euler approach accounting for the effect of surface structure modification on the flow field applying a continuous mesh adaptation algorithm.

Some of the most common single-particle erosion models are summarised, following the work of Novelletto Ricardo (2020) and Novelletto Ricardo and Sommerfeld (2022).

7.4.1 THE FINNIE MODEL

The first single-particle erosion model for ductile materials was proposed by Finnie (1958, 1960). The model assumes that a particle impacting on a surface with an impingement angle α removes material from the surface by displacement or cutting action of the abrasive particle. Therefore, the surface of the material is assumed to deform plastically, and the particle does not break upon impact. The resulting equation is

$$E = \begin{cases} \dfrac{M\bar{u}_p^2}{p\psi K}\left(\sin(2\alpha) - \dfrac{6}{K}\sin^2(\alpha)\right) & for \quad \alpha \le \dfrac{6}{K} \\[4mm] \dfrac{M\bar{u}_p^2}{p\psi K}\left(\dfrac{K\cos^2(\alpha)}{6}\right) & for \quad \alpha > \dfrac{6}{K} \end{cases} \tag{7.47}$$

where M is the total mass of particles, $\vec{u}_p$ is the particle impact velocity, p is flow stress of the material surface, ψ is the ratio of the depth of contact to the depth of cut, and K is the ratio of normal to tangential force. The first equation, that is, for $\alpha \leq 6/K$, corresponds to the case in which the particle leaves the material surface while still cutting, whereas the second equation, that is, for $\alpha > 6/K$, corresponds to the case in which the horizontal motion of the particle stops while cutting. In his work, Finnie (1960) assumed $K = 2$ and $\psi = 2$, that many particles are not as effective as the idealized particle and, therefore, arbitrarily chose a value of 50% off the predicted erosion. For erosion predictions with the Finnie model, a value of 2.7 GPa is assumed for p.

7.4.2 THE NEILSON AND GILCHRIST MODEL

Neilson and Gilchrist (1968) simplified the model proposed by Bitter (1963a, 1963b), which resulted in the following predictive equation for the total erosion

$$E = E_C + E_D \tag{7.48}$$

in which E_C and E_D represent the combined contributions from cutting and deformation wear, respectively. As usual, the cutting wear is computed as a function of the particle impact angle in the following form

$$E_C = \begin{cases} \dfrac{M\vec{u}_p^2 \cos^2(\alpha)\sin(n\alpha)}{2\phi_C} & for \quad \alpha < \alpha_0 \\[4mm] \dfrac{M\vec{u}_p^2 \cos^2(\alpha)}{2\phi_C} & for \quad \alpha > \alpha_0 \end{cases} \tag{7.49}$$

where M is the total mass of particles impacting the surface at a particle impact angle α and impact velocity $\vec{u}_p$. ϕ_C and α_0 are the cutting coefficient and the transition angle, respectively. The transition angle is normally set as 45°, and the deformation wear is given by the expression

$$E_D = \frac{M\left[\vec{u}_p \sin(\alpha) - K\right]^2}{2\varepsilon_C} \tag{7.50}$$

with ε_C being the deformation coefficient and K the cutoff velocity below which no deformation occurs. The authors also point out that the cutting wear predominates in the erosion of aluminum and that K is often negligible, as it is generally small with respect to the particle velocity. Therefore, the cutoff velocity is set to zero. Finally, ϕ_C and ε_C are specified as $3.332 \cdot 10^7$ and $7.742 \cdot 10^7$, respectively.

7.4.3 THE CHEN MODEL

The predictive equation proposed by Chen et al. (2004) assumes the same form as expressed by Finnie (1960):

$$E = K\vec{u}_p^n f(\alpha) \tag{7.51}$$

TABLE 7.2
Parameters for the Chen (2004) Model According to Wong et al. (2013)

K	n	A	B	W	X	Y	Z	α_0
$1.44 \cdot 10^{-8}$	2.2	−7	5.45	−3.4	0.4	−0.9	1.556056	23°

where E is measured as mass of material eroded per unit mass of particle impacting the surface, $\bar{u}_p$ is the particle impact velocity, n is the velocity exponent, α is the particle impact angle, K is a scaling parameter, and $f(\alpha)$ is a function which accounts for the influence of the impact angle on erosion. The function $f(\alpha)$ is expressed as follows:

$$f(\alpha) = \begin{cases} A\alpha^2 + B\alpha & \text{for} \quad \alpha \leq \alpha_0 \\ X\cos^2(\alpha)\sin(W\alpha) + Y\sin^2(\alpha) + Z & \text{for} \quad \alpha > \alpha_0 \end{cases} \quad (7.52)$$

The parameters in Eq. (7.52), that is, A, B, X, W, Y, Z, α_0, along with K and n, need to be experimentally determined. The values of these parameters were determined for sand particles impacting on aluminum by Wong et al. (2013) and are shown in Table 7.2.

7.4.4 THE ZHANG MODEL

Zhang et al. (2007) proposed the following predictive erosion equation:

$$E = C(BH)^{-0.59} F_s V_p^n F(\alpha) \quad (7.53)$$

where C and n are empirical constants, BH is the Brinell hardness of the target material, F_s is a particle shape coefficient, V_p is the particle impact velocity, α is the impact angle, and $F(\alpha)$ is a function of the impact angle. $F(\alpha)$ is modelled as follows:

$$F(\alpha) = 5.4\alpha - 10.11\alpha^2 + 10.93\alpha^3 - 6.33\alpha^4 + 1.42\alpha^5 \quad (7.54)$$

C and n are assumed to have values equal to 2.17×10^{-7} and 2.41, respectively. The particle shape coefficient F_s takes a value of 0.2 for spherical particles, 0.53 for semi-rounded particles, and 1 for angular (sharp) particle. This predictive equation depends entirely on flow information and eroded material properties and, consequently, like the Oka et al. (2005) model, is very robust.

7.4.5 THE OKA ET AL. MODEL

The predictive erosion model proposed by Oka et al. (2005) and Oka and Yoshida (2005) is one of the most general models for erosion predictions. This erosion model is applicable to many types of materials under various conditions involving particle impact angles and velocities as well as particle sizes and properties and is therefore considered in this work and briefly described in the next section. The predictive equation can be expressed as

$$E_V(\alpha) = g(\alpha) E_{90} \qquad (7.55)$$

where $E_V(\alpha)$ and E_{90} represent the volume of eroded material per mass of impinging particles, in which E_{90} expresses the erosion damage at normal impact angle. The mass of eroded mass per mass of impinging particles is then obtained from the expression:

$$E = 10^{-9} \rho_w E_V(\alpha) \qquad (7.56)$$

where the constant 10^{-9} has the dimension [m³/mm³], and ρ_w is the density of the wall material. $g(\alpha)$ is a function of the impact angle that describes the simultaneous phenomena of cutting (ductile material) and repeated deformation wear (brittle material), as shown in Figure 7.9:

$$g(\alpha) = (\sin \alpha)^{n_1} \left[1 + Hv(1 - \sin \alpha) \right]^{n_2} \qquad (7.57)$$

where H_v is initial eroded material Vickers hardness. The term $(\sin \alpha)^{n_1}$ is associated with repeated plastic deformation reaching its maximum at normal impact angle. On the other hand, the term $\left[1 + Hv(1 - \sin \alpha) \right]^{n_2}$ characterizes the cutting wear, which is more pronounced at shallower impact angles. The exponents n_1 and n_2 are determined by the eroded material hardness and other properties such as particle shape as follows:

$$\begin{cases} n_1 = s_1 (Hv)^{q_1} \\ n_2 = s_2 (Hv)^{q_2} \end{cases} \qquad (7.58)$$

The erosion at normal angle E_{90} depends on the impact velocity, particle diameter, and eroded material hardness accordingly to the expression:

$$E_{90} = K(aHv)^{k_1 b} \left(\frac{u_p}{u_{ref}} \right)^{k_2} \left(\frac{D_p}{D_{ref}} \right)^{k_3}. \qquad (7.59)$$

In Eq. (7.59), u_p and D_p are, respectively, the impact velocity and particle diameter, and u_{ref} and D_{ref} are the impact velocity and diameter used as reference in the experiments conducted by Oka et al. (2005) and Oka and Yoshida (2005). K is a particle property factor that considers the particle shape, for example, angularity, and particle hardness and is assumed to be in arbitrary units. The exponents k_1 and k_3 are calculated based on the properties of the particle and are also expressed in arbitrary units. k_3 is found by the authors to be roughly constant. k_2 depends on the Vickers hardness of the eroded material and on particle properties and can be expressed by

$$k_2 = r(Hv)^p. \qquad (7.60)$$

TABLE 7.3
Parameters of the Oka Predictive Erosion Model (Oka et al., 2005; Oka and Yoshida, 2005)

Particle Type	K	k_1	k_2	k_3	u_{ref}	D_{ref}	n_1	n_2
Quartz Sand	65	−0.12	$2.3(Hv)^{0.038}$	0.19	104	326	$0.71(Hv)^{0.14}$	$2.4(Hv)^{-0.94}$
Glass Beads	27	−0.16	2.1	0.19	100	200	$2.8(Hv)^{0.41}$	$2.6(Hv)^{-1.46}$

The term $K\left(aHv\right)^{k_1b}$ in Eq. (7.59) highly depends on the Vickers hardness of the eroded material and the type of particles, which are not correlated with the impact conditions and others factor. The relationship between the Vickers hardness of the eroded material and E_{90} at the reference impact velocity for the material investigated was derived from the experimental data from the authors and is presented in Table 7.3.

In order to demonstrate the performance of the different single-particle impact erosion models, numerical computations were conducted based on the Euler/Lagrange approach and combined with the standard k-ε turbulence model (Novelletto Ricardo and Sommerfeld, 2022). Particles are tracked considering both the translational and rotational motion and all relevant forces acting on the particles, such as gravity/buoyance, drag and transverse lift due to shear and particle rotation are accounted for. Particle dispersion due to turbulence is predicted stochastically by means of a single step Langevin equation, and the influence of wall roughness on particle-wall interactions is considered stochastically, in terms of a roughness angle distribution (Sommerfeld and Huber, 1999).

For validating the different erosion models, the experiments of Mazumder et al. (2008) were considered, where the magnitude of erosion at different locations along a pipe elbow was measured. These experiments were performed for a vertical to horizontal particle-laden two-phase flow through a pipe system. The elbow used in the experiments was made of aluminum 6061-T6, which has density and Vickers hardness of 2,700 kg/m³ and 1.049 GPa, respectively. The bulk velocity of air in the experiments was 34.1 m/s, which results in a Reynolds number of about 57,750. Sand particles with a mean diameter of 182 μm and a mass loading of 0.013 (kg of particles/kg of air) were injected into the vertical pipe at about 1.22 m below the test elbow. The test specimen was a 90° elbow with a diameter of 25.4 mm and a curvature radius of 38.1 mm, as illustrated in Figure 7.10. A horizontal pipe with a length of 10 diameters is connected to the elbow. Details of the numerical mesh used in the calculations are presented in Figure 7.10 (Novelletto and Sommerfeld, 2022).

The numerical domain of the calculations was discretized into approximately 1,000,000 hexahedral elements. In the near-wall region, the elements were gradually refined due to the existence of high-velocity gradients and the boundary layer. A total of 4,953,000 parcels per second were tracked through the flow field using the considered particle forces and a dynamic Lagrangian time step procedure (see Chapter 8.7.2). The simulations were conducted in an unsteady mode, where the Eulerian time step was set to $1.0 \cdot 10^{-5}$ s. Inter-particle collisions as well as flow modification by the particles (two-way coupling) were neglected in the calculations as the experiments were conducted with a low mass loading of 0.013. Particles were

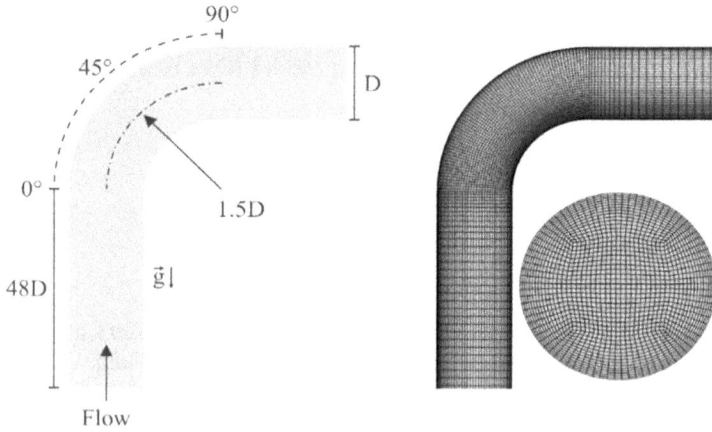

FIGURE 7.10 Experimental configuration of Mazumder *et al.* (2008) and numerical grid used for the flow simulations by Novelletto Ricardo and Sommerfeld (2022); the test element is a 90° elbow with a diameter of D = 0.0254 m and a radius of 0.0381 m (vertical pipe length, 1.22 m, horizontal outlet pipe length, 0.254 m).

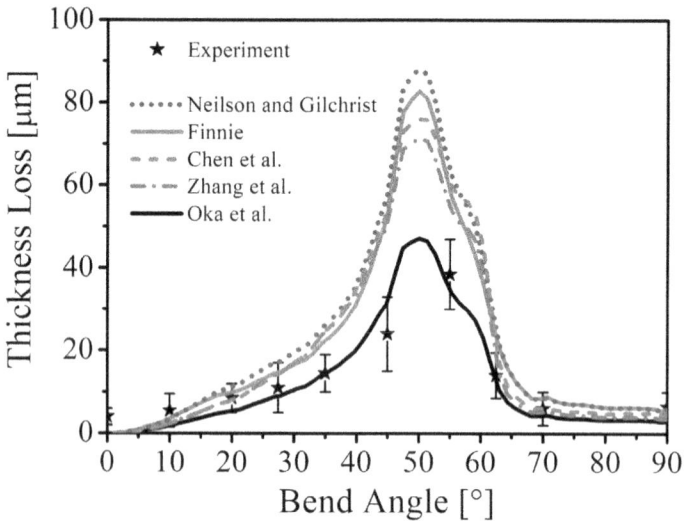

FIGURE 7.11 Comparison of the predicted thickness loss by different erosion models (Novelletto Ricardo and Sommerfeld, 2022) with the experimental data from Mazumder *et al.* (2008) for a profile along the middle plane of the 90° bend of Figure 7.10. The order of the labels for the different models correspond to the decreasing maxima in the erosion profiles.

injected into the system with the mean fluid velocity of 34.1 m/s and a fluctuating velocity for all three components of 4.5% of the bulk flow velocity was considered.

A quantitative comparison of the numerical results predicted by different erosion models with the measured erosion thickness loss along the central line of the outer wall of the bend is presented in Figure 7.11. The vertical bars indicate the standard

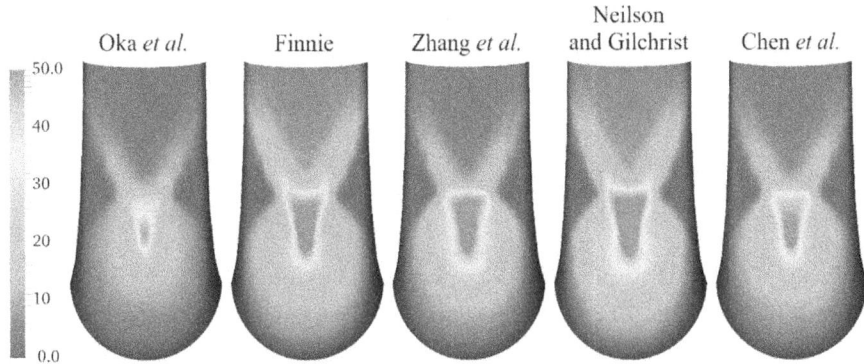

FIGURE 7.12 Predicted erosion scar by different erosion models for the Mazumder *et al.* (2008) two-phase bend flow. The gray scale is identical for all individual graphs.

deviation of the measurements. The bend inlet corresponds to a bend angle of $0°$, whereas a bend angle of $90°$ corresponds to the bend outlet. All erosion models predicted similar thickness loss profiles, although their magnitudes are different. As the flow field is the same for all erosion correlations and most of them depend mainly on particle impact velocity and angle, the shape of the curves is expected to be similar. Additionally, the thickness loss magnitude depends mainly on material and particle properties as well as other parameters, such as velocity exponent. Therefore, the models performed well in predicting the location of maximum thickness loss, which is between bend angles of $50°$ to $55°$ in this case. The slight difference in the location of maximum erosion between measured and calculated values can be attributed to the predictions of the two-phase flow field rather than to the erosion correlations, as all of them predicted similar erosion peaks along the bend. Considering the complexity of the phenomena involved in multiphase flows, the quality of the results is reasonably good. Among the models analyzed, the Oka model (Oka *et al.*, 2005; Oka and Yoshida, 2005) provides better agreement with the experimental data. All the other models overpredicted the maximum thickness loss. It must be noted that the results obtained with the numerical approach by Novelletto Ricardo and Sommerfeld (2022) with the Oka model (Oka *et al.*, 2005; Oka and Yoshida, 2005) are very similar to the numerical results obtained by Lain and Sommerfeld (2019) using a different numerical code.

Figure 7.12 illustrates the predicted erosion scar on the outer wall of the bend by the different erosion models. The depicted view is in flow direction from the bend entrance, and the bend outlet is the upper edge of the individual graphs. The erosion scars are characterized by a narrow, elongated region of high thickness loss, which is surrounded by a wide, almost circular area of moderate erosion. The region of high thickness loss arises mainly due to primary particle-wall collisions with high particle impact velocity. Downstream of the location of high erosion, a V-shaped structure is observed, which is caused mainly by secondary particle-wall collisions. This behavior has been numerically observed by several other studies (Chen *et al.*, 2004; Parsi *et al.*, 2014; Pereira *et al.*, 2014; Solnordal *et al.*, 2015; Duarte *et al.*, 2015,

2020; Lain and Sommerfeld, 2019). However, according to Solnordal *et al.* (2015), a V-shaped structure is normally not observed experimentally.

REFERENCES

Almohammed, N. and Breuer, M., 2016b, Modelling and simulation of particle-wall adhesion of aerosol particles in particle-laden turbulent flows, *Int. J. of Multiphase Flow*, **85**, 142–156.

Bitter, J.G.A., 1963a, A study of erosion phenomena part 1, *Wear*, **6**, 5–21.

Bitter, J.G.A., 1963b, A study of erosion phenomena Part 2, *Wear*, **6**, 169–190.

Brach, R.M. and Dunn, P.F., 1992, A mathematical model of the impact and adhesion of microspheres, *Aerosol Science and Technology*, **16**, 51–64.

Chen, X., McLaury, B.S. and Shirazi, S.A., 2004, Application and experimental validation of a computational fluid dynamics (CFD)-based erosion prediction model in elbows and plugged tees, *Comput. Fluids*, **33**, 1251–1272.

Dahneke, B., 1995, Particle bounce or capture: Search for an adequate theory: I. Conservation-of-energy model for a simple collision process, *Aerosol Science and Technology*, **23**, 25–39.

Duarte, C.A.R., de Souza, F.J. and dos Santos, V.F., 2015, Numerical investigation of mass loading effects on elbow erosion, *Powder Technol.*, **283**, 593–606.

Duarte, C.A.R., de Souza, F.J., Venturi, D.N. and Sommerfeld, M., 2020, A numerical assessment of two geometries for reducing elbow erosion, Particuology, **49**, 117–133.

El Togby, M.S., Ng, E. and Elbestawi, M.A., 2005, Finite element modelling of erosive wear, *Int. J. Machine Tools & Manufacture*, **45**, 1337–1346.

Feng, Z.-G. and Michaelides, E. E., 2002b, Hydrodynamic force on spheres in cylindrical and prismatic enclosures, *Int. J. Multiphase Flow*, **28**, 479–496.

Feng, Z.-G. and Michaelides, E.E., 2002a, Inter-particle forces and lift on a particle attached to a solid boundary in suspension flow, *Physics of Fluids*, **14**, 49–60.

Finnie, I., 1958, The mechanism of erosion of ductile metals, in *Proceedings of the 3rd U.S. National Congress of Applied Mechanics* (Ed. R.M. Hagthormthwaite), pp. 527–532, American Society of Mechanical Engineers, New York, N.Y.

Finnie, I., 1960, Erosion of surfaces by solid particles, *Wear*, **3**, 87–103.

Grant, G. and Tabakoff, W., 1975, Erosion prediction in turbomachinery resulting from environmental solid particles, *J. Aircraft*, **12**, 471–478.

Hertz, H., 1882, Über die Berührung fester elastischer Körper, *J. Reine Angew. Math.*, **92**, 156–171.

Hiller, R.B., 1981, *Der Einfluss von Partikelstoß und Partikelhaftung auf die Abscheidung in Faserfiltern*. Dissertation, Universität Karlsruhe, VDI-Verlag GmbH Düsseldorf.

Hutchings, I.M. and Winter, R.E., 1974, Particle erosion of ductile metals: A mechanism of material removal, *Wear*, **27**, 121–128.

Johnson, K.L., Kendall, K. and Roberts, A.D., 1971, Surface energy and the contact of elastic solids, *Proc. R. Soc. Lond.*, **A324**, 301–313.

Kleis, I. and Kulu, P., 2008, *Solid Particle Erosion Occurrence, Prediction and Control*, Springer-Verlag, Heidelberg, Germany.

Konan, N.A., Kannengieser, O. and Simonin, O., 2009, Stochastic modeling of the multiple rebound effects for particle-rough wall collisions, *Int. J. of Multiphase Flow*, **35**, 933–945.

Kosinski, P., Hoffmann, A.C., 2009, Extension of the hard-sphere particle-wall collision model to account for particle deposition, *Physical Review E*, **79** (061302).

Kussin, J. and Sommerfeld, M., 2002, Experimental studies on particle behaviour and turbulence modification in horizontal channel flow with different wall roughness, *Experiments in Fluids*, **33**, 143–159.

Lain, S. and Sommerfeld, M., 2008, Euler/Lagrange computations of pneumatic conveying in a horizontal channel with different wall roughness, *Powder Technology*, **184**, 76–88.

Lain, S. and Sommerfeld, M., 2019, Numerical prediction of particle erosion of pipe bends, *Advanced Powder Technology*, **30**, 366–383.

Levy, A., 1995, *Solid Particle Erosion and Erosion-Corrosion of Materials*, 2nd edition, ASM Int., Materials Park, OH, USA.

Levy, A. and Chik, P., 1983, The effect of erodent composition and shape on the erosion of steel, *Wear*, **89**, 151–162.

Levy, A. and Hickey, G., 1982, Surface degradation of metals in simulated synthetic fuels plant environments, *NACE Corrosion/82, Int. Corrosion Forum*, p. 154, Houston TX.

Matsumoto, S. and Saito, S., 1970a, On the mechanism of suspension of particles in horizontal pneumatic conveying: Monte Carlo simulation based on the irregular bouncing model, *J. Chem. Engng Japan*, **3**, 83–92.

Matsumoto, S. and Saito, S., 1970b, Monte Carlo simulation of horizontal pneumatic conveying based on the rough wall model, *J. Chem. Engin. Japan*, **3**, 223–230.

Mazumder, Q.H., Shirazi, S.A. and McLaury, B., 2008, Experimental investigation of the location of maximum erosive wear damage in elbows, *J. Pressure Vessel Technol.*, **130**, 1–7.

Meng, H.C., 1994, *Wear Modelling: Evaluation and Categorization of Wear Models*, University of Michigan, Ann Arbor, MI, USA.

Meng, H.C. and Ludema, K.C., 1995, Wear models and predictive equations-their form and content, *Wear*, **181**, 443–457.

Michaelides, E.E., 1987, Motion of particles in gases: Average velocity and pressure loss, *J. Fluids Engineering*, **109**, 172.

Neilson, J.H. and Gilchrist, A., 1968, Erosion by a stream of solid particles, *Wear*, **11**, 111–122.

Novelletto Ricardo, G.A., 2020, *Studies of Erosion in Gas-Solid Flows by Using Experimental and Numerical Techniques*. Dissertation Fakultät für Verfahrens- und Systemtechnik der Otto-von-Guericke-Universität Magdeburg.

Novelletto Ricardo, G.A. and Sommerfeld, M., 2020, Experimental evaluation of surface roughness variation of ductile materials due to solid particle Erosion, *Advanced Powder Technology*, **31**, 3790–3816.

Novelletto Ricardo, G.A. and Sommerfeld, M., 2022, Comprehensive euler/lagrange modelling including particle erosion for confined gas-solid flows, *Submitted to Wear (J.)*, March.

Oka, Y.I., Okamura, K. and Yoshida, T., 2005, Practical estimation of erosion damage caused by solid particle impact: Part 1: Effects of impact parameters on a predictive equation, *Wear*, **259**, 95–101.

Oka, Y.I. and Yoshida, T., 2005, Practical estimation of erosion damage caused by solid particle impact: Part 2: Mechanical properties of materials directly associated with erosion damage, *Wear*, **259**, 102–109.

Parsi, M., Najmi, K., Najafifard, F., Hassani, S., McLaury, B.S. and Shirazi, S.A., 2014, A comprehensive review of solid particle erosion modeling for oil and gas wells and pipelines applications, *J. Natural Gas Science and Engineering*, **21**, 850–873.

Pereira, G.C., de Souza, F.J. and Martins, D.A., 2014, Numerical prediction of the erosion rate due to particles in elbows, *Powder Technology*, **261**, 105–117.

Quintero, B., Laín, S. and Sommerfeld, M., 2021, Derivation and validation of a hard-body particle-wall collision model for non-spherical particles of arbitrary shape, *Powder Technology*, **380**, 526–538.

Radenkovic, D. and Simonin, O., 2018, Stochastic modelling of three-dimensional particle rebound from isotropic rough wall surface, *Int. J. of Multiphase Flow*, **109**, 35–50.

Rizkalla, P.A., 2007, *Development of a Hydroerosion Model using a Semi-Empirical Method Coupled with an Euler-Euler Approach*. Dissertation School of Aerospace, Mechanical & Manufacturing Engineering Royal Melbourne Institute of Technology (RMIT) November.

Sgrott Junior, O.L. and Sommerfeld, M., 2019, Influence of inter-particle collisions and agglomeration on cyclone performance and collection efficiency, *Canadian J. Chemical Engineering*, **97**, 511–522.

Siegel, W., 1991, *Pneumatische Förderung: Grundlagen, Auslegung, Anlagenbau, Betrieb*, Vogel Verlag, Würzburg, Germany.

Solnordal, C.B., Wong, C.Y. and Boulanger, J., 2015, An experimental and numerical analysis of erosion caused by sand pneumatically conveyed through a standard pipe elbow, *Wear*, **336–337**, 43–57.

Sommerfeld, M., 1992, Modelling of particle/wall collisions in confined gas-particle flows, *Int. J. Multiphase Flow*, **18**, 905–926.

Sommerfeld, M., 1996, *Modellierung und numerische Berechnung von partikelbeladenen turbulenten Strömungen mit Hilfe des Euler/Lagrange-Verfahrens*. Habilitationsschrift (habilitation thesis), Universität Erlangen-Nürnberg, Shaker Verlag, Aachen.

Sommerfeld, M., 2002, Kinetic simulations for analysing the wall collision of non-spherical particles, *Joint US ASME/European Fluids Engineering Summer Conference*, Montreal, Paper No. FEDSM, 2002–31239.

Sommerfeld, M., 2003, Analysis of collision effects for turbulent gas-particle flow in a horizontal channel: Part I. Particle transport, *Int. J. Multiphase Flow*, **29**, 675–699.

Sommerfeld, M. and Huber, N., 1999, Experimental analysis and modelling of particle-wall collisions, *Int. J. of Multiphase Flow*, **25**, 1457–1489.

Sommerfeld, M. and Kussin, J., 2003, Analysis of collision effects for turbulent gas-particle flow in a horizontal channel: Part II. Integral properties and validation, *Int. J. Multiphase Flow*, **29**, 701–718.

Sommerfeld, M. and Kussin, J., 2004, Wall roughness effects on pneumatic conveying of spherical particles in a narrow horizontal channel, *Powder Technology*, **142**, 180–192.

Sommerfeld, M. and Lain, S., 2009, From elementary processes to the numerical prediction of industrial particle-laden flows, *Multiphase Science and Technology*, **21**, 123–140.

Sommerfeld, M. and Lain, S., 2015, Parameters influencing dilute-phase pneumatic conveying through pipe systems: A computational study by the Euler/Lagrange approach, *The Canadian J. of Chemical Engineering*, **93**, 1–17.

Sommerfeld, M. and Zivkovic, G., 1992, Recent advances in the numerical simulation of pneumatic conveying through pipe systems, in *Computational Methods in Applied Science* (Eds. Hirsch, et al.), pp. 201–212, First European Computational Fluid Dynamics, Brussels.

Thornton, C. and Ning, Z., 1998, A theoretical model for the stick/bounce behaviour of adhesive, elastic-plastic spheres, *Powder Technology*, **99**, 154–162.

Tsuji, Y., Morikawa, Y., Tanaka, T., Nakatsukasa, N. and Nakatani, M., 1987, Numerical simulation of gas-solid two-phase flow in a two-dimensional horizontal channel, *Int. J. Multiphase Flow*, **13**, 671–684.

Tsuji, Y., Oshima, T. and Morikawa, Y., 1985, Numerical simulation of pneumatic conveying in a horizontal pipe, *KONA*, **3**, 38–51.

Venturini, P., 2010, *Modelling of Particle-Wall Deposition in Two Phase Gas-Solid*. Ph.D. Thesis, Sapienza Università di Roma, Italy.

Verma, R., Agarwal, V.K., Pandey, R.K. and Gupta, P., 2018, Erosive wear reduction for safe and reliable pneumatic conveying systems: Review and future directions, *Life Cycle Reliability and Safety Engineering*, **7**, 193–214.

Volk, R., 2005, *Rauheitsmessung: Theorie und Praxis*, Beuth Verlag GmbH, Berlin, Germany.

Wall, S., John, W. and Wang, H.C., 1990, Measurements of kinetic energy loss for particles impacting surfaces, *Aerosol Science and Technology*, **12**, 926–946.

Weickert, M., Sommerfeld, M., Teike, G. and Iben, U., 2011, Experimental and numerical investigation of the hydroerosive grinding, *Powder Technology*, **214**, 1–13.

Westman, M.A., Michaelides, E.E. and Thompson, F.A., 1987, Pressure losses due to bends in pneumatic conveying, *J. of Pipelines*, **7**, 15.

Wong, C.Y., Solnordal, C., Swallow, A. and Wu, J., 2013, Experimental and computational modelling of solid particle erosion in a pipe annular cavity, *Wear*, **303**, 109–129.

Xie, J., Dong, M., Li, S., Shang, Y. and Fu, Z., 2018, Dynamic characteristics for the normal impact process of micro-particles with a flat surface, *Aerosol Sience and Technology*, **52**, 222–233.

Zeng, L., Najjar, F., Balachandar, S. and Fischer, P., 2009, Forces on a finite-sized particle located close to a wall in a linear shear flow, *Physics of Fluids*, **21** (033302).

Zhang, Y., Reuterfors, E.P., McLaury, B.S., Shirazi, S.A. and Rybicki, E.F., 2007, Comparison of computed and measured particle velocities and erosion in water and air flows, *Wear*, **263**, 330–338.

8 Numerical Methods and Modelling Approaches

8.1 SUMMARY OF NUMERICAL METHODS FOR SINGLE-PHASE FLOWS

The numerical calculation of single-phase flows may be broadly separated into three classes depending on their spatial resolution, as illustrated in Figure 8.1. An introduction to these methods may be found in Ferziger and Peric (2002) together with details about discretization schemes and solution procedures. The first approach, direct numerical simulations (DNS), in principle resolves all relevant flow scales down to the Kolmogorov scales (i.e. the turbulent dissipation scales). Consequently, the numerical grid should be smaller than these scales to ensure proper resolution. This, however, implies that the size of the computational domain is limited due to the limited number of numerical control volumes that can be handled by today's computers. Consequently, DNS is not able to resolve the entire spectrum of vortex scales or wave numbers, and hence, there is a cutoff at larger scales. Moreover, the ratio of macro-scale vortices to the Kolmogorov scale is proportional to $Re^{3/4}$. As a result, the flow Reynolds number considered in DNS has a limitation too. Typically, the maximum Re is between 5,000 and 10,000. Therefore, the application of DNS is limited to simple flow conditions, such as homogeneous isotropic turbulence, shear flows, and channel flows. Nevertheless, the DNS method is very helpful in analyzing and understanding basic phenomena in turbulence research.

The second class of numerical methods, which is rapidly growing in importance, is large eddy simulations (LES) where the filtered Navier-Stokes equations are solved, as shown in Figure 8.1. The numerical grid may be much coarser compared to DNS, but it requires a sub-grid-scale (SGS) turbulence model. With such a coarse grid, processes with larger dimensions can be simulated. In LES the vortex spectrum is fully resolved except for the dissipation regime (i.e. the Kolmogorov scale). These unresolved scales (SGS) are close to being isotropic and therefore require relatively simple turbulence modelling (Lesieur *et al.*, 2005; Fröhlich, 2006).

The third class of numerical methods is based on the Reynolds-averaged Navier-Stokes (RANS) equations combined with an appropriate turbulence model. A number of turbulence models have been developed in the past (Wilcox, 2006) with a variety of derivatives and improvements. Most of them are based on the solution of additional transport equations such as the two-equation k-ε and k-ω turbulence models, as well as the full Reynold-stress turbulence model.

A turbulence model is based on the integration of the entire energy spectrum. The k-ε model solves additional transport equations for the turbulent kinetic energy k and its dissipation rate ε, which is related to the Kolmogorov scale. These values provide the turbulent viscosity that is introduced in the transport equations. In

DOI: 10.1201/9781003089278-8

239

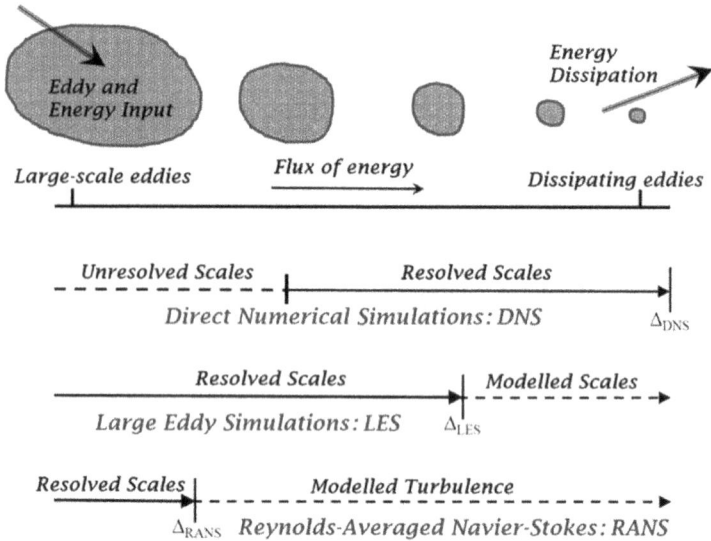

FIGURE 8.1   A diagram of numerical methods for turbulent single-phase flows indicating the respective resolution of eddies and turbulence; the spectrum of eddies ranges from the system-induced large scales to the Kolmogorov scales; direct numerical simulations (DNS) resolving the small turbulent scales up to the large scales on the order of the domain size; large eddy simulations (LES) resolving all scales down to the dissipation length scale (filter size); calculations based on Reynolds-Averaged Navier-Stokes (RANS) equations resolving only large-scale structures, with small scales, are described by turbulence models (Bakker, 2002).

the Reynolds stress model, seven additional transport equations have to be solved, making this method more complex, and sometimes it is difficult to obtain converged results. Such an approach is suggested for complex anisotropic flows—for example, swirling flows in combustors or cyclones or flows with recirculation and separations. The resulting RANS-equations may be solved in a steady-state mode, but also in an unsteady way, depending on the application.

8.2   HIERARCHY OF NUMERICAL METHODS FOR MULTIPHASE FLOWS

Multiphase flows are very complex due to their appearance in quite different forms. Globally one may distinguish between discontinuous and separated multiphase flows, where clear interfaces exist between the different phases, and dispersed multiphase flows, where the dispersed phase elements (e.g. solid particles and droplets) are homogeneously distributed in the flow domain. Due to these different structures, it is also not possible to use only one kind of numerical method that allows the complete numerical description and parameter determination of industrial processes. Whether separated or dispersed multiphase flows are considered requires different numerical approaches and the consideration of different physical phenomena and parameters. Naturally, a mix of separated and dispersed two-phase flow may occur, for example, due to the entrainment of fine droplets from the liquid phase into the gas phase and vice versa, as it is typical for annular flows. In addition, the liquid fraction in separated flows may include droplets, bubbles, or solid particles, as for example in oil and gas recovery. For a numerical calculation of such rather complex and mostly time-dependent multiphase flows, different methods need to be combined. The most complex type of multiphase flow is the transient multiphase flow, which evolves from pure liquid to pure vapor, with different multiphase flow states in between calling for different numerical methods. On the other hand, more simple one-dimensional methods may be used for large-scale industrial equipment, supported by numerous correlations and closures, as for example for long pipelines in oil and gas transportation (Lurie and Sinaiski, 2008; Issa, 2009; Kjeldby et al., 2011). For the numerical calculations of separated and slug-type multiphase flows, one may use a two-fluid approach, including an appropriate treatment of the interfaces, wherefore different numerical methods may be used (Prosperetti and Tryggvason, 2009).

For dispersed flows of industrial scale—including solid particles, droplets, and bubbles—an interface resolving simulation is, even with today's computational power, not possible. This type of two- or multiphase flows in a technical or industrial scale are calculated numerically by treating the dispersed phase elements as point-particles or point-masses using appropriate correlations for describing their fluid dynamic transport, as for example drag and lift coefficients. In most technical cases, the particle Reynolds number is larger than unity so that a theoretical derivation of such correlations is not possible, and therefore, experimentally based correlations are mostly used (Michaelides and Roy, 1987; Sommerfeld et al., 2008; Sommerfeld, 2010). In support of, particle-resolved numerical simulations are increasingly applied for deriving resistance coefficients, for example, for non-spherical particles (Hölzer and Sommerfeld, 2009; Zastawny et al., 2012; Sommerfeld and Qadir, 2018). In addition, a large number of other particle-scale phenomena influence dispersed two-phase flows, such as inter-particle collisions, agglomeration, wall collisions, coalescence, and breakup of droplets and bubbles, to name only a few. In a point-particle approach, all these elementary processes need additional modelling and closures. This is the most important task in the development of macro-scale numerical methods for industrial and technical processes.

As a result, one encounters the so-called *multi-scale phenomenon*, which is especially important for multiphase flows, as shown in Figure 8.2. Industrial-scale processes can only be simulated with a limited number of grid cells and dispersed particles, even if treated as point-masses. This is illustrated in the left part of Figure 8.2 for an industrial-scale fluidised bed simulated by a two-fluid approach (Ozel et al., 2013). Using a too-coarse mesh, which, however, yields a reasonable computational time, the fine-scale structures of particle clustering during their transport through the riser cannot be resolved. Resolution can only be improved with a very fine mesh, which consequently results in unrealistically high computational time and storage requirements. Therefore, a sub-grid-scale drift velocity model was developed to allow for coarser meshes (Igci et al., 2008; Parmentier and Simonin, 2012). This shows that all phenomena on scales smaller than the numerical grid require modelling, which is first of all turbulence modelling for the fluid flow when using Reynolds-averaged Navier-Stokes (RANS) equations or LES methods.

The modelling of particle-scale phenomena was mainly based on detailed experiments, for example, in the case of collisions between droplets, which was thoroughly analyzed by imaging techniques. However, due to the increasing computational power, more and more fully resolved DNS methods (i.e. resolving particles or bubbles and droplets by the numerical grid, including the flow around them) are being applied to evaluate models to be used for point-particle approximations. Such simulations are called *micro-scale simulations*, as shown in right part of Figure 8.2. The hydrodynamic interaction between colliding solid particles was simulated by the Lattice-Boltzmann method (LBM) in the frame of agglomeration studies (Ernst and Sommerfeld, 2015; Ernst, 2016). In the middle of Figure 8.2, meso-scale simulations are shown, conducted by LBM, but considering particles as point masses (Ernst and Sommerfeld, 2012; Ernst *et al.*, 2019). Different measures were used for characterizing particle segregation in homogeneous isotropic turbulence and accounting for inter-particle collisions. DNS and LES that consider point-particles for analyzing particle clustering or particle transport in channels or pipes fall into this category of simulations. In the case of DNS, simulations can be conducted for simple geometries up to flow Reynolds numbers of about 5,000 to 10,000. LES allows the consideration of much more complex multiphase flows, due to the coarser grid resolution compared to DNS, but the dispersed phase is still treated by the point-particle approximation

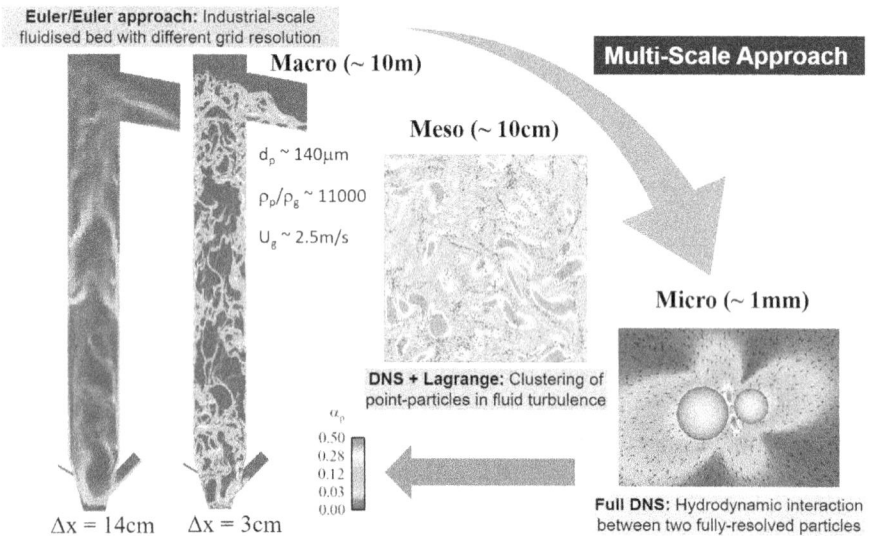

FIGURE 8.2 Illustration of multiscale approach for the numerical calculation of dispersed two-phase flows. Left: Industrial-scale fluidized bed simulation by a two-fluid approach with different grid resolution (Parmentier and Simonin, 2012; Ozel *et al.*, 2013). Middle: Analysis of particle segregation and accumulation in homogeneous isotropic turbulence accounting for inter-particle collisions using LBM (Ernst and Sommerfeld, 2012). Right: Hydrodynamic interaction between colliding solid particles simulated by the LBM in the frame of agglomeration studies (Ernst and Sommerfeld, 2015; Ernst, 2016).

as it has been shown in the reviews by Dhotre *et al.* (2013) for bubbly flows and by Kuerten (2016) for particle-laden flows.

The hybrid Euler-Lagrange point-mass approach is not only applied in the frame of RANS, but also when using DNS and LES. In these methods, the fluid flow is solved on a Eulerian mesh, and particle motion is calculated in a Lagrangian frame-work. In such hybrid approaches, the coupling between point-particles and fluid flow needs special attention. This concerns all coupling mechanisms, including momentum, thermal energy, mass, and turbulence. Due to the use of point-particles, numerical simulations of millions of small particles in a flow are feasible. In the point-particle approach, the presence of the particle is represented by a point: the center of the particle and the interactions between the surface of the particle and the fluid must be accounted for empirically. The main assumption of the point-particle approach is that the particle diameter must be smaller than the Kolmogorov micro-scale, η_κ, and significantly smaller than the grid size on which the fluid behav-ior is predicted.

For a further classification of the coupling mechanisms, the particle concentra-tion or volume fraction and, consequently, the averaged inter-particle spacing may be used. The proximity of particles can be easily determined for regular and homo-geneous arrangements of the particles. The particle volume fraction is the volume of a single spherical particle divided by the box volume (edge length L), including one particle only. For such a cubic arrangement, the inter-particle spacing, that is, the distance between the centers of particles, is obtained from (Sommerfeld, 2017)

$$\frac{L}{D_p} = \left(\frac{\pi}{6\,\alpha_P} \right)^{1/3} \tag{8.1}$$

The result is depicted in Figure 8.3 by comparing the inter-particle spacing and the dispersed phase volume fraction. As an example, considering a volume fraction of 1%, the spacing is 3.74 diameters and for 10% only 1.74. Hence, for such high volume fractions, the particles cannot be treated to move isolated since fluid dynamic inter-actions become of importance. In other practical fluid-particle systems however, the particle volume fraction is much lower, as found in pneumatic conveying. Consider, for example, a gas-solid flow (particle density $\rho_p = 2{,}500$ kg/m^3, gas density of $\rho_F = 1.18$ kg/m^3) with a mass loading of one (i.e. $\eta = 1$) and assume no slip between the phases, then the volume fraction is about 0.05% (i.e. $\alpha_p = 5 \times 10^{-4}$). This results in an inter-particle spacing of about 10 particle diameters. Hence, under such conditions, a fluid dynamic interaction may be neglected, and the particles can be regarded as moving independently.

In Figure 8.3, one may distinguish between dilute and dense dispersed two-phase flows. In dilute systems, particles move independently, and their transport is mainly governed by the relevant forces, including external and interfacial forces, such as drag and lift. Such a dilute system exists for volume fractions up to $\alpha_p = 5 \times 10^{-4}$ and an inter-particle spacing up to $L/D_p \approx 10$. In denser systems when the inter-particle spacing becomes $L/D_p < 10$, fluid dynamic interactions between particles occurring upon approach (Figure 8.2 right) and eventually inter-particle collisions become of

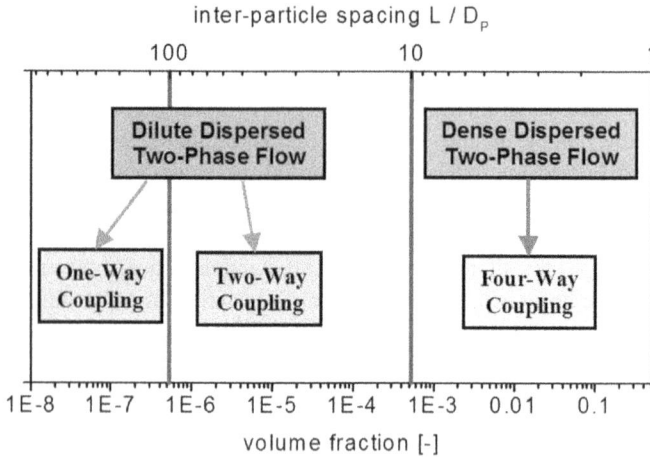

FIGURE 8.3 Regimes of dispersed two-phase flows as a function of particle volume fraction and inter-particle spacing for a regular cubic arrangement of particles and coupling regimes (Sommerfeld *et al.*, 2008).

greater importance relative to the fluid dynamic transport. Ultimately, when the particle volume fraction becomes larger than about 10%, one refers to this regime as "contact dominated," where multiple inter-particle contacts and adhesion forced become very important.

In terms of coupling between the phases, that is, the effect of the flow on particle transport and how the particles may modify the flow field, again the volume fraction is an important property, and four regimes may be identified. For very low particle phase volume fraction (below 10^{-6} and for $L/D_p > 100$), a *one-way coupled calculation* is sufficient, implying that the particles are tracked through the previously computed and fully converged flow field as a passive scalar, and their properties are ensemble averaged. For higher volume fractions, the influence of the particle phase on the flow field and turbulence properties must be taken into account, also called *two-way coupling*. The associated computational approach was first introduced by Crowe *et al.* (1977) and was called *particle-source-in-cell* (PSIC) method. In this method, particles are tracked in the flow field as predicted by the carrier flow-solver, and the particle phase properties as well as the source terms in each computational cell of the flow domain are collected and ensemble averaged. The source terms are sampled for kinetic interaction (i.e. for the momentum equations and the transport equations for the turbulence properties) as well as for heat and mass transfer across the phase boundaries, if required. It should be emphasized that only transport processes occurring at the interface of the particle/droplet contribute to the coupling. Natural external forces, or body forces, such as gravity are not relevant. Particles contribute to the cell-based source terms only if they are residing in the control volume that is considered.

The term *four-way coupling* is used when in connection with two-way coupling also inter-particle collisions, turbulence modulation by the particles, and fluid-dynamic interaction upon collision are considered. This is relevant for particle volume fractions larger than about 10^{-3} (i.e. $L/D_p < 10$), as illustrated in Figure 8.3. In this type of modelling, the particle phase properties are not only influenced by the modified flow field but also by the alteration of the particle phase velocities through inter-particle collisions. Finally, *three-way coupling* is only associated with fluid-dynamic interactions between approaching or nearby moving particles (Chapter 2 in Michaelides *et al.*, 2017). These effects, among others, result in a modification of the fluid dynamic forces acting on particles, such as drag and lift (Prahl *et al.*, 2007). Also, the so-called swarm drag, often applied in fluidization, is a kind if three-way coupling and mimics, on the average, the fluid dynamic interactions between particles (Loha *et al.*, 2012). Such a three-way coupling mode, however, may be difficult to observe experimentally since at these high-volume fractions, inter-particle collisions occur simultaneously in practical two-phase flow systems. For particles in homogenous isotropic turbulence, limits of validity for the various coupling approaches have been presented by Elghobashi (1994).

8.3 PARTICLE-SCALE SIMULATION METHODS

Fully resolved simulations, which are feasible for a small group of particles, may considerably support the modelling of particle-scale phenomena needed for macro-scale simulations. The requirements for such kind of resolved simulations first depend on the type of particles considered. Resolved simulations for rigid solid particles (as illustrated in the right part of Figure 8.2) do not need any special treatment of the interface and only require a no-slip wall boundary condition or a suitable slip model on the particle surface. Special interpolation methods are often needed if the particle moves across a fixed Eulerian grid, and one may use adaptive grids that follow the resolved particles. A special type of method is the immersed boundary method (IBM), which has proven to be computationally very efficient. The no-slip boundary condition is not explicitly enforced in IBM but simulated by a source term distribution in the region of the particle. The second class of particles are those with non-rigid interface and internal flows, namely, droplets or bubbles. In this case, the major problem is resolving the state discontinuities (jumps) across the interface accurately. In most cases, the flow inside and outside these viscous particles is calculated with a one-field formulation with appropriate boundary or jump conditions at the interface.

8.3.1 SUMMARY RESOLVED RIGID PARTICLES

Numerical simulations of the flow around resolved rigid particles, which are either stationary in space or immersed in a certain flow system, require that a no-slip condition at the particle surface is satisfied. For such kind of simulations, referred to as PR-DNS (particle-resolved DNS), several numerical approaches may be used (e.g. Tenneti and Subramaniam, 2014; Luo *et al.*, 2016). Finite element methods with body-fitted numerical grids require a re-generation of the grid at each time step as the particles are moving through the flow (see e.g. Hu *et al.*, 2001). Such an approach is numerically very costly and, therefore, limited to systems with only a few particles or to two-dimensional simulations, as shown in Figure 8.4a. However,

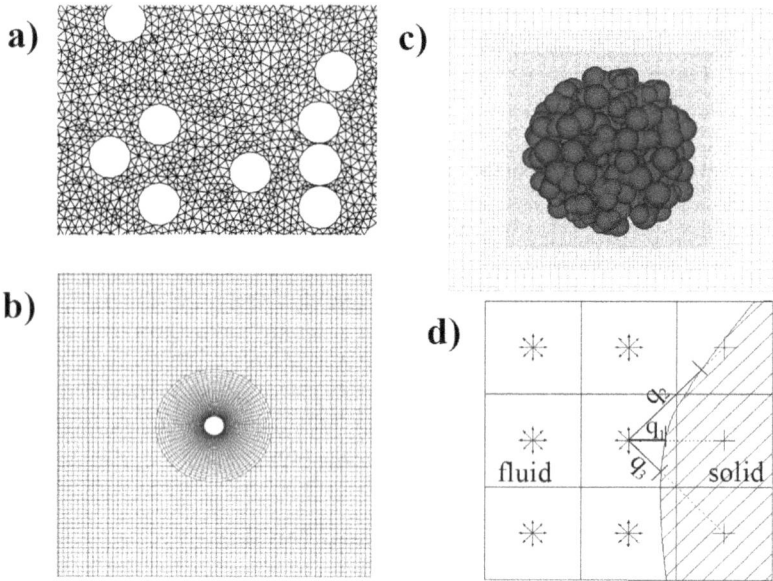

FIGURE 8.4 Illustrations of the different methods for conducting fully resolved particle simulations; (a) finite element methods with body-fitted numerical grid (Hu *et al.*, 2001); (b) overlay grid fitted to the surface of a particle (Burton and Eaton, 2005); (c) grid refinement around an agglomerate for LBM; (d) curved wall boundary condition in the bounce-back method for LBM simulations (Dietzel *et al.*, 2016).

there exists a broad range of different methods, which are based on a structured Eulerian grid for calculating the fluid flow. Based on the particles' treatment, these methods may be classified as follows:

1. Methods using an overlay grid attached to the particles and a structured base-grid for the flow simulations. These methods require interpolation of the flow field between both grids.
2. Methods using a structured grid for the flow simulation and enforcing the no-slip condition on the particle surface by introducing some kind of force field distribution in the region of the particle object. These methods may be grouped under the term *fictitious domain* methods, each using different approaches for emulating the presence of the particles and enforcing the no-slip condition on the surface.
3. The Lattice-Boltzmann method (LBM) normally uses a regular base-grid, and embedded solid bodies or particles are considered as being rigid with a bounce-back boundary condition on the surface.

An overlay grid may perfectly fit the surface of a particle, as shown in Figure 8.4b, but the results between base-grid and particle grid have to be interpolated (Burton and Eaton, 2005; Vreman, 2016). Such a method is suitable for a few fixed particles, which might even have a complex geometry, but becomes more difficult to handle if the particles move, interact, and collide.

In the fictitious domain method, a field of Lagrange multipliers is applied over the particle volume so that the no-slip boundary condition is enforced on the particle surface (Glowinski *et al.*, 2001). This approach allows handling a higher number of particles and, also, allows collisions between them. Glowinski *et al.* (2001) considered the two-dimensional sedimentation of 6,400 particles and the fluidization of 1,204 particles in a small domain having the thickness of the particle diameter. With the development of significant computational power, recent simulations using the fictitious domain method handle computations of millions of particles.

The force-coupling method allows the coupling between resolved particles and fluid flow by adding a finite and localized (in the region of the particle) forcing to the Navier-Stokes equations as a spatially distributed source term (Lomholt and Maxey, 2003; Climent and Maxey, 2009). The particles are tracked in a Lagrangian way based on the interaction force between fluid and particle and with the relevant external forces. This method does not need a numerical grid fitted to the surface of the particle, and the two-phase system may be simulated on a regular grid.

The immersed boundary method (IBM), which originated with the work of Peskin (1972), is also an approximate method to account for the presence of any stationary or moving object in fluid flows using a regular Cartesian grid. The no-slip boundary condition on the surface of the object is realized by introducing a localized force field, which appears in the fluid momentum equation as a distributed source term. The IBM has become very popular in any flow system with fluid-solid boundaries, moving or stationary. The method was adapted for allowing simulations for a large number of resolved solid particles (on the order of 1,000) by Feng and Michaelides (2004, 2005) and then by Uhlmann (2005). Feng and Michaelides (2008a, 2008b) extended the IBM to the energy equation of particles by including a heat source/sink field on the surface of the particles and computed the interacting momentum and energy exchanges of a few hundred particles that interact mechanically and thermally with the carrier fluid. Forces acting on non-spherical particles have also been obtained by this method by Zastawny *et al.* (2012). The main advantage of the IBM is its conceptually simple structure, which is easy to implement and numerically quite efficient. As with all the other fictitious domain methods, a flow field is also produced inside the objects/particles that have a rigid outer surface. However, it has been proven that the effect of this field on the external flow to the objects/particles is negligible.

A powerful method, which also allows simulations of resolved particles moving in a flow, is the Lattice-Boltzmann method (LBM). The Navier-Stokes equations are not solved in the LBM, but the Lattice-Boltzmann equation describing the change of state of a fluid through a probability distribution function. A bounce-back rule is typically used to simulate the particle surfaces (Ladd, 1994; Feng and Michaelides, 2009). This approach is numerically efficient and easy to parallelize, but it has the limitation of a maximum flow Mach number because the method is only applicable to incompressible flows. Normally, this method uses an equidistant regular grid for discretizing flow domain and particles. However, a local grid refinement in several steps may be used to improve the spatial resolution (Filippova and Hänel, 1998; Dietzel and Sommerfeld, 2013), as illustrated in Figure 8.4c.

The LBM has been applied to numerous problems of particle-laden flows as well as for studying the flow around non-spherical particles (Hölzer and Sommerfeld, 2009) and agglomerates (Davis *et al.*, 2012; Dietzel and Sommerfeld, 2013; Dietzel *et al.*, 2016) because of its numerical efficiency and good spatial resolution. The interaction of a flow with embedded objects is realized by using a bounce-back boundary condition on the actual surface of the particle (Figure 8.4 d) by applying an extrapolation method (Guo *et al.*, 2002). Consequently, the fluid dynamic forces acting on complex aggregate structures are directly obtained through the bounce-back boundary condition of the LBM simulations.

Moving particles in turbulent (Gao *et al.*, 2013) and laminar flows (Feng and Michaelides, 2002; Ernst and Sommerfeld, 2015) have also been studied by the LBM. The resolved particles move across a regular Eulerian grid using the forces and moments acting on the particles. The fluid nodes in front of the particle are switched to solid nodes and solid nodes behind the particle to fluid nodes by an extrapolation approach. In the work of Ernst *et al.* (2013), the agglomeration of resolved falling particles and the evolving structures were analyzed, and in the second study, additionally, agglomeration in a shear flow was considered (Ernst and Sommerfeld, 2015), emphasizing the effect of hydrodynamic interaction on agglomeration. In the LBM simulations, there is a problem when the gap between contacting particles becomes smaller than the mesh size. Therefore, some nodes inside contacting particles may be switched to fluid nodes using the equilibrium distribution function together with the known particle velocity. In agglomeration studies, contacting particles were assumed to stick together; however, one may account for the interaction forces between contacting resolved particles, for example, by a square well potential (Derksen, 2012) for the analysis of resolved particle agglomeration in homogeneous isotropic turbulence. In this study, the LBM was used in combination with the IBM to mimic the presence of solid particles.

8.3.2 LATTICE-BOLTZMANN METHOD

The LBM originated from molecular dynamics models, such as Lattice Gas Automata (Higuera *et al.*, 1989; Benzi *et al.*, 1992; Ladd and Frenkel, 1990; Crouse, 2003; Feng and Michaelides, 2002; Aidun and Clausen, 2010). In contrast to other approaches of CFD methods that solve the conservation equations of the macroscopic properties, such as the Navier Stokes equations, the LBM describes the fluid behavior on a mesoscopic scale (He and Luo, 1997; Chen and Doolen, 1998). The method typically uses regular numerical grids for the spatial discretization and has the advantage of being capable to consider the flow about arbitrary and complex-shaped objects. The basic features of the LBM with respect to complex-shaped particles, such as agglomerates or irregular-shaped particles, fixed in a stationary flow or moving through certain flow systems were described (Dietzel and Sommerfeld, 2013; Sommerfeld and Qadir, 2018; Ernst *et al.*, 2013; Ernst and Sommerfeld, 2015).

The main parameter in the Boltzmann statistics is the distribution function $f = f(x, \xi, t)$, which represents the number of fluid elements having velocity ξ at the location x and time t (vector parameters are denoted with bold letters). The fluid elements represented by a probability distribution function move along a lattice mesh and collide at the lattice nodes. The temporal and spatial development of the distribution function is described by the Boltzmann equation considering the collisions between fluid elements

$$\frac{\partial f}{\partial t} + \xi \frac{\partial f}{\partial x} + F_{ext} \cdot \frac{\partial f}{\partial \xi} = \Omega(f). \tag{8.1}$$

Eq. (8.1) is derived under the assumption that any external forces, F_{ext}, are neglected. Typically, the Boltzmann equation is solved with the help of a single relaxation time collision operator approximated by the Bhatnagar-Gross-Krook (BGK) approach (Bhatnagar et al., 1954). The relaxation of the distribution function to an equilibrium distribution is assumed to occur during a single constant relaxation parameter τ. Consequently, the collision operator, $\Omega(f)$, approximated by the Bhatnagar-Gross-Krook (Bhatnagar et al., 1954) approach, is written as

$$\Omega(f) = -\frac{1}{\tau}\left(f - f^{(0)}\right), \tag{8.2}$$

leading to the following form of the Boltzmann equation

$$\left(\frac{\partial}{\partial t} + \xi \cdot \nabla\right) f(x, \xi, t) = -\frac{1}{\tau}\left(f(x, \xi, t) - f^{(0)}(x, \xi, t)\right) \tag{8.3}$$

The equilibrium distribution $f^{(0)}$ is obtained from Maxwell's equation using the speed of sound

$$f^{(0)}(x, \xi, t) = \frac{\rho}{\left(2\pi c_s^2\right)^{\frac{3}{2}}} exp\left(-\frac{(\xi - u)^2}{2c_s^2}\right) \tag{8.4}$$

The density and momentum per unit volume can be derived as moments of the distribution function

$$\rho = \int_{\xi_i = -\infty}^{\infty} f(x, \xi, t)\, d^3\xi_i \tag{8.5}$$

$$\rho u = \int_{\xi_i = -\infty}^{\infty} \xi f(x, \xi, t)\, d^3\xi_i \tag{8.6}$$

Substitution of the continuous velocities in Eq. (8.3) by discrete ones, indicated with subscript σi, leads to the discrete Boltzmann equation. The number of available discrete velocity directions σi that connect the lattice nodes with each other depends on the applied model. One of them, the D3Q19 model, is shown in Figure 8.5. The grid applies to a three-dimensional grid (D3) that entails 19 discrete propagation directions (Q19), which are distinguished in the 6 horizontal and vertical directions ($\sigma = 1$), 12 diagonal directions ($\sigma = 2$), and one corresponding to zero

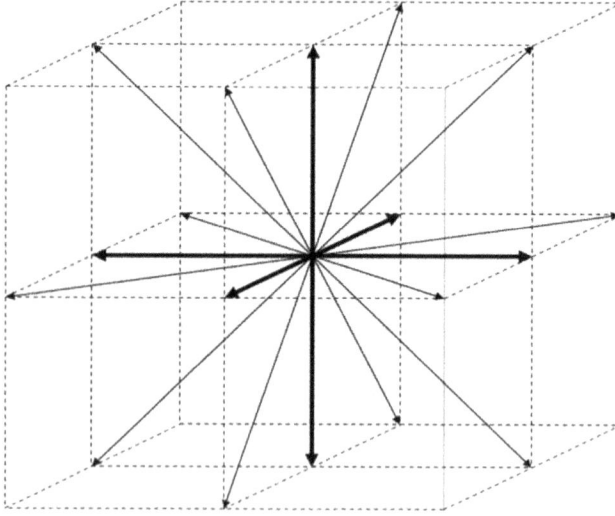

FIGURE 8.5 Schematic diagram of the discrete velocity directions of the D3Q19 model in the Lattice-Boltzmann method.

velocity ($\sigma = 0$). Information is transported along these lattice directions followed by the collision or relaxation step. Discretization in space and time yields the Lattice Boltzmann equation in the form

$$f_{\sigma i}\left(x + \xi_{\sigma i}\Delta t, t + \Delta t\right) - f_{\sigma i}\left(x, t\right) = -\frac{\Delta t}{\tau}\left(f_{\sigma i}\left(x, t\right) - f_{\sigma i}^{(0)}\left(x, t\right)\right) \qquad (8.7)$$

The discretized equilibrium distribution is obtained from a Taylor series expansion for u, limiting the application of the equation to incompressible flows with low Mach number (Feng and Michaelides, 2002). The appropriate weighting factors for the D3Q19 model are as follows

$$f_{\sigma i}^{(0)}\left(x, t\right) = \omega_{\sigma i}\rho\left(1 + \frac{3\xi_{\sigma i}\cdot u}{c^2} + \frac{9\left(\xi_{\sigma i}\cdot u\right)^2}{2c^4} - \frac{3u^2}{2c^2}\right) \qquad (8.8)$$

$$\omega_{\sigma i} = \begin{cases} 1/3; & \sigma = 0, \ i = 1 \\ 1/18; & \sigma = 1, \ i = 1\ldots6 \\ 1/36; & \sigma = 2, \ i = 1\ldots12 \end{cases} \qquad (8.9)$$

A grid constant c can be defined as the ratio of the spatial to temporal discretization, which is also related to the speed of sound c_s

$$c = \frac{\Delta x}{\Delta t} = \sqrt{3}c_s \qquad (8.10)$$

The moments of the discrete distribution function, from Eqs. (8.5) and (8.6), are as follows

$$\rho(x,t) = \sum_{\sigma} \sum_i f_{\sigma i}(x,t) \tag{8.11}$$

$$\rho(x,t) u(x,t) = \sum_{\sigma} \sum_i \xi_{\sigma i} f_{\sigma i}(x,t) \tag{8.12}$$

The local pressure may be derived from the local density and the speed of sound using Eq. (8.10)

$$p(x,t) = c_s^2 \rho(x,t) = \frac{1}{3}\left(\frac{\Delta x}{\Delta t}\right)^2 \rho(x,t) \tag{8.13}$$

The relaxation parameter τ is related to the dynamic viscosity of the fluid

$$\mu = \frac{1}{6}\rho c^2 (2\tau - \Delta t) \tag{8.14}$$

Eq. (8.14) is typically used for the determination of the relaxation parameter, τ. The numerical stability of the LBM is influenced by several criteria, such as the upper limit of the Mach number ($Ma = u/c_s < 0.3$) and the lower bound of the relaxation time parameter ($t/\Delta t > 0.5$). This places a limitation on the Reynolds number that can be simulated, depending however on the resolution by the numerical grid. Local grid refinement may be applied for accurately resolving complex shaped objects embedded in the flow as well as for strong velocity gradients. Naturally, in regions of lower velocity gradients, only coarser grids are required. This results in significant reduction of the computational effort (Crouse, 2003, Feng and Michaelides, 2003).

8.3.2.1   Treatment of Solid-Fluid Boundaries

In order to resolve the surface of complex-shaped objects inside the flow field with a regular structured mesh, the exact position of the particle surface within a cell is required. There are several methods for describing a complex surface with a no-slip boundary condition in the frame of the LBM, for example, Bouzidi et al. (2001); Guo et al. (2002); Mei et al. (2002). Due to its applicability with moving boundaries, the extrapolation scheme proposed by Guo et al. (2002) is presented in this section.

The basic premise of this scheme is to decompose the numerical grid into two parts, as shown in Figure 8.6. All nodes that lay within a particle structure are marked as solid nodes (gray in the figure), and nodes outside the particle are marked as fluid nodes (white). It is important to note that in the LBM, the information and particle properties are stored at the grid nodes, while the grid cells represent their associated control volumes. The position for any type of grid node is always located at the front left bottom vertex within the cubic grid cells. According to Figure 8.6, the link between the fluid node x_F and the solid node x_S

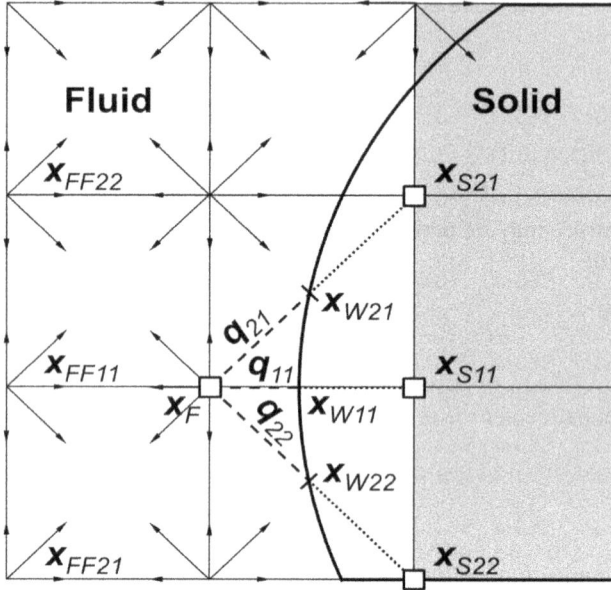

FIGURE 8.6 Schematic two-dimensional diagram of the boundary condition in the LBM: The numerical grid is divided into fluid cells (white grid cells) and solid cells (gray grid cells). The effective contour of the curved object (bold solid line) is discretized by the distribution functions (black arrows). The relative distance between the fluid node and the curved particle surface is weighted by the parameter $q_{\sigma i}$ (dashed lines). Note that the grid nodes (small open squares) are always located at the front left bottom vertex within the cubic grid cells.

intersects the physical boundary at the point x_W. The fraction of the intercept on the fluid side is

$$q = \frac{|x_F - x_W|}{|x_F - x_S|} \qquad (8.15)$$

The parameter q indicates the relative distance between the particle surface and the next fluid node in the σi-direction, as shown in Figure 8.6.

For completing the propagation step of the fluid at x_F, the post-relaxation distribution function at the solid node $f_{\sigma i}^{*}(x_S, t)$ is needed. This is defined as

$$f_{\sigma i}^{*}(x_S, t) = f_{\sigma i}^{eq}(x_S, t) + \left(1 - \frac{\Delta t}{\tau}\right) f_{\sigma i}^{neq}(x_S, t) \qquad (8.16)$$

where $f_{\sigma i}^{eq}(x_S, t)$ and $f_{\sigma i}^{neq}(x_S, t)$ are the equilibrium and non-equilibrium parts of $f_{\sigma i}(x_S, t)$, respectively. The equilibrium part $f_{\sigma i}^{eq}(x_S, t)$ is approximated by the expression

$$f_{\sigma i}^{eq}(x_S, t) = \omega_{\sigma i}\left[p_P + \rho\left(\frac{3\xi_{\sigma i} u_S}{c^2} + \frac{9(\xi_{\sigma i} u_S)^2}{2c^4} - \frac{3u_S^2}{2c^2} \right) \right], \qquad (8.17)$$

where ρ_P and u_S are the density and velocity at the solid nodes x_S, and ρ is the fluid density at x_F. The velocity u_S is determined by a linear extrapolation as a function of q

$$u_S = u(x_W) + (q-1)u(x_F) + (1-q)\frac{2u(x_W) + (q-1)u(x_F)}{1+q}, \quad q < 0.75$$

$$u_S = \frac{u(x_W) + (q-1)u(x_F)}{q}, \quad q \geq 0.75$$

(8.18)

The non-equilibrium part $f_{\sigma i}^{\mathrm{neq}}(x_S,t)$ can be approximated by the non-equilibrium part of the distribution function at the fluid nodes x_F and x_{FF} as follows

$$f_{\sigma i}^{\mathrm{neq}}(x_S,t) = q_{\sigma i}f_{\sigma i}^{\mathrm{neq}}(x_F,t) + (1-q_{\sigma i})f_{\sigma i}^{\mathrm{neq}}(x_{FF},t), \quad q < 0.75$$

$$f_{\sigma i}^{\mathrm{neq}}(x_S,t) = f_{\sigma i}^{\mathrm{neq}}(x_F,t), \quad q \geq 0.75$$

(8.19)

Using the expression for $q_{\sigma i}$, the particles are fully resolved by the numerical grid. Consequently, details of the flow around the particles are captured, and the forces on the particles follow from the surface boundary condition of the LBM (i.e. the bounce-back boundary condition).

8.3.2.2 Description of the Particle Motion

The particle motion is described by two governing equations: Newton's second law for the translation and the Euler's equations for rotation. The forces and torques, which cause the change of position and orientation, are computed from the fluid-particle interactions. The flow-induced forces and torques acting on the particles (moving or fixed in space) are determined by balancing the momentum of the fluid at the particle surface. The difference of the fluid momentum before and after a surface contact yields the local forces $F_{\sigma i}$. The local torques $T_{\sigma i}$ are derived from the local force and their distance from the center of rotation x_R

$$F_{\sigma i}(x,t+\Delta t/2) = \frac{\Delta V}{\Delta t}\left(f_{\sigma i}(x,t+\Delta t) - f_{\sigma i}^*(x,t)\right)\xi_{\sigma i}$$

(8.20)

$$T_{\sigma i}(x,t+\Delta t/2) = (x - x_R) \times F_{\sigma i}(x,t+\Delta t/2)$$

(8.21)

where $f_{\sigma i}^*(x,t)$ is the post-relaxation discrete distribution function before the propagation step. Summarizing all local forces and torques along the object's surface yields the total force and torque

$$F(t+\Delta t/2) = \sum_x \sum_{\sigma i} F_{\sigma i}(x,t+\Delta t/2)$$

(8.22)

$$T(t+\Delta t/2) = \sum_x \sum_{\sigma i} T_{\sigma i}(x,t+\Delta t/2)$$

(8.23)

The particle motion is calculated in a Lagrangian frame of reference, which considers a discrete resolved particle travelling in a continuous fluid medium. The changes of the position and angular displacement of the particle, x_P and φ_P, as well as the translational and angular components of the particle velocity, $\boldsymbol{u}_P$ and $\boldsymbol{\omega}_P$, are calculated by solving a set of ordinary differential equations (ODEs) along the particle trajectory

$$\boldsymbol{u}_P = \frac{d\boldsymbol{x}_P}{dt}; \quad m_P \frac{d\boldsymbol{u}_P}{dt} = \boldsymbol{F} + \boldsymbol{F}_{\text{ext}} \tag{8.24}$$

$$\boldsymbol{\omega}_P = \frac{d\boldsymbol{\varphi}_P}{dt}; \quad \boldsymbol{J}_P \frac{d\boldsymbol{\omega}_P}{dt} + \boldsymbol{\omega}_P \times \left(\boldsymbol{J}_P \cdot \boldsymbol{\omega}_P\right) = \boldsymbol{T} \tag{8.25}$$

where m_P is the particle mass, $\boldsymbol{J}_P$ the moment of inertia of the particle, and dt the time step of the Lattice-Boltzmann scheme. Therefore, the fluid flow and the particle movement are calculated with the same temporal discretization. The particle variables are updated during each LBM iteration step, realising a transient simulation. In addition to the flow-induced forces and torques on a particle, $\boldsymbol{F}$ and $\boldsymbol{T}$, external forces, $\boldsymbol{F}_{\text{ext}}$, are also considered in the equation of motion. For example, when the gravity/buoyancy force acts on the particles, Eq. (8.24) is modified to the following

$$m_P \frac{d\boldsymbol{u}_P}{dt} = \boldsymbol{F} + \left(\rho_P - \rho\right)V_P\boldsymbol{g} \tag{8.26}$$

where V_P is the volume of the particle, and $\boldsymbol{g}$ is the gravitational acceleration. In order to calculate the particle motion, the integration of Eqs. (8.24) to (8.26) may be performed with the help of a forward Euler scheme. The linearized ODEs for translational motion are

$$\boldsymbol{u}_P\left(t\right) = \frac{\boldsymbol{F}\left(t - \Delta t\right) + \boldsymbol{F}_{\text{ext}}\left(t - \Delta t\right)}{m_P} \Delta t + \boldsymbol{u}_P\left(t - \Delta t\right) \tag{8.27}$$

and

$$\boldsymbol{x}_P\left(t\right) = \frac{\boldsymbol{u}_P\left(t - \Delta t\right) + \boldsymbol{u}_P\left(t\right)}{2} \Delta t + \boldsymbol{x}_P\left(t - \Delta t\right) \tag{8.28}$$

For the rotation of rigid particles, the mathematical concept of quaternions is used to represent the angular displacement of the particle within the three-dimensional space (Kuipers, 2002; Feng and Michaelides, 2004). In addition to the higher computational accuracy and efficiency, an advantage over ordinary rotation matrices is that the solution of quaternions allows an enhanced stability of the numerical solution as a result of the absence of singularities. The change of the particle angular velocity and angular displacement are obtained from the following expressions

$$\omega_{P,x}\left(t\right) = \frac{1}{J_{P,xx}}\left[T_x + \left(J_{P,yy} - J_{P,zz}\right)\omega_{P,y}\left(t\right)\omega_{P,z}\left(t\right)\right]\Delta t + \omega_{P,x}\left(t - \Delta t\right)$$

$$\omega_{P,y}\left(t\right) = \frac{1}{J_{P,yy}}\left[T_y + \left(J_{P,zz} - J_{P,xx}\right)\omega_{P,z}\left(t\right)\omega_{P,x}\left(t\right)\right]\Delta t + \omega_{P,y}\left(t - \Delta t\right)$$

$$\omega_{P,z}\left(t\right) = \frac{1}{J_{P,zz}}\left[T_z + \left(J_{P,xx} - J_{P,yy}\right)\omega_{P,x}\left(t\right)\omega_{P,y}\left(t\right)\right]\Delta t + \omega_{P,z}\left(t - \Delta t\right) \tag{8.29}$$

$$\varphi_P(t) = \frac{\omega_P(t - \Delta t) + \omega_P(t)}{2} \Delta t + \varphi_P(t - \Delta t) \qquad (8.30)$$

It must be noted that for the linearization of the ODEs, in this case the time step, Δt, must be several times smaller than the particle response time, τ_v.

8.3.2.3 Moving Solid-Fluid Boundaries

Because of the Lagrangian particle treatment, a change of particle position involves state modifications of the numerical grid and of the boundary conditions. When a particle is moving over a stationary fluid grid, nodes which during a time step are overlapped by the particle, are switched from fluid to solid status. Nodes left behind a particle and previously marked as solid are switched to fluid nodes. Figure 8.7 demonstrates the procedure of switching the cell state from fluid to solid and vice versa for the boundary of a disk moving over a two-dimensional grid.

For the initialization of new fluid nodes, the 19 equilibrium and non-equilibrium distribution functions of each node may be reconstructed through an extrapolation method, proposed by Caiazzo (2008). The so-called refill method consists of four steps. At first, the fluid velocity and density of the new fluid node are obtained by extrapolation from the closest fluid neighbor nodes. Secondly, the corresponding

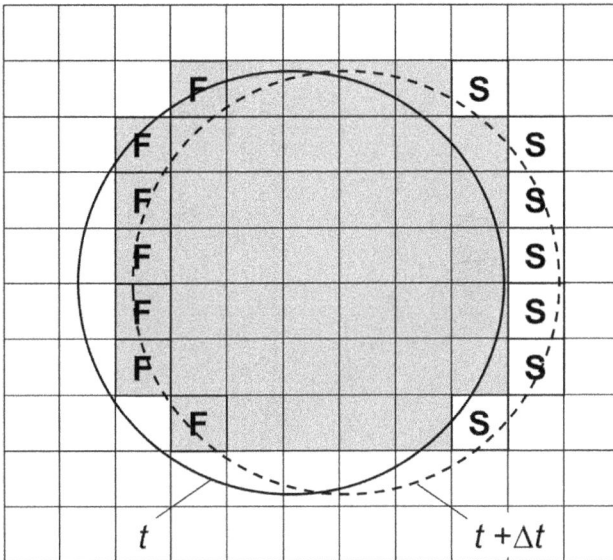

FIGURE 8.7   Motion of a spherical particle over the structured regular mesh from time t to $t + \Delta t$: The switching from fluid to solid status, denoted by S, and from solid to fluid status, denoted by F, are illustrated in a two-dimensional diagram.

equilibrium distribution is calculated in a second step using the expression (Lallemand and Luo, 2003)

$$f_{\sigma i}^{eq} = \omega_{\sigma i} \tilde{\rho} \left[1 + \frac{3\xi_{\sigma i}\tilde{u}}{c^2} + \frac{9(\xi_{\sigma i}\tilde{u})^2}{2c^4} - \frac{3\tilde{u}^2}{2c^2} \right]. \tag{8.31}$$

where $\tilde{\rho}$ and $\tilde{u}$ are extrapolations of the fluid density and velocity at the reconstructed fluid node. Thirdly, the non-equilibrium part of the distribution function at the new fluid nodes is determined from the non-equilibrium part of the neighbour nodes by extrapolation (Caiazzo, 2008). Fourthly, both the equilibrium and non-equilibrium parts are used to initialize the missing distributions according to Eqs. (8.1 and 8.2). Thömmes et al. (2009) demonstrated the applicability of this procedure by simulating multiphase flows with free surfaces.

8.3.2.4 Solid Boundaries in Close Contact

During the approach of two particles or particle-wall interaction, the distance between the solid surfaces may fall below the resolution limit of the finite grid. This leads to a non-physical force acting on the particle because fluid nodes are missing in the gap between the two surfaces, and hence, the momentum balance of the fluid is incomplete at the particle surface. Figure 8.8 illustrates the approach stage of two

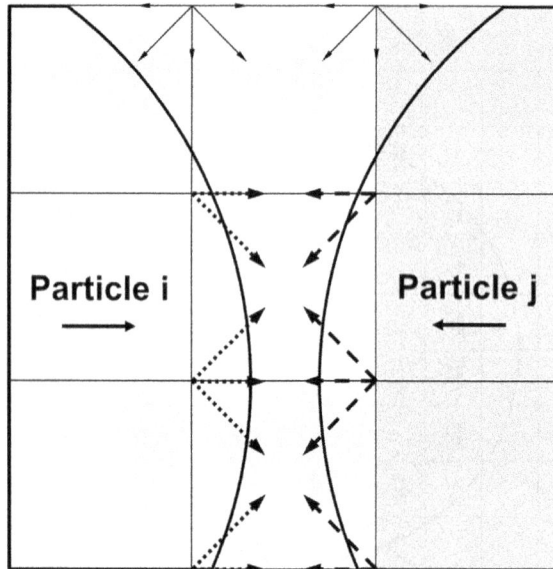

FIGURE 8.8 Schematic diagram of two approaching particles, i and j, where the flow in the gap between both particles is not fully resolved by the numerical grid. The dashed arrows represent the distribution functions at the nodes of particle j that are missing for the momentum balance of the fluid along the surface of particle i. Dotted arrows belong to particle j. Note that the grid nodes are always located at the front left bottom vertex within the cubic grid cells.

particles, i and j. It is apparent that fluid nodes are missing between particle nodes (dashed and dotted arrows in the figure). A realistic flow-induced force acting on each particle and acting against the direction of the approaching other particle cannot be obtained. This has significant consequences for any interaction event (collision, agglomeration, etc.) taking place by the particles.

To address this problem, a reconstruction scheme for the missing fluid distribution functions is applied to avoid the non-physical approach of the two particles. While the particles are moving in close neighbourhood, temporary so-called non-fluid nodes with fluid properties are dynamically generated at the solid nodes within the opposite collision partner, if the gap width between the particle surfaces becomes smaller than the spatial discretization Δx (Ladd, 1997; Xu and Michaelides, 2003). The positions of these nodes correspond to the positions of the missing fluid distribution functions, which are depicted in Figure 8.8. Non-fluid nodes are generated on demand prior to the computation of the momentum balance of the fluid at the particle surface for each time step. Hence, the temporary switchover from solid nodes to non-fluid nodes (i.e. solid nodes with fluid properties) has no effect on the effective particle shape. The initialization of these artificial nodes is accomplished by reconstructing the fluid velocity at the particle surface. On this surface, the fluid velocity is equal to the velocity of the particle, $u_F = u_P$, due to the no-slip condition. Accordingly, the missing distribution functions can be determined using an equilibrium distribution in analogy to Eq. (8.8)

$$f_{\sigma i} = \omega_{\sigma i}\rho \left[1 + \frac{3\xi_{\sigma i}u_P}{c^2} + \frac{9\left(\xi_{\sigma i}\,u_P\right)^2}{2c^4} - \frac{3u_P^{\,2}}{2c^2} \right] \qquad (8.32)$$

The momentum balance of the fluid distribution functions at the particle surface is now fulfilled.

It must be noted that even with this approach, the resolution of the fluid is still limited and that a physical lubrication layer cannot be accurately simulated when particles are in very close proximity. The portion of lubrication, which is not directly resolved by the numerical grid, must be modelled by an empirical approximation. Such an approximation is the introduction of a pair-wise sub-grid repulsion force acting on all involved particles (Nguyen and Ladd, 2002; Feng and Michaelides, 2003).

The LBM is an attractive and efficient numerical approach for complex objects embedded in a flow or moving through a flow domain. Using local grid refinement, even highly structured agglomerates may be accurately resolved using a curved wall boundary condition (Dietzel et al., 2016). Compared to the finite element method (FEM), the requirements on the mesh structure and number of control volumes are significantly lower, as may be seen in Figure 8.9.

8.3.2.5 Examples of LBM Applications

The modelling of dry powder inhalers entails numerous elementary processes, which may be analysed by particle-resolved direct numerical simulations (PR-DNS). One

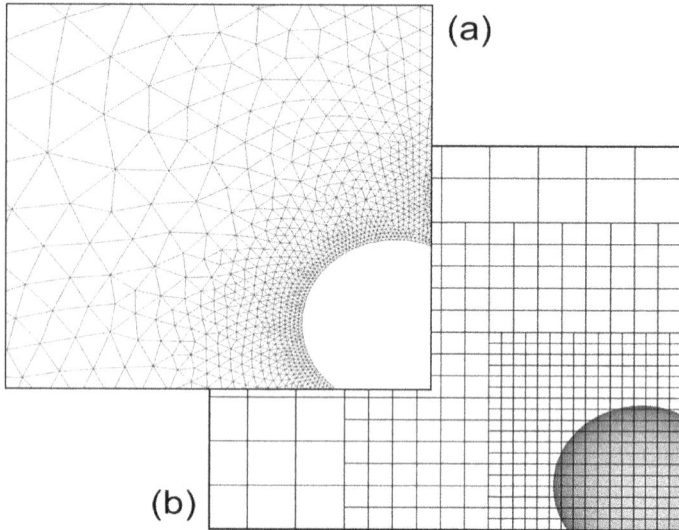

FIGURE 8.9 Illustration of the discretized fluid domain in the region of an identical particle under the application of local grid refinements: (a) FEM and (b) LBM.

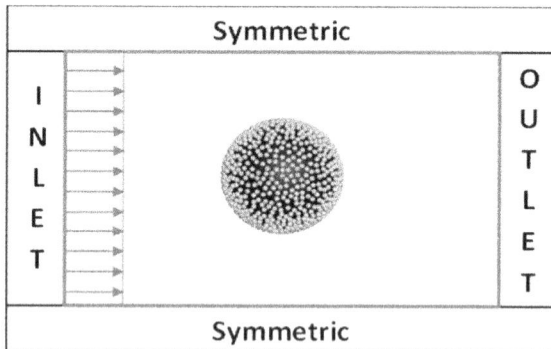

FIGURE 8.10 Computational domain with the applied boundary conditions; a cuboid domain where the particle cluster is centrally fixed and exposed to a plug flow.

challenging example for the application of LBM to complex geometries is the detachment of fine spherical drug particles from a larger carrier particle through the flow stresses (Cui *et al.*, 2014). Typically, the larger spherical carrier has a diameter of 100 μm, and the drug particle diameter is up to 5 μm. When considering a surface coverage of 50% for the carrier, then 882 spherical drug particles are randomly paced on the carrier surface in a single layer. This requires an extremely fine discretization for simulating the flow about such a cluster. For the simulation of such a case, the particle cluster was centrally fixed (i.e. also, no rotation was considered) in a cuboid flow domain and exposed to a laminar plug flow, as shown in Figure 8.10. At the outlet, a gradient-free condition is applied and at all side faces symmetry boundary conditions are used. In these simulations, first the carrier particle Reynolds number was varied, being defined with the diameter of the carrier particle, $D_{carrier}$, and the inflow velocity, U_0

$$\mathrm{Re} = \frac{\rho\, D_{carrier}\, U_0}{\mu} \qquad\qquad (8.33)$$

The selection of the grid dimension (the resolution of drug particles) and the size of the computational domain must be first evaluated for reliable numerical computations. With four grid refinement regions around the cluster, sufficient resolution of each fine drug particle was obtained, namely, six grid cells per diameter, as shown in Figure 8.11. This was the conclusion of a detailed study with continuously changing domain size and grid resolution and comparing the resulting forces on a drug particle (Cui and Sommerfeld, 2015).

The total number of grid cells and the number of refinement regions were selected according to the case considered and the carrier Reynolds number. For smaller Reynolds numbers, $\mathrm{Re} < 100$, the total mesh had 3.8 million control volumes, and for $\mathrm{Re} > 100$, the domain was increased, and 5.7 million control volumes were used.

In the simulations, the forces acting on the fine drug particles in dependence of their angular location on the carrier surface (position angle) and the Reynolds number, Re, were evaluated considering different flow situations. The relevant forces to possibly detach the fine particles from the carrier are the total fluid force, the van der Waals adhesion force, and the friction force. The total fluid force can be separated into a normal and a tangential component, that is, F_n and F_t. Both fluid forces are relevant for a potential detachment by liftoff, sliding, or rolling (Ibrahim *et al.*, 2003). The gravity force is about eight times smaller in magnitude than these forces and therefore may be neglected.

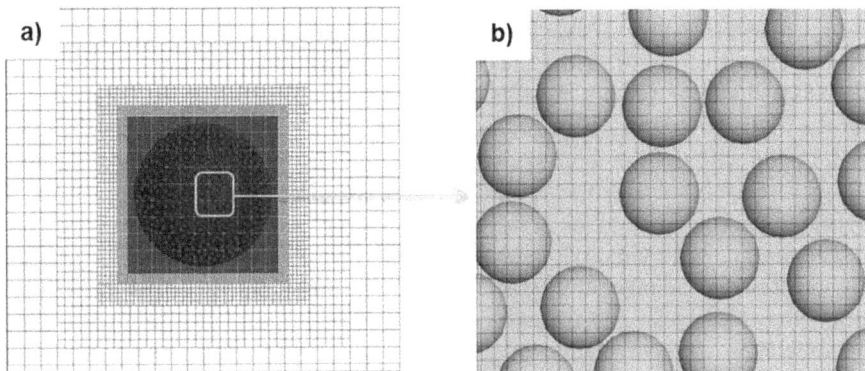

FIGURE 8.11 Multigrid approach for resolving all the relevant length scales, (a) different grid levels or refinement regions around the particle cluster, (b) zoom into the finest mesh region showing the resolution of the fine 5μm spherical particles (from Cui and Sommerfeld, 2015).

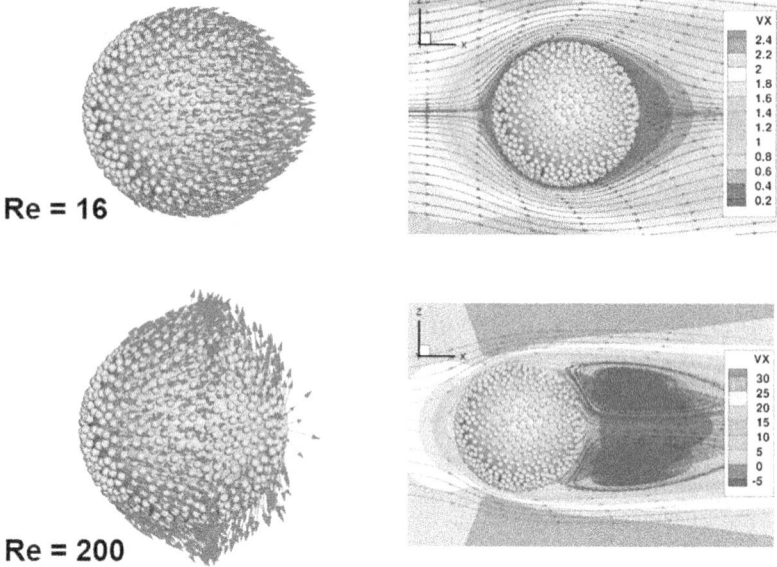

FIGURE 8.12 Flow field about the particle cluster and the resulting directions of the total fluid forces on the fine particles for two Reynolds numbers (Re = 16 and 200). Left: Distribution of particles with vectors indicating the fluid force direction. Right: Color contours for the stream-wise velocity (scale in m/s) and streamlines (from Cui and Sommerfeld, 2015).

All total force vectors on the fine particles are illustrated in Figure 8.12 for two carrier particle Reynolds numbers, together with the flow field about the cluster. It should be emphasized that the force vectors only indicate the direction and not the magnitude for clarity (i.e. otherwise many force vectors would not be visible). The directions of the forces are strongly correlated with the flow structure about the cluster. For the lower Reynolds numbers, no wake separation occurs, and consequently the force vectors in the wake region are mostly directed away from the carrier in the stream-wise direction. Near the equator region (i.e. plane perpendicular to the flow direction), the force vectors are almost parallel to the carrier surface for small *Re*. For the highest Reynolds number, the region of wake separation is clearly visible from the direction of the force vectors. Within the wake, with the circulating flow along the cluster surface, the velocity vectors are directed radially outward with respect to the centerline of the geometry.

The forces on individual drug particles sitting on the carrier surface depend on the flow Reynolds number and associated temporal variations as well as the angular location on the carrier and the coverage. Therefore, in order to obtain reasonable approximations of the hydrodynamic forces on the drug particles over the surface of the carrier, that is, in dependence of angular position, at least four simulation runs were conducted for each condition with a different randomized fine particle distribution. In addition, for each drug particle and case, a temporal averaging over the last 10,000 time steps was conducted.

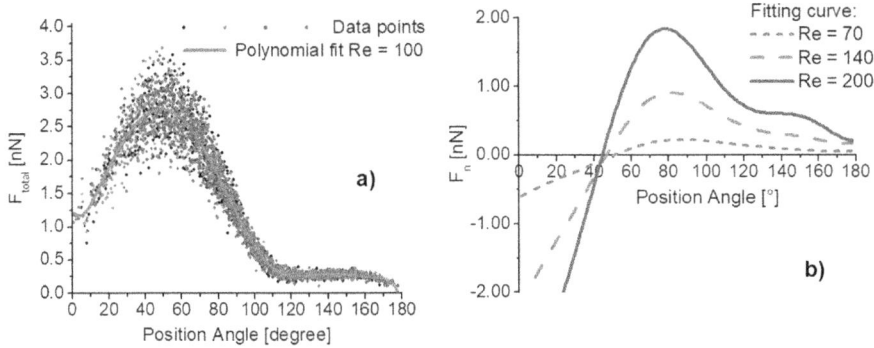

FIGURE 8.13 (a) Total fluid force on all the fine drug particles situated on the carrier particle as a function of the position angle for four different random distributions (dots) and the resulting polynomial fitting curve (Re = 100, coverage degree 50%, $D_{fine}/D_{carrier}$ = 5/100); (b) Fitting curves for the normal force on the fine particles as functions of the position angle for different Reynolds numbers, that is, 70, 140, and 200 (coverage degree 50%, $D_{fine}/D_{carrier}$ = 5/100).

Since the flow about the particle cluster is axisymmetric for a plug flow, the forces on all the fine particles depend on the position angle only, as shown in the plot of Figure 8.13a. All the data points in this figure emanate from four simulation runs for Re = 100. Due to the large scatter, it is more convenient to develop a correlation curve, which is depicted in Figure 8.13a. The maximum of the total force in this case is at an angle of about 50° from the front stagnation point, where the boundary layer on the cluster is very thin, and the streamlines are squeezed together. This is also the region with the largest scatter of the forces. Within the wake region on the back side of the cluster, i.e. beyond an angle of about 110° at Re = 100, the total force on the fine particles is much smaller. In addition, the scatter of the total force is considerably lower in this region.

The magnitude of the forces exerted by the fluid and the angular location of the force maximum largely depend on the Reynolds number. The fitting curves for the normal force on the fine particles are shown as a function of the position angle for different Reynolds numbers in Figure 8.13b. The magnitude of the force is increasing with free stream velocity (i.e. Reynolds number). In the front section of the cluster, the normal force is negative (i.e. for position angles smaller than 40° for the considered Reynolds numbers), implying that these fine particles cannot be detached by liftoff as they are pushed toward the carrier particle. Beyond this stagnation region, the normal force becomes positive and continuously increases to a maximum, allowing for liftoff and detachment of the fine particles, when it becomes larger than the van der Waals force. As a result of the wake separation behind the particle, the maximum of the normal force is continuously shifted toward the front side of the particle. For higher Reynolds numbers (Figure 8.13b), the maximum in the normal force is found just in front of the equator, i.e. between 75° to 85°. Even for the highest Reynolds number, the normal force reaches only about 2 nN. This is far below measured adhesion forces for the different carrier particle surface treatment, which is in the range between 63 nN and 257 nN (Sommerfeld et al., 2019). Consequently, the drug particles cannot be detached by direct liftoff but possibly with a certain probability through sliding or rolling (Cui and Sommerfeld, 2015). These results demonstrate the usefulness of PR-CFD for a detailed analysis of particle-scale processes, which are hardly accessible through experiments.

8.3.3 Immersed Boundary Methods

The prescription of the interfacial boundary conditions is a rather unusual feature within the framework of the governing equations the numerical codes solve. In the Newtonian scientific framework, the governing equations that are solved for the transport of momentum are given in terms of forces (rates of change of momentum), while velocity interfacial conditions need to be prescribed at the interface by a commonly agreed convention or experimental data. A more consistent approach within the Newtonian framework would be to describe the interfaces and other boundaries in terms of interfacial forces. Peskin (1972, 1977, 2003) used such an approach and developed the immersed boundary method (IBM) to model the flow of blood in the heart. The method uses a fixed Cartesian mesh for the fluid, which is composed of Eulerian nodes. For the solid boundaries, which are immersed in the fluid, the IBM uses a set of Lagrangian points that are advected according to the rules of fluid-solid interactions. The boundaries are modeled as continuous force fields acting on the fluid. The advantage of the IBM is that, instead of re-meshing the fluid domain to account for the flow of particles, the method uses a fixed mesh to represent the fluid field.

Feng and Michaelides (2004) combined the IBM and the Lattice Boltzmann Method (LBM) for the first time to model the flow and interactions of large groups of particles in flow systems. They computed the boundary surface force at the interfaces using a numerical penalty method. The moving boundaries of the particles are represented by a set of Lagrangian boundary points, which are advected by the fluid. Feng and Michaelides (2005) also developed the *Proteus* code using a numerical method that combines the direct forcing scheme (Mohd-Yusof, 1997) and the IBM. *Proteus* makes use of Eulerian lattice nodes for the fluid flow field and Lagrangian boundary nodes to represent particles or moving-boundary surfaces. A few months later, Uhlmann (2005) also developed a similar computational method that combines the IBM with a finite-difference based fluid solver. A more recent extension of the IBM has been developed in combination with the resolved discrete particle model (RDPM) to be used primarily with heat and mass interaction in the particulate flow systems. This numerical scheme has also received significant attention (Gan *et al.*, 2003; Yu and Shao, 2007; Feng and Michaelides, 2008a, 2008b, 2009; Tavassoli *et al.*, 2013; Tenneti *et al.*, 2013) and is briefly described at the end of this section.

8.3.3.1 Fundamentals of the Immersed Boundary Methods

The key characteristic of the IBM is that the existence of a solid body is represented by its effect on the fluid as a force field. This is implemented by introducing a *body force density* term into the fluid momentum equations. Let us consider a particle with a boundary surface, Γ, immersed in a three-dimensional incompressible viscous fluid with a domain, Ω, as detailed in Feng and Michaelides (2004). The particle boundary surface, Γ, is represented by the Lagrangian parametric coordinates, s, and the flow domain, Ω, is represented by the Eulerian coordinates x. Hence, the positions of the surface may be written as $x = X(s,t)$. The no-slip boundary condition at the particle surface is satisfied by enforcing the velocity at all interfaces to be equal to the velocity of the fluid at the same location

$$\frac{\partial X(s,t)}{\partial t} = u\big(X(s,t),t\big), \tag{8.34}$$

where u is the fluid velocity. Let $F(s,t)$ and $f(x,t)$ represent the particle surface force density and the fluid body force density, respectively. In this treatment of the particulate system, the entire domain, Ω, can be regarded to be filled with fluid, and its velocity field is described by

$$\rho\left(\frac{\partial u}{\partial t} + u\cdot\nabla u\right) = \mu\nabla^2 u - \nabla p + f. \tag{8.35}$$

$$\nabla\cdot u = 0. \tag{8.36}$$

with

$$f(x,t) = \int_\Gamma F(s,t)\delta\big(x - X(s,t)\big)ds, \tag{8.37}$$

and

$$\frac{\partial X}{\partial t} = \int_\Omega u(x,t)\delta\big(x - X(s,t)\big)dx, \tag{8.38}$$

where $p(x,t)$ is the fluid pressure, ρ is the fluid density, and μ the fluid viscosity.

Eqs. (8.35) and (8.36) are the governing equations for viscous, isothermal, incompressible flow, which include the effect of the particle interface as a force density. Eq. (8.37) shows how the force density of the fluid, $f(x,t)$, may be calculated from the immersed boundary surface force density, $F(s,t)$. Eq. (8.38) is essentially the no-slip condition at the interface, since the particle moves at the same velocity as the neighboring fluid.

In its original development, the IBM theory was tailored for solving the interaction between elastic, deformable fibers, and fluid. To model a particle, the density force is only defined on the particle surface, as indicated by Eq. (8.37). This force field is set to be zero inside the particle. In the numerical implementation of the IBM, the entire fluid domain, including the parts that are occupied by immersed bodies, is divided into a set of fixed regular nodes. The fixed nodes are not moving with the flow and are simple Eulerian nodes. The immersed fiber is discretized by a set of segments with each segment represented by a boundary point that moves under the action of the fluid. We will call these boundary nodes Lagrangian nodes. It must be noted that the Lagrangian nodes do not necessarily coincide with the Eulerian nodes.

There are several schemes to compute the surface force density associated with the Lagrangian nodes that represent the fluid-particle interface. For example, the "penalty method" considers the particle boundary as an elastic fiber with high stiffness in the case of modelling solid particles (Feng and Michaelides, 2004, 2005). In this case, we define a set of reference nodes and a set of boundary nodes for each particle. We use a template that is undergoing a rigid body motion consisting of the reference nodes. Before the particle deforms, the reference nodes are coincident with the particle boundary nodes. Once the particle has deformed under particle-fluid interaction, the particle boundary nodes are compared to the reference nodes. During a single time step, the reference nodes following the motion of a template, which is a rigid particle while the boundary nodes following the motion of fluid, thus resulting in a distortion. Such a distortion between the boundary nodes and reference nodes causes an elastic restoration force on the particle boundary node. This method requires the input in the computational scheme the stiffness of the fiber or particle. A very large value for the stiffness is used for modelling solid particles.

Figure 8.14 is a schematic diagram of the main steps that constitute the IBM. Part (a) shows the boundary of an ellipsoidal particle, which is carried by the fluid within the Eulerian grid, where the fluid computations take place. Part (b) shows the discretization of the fluid-particle interface using a fixed number of points. If the interface is deformable, these interfacial points are connected by springs with finite spring constants (Feng and Michaelides, 2004). If the interface is rigid, the spring constant value is set to a very large value. Part (c) depicts the substitution of the ellipsoidal particle surface and its effect on the fluid domain by a system of forces acting on the nodes of the Lagrangian grid. Part (d) is the final step in the interface modelling and shows the spreading of the forces from the Lagrangian points that move with the particle to the nodes of the Eulerian grid, which are stationary and where the flow computations take place.

In modelling the interactions between a rigid body and a fluid, the IBM can be extended by replacing the surface force F with a body force f, which acts only in the

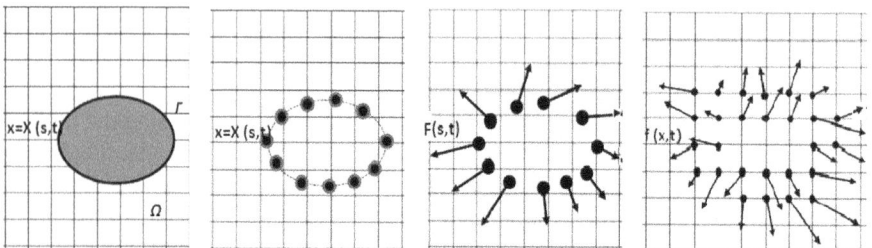

FIGURE 8.14 The four main steps in the implementation of the IBM. From left to right: (a) The Eulerian grid and the (ellipsoidal) particle whose surface is described by the parameter s, $X(s,t)$; (b) discretization of the particle surface into Lagrangian points; (c) substitution of the particle surface with a system of forces at the Lagrangian points that enforces the no-slip velocity condition; (d) spreading of the forces in the Eulerian grids in the form of force density.

region occupied by the solid. The body force in the region of the particles is determined in a way to enforce that this region moves exactly like a solid. This method is called direct-forcing scheme and provides a very good way to compute the velocity field. Unlike the penalty method, this method does not require another artificial input. Since we consider the entire geometric domain as being occupied by fluid, with the body density force acting in regions that are known to be occupied by a solid particle, the fluid motion is described by Eqs. (8.34) and (8.35).

Let us assume that the velocity in the regions occupied by a solid particle that represents/mimics the motion of a solid particle is u. Then the force density must satisfy the following equation

$$f = \rho \left(\frac{\partial u}{\partial t} + u \cdot \nabla u \right) - \mu \nabla^2 u + \nabla p. \tag{8.39}$$

Therefore, the body force at a point X_P can be computed by replacing the fluid velocity u in Eq (8.39) with the expected velocity of that point attached to the particle, U_P. With known velocity and pressure fields at the time step $t = t_n$, we then have an explicit scheme to determine the force term at the Lagrangian boundary point L at time $t = t_{n+1}$, which is as follows

$$f_P^{(n+1)} = \rho \left(\frac{U_P^{(n+1)} - u_P^{(n)}}{\Delta t} + u_P^{(n)} \cdot \nabla u_P^{(n)} \right) - \mu \nabla \cdot \nabla u_P^{(n)} + \nabla p_P^{(n)}. \tag{8.40}$$

For a rigid particle, some of the terms in Eq. (8.40) can be further simplified. For example, the viscous force can be dropped because of the zero strain rates when the fluid is within the solid boundary and undergoes a rigid body motion. However, and in order to reduce computational complexity, the force density is typically set to zero inside the boundaries of particles, and the inside fluid is not strictly in rigid body motion.

In contrast to the smooth-interface approach, a sharp-interface formulation imposes a velocity corresponding to the no-slip condition directly at the Lagrangian marker points, by constraining the nodal variables of the nearest neighbouring Eulerian cells directly, using an interpolation function, typically a linear one (Majumdar et al., 2001; Mark and van Wachem, 2008; Mittal et al., 2008). This results in the "sharp" representation of the IB, which can show a second-order spatial convergence behavior on some specific features of the flow. Although this is favorable for some applications, since the fictitiously forced layer around the IB is typically much thinner and potentially less stable, an important drawback of the sharp-interface formulation is that with transient flows, the particle behavior can become bizzare with spurious flow patterns for cases with moving boundaries, especially when a coarse mesh and/or small time steps are used, due to geometric inconsistencies in the discretization of the continuity constraint (Seo and Mittal, 2011) and to the sudden transfer of cells from the solid to fluid phase (and vice versa) between consecutive time steps (Lee et al., 2011). Yang and Balaras (2006) proposed a robust field-extension procedure that extends the pressure and velocity fields into the solid body in order to implicitly treat problematic grid cells. The extension is shown to significantly reduce the occurence of spurious oscillation. A special type of the sharp-interface IBM is the so-called "cut-cell" approach, where the IB cuts the Eulerian mesh into locally unstructured cells (Seo and Mittal, 2011; Zastawny *et al.*, 2012). Although such algorithms are complex, they can substantially reduce the oscillations when the IB moves in time.

8.3.3.2 Applications of the Immersed Boundary Methods

The IBM has been frequently used to study the behavior of fluid-solid flows, fully resolving the flow between the particles. There are numerous examples in studying turbulent fluid-particle channel flow (Costa *et al.*, 2016; Fornari *et al.*, 2016) and particle-laden homogeneous isotropic turbulence (Chouippe and Uhlmann, 2015; Schneiders *et al.*, 2019; Uhlmann and Chouippe, 2017). An overview of some applications can be found in Brändle de Motta *et al.* (2019). However, applying the IBM to flows with thousands or more particles remains computationally very expensive and is usually only of interest to study specific physical phenomena or trends in flows of a relatively small scale and up to moderate Reynolds number.

There are also a number of studies studying the behavior of one or a few particles in flows to derive or fit the fluid-particle interaction models, to study the effect of particle shape, particle rotation, or fluid Reynolds number (Hölzer and Sommerfeld, 2009; Sanjeevi *et al.*, 2018; Zastawny *et al.*, 2012). Figure 8.15 shows the instantaneous velocity field past an oblate particle, with an aspect ratio of $\frac{b}{a} = 0.2$, with an equivalent diameter set to unity. The incidence angle of attack is 30°, and the particle Reynolds number is 300. To speed up the computation, an adaptive mesh refinement

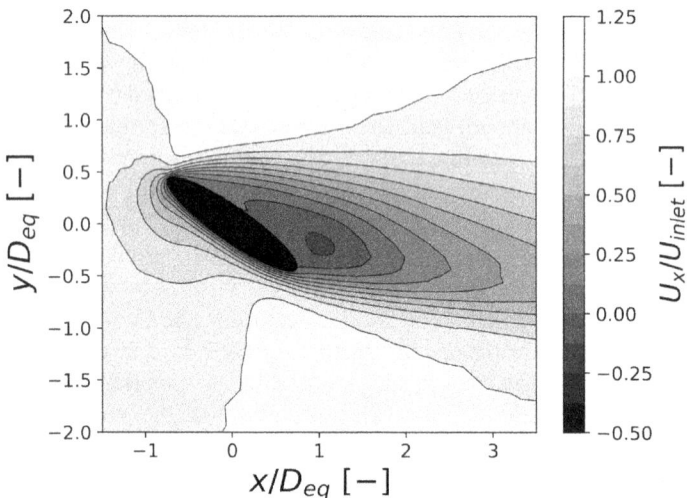

FIGURE 8.15 The IBM simulation of the flow passing an ellipsoid. The color map represents the magnitude of the velocity field, and the length is made dimensionless by the equivalent diameter of the ellipsoid.

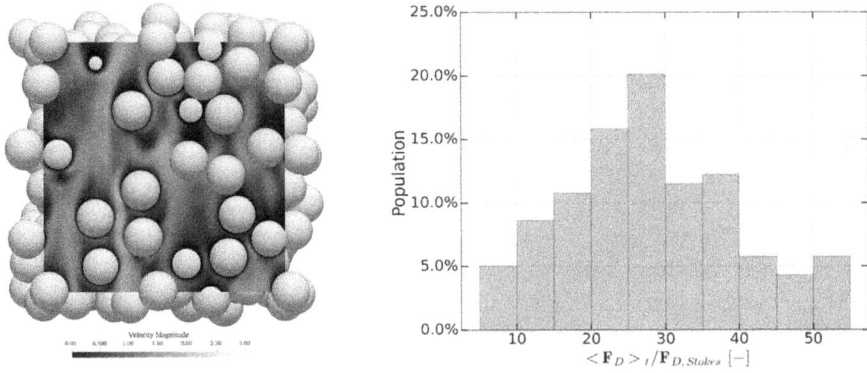

FIGURE 8.16 An IBM simulation of a flow through an array of spheres, with a solid volume fraction of 30%. On the left is the representation of the flow field and on the right the histogram of the averaged drag forces exerted on the particles, scaled by the Stokes drag force on a single particle in a uniform flow.

(AMR) is used, where the finest resolution is located at the fluid-solid interface ($Deq/\Delta x = 48$). This allows the generation of a substantial computational domain in order to converge with a reasonable computational cost toward a solution independent of the boundary conditions. The force on the ellipsoid can be easily found by adding the IBM forces within the flow domain.

Another recent application of the IBM is to study the interaction of a flow with an array of spheres. An example of a flow past a random arrangement of monodisperse spheres is shown in Figure 8.16. The volume fraction of particles is 30%, which results in the placement of 140 particles in the domain. The average particle Reynolds number of the simulation is approximately 100, based on the superficial velocity across the entire domain. The numerical resolution in the domain comprises 32 cells across the diameter of each particle. The right part in Figure 8.16 presents the histogram of the temporal averaged drag force exerted on the particles, scaled by the Stokes drag force. These results show that there is a large variation in drag force on particles in an array, which is not typically taken into account in Eulerian-Eulerian or Eulerian-Lagrangian simulation. The force may be incorporated in a stochastic way (Hardy *et al.*, 2022).

By performing a large number of direct numerical simulations, using the IBM, the behavior of a single or a few complex particles can be studied, and the results can be used at larger scales, such as Eulerian-Lagrangian particle tracking (Davis *et al.*, 2012; van Wachem *et al.*, 2015).

8.4 POINT-PARTICLE DNS

The term DNS typically refers to spatially fully resolved numerical simulations of typically simple flow fields since the large-scale vortical structures, which are sources of turbulence, cannot be resolved. Consequently, the DNS does not capture the entire energy spectrum for moderately high Reynolds numbers. Therefore, typical applications of DNS methods are related to homogeneous isotropic turbulence (HIT) and channel flows, applying periodic boundary conditions. For simulating particle-laden flows with such a very good turbulence resolution, the dispersed particles are treated as point-masses (Kuerten, 2016) subject to the conditions specified in Section 8.2. Consequently, the DNS/Lagrangian point-particle approach has been applied for very small particles for a long time to analyze phenomena such as preferential concentration in HIT and channel flows. The accumulation of particles in certain regions of turbulence structures strongly depends on the particle Stokes number.

When modelling a particle-laden flow with point-particles, the equations governing the fluid phase are the Navier-Stokes equations

$$\nabla \cdot \mathbf{u}_f = 0$$

$$\rho_f \left[\frac{\partial \mathbf{u}_f}{\partial t} + \nabla \mathbf{u}_f \mathbf{u}_f \right] = -\nabla p + \nabla \cdot \left(\bar{\bar{\tau}}_f \right) + \rho_f \mathbf{g} + \mathbf{M} \qquad (8.41)$$

where $\mathbf{u}_f$ is the fluid velocity vector, ρ_f the fluid density, $\bar{\bar{\tau}}_\phi$ the fluid stress tensor, p the static pressure, and to which $\mathbf{M}$, a term corresponding to the transfer of momentum from the particles to the fluid, is added in case two-way coupling (i.e. the particles exchange momentum with the fluid) is considered. When the particle volume fraction is significant ($\alpha_p > 0.005$), the Navier-Stokes equations are modified to include a volume fraction, which can be derived with a filtering approach

$$\nabla \cdot \alpha_f \mathbf{u}_f = 0$$

$$\rho_f \left[\frac{\partial \alpha_f \mathbf{u}_f}{\partial t} + \nabla \alpha_f \mathbf{u}_f \mathbf{u}_f \right] = -\alpha_f \nabla p + \nabla \cdot \left(\alpha_f \bar{\bar{\tau}}_f \right) + \alpha_f \rho_f \mathbf{g} + \mathbf{M} \qquad (8.42)$$

Note that the unknowns corresponding to the fluid flow, that is, the fluid velocity and the fluid pressure, are solved in the *Eulerian framework*, typically a fixed mesh (at least during each time step).

In the point-particle approach, the motion of a spherical particle is described by Newton's second law of motion

$$\rho_p \frac{\pi d_p^3}{6} \frac{d\mathbf{u}_p}{dt} = \left(\mathbf{F}_{p,fluid} + \mathbf{F}_{p,body} \right) \qquad (8.43)$$

where $\dfrac{d\mathbf{u}_p}{dt}$ is the acceleration of each particle, $\mathbf{F}_{p,fluid}$ is the resultant of the fluid forces acting on the particle surface, and $\mathbf{F}_{p,body}$ is the resultant of the body forces acting on the particle. When particle-particle collisions are considered, a possible third type of force is added to the equation when using a soft-sphere interaction model to represent the effects of collisions on that particle. This is not the case when using a hard-sphere particle collision model. Note that the particles are tracked in the Lagrangian framework, and their positions and velocities are updated each time step.

When the particles are significantly denser than the fluid ($\rho_p \gg \rho_f$), the drag is the dominant fluid force contribution, and other forces such as the Basset history force and the added-mass force can be neglected ($\mathbf{F}_{p,fluid} \approx \mathbf{F}_{p,drag}$). The equation for the drag acting on a spherical particle is

$$\mathbf{F}_{p,drag} = 3\pi\mu_f d_p f\left(\alpha_p, Re_p\right)\left(\tilde{\mathbf{u}}_{f@p} - \mathbf{u}_p\right) \tag{8.44}$$

which corresponds to Stokes's law augmented with an empirical factor $f(\alpha_p, Re_p)$ in order to consider finite values of the particle Reynolds number, Re_p and the particle volume fraction, α_p. The drag force depends on $\tilde{\mathbf{u}}_{f@p}$, the *undisturbed* fluid velocity associated with the particle. This is defined as the local fluid velocity at the particle center, as though the particle had been taken out of the flow.

In many point-particle approaches, the undisturbed fluid velocity at the location of the particle is simply found by interpolation from the existing fluid velocities computed on the Eulerian mesh. It is then required to transform the Eulerian fluid properties to the Lagrangian framework at the particle's center by an interpolation technique. The required accuracy of the interpolation depends on the type of fluid-particle flow under consideration. For a turbulent flow, typically a spline, or a third order Lagrangian polynomial, is required to minimize the filtering errors associated with the interpolation (Yeung and Pope, 1988). Such a scheme is also important to be able to predict two-particle dispersion and coagulation.

For two-way coupling, assuming that the disturbed fluid velocity is the same as the undisturbed fluid velocity may not be accurate, depending on the fluid mesh spacing and the particle size. However, the estimation of the undisturbed velocity is notoriously difficult as it involves access to an equivalent, conceptual flow where the particle under consideration does not exist (Evrard *et al.*, 2021). In a two-way coupled point-particle framework, specifically, the fluid velocity at the location of each particle features a disturbance that is induced by the momentum transferred from the particle to the fluid. Determining the undisturbed velocity would thus formally require solving the governing flow equations without transfer of momentum from the particle under consideration as many times as there are particles in the flow. Since this is, in practice, impossible, the undisturbed velocity is almost always approximated by the *disturbed* or *actual* fluid velocity interpolated to the position of the particle. The errors associated with the two-way coupled point-particle approach and approaches for determining the undisturbed fluid velocity are further discussed in Section 8.7.5.

Assumptions are required to determine the transfer of momentum from the particles to the fluid. The particle-source-in cell (PSIC) model (Crowe *et al.*, 1977) is widely used to address this issue. In a Eulerian fluid cell K, the momentum transfer term as defined in the PSIC model reads as

$$\mathbf{M}_k = \sum_{p \in K} -\mathbf{F}_{p,fluid}\delta_K\left(\mathbf{x}_p\right) \tag{8.45}$$

where $\bar{\mathbf{x}}_p$ is the center of the particle p, and δ_K is a piecewise-constant approximation of the Dirac delta function whose compact support is the computational cell K where the particle is located. If the cell has a volume V_K, then δ_K is defined as

$$\delta_K(\mathbf{x}) = \begin{cases} \dfrac{1}{V_k}, & \mathbf{x} \in cell \\ 0, & elsewhere \end{cases} \tag{8.46}$$

In the PSIC-EL numerical framework, it is recognized that the errors due to a particle's self-induced flow disturbance grow proportionally with the ratio between the particle diameter, d_p, and the mesh dimension, $h = V_K^{1/3}$. To mitigate these errors, that is, keep the magnitude of the local velocity disturbance small, a stringent constraint is usually imposed: The tracked Lagrangian particles have to be much smaller than the computational cells in which the governing flow equations are solved. The consensus in the community seems to be $\dfrac{d_p}{h} \leq 0.1$. This will be further discussed in Section 8.7.5.

8.4.1 EXAMPLES OF POINT-PARTICLE DNS

A widely found application of point-particle DNS is related to the analysis of particle behavior in turbulent flows. For that purpose, mostly a **homogeneous isotropic turbulence** structure (HIT) is generated, and the behavior of different-sized inertial particles, or better particles having different Stokes numbers, is analyzed using different approaches for characterizing particle concentration distributions. Because of particle inertia, the movement of particles create inhomogeneities in their spatial distribution (Bec *et al.*, 2005; Monchaux *et al.*, 2012). This phenomenon, referred to as *preferential concentration*, has been observed in experiments (Eaton and Fessler, 1994) and is a consequence of the interaction of particles with the turbulence structure of the carrier fluid. As a result, heavy particles tend to accumulate in regions of low vorticity and high strain rate (Wang and Maxey, 1993), leading under certain conditions to the formation of clusters and regions with significantly higher concentration of particles surrounded by areas of low concentration (Squires and Eaton, 1991). Clusters may be formed in both homogeneous and inhomogeneous turbulent flows. The phenomenon of clustering of heavy particles in inhomogeneous turbulent flows is explained by the effect of turbulent migration (turbophoresis) from regions of high intensity of turbulent velocity fluctuations to regions of low turbulence (Reeks, 1983). Clustering of inertial particles also takes place in homogeneous turbulence, where the gradients of the mean velocity fluctuations of the carrier flow are zero, and consequently, particle transport via turbophoresis does not take place in the conventional sense. However, despite the stochastic character of turbulence, the distribution of heavy inertial particles in turbulent flows is not random, and their interaction with the coherent vortex structures of the turbulent flow may give rise to significant clustering.

For demonstrating such clustering behavior, direct numerical simulation (DNS) results are presented obtained by a three-dimensional lattice Boltzmann method (LBM) extended by an Lagragian point-particle approach (Ernst and Sommerfeld,

2012; Ernst *et al.*, 2019). The discretization of velocity directions is based on the D3Q19 model, which applies to a three-dimensional grid, provides 19 propagation directions on each single node, and is described in Section 8.3.2. The relaxation of the local distribution function to the equilibrium distribution is assumed to occur at a constant single relaxation time collision operator approximated by the BGK approach (Bhatnagar *et al.*, 1954), which showed a suitable stability in terms of the also applied spectral forcing scheme. The computational domain is a cube of side L = 0.128 m with 128^3 grid points and with fully periodic boundaries. The fluid density ρ is 1.17 kg/m^3. Homogeneous isotropic turbulence is realized by generating a force in spectral space and introducing it as a change of velocity in the turbulent flow field. The applied spectral forcing scheme used in this study was proposed by Eswaran and Pope (1988). Turbulence characteristics resulting from the simulation are summarized in Table 8.1.

At the beginning of a simulation, all particles are introduced in random non-overlapping positions inside the entire computational domain. The initial particle velocity is set to the local instantaneous fluid velocity. In order to have particle response times, which are on the order of the relevant timescales of the turbulence, the ratio ρ_P/ρ was varied between 47.3 and 4,730 keeping ρ constant. The particle diameter was constant at $d_P/\eta_K = 0.2$. Under these conditions, the particle Stokes numbers are in the range 0.1 to 10. The volume fraction of the particle phase, α_P, is varied between 0.01% and 0.1% to study the influence of inter-particle collisions in particle variables and on the particle segregation. The detection and modelling of fully elastic particle-particle collisions is computed using the deterministic collision model proposed by Sundaram and Collins (1997), which assumes that the collisions are binary and quasi-instantaneous. Modification of flow and turbulence (two-way coupling) by the inertial particles as well as the effects of gravity is not considered at the moment, implying that only the drag is acting on the particles.

The movement of particles immersed in turbulent flow is controlled by the local velocity fluctuations and by the ordered motion of large-scale turbulent structures. Since the dynamics of inertial particles is dissipative, this leads to an inhomogeneous

---

TABLE 8.1
Turbulence Characteristics of the LBM-based Direct Numerical Simulations with Dispersed Point-like Particles

Parameter	Symbol	Value
Kinematic fluid viscosity	ν	1.47×10^{-5} m^2/s
Fluctuating velocity	u'	2.75×10^{-2} m/s
Dissipation rate	ε	2.92×10^{-3} m^2/s^3
Integral length scale	L_{int}/L	0.11
Integral timescale	T_{int}/τ_K	4.83
Kolmogorov length scale	$\eta_K/\Delta x$	1.02
Kolmogorov timescale	$\tau_K/\Delta t$	208.5
Taylor Reynolds number	Re_T	18.0

particle distribution within the flow field. Interactions of particles with the turbulence features of the flow are described by the turbulent Stokes number $St_K = \langle \tau_p \rangle / \tau_K$, which is equal to the ratio between the number average response time of particles and the Kolmogorov time microscale, τ_K, which is expressed as $\tau_K = (v/\varepsilon)^{1/2}$. For $St_K \ll 1$, the response time of the particle is much less than the characteristic time of the smallest vortical flow structures, and the particle will easily adapt to the changes in the flow velocity. In this case, particles behave as fluid tracers, and as such, they distribute uniformly over the flow field. On the other hand, for $St_K \gg 1$, the particles will have not sufficient time to respond to the fluid velocity changes as a result of their inertia, and their velocity will be little affected by the turbulent structures. In this limit, particle motion is essentially ballistic, the particles distribute uniformly, but they may reach the same position with very different velocities, a fact known as the *sling effect* (Falkovich *et al.*, 2002). However, for intermediate inertia, $St_K \sim 1$, particles respond partially to the fluid fluctuations in such a way that they tend to be centrifuged out of the coherent eddies and accumulate in regions of high strain-rate, generating a local rise of concentration of heavy particles in zones of low vorticity. These effects are more pronounced when particle response time approaches the Kolmogorov time microscale of turbulence. This different particle behavior and the resulting segregation and clustering is illustrated in Figure 8.17 showing an instantaneous slice through the flow field and the particles distributed therein with $St_K = 1.25$ and $St_K = 9.67$, respectively.

The main objective in the research of Ernst *et al.* (2019) was to reveal the influence of inter-particle collisions on particle distribution and clustering in HIT. The resulting effect is illustrated in Figure 8.18 for $St_K \approx 1$, which is a situation of maximal clustering. Results are presented for a slice, of thickness $2\eta_K$, along the x-axis.

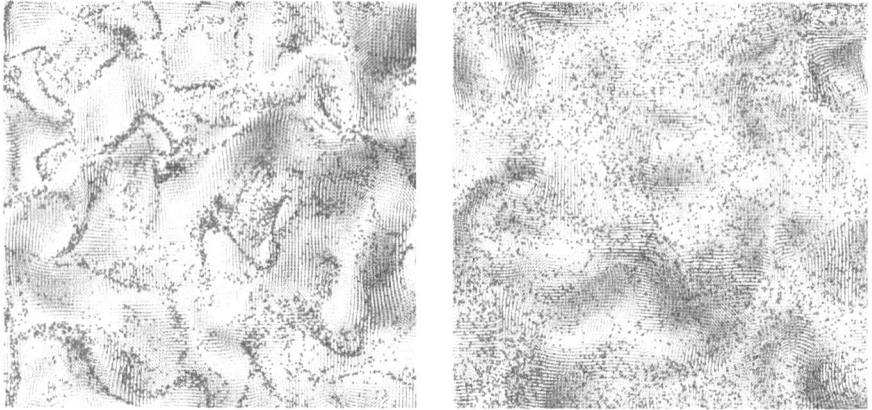

FIGURE 8.17 Snapshot of a 2D slice through the fluid turbulent field laden with particles. Illustration of particle segregation having intermediate ($St_K = 1.25$) and large ($St_K = 9.67$) inertia (left and right, respectively). The particle volume fraction is 0.1%; computations with inter-particle collisions; edge lengths l and thickness s of the image sections: $l/\eta_K = 128$ and $s/\eta_K = 2$, respectively (from Ernst *et al.*, 2019).

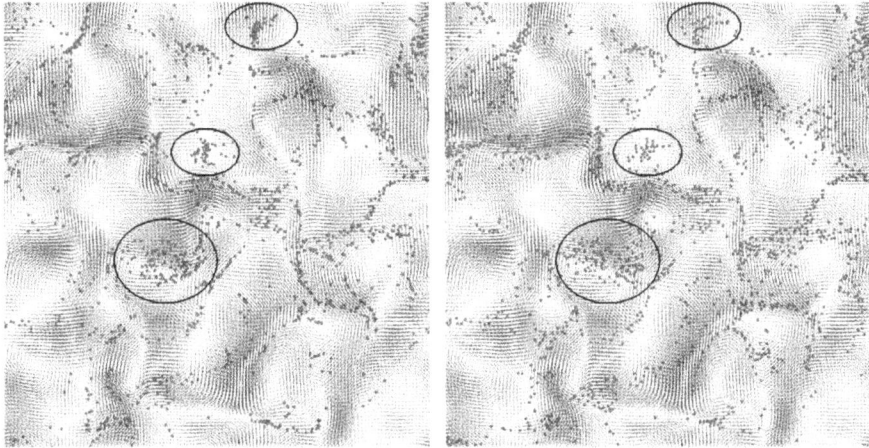

FIGURE 8.18 Instantaneous positions of solid particles in a slice of the HIT considering a frozen turbulence field. The computations are without (left) and with (right) inter-particle collisions ($l/\eta_K = 128$, $s/\eta_K = 2$, $St_K = 1.15$, $\alpha_p = 1.28\%$) (from Ernst *et al.*, 2019).

The left part of Figure 8.18 shows the case where collisions are disregarded while in the right plot inter-particle collisions are considered. In this case, a *frozen* turbulence field (this implies that particle tracking is conducted on the same flow field changing not in time) is considered, and a direct comparison between colliding and non-colliding particle distributions can be carried out. For non-colliding particles, they tend to accumulate in specific locations of the flow forming filaments (in the 2D view) where the same volume can be occupied by several particles at the same time (as in Figure 8.17 left). When inter-particle collisions are considered, this situation is no longer possible, since in these clustering regions with higher concentrations, inter-particle collisions are likely to occur yielding a dispersion of the particles. The specific regions where these effects may be identified are indicated by the ellipses in both parts of Figure 8.18.

There are several ways to characterise the spatial distribution of particles in flow structures (Monchaux *et al.*, 2012; Ernst *et al.*, 2019) and to identify the occurrence of clustering. The *segregation parameter* (Soldati and Marchioli, 2009) or clustering index (Monchaux *et al.*, 2012) Σ_p compares the standard deviation for the actual particle distribution with that of a Poisson distribution; positive values of Σ_p indicate segregation of particles, and zero Σ_p implies a homogeneous distribution of particles. The results of numerous simulations with varying particle diameters and volume fraction by comparing the cases without (open symbols) and with (half-closed symbols) inter-particle collisions are summarised in Figure 8.19, where it is clearly shown that Σ_p increases with the particle volume fraction for all the computed Stokes numbers. In agreement with previous results (Fede and Simonin, 2010), the particle segregation parameter tends to zero for the cases of very low and high inertia particles, indicating a random particle distribution. For particles with intermediate inertia, the Σ_p-values have a clear maximum, indicating particle

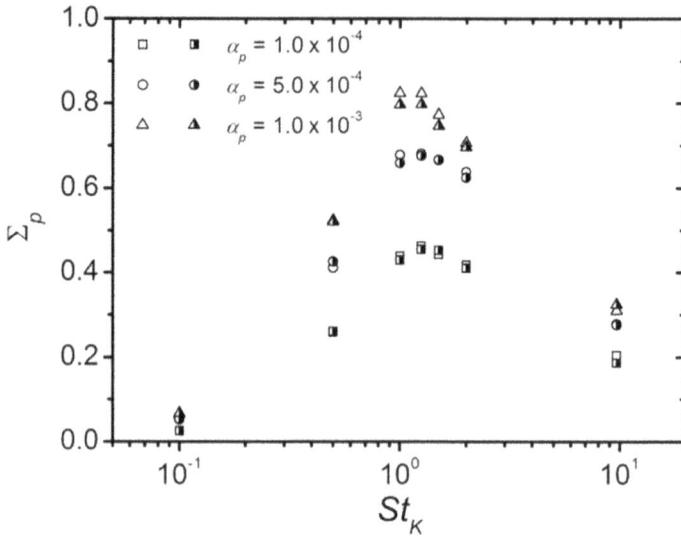

FIGURE 8.19 Particle segregation parameter Σ_p obtained from unsteady simulations as a function of the turbulent particle Stokes number for different particle volume fractions. Open symbols indicate no particle collisions, and half-closed symbols indicate inter-particle collisions: $St_K = 1.25$ (from Ernst *et al.*, 2019).

clustering or segregation to occur, as shown in Figure 8.19. Also, in agreement with Fede and Simonin (2006), the maximum value is obtained for $St_K \approx 1$, indicating that particles tend to experience *maximal* clustering for response timescales close to the Kolmogorov timescale of the turbulence. Such maximal clustering was found for $St_K = 1.25$, regardless of the volume fraction considered. The value of the segregation parameter augments significantly with increasing solids volume fraction, independently of whether inter-particle collisions are taken into account or not. Actually, the influence of collisions is marginal in the plots of Σ_p versus St_K, and only at the higher volume fraction of 0.1% is a slight decrease of particle segregation noticeable. This result is consistent with the visualizations depicted in Figure 8.18, where inter-particle collisions reduce clustering and yield a somewhat better particle dispersion, thus, decreasing the value of Σ_p. This effect is also clearly increasing with particle volume fraction and is expected to be much more important if the volume fraction exceeds 10%.

The effect of increasing α_p on the segregation parameter Σ_p can be better illustrated using simulations obtained with a frozen turbulent field. In this case, the differences of the particle segregation parameter between non-colliding and colliding particles are much more evident than for the case of an evolving turbulent field. Hence, Figure 8.20 shows such results, where Σ_p increases continuously with α_p when inter-particle collisions are not taken into account. In this case, $\Sigma_p \sim \alpha_p^{0.27}$. When inter-particle collisions are considered, Σ_p is clearly lower and shows a very moderate increase. Figure 8.20 also includes the segregation parameter obtained for

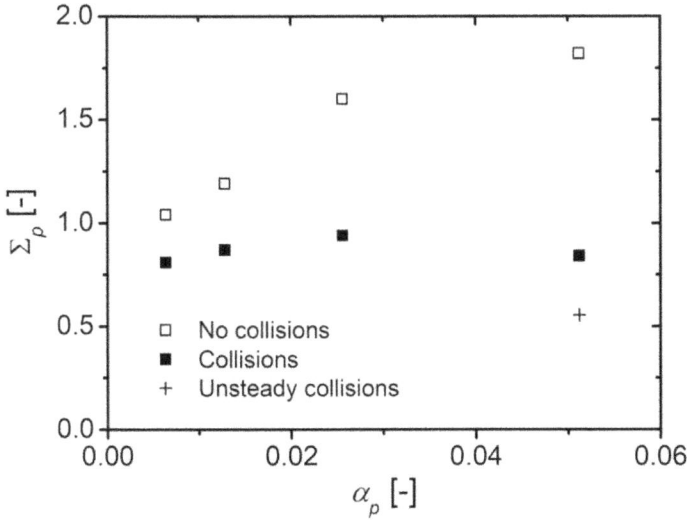

FIGURE 8.20 Particle segregation parameter Σ_p as a function of the particle volume fraction α_p for a frozen turbulence field ($St_K = 1.15$) considering non-colliding and colliding particles. The value of Σ_p for the case of unsteady, evolving turbulence with inter-particle collisions at $\alpha_p = 5.12\%$ (symbol: +) is also shown for comparison (from Ernst *et al.*, 2019).

the case of unsteady turbulence with inter-particle collisions and $\alpha_p = 5.12\%$. In this case, the parameter is noticeably lower than for the frozen turbulence cases.

Small solid particles suspended in a **turbulent channel flow** are commonly found in numerous industrial and environmental applications. This problem has been widely studied numerically and is a good benchmark case for a second example. The DNS of single-phase channel flow has been described by Kim *et al.* (1987), and DNS of several point-particle approaches has been described by Marchioli *et al.* (2008). The computational domain consists of two "infinitely" sized parallel plates, spaced $2h$ apart in the y-direction, see Figure 8.21. To approximate the infinitely sized plates in the x-z-plane, periodic boundary conditions are applied in both x and z directions. The computational domain has the size $4\pi h \times 2\pi h \times 2h$. The flow is predicted by solving the Navier-Stokes equation with an additional source term in the direction of the flow that induces the flow

$$S_{p,x} = \frac{\rho_f v_f^2 Re_\tau^2}{h^3}$$

(8.47)

Where h is the half channel height, v_f is the fluid kinematic viscosity, and Re_τ is the friction Reynolds number

$$Re_\tau = \frac{u_\tau h}{v_f}$$

(8.48)

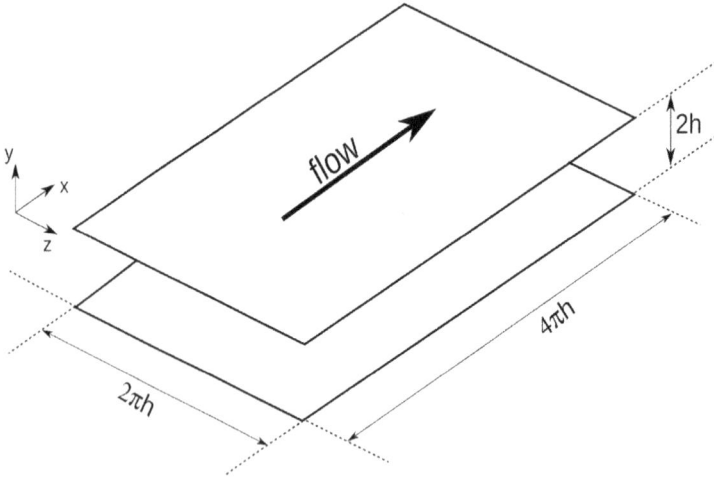

FIGURE 8.21 Setup of a turbulent flow in a channel. The flow is in in the x-direction between the plates.

where u_τ is the *friction velocity*, defined as

$$u_\tau = \sqrt{\frac{\tau_0}{\rho_f}} \tag{8.49}$$

where τ_0 is the wall shear stress. Imposing this source term fixes the wall shear stress and the friction velocity for a given case. In the current example, the friction Reynolds number is taken as 150.

In the wall normal direction, y, no slip conditions are applied on the boundaries. For a friction Reynolds number of 150, the number of grid points for a second order accurate method are 159 × 169 × 159. The grid spacing is uniform in the x- and y-directions but is strongly refined in the wall normal direction, using a *tanh* distribution with a growth factor of 1.6. The refinement ensures there are at least 4 grid points within $y^+ \leq 5$ layers of the wall. For a two-way coupled simulation, it should be noted that the cells near the two walls may be very small, and a standard particle-source-in-cell treatment of the fluid source terms may violate the assumptions of the particle diameter being smaller than the fluid cell. The spatial and temporal discretization should be at least second order accurate, and the interpolation of the fluid flow properties to the particle positions should be done by a spline interpolation or better.

In the work of Zhao and van Wachem (2013a, 2013b) and Zhao *et al.* (2015), a series of one-, two-, and four-way coupled point-particle DNS were carried out with spherical and spheroidal particles of varying particle Stokes number. Some of the results in these simulations are shown in Figure 8.22, where the single-phase fluid velocity is compared with the fluid velocity resulting from adding spheres or spheroidal particles.

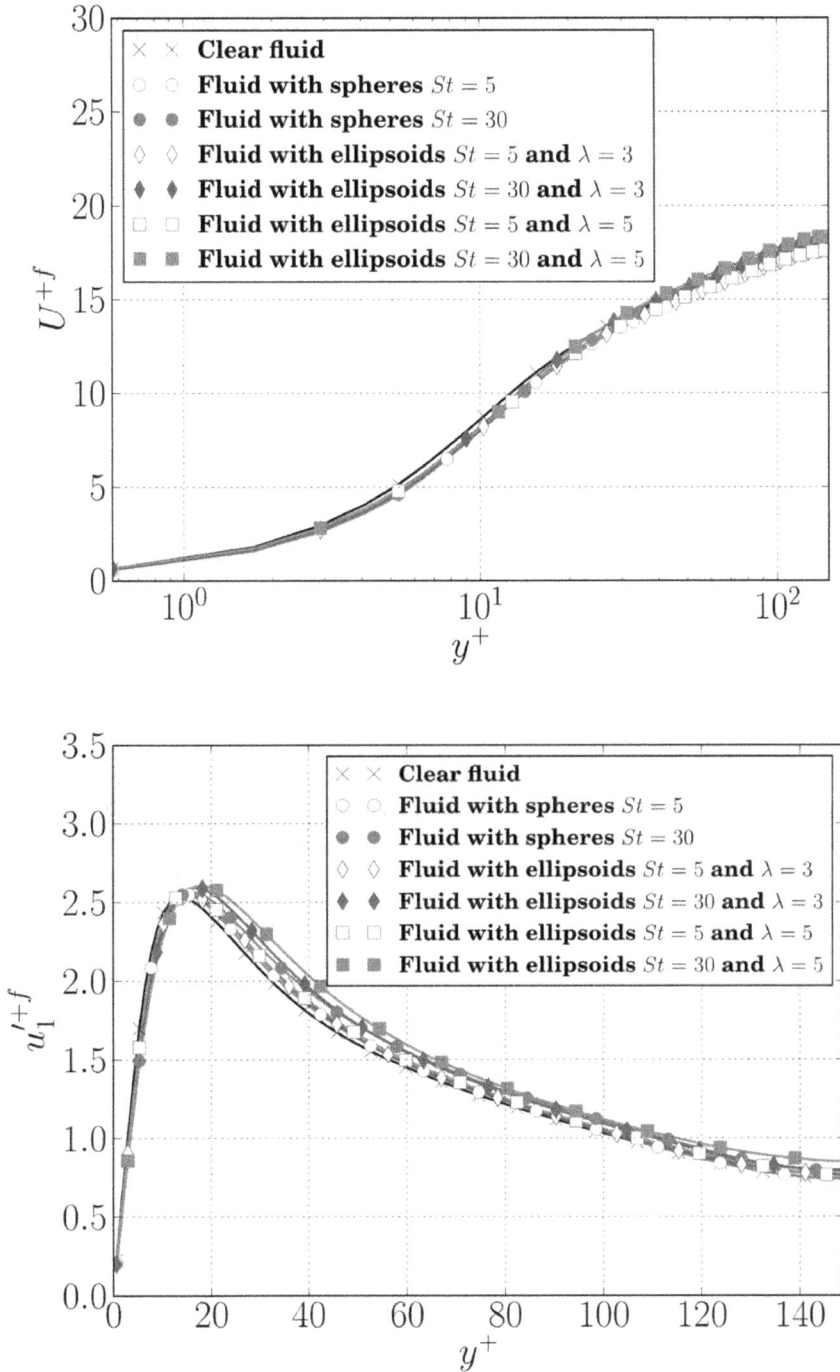

FIGURE 8.22 Fluid velocity profile (top) and RMS fluid velocity (bottom) in the span-wise direction as functions of the dimensionless distance to the wall.

Although the number of particles added to the channel flow is very small ($\alpha_p \approx 2 \times 10^{-4}$), and the particles have little effect on the average fluid velocity in the channel, as can be seen from Figure 8.22, the particles still change the turbulence behavior of the fluid. The dissipation of the fluid turbulence kinetic energy (TKE) decreases, and the particles act as an additional source of TKE dissipation, mainly in the near-wall region. The particle concentration is non-uniform, and most of the particles are present in the near wall region. As the Stokes number of the particle increases, the particles increasingly attenuate the flow turbulence intensity, an effect that is enhanced as the aspect ratio of spheroids increases. The previous test case shows that point-particle DNS has the ability to elucidate the behavior of complex turbulent fluid-particle systems. Equilibrium can be obtained relatively easily by applying periodic boundary conditions.

8.5   POINT-PARTICLE LES

The advantages offered by particle-laden large eddy simulations (LES) are the reduced computational effort compared to DNS and the higher accuracy concerning the particle-turbulence interactions, compared to RANS methods. Such advantages have made this framework an important tool to predict the behavior of turbulent fluid-particle systems. Most of the points discussed in the previous section also apply to tracking particles as point-particles in the framework of LES. However, one additional complexity arises, namely, determining $\tilde{u}_{f@p}$, the *undisturbed* fluid velocity at the center of the particle. In an LES simulation, the actual, instantaneous fluid velocity is not available, but only a filtered velocity field is computed, denoted as $\bar{u}_f$. A typical spectrum of a turbulent flow is shown in Figure 8.23, where the LES filter-size is shown as Δ_{fluid}. The resolved part of the spectrum gives a filtered fluid velocity, $\bar{u}_f$. However, the unresolved part of the spectrum is not resolved in the simulation and is only accounted for indirectly using a model. Therefore, part of the fluid velocity is not accessible in a deterministic, instantaneous way.

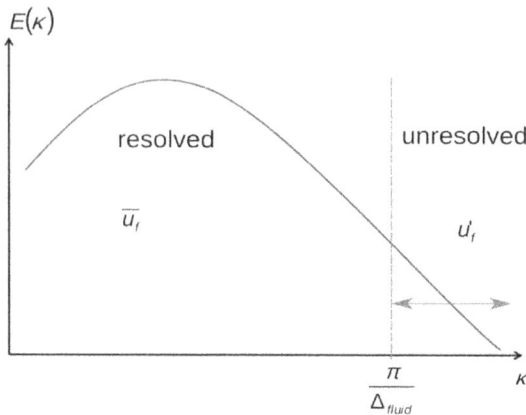

FIGURE 8.23   A typical spectrum of a turbulent flow, showing the filter length and the resolved and unresolved turbulence scales.

To obtain the correct fluid velocity at the particle position, one may use the equation

$$\tilde{u}_{f@p} = \overline{\tilde{u}}_{f@p} + u'_{f@p} \tag{8.50}$$

However, the instantaneous prediction of the unresolved fluid velocity, $u'_{f@p}$, in a deterministic way is not possible. Neglecting the unresolved fluid fluctuating velocity, that is, $\tilde{u}_{f@p} \approx \overline{\tilde{u}}_{f@p}$, may influence the evolution of the particle position as a function of time, as shown in Figure 8.24. This may affect the velocity of the particle, the position of the particle, and the residence time of the particle in a certain area.

There are several models to determine the unresolved fluid fluctuating velocity, $\tilde{u}_{f@p}$, in a *stochastic* framework. Bini and Jones (2008) describe cases in which, especially for light and small particles, a sub-grid scale (SGS) model for the unresolved fluid velocity is needed to mimic the effect of the unresolved scales. Models for the unresolved fluid velocity typically adopt a Fokker-Plank equation, which is approximated by a Wiener process that has the form

$$W(t_n) = \sqrt{\delta t} \sum_{i=1}^{n} \xi_i \tag{8.51}$$

Where n indicates the number of the time step, δt the time step, and ξ_i is a random variable at time-level i, with a zero mean and a variance of unity. Adding the unresolved velocity fluctuations to the filtered velocity leads to the addition of a diffusion-like term to the particle equation of motion as given by

$$\rho_p \frac{\pi d_p^3}{6} \frac{du_p}{dt} = \left(F_{p, fluid}\left(\tilde{u}_{f@p}\right) + F_{p, body} \right) + \overline{\overline{B}} d\overline{W} \tag{8.52}$$

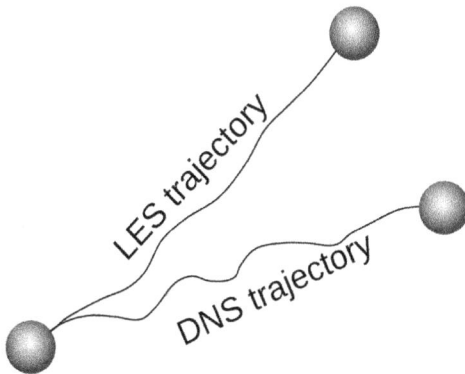

FIGURE 8.24 Schematic diagram of updating the position of a particle using LES (with the filtered fluid velocity), top trajectory, and using DNS (the actual, instantaneous fluid velocity), bottom trajectory.

where $\overline{W}$ is the vectorized Wiener process, and the tensor $\overline{\overline{B}}$ represents the net diffusion. In the most rudimentary models, the diffusion process is assumed isotropic and homogeneous, and the tensor $\overline{\overline{B}}$ then simplifies to $\overline{\overline{B}} = b\delta_{ij}$, where the local constant b depends on the time and length scales of the unresolved fluid velocity and on the particle properties. An estimate of these can be made by

$$b = \sqrt{C_0 \frac{k_{sgs}}{\tau_p}} \tag{8.53}$$

where C_0 is a model constant, k_{sgs} is the sub-grid scale turbulence kinetic energy, and τ_p is a timescale that determines the rate of interaction between the particle and the turbulence dynamics, which is related to the local turbulence Stokes number. If the particle timescale (and, thus, the particle Stokes number) is very large, the effect of the unresolved velocity fluctuations on the behavior of the particle is small, and the SGS model can be neglected. There are various models for the diffusion of the particles arising from the unresolved fluid velocity field—and extensions to non-isotropic flows can be made (Bini and Jones, 2008).

Another, more complex approach, for generating the SGS fluctuating velocity seen by the particles is based on a single-step correlated Langevin model (Lipowsky and Sommerfeld, 2005; Sgrott Júnior and Sommerfeld, 2019). The model assumes that the SGS fluctuations are isotropic, which in the frame of LES is a reasonable assumption. The new fluid velocity fluctuation components seen by the particle at the next location are determined using a correlated and a random part, both depending on Lagrangian and Eulerian correlation functions

$$u_i'^{n+1} = R_{P,i}\left(\frac{\Delta t_L}{T_L}, \frac{\Delta r}{L_E}\right) u_i''' + \sigma_F \sqrt{1 - R_{P,i}^2\left(\frac{\Delta t_L}{T_L}, \frac{\Delta r}{L_E}\right)}\, \xi_i \tag{8.54}$$

where the superscripts denote the time step and the subscripts the spatial component. The correlation function $R_{P,i}$ has Lagrangian and Eulerian components, depending on Δr, the spatial separation between the virtual fluid element, and the particle during the time Δt_L. The SGS turbulence may be considered isotropic so that σ_F represents the rms value of the fluid velocity fluctuation, and ξ_i denotes a random number with Gaussian distribution having zero mean and unit variance. The first term on the right-hand side of the equation represents the correlated part, relevant for small Lagrangian time steps, while the second term is the random contribution to the velocity fluctuation active for larger time steps. More details on the method are given in Section 8.7.3.

The required timescale, $T_L = c_T \sigma_F^2 / \varepsilon_{SGS}$, and the length scale of SGS turbulence, $L_E = c_L \sigma_F T_L$, are estimated with $\sigma_F = \sqrt{2/3 k_{SGS}}$ and the constants $c_T = 0.24$ and $c_L = 3.0$, obtained from a validation (Sommerfeld, 1996). From the LES results, it is possible to estimate the unresolved turbulent kinetic energy and the dissipation rate as follows (Lilly, 1967)

$$k_{SGS} = \frac{\mu_{t,c}^2}{\rho_F^2 (0.094\,\Delta)^2} \tag{8.55}$$

$$\varepsilon_{\text{SGS}} = \frac{C_\varepsilon k_{\text{SGS}}^{3/2}}{\Delta} \qquad (8.56)$$

It should be noted that SGS models only reconstruct the unresolved fluid velocity in a stochastic framework. A deterministic prediction of the unresolved fluid velocity is not possible since the information that was lost due to filtering cannot be regained *a posteriori*.

8.5.1 EXAMPLES OF A POINT-PARTICLE LES

The advantage of applying LES instead of DNS is that the behavior of flows of much higher Reynolds numbers can be investigated. For instance, to compute the flow in the horizontal channel experimentally described in Kussin and Sommerfeld (2002), a DNS approach is computationally very expensive, and an LES approach must be adopted. The Reynolds number of this channel flow is close to $Re = 42,585$ (gas phase bulk velocity 19.7 m/s) and the friction Reynolds number is $Re_\tau \approx 600$. The experimental channel (height 35 mm and width 350 mm) is 6 m in length, which ensures the turbulence is statistically steady. For the computations, a shorter channel can be used, where periodic boundary conditions are employed in the stream-wise direction. In Mallouppas and van Wachem (2013), point-particle LES with spherical particles is carried out, in which different LES models as well as the hard-sphere and soft-sphere collision models are compared. Although the average volume fraction is very low ($\alpha_p \approx 5 \times 10^{-4}$), the local volume fraction can become much higher near the bottom wall because gravitation and particle collisions need to be considered. The glass particles considered here have a diameter of 195 μm and a density of $\rho_P = 2500$ kg/m^3.

The computational domain consists out of 1 million cells. In the wall-normal direction, the mesh is strongly refined, using a *tanh* growth with a coefficient of 7, which enables a resolution corresponding to a DNS in the very near wall region. This will be automatically detected by a dynamic LES model, but for a static LES model, such as the Smagorinsky model, van Driest dampening in the near wall region must be applied. A suitable model for dealing with particle-particle and particle-wall collisions, including the effects of wall roughness, is crucial to obtaining accurate results.

The results of the fluid velocity predictions with the two LES approaches are shown in Figure 8.25. It is shown that both LES approaches predict the flow accurately. It is also apparent that, even though the flow is dilute, there is still a significant effect of two-way coupling as the flow is slowed down in the particle-laden cases. The fluid turbulence is dampened somewhat by the behavior of the particles.

The particle velocity profiles are shown in Figure 8.26. Although there is very little difference between the hard-sphere and the soft-sphere model predictions, taking into account particle-particle interactions and even more so particle-wall interactions are very important. Taking into account the rough nature of the walls in the simulations is also very important, as can be seen in Figure 8.26.

This case shows that point-particle LES can be a very accurate tool to predict turbulent gas-solid flows, although assumptions need to be made on the particle

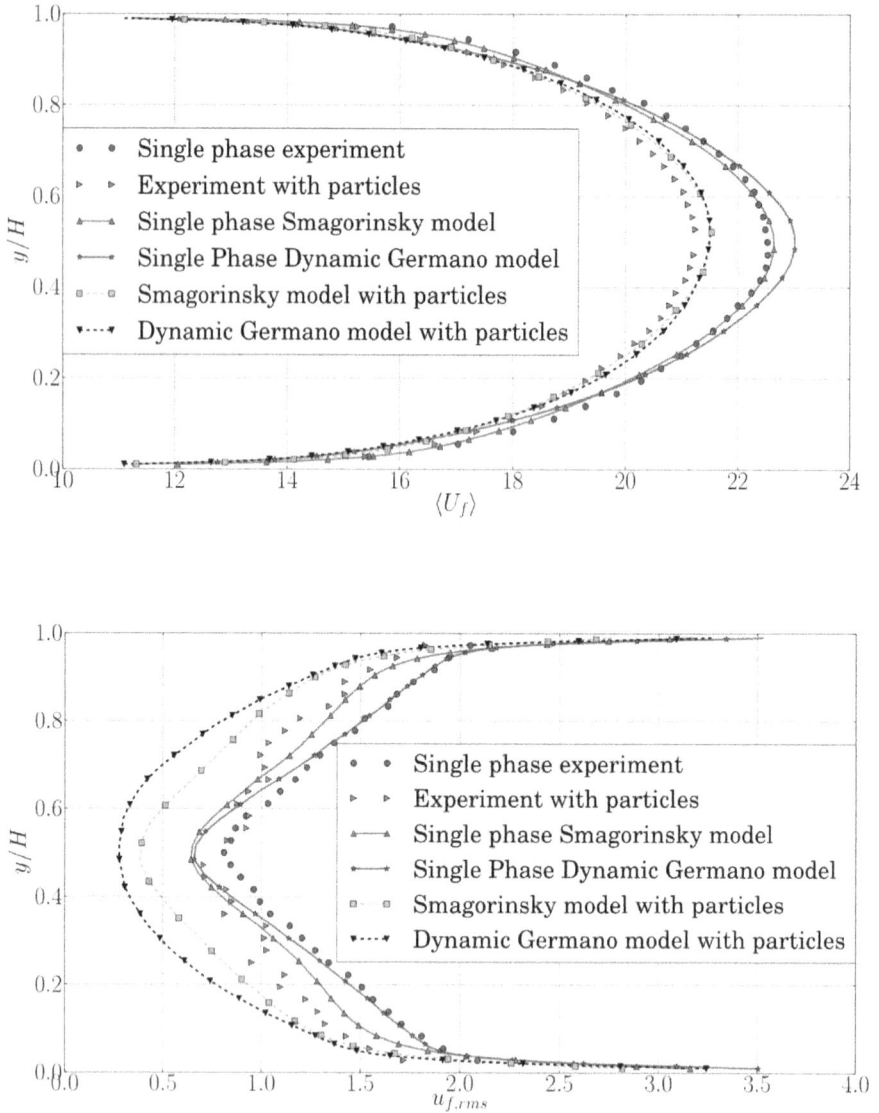

FIGURE 8.25 The horizontal mean (top) and RMS (bottom) fluid velocity for the single-phase fluid and the particle-laden fluid as a function of the dimensionless channel height. Comparison of experiment with different LES approaches.

size (they need to be small) and on the interactions of the particles with the unresolved or sub-grid scales. In simple geometries, there seems to be little effect of the type of LES model. Moreover, for such flows, the hard-sphere model yields almost identical results compared to the soft-sphere model since most particle-particle and particle-wall interactions are of a very short time. The advantage of using the point-particle approach is that complex physical phenomena, such as rough walls, can be

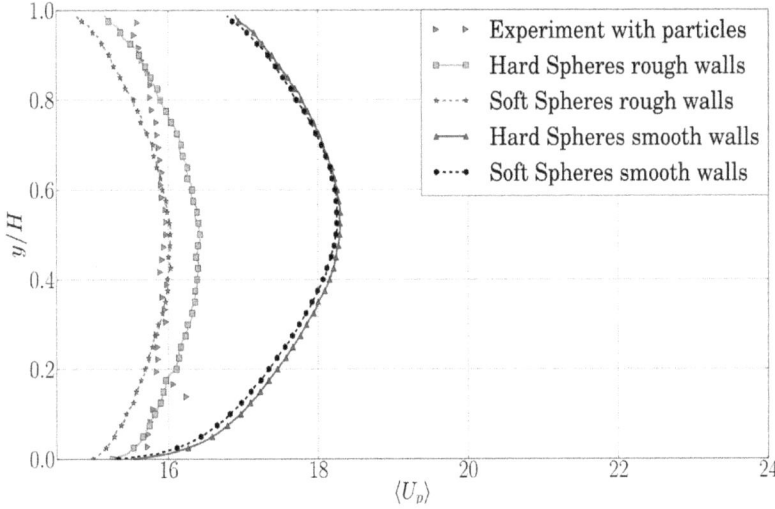

FIGURE 8.26 The horizontal mean particle velocity as a function of the dimensionless channel height for the hard-sphere and soft-sphere (DEM) models with rough and smooth walls and comparison with experiments (LES based on the Dynamic Germano model).

accounted for relatively easily. Including such physical phenomena is, usually, far more complicated in the Eulerian-Eulerian approach.

8.6   EULER/EULER OR MULTI-FLUID APPROACH

The Eulerian-Eulerian model describes the evolution of the *statistical behavior* of a multiphase flow in an averaged way and uses the concept of a volume fraction to account for the local number of droplets or particles that occupy a point in space and time. In these models, the dispersed phase, as well as the continuous phase, are described as a continuous fluid with appropriate closure equations. Hence, one calculates only the average local volume fraction, velocity, fluctuations, etc., but not the properties of each individual dispersed particle or droplet. The averaging process to derive the Eulerian-Eulerian model can be performed in three different frameworks: volume averaging, ensemble averaging, and using a probability density function (PDF).

8.6.1   Volume Averaging Over an Indicator Function

The volume averaging procedure of the Eulerian-Eulerian model is usually performed by volume averaging over an indicator function. In a fluid-fluid formulation, both phases can be volume averaged over a fixed volume, as proposed by Ishii (1975) and Ishii and Hibiki (2006). This volume has to be relatively large compared to the size of individual particles or droplets. A phase indicator function is introduced, $I_k(\mathbf{x})$, which is equal to one when the point $\mathbf{x}$ is occupied by phase k, and zero if it is not. Averaging over the indicator function leads to the volume fraction of both phases resulting in

$$\alpha_k = \frac{1}{V} \int_V I_k(\mathbf{x}) \, dV \tag{8.57}$$

where V is the averaging volume. Since both the continuous and dispersed phases are considered fluids, they are treated in the same way in the averaging process. Hence, the origin of the name *two-fluid* model for the Eulerian-Eulerian model. The form of the momentum balances for both phases are the same and are written as follows

$$\frac{\partial \alpha_k \rho_k \langle \mathbf{u}_k \rangle}{\partial t} + \nabla \cdot \left(\alpha_k \rho_k \langle \mathbf{u}_k \rangle \langle \mathbf{u}_k \rangle \right) = -\nabla \left(\alpha_k \langle p_k \rangle \right) + \nabla \cdot \left(\alpha_k \langle \overline{\overline{\tau}}_k \rangle \right)$$
$$+ \alpha_k \rho_k \mathbf{g} + \mathbf{M}_k \tag{8.58}$$

where k is the phase number and $\mathbf{M}_k$ is the interphase momentum exchange between the phases, with $\Sigma_i \, \mathbf{M}_i = 0$. The density terms ρ_k are averaged in the same way as the velocity. The distribution of stress within both phases is important since the dispersed phase is considered as a fluid. Hence, the "jump" conditions are used to determine the precise formulation of $\mathbf{M}_k$. The interphase momentum transfer is defined as

$$\mathbf{M}_k = -\sum_j \frac{1}{L_j} \left(p_k \mathbf{n}_k - \mathbf{n}_k \cdot \overline{\overline{\tau}}_k \right)$$
$$= \sum_j \frac{1}{L_j} \left[\left(\langle p_{ki} \rangle - p_k \right) \mathbf{n}_k - \langle p_{ki} \rangle \mathbf{n}_k - \mathbf{n}_k \cdot \left(\langle \overline{\overline{\tau}}_{ki} \rangle - \overline{\overline{\tau}}_k \right) + \mathbf{n}_k \cdot \langle \overline{\overline{\tau}}_{ki} \rangle \right] \tag{8.59}$$

where $1/L_j$ is the interfacial area per unit volume, p_k is the pressure in the bulk of phase k, $\langle p_{ki} \rangle$ is the average pressure of phase k at the interface, $\overline{\overline{\tau}}_k$ denotes the shear stress in the bulk, and $\langle \overline{\overline{\tau}}_{ki} \rangle$ represents the average shear stress at the interface between the phases. Mass transfer between the phases is not included in the previous equation but can be easily accounted for. The terms $(\langle p_{ki} \rangle - p_k) \mathbf{n}_k$ and $\mathbf{n}_k \cdot \left(\langle \overline{\overline{\tau}}_{ki} \rangle - \overline{\overline{\tau}}_k \right)$ are identified (Ishii, 1975; Ishii and Hibiki, 2006) as the form drag and the viscous drag, respectively, making up the total drag force. The other terms can be written out as

$$\mathbf{M}_k = \mathbf{I} + \langle p_k \rangle \nabla \alpha_k + \left(\langle p_{ki} \rangle - \langle p_k \rangle \right) \nabla \alpha_k - \left(\nabla \alpha_k \right) \cdot \langle \overline{\overline{\tau}}_{ki} \rangle \tag{8.60}$$

where $\mathbf{I}$ represents the total form and viscous drag. According to Ishii and Mishima (1984), the last term on the right-hand side is an interfacial shear term, which is important in separated flows. According to Ishii (1975), the term $(\langle p_{ki} \rangle - \langle p_k \rangle)$ only plays an important role when the pressure at the bulk is significantly different from that at the interface, as in stratified flows. For many applications, both terms are negligible, and the last equation is reduced to

$$\mathbf{M}_k = \mathbf{I} + \langle p_k \rangle \nabla \alpha_k \tag{8.61}$$

where the form and viscous drag, **I**, is written as the interphase drag times the local velocity difference between the phases

$$\rho_c \alpha_c \left[\frac{\partial \mathbf{u}_c}{\partial t} + \mathbf{u}_c \cdot \nabla \mathbf{u}_c \right] = -\alpha_c \nabla p + \nabla \cdot \left(\alpha_c \overline{\overline{\tau}}_c \right) + \alpha_c \rho_c g - \beta \left(\mathbf{u}_c - \mathbf{u}_p \right) \quad (8.62)$$

$$\rho_p \alpha_p \left[\frac{\partial \mathbf{u}_p}{\partial t} + \mathbf{u}_p \cdot \nabla \mathbf{u}_p \right] = -\alpha_p \nabla p + \nabla \cdot \left(\alpha_p \overline{\overline{\tau}}_p \right) + \alpha_p \rho_p g + \beta \left(\mathbf{u}_c - \mathbf{u}_p \right) \quad (8.63)$$

where p is the actual pressure and β is the interphase drag constant, used to model the combined form drag and skin drag of the particle or droplet. This constant is typically obtained experimentally from pressure drop measurements in fixed, fluidized, and/or settling beds (Ergun, 1952; Wen and Yu, 1966) or through direct numerical simulations (Tenneti *et al.*, 2011).

8.6.2 AVERAGING OVER AN ENSEMBLE OF PARTICLES

When averaging over a number of particles, ensemble averaging can also be applied. Anderson and Jackson (1967) and Jackson (1997) used a formal mathematical definition of the local mean variables to translate the point Navier-Stokes equations for the fluid and the Newton's equation of motion for a single particle directly into the continuum equations representing momentum balances for the fluid and solid phases. The point variables are averaged over regions large enough with respect to the particle diameter but small with respect to the characteristic dimension of the complete system, and this makes necessary the separation of scales. A weighting function, $g(\|\mathbf{x}-\mathbf{y}\|)$, is introduced in forming the local averages of system point variables, where $\|\mathbf{x}-\mathbf{y}\|$ denotes the distance between two arbitrary points in space. An example of such a weighting function is shown in Figure 8.27. This weighting

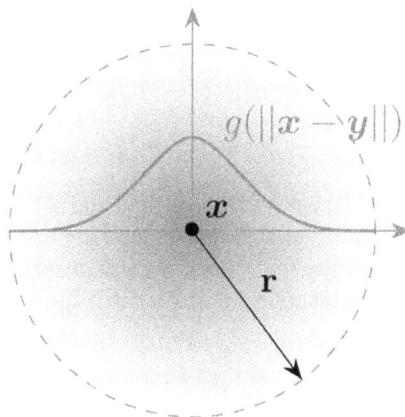

FIGURE 8.27 An example of the weighting function, g, representing the influence of a particle at the location **x**.

function represents the location and influence of one particle. The integral of g over the total space is normalized to one

$$\int_{\mathbb{R}^3} g\left(\|\mathbf{x}\|\right) \mathrm{d}\mathbf{x} = 4\pi \int_0^\infty g(r) r^2 \mathrm{d}r = 1 \tag{8.64}$$

The radius a of the function g is defined by the equation

$$\int_0^a g(r) r^2 \mathrm{d}r = \int_a^\infty g(r) r^2 \mathrm{d}r \tag{8.65}$$

Provided g is chosen so that a satisfies $r_p \ll a \ll L$, where r_p is the particle radius, and L is the shortest macroscopic length scale, the averages so defined will not depend significantly on the particular functional form of g or its radius. This means that, if the function for averaging is used, the precise form of the function does not matter.

The carrier fluid phase volume fraction $\alpha_c(\mathbf{x})$ and the particle number density $n(\mathbf{x})$ at the point $\mathbf{x}$ are directly related to the weighting function g

$$\alpha_c\left(\mathbf{x}\right) = \int_{V_c} g\left(\|\mathbf{x}-\mathbf{y}\|\right) \mathrm{d}V_\mathbf{y} \tag{8.66}$$

$$n\left(\mathbf{x}\right) = \sum_p g\left(\|\mathbf{x}-\mathbf{x}_p\|\right) \tag{8.67}$$

where V_c is the carrier fluid phase volume, and $\mathbf{x}_p$ is the position of the center of particle p. The local mean value of the fluid phase point properties, $\langle\Phi\rangle_c$ is defined by the expression

$$\alpha_c\left(\mathbf{x}\right)\langle\Phi\rangle_c\left(\mathbf{x}\right) = \int_{V_c} \Phi(\mathbf{y}) g\left(\|\mathbf{x}-\mathbf{y}\|\right) \mathrm{d}V_\mathbf{y} \tag{8.68}$$

The solid phase averages are not defined in a way analogous to the continuous fluid phase averages since the motion of the solid phase is determined with respect to the center of the particle, and average properties need only depend on the properties of the particle as a whole. Hence, the local mean value of the solid phase point properties is defined as

$$n\left(\mathbf{x}\right)\langle\Phi\rangle_p\left(\mathbf{x}\right) = \sum_p \Phi_p\, g\left(\|\mathbf{x}-x_p\|\right) \tag{8.69}$$

The average space and time derivatives for the fluid and solid phases follow from the preceding definitions. The averaging rules are then applied to the point continuity and momentum balances for the fluid. For the solid phase, the averaging rules are applied to the equation of motion of a single particle p

$$\rho_p V_p \frac{\partial \mathbf{u}_p}{\partial t} = \int_{S_p} \bar{\bar{\sigma}}_c(\mathbf{y}) \cdot \mathbf{n}(\mathbf{y}) \mathrm{d}A_\mathbf{y} + \sum_{q\neq p} \mathbf{F}_{qp} + \rho_p V_p \mathbf{g} \tag{8.70}$$

where $\mathbf{u}_p$ is the particle velocity, ρ_p is the particle density, V_p is the volume of particle p, $\bar{\bar{\sigma}}_c$ is the fluid phase stress tensor, S_p denotes the surface of particle p, and $\mathbf{F}_{qp}$

represents the resultant force exerted on the particle p from contacts with other particles.

The resulting momentum balances for the fluid and solid phases, after dropping the averaging brackets $<>$ on the variables, are written as follows

$$\rho_c \alpha_c \left[\frac{\partial \mathbf{u}_c}{\partial t} + \mathbf{u}_c \cdot \nabla \mathbf{u}_c \right] = \nabla \cdot \left(\alpha_c \bar{\bar{\sigma}}_c \right)$$

$$- \sum_p \int_{S_p} \bar{\bar{\sigma}}_c \cdot \mathbf{n}(\mathbf{y}) g \left(\| \mathbf{x} - \mathbf{y} \| \right) dA_y + \rho_c \alpha_c \mathbf{g} \tag{8.71}$$

$$\rho_p \alpha_p \left[\frac{\partial \mathbf{u}_p}{\partial t} + \mathbf{u}_p \cdot \nabla \mathbf{u}_p \right] = \sum_p g \left(\| \mathbf{x} - \mathbf{x}_p \| \right) \int_{S_p} \bar{\bar{\sigma}}_c \cdot \mathbf{n}(\mathbf{y}) dA_y + \nabla \cdot \bar{\bar{\sigma}}_p + \rho_p \alpha_p \mathbf{g} \tag{8.72}$$

The first term on the right-hand side of the fluid phase equation of motion represents the effect of stresses in the fluid phase, the second term on the right-hand side represents the traction exerted on the fluid phase by the particle surfaces, and the third term represents the gravity force on the fluid. The first term on the right-hand side of the solid phase equation of motion represents the forces exerted on the particles by the fluid, the second term on the right-hand side represents the force due to solid-solid contacts, which can be described using concepts from kinetic theory, and the third term represents the gravity force on the particles. The averaged shear tensor of the fluid phase can be rewritten with the Newtonian definition as

$$\bar{\bar{\sigma}}_c = -p_f \bar{\bar{\mathbf{I}}} + \frac{\mu_c}{\alpha_c} \left(\nabla \mathbf{u}_c + \left(\nabla \mathbf{u}_c \right)^T \right) \tag{8.73}$$

where the fluid phase volume fraction is introduced in the volume averaging process.

It must be noted that the forces due to fluid traction are treated differently in the fluid phase and solid phase momentum balances. In the solid (particle) phase, only the resultant force acting on the center of the particle is relevant; the distribution of stress within each particle is not needed to determine its motion. Hence, in the solid phase momentum balance, the resultant forces due to fluid traction acting everywhere on the surface of the particles are calculated first, then these are averaged to the particle centers. In the fluid phase momentum balance, the traction forces fluid-solid interaction are calculated at the particle surface and are employed there. Hence, the fluid phase traction term is given as

$$\sum_p \int_{S_p} \bar{\bar{\sigma}}_c \cdot \mathbf{n}(\mathbf{y}) g \left(\| \mathbf{x} - \mathbf{y} \| \right) dA_y = \sum_p g \left(\| \mathbf{x} - \mathbf{x}_p \| \right) \int_{S_p} \bar{\bar{\sigma}}_c \cdot \mathbf{n}(\mathbf{y}) dA_y$$

$$- \nabla \cdot \left[r_p \sum_p g \left(\| \mathbf{x} - \mathbf{x}_p \| \right) \int_{S_p} \left(\bar{\bar{\sigma}}_c \cdot \mathbf{n}(\mathbf{y}) \right) \mathbf{n}(\mathbf{y}) dA_y \right] + \mathcal{O} \left(\nabla^2 \right) \tag{8.74}$$

which is a result of a Taylor series expansion in $g(\|\mathbf{x} - \mathbf{y}\|)$ about the center of the particle with radius r_p. Terms of $\mathcal{O}\left(\nabla^2 \right)$, where the ∇ represents the gradients in the expansion, and higher have been neglected. Note that the first term on the right-hand side of Eq. (8.74) is the same as the fluid traction term in the particle phase momentum balance. The difference in the manner in which the resultant forces due to fluid traction act on the surfaces of the particles is a key distinction between the Jackson [Eq. (8.74)] and Ishii [Eq. (8.59)] formulations. In the formulation by (Ishii, 1975), which is applicable to fluid droplets, the fluid-droplet traction term is the same in the fluid phase and in the dispersed phase governing equations.

The integrals involving the traction on a particle surface have been derived by Nadim and Stone (1991) and are given in Jackson (1997) as

$$\sum_p g\left(\parallel \mathbf{x} - \mathbf{x}_p \parallel\right)\int_{S_p} \bar{\bar{\sigma}}_c \cdot \mathbf{n}(\mathbf{y}) \mathrm{d}A_y = \frac{\mathbf{I}}{\alpha_p} + \rho_c \alpha_p g + \rho_c \alpha_p \frac{D\mathbf{u}_c}{Dt} \tag{8.75}$$

$$\nabla \cdot \left[r_p \sum_p g\left(\parallel \mathbf{x} - \mathbf{x}_p \parallel\right)\int_{S_p} \left(\bar{\bar{\sigma}}_c \cdot \mathbf{n}(\mathbf{y})\right)\mathbf{n}(\mathbf{y}) \mathrm{d}A_y \right] = -\nabla\left(\alpha_p p_c\right) \tag{8.76}$$

where $\mathbf{I}$ is the interphase momentum transfer coefficient, including viscous and form drag, lift force, etc.

The resulting governing equations are averaged as follows

$$\rho_c\left[\frac{\partial \mathbf{u}_c}{\partial t} + \mathbf{u}_c \cdot \nabla \mathbf{u}_c \right] = \nabla \cdot \bar{\bar{\tau}}_c - \nabla p_c - \frac{\beta}{\alpha_c}\left(\mathbf{u}_c - \mathbf{u}_p\right) + \rho_c g \tag{8.77}$$

$$\rho_p \alpha_p \left[\frac{\partial \mathbf{u}_p}{\partial t} + \mathbf{u}_p \cdot \nabla \mathbf{u}_p \right] - \rho_c \alpha_p \left[\frac{\partial \mathbf{u}_c}{\partial t} + \mathbf{u}_c \cdot \nabla \mathbf{u}_c \right]$$
$$= \frac{\beta}{\alpha_c}\left(\mathbf{u}_c - \mathbf{u}_p\right) + \alpha_p \left(\rho_p - \rho_c\right)g + \nabla \cdot \bar{\bar{\tau}}_p - \nabla p_p \tag{8.78}$$

The pressures of the two phases are typically assumed to be the same, $p = p_c = p_p$. As can be seen, these momentum equations have a very similar form as Eqs. (8.62) and (8.63), but there are small differences. These differences arise from the fact that for droplets, as considered by Ishii (1975), the stresses are continuous throughout the medium, whereas in a particle there are no stresses. This has been described and confirmed by van Wachem et al. (2001a), and in most practical cases, these differences are small.

8.6.3 PROBABILITY DENSITY FUNCTION

Averaging over the dispersed phase may also occur in terms of a probability density function (PDF). Consider the distribution of velocities among a large number of particles in a volume element $\mathrm{d}\mathbf{x}$, where $\mathbf{x}$ is a spatial coordinate, can be represented by the distribution of their velocity points $\mathbf{u}$ in the velocity space. The number density of this volume element will generally be a function of the location in space, $\mathbf{x}$, of the time, t, as well of the velocity $\mathbf{u}$. Therefore, the number density of the particles at volume $\mathbf{x}$ with velocity $\mathbf{u}$ at time t is denoted by $f(\mathbf{u}, \mathbf{x}, t)$. This definition implies that the probable number of particles, which at time t are situated in the volume element $(\mathbf{x}, \mathbf{x} + d\mathbf{x})$ and have velocities lying in the range $(\mathbf{u}, \mathbf{u} + d\mathbf{u})$ is

$$f(\mathbf{u}, \mathbf{x}, t)\, d\mathbf{u}\, d\mathbf{x}$$

The normalised function of f is called the *probability density function*. This definition involves a probability concept; the results in which function f appears will be a result of the probable, or average, behavior of the particles.

The total number of particles in the space at time t is obtained by integrating f over all possible velocities and locations. The number density is thus defined as

$$n(\mathbf{x}, t) = \int f(\mathbf{u}, \mathbf{x}, t)\, d\mathbf{u} \tag{8.79}$$

The function f can never be negative, must tend to zero as $\mathbf{u}$ becomes infinite, and is assumed to be finite and continuous for all values of t.

When the behavior or number of the particles also depends upon the temperature, T, or the mass, m, the definition of the probability density function is extended to the following expression

$$f(\mathbf{u}, \mathbf{x}, T, m, t)\, d\mathbf{u}\, d\mathbf{x}\, dT\, dm$$

This equation implies that there is a number of particles at time t, in volume element $(\mathbf{x}, \mathbf{x} + d\mathbf{x})$, with velocities in the range $(\mathbf{u}, \mathbf{u} + d\mathbf{u})$, temperatures in the range $(T, T + dT)$, and masses in the range $(m, m + dm)$. The definitions from this probability density function can be derived analogously. In most cases in this work, however, this extension is not necessary.

8.6.4 THE BOLTZMANN EQUATION

When each particle is subjected to an external force, $\mathbf{F}$, with acceleration $\mathbf{F}/m = \mathbf{a}$, and considering the time interval between t and $t + dt$, the velocity $\mathbf{u}$ of any particle that does not collide with another particle will change to $\mathbf{u} + \mathbf{a}\, dt$, and its position vector $\mathbf{x}$ will change to $\mathbf{x} + \mathbf{u}\, dt$. The number of particles $f(\mathbf{u}, \mathbf{x}, t)\, d\mathbf{u}\, d\mathbf{x}$ at time t is equal to the number of particles $f(\mathbf{u} + \mathbf{a}\, dt, \mathbf{x} + \mathbf{u}\, dt, t + dt)\, d\mathbf{u}\, d\mathbf{x}$ if external effects and collisions between particles are neglected. The change of f over dt is caused only by external effects or the collisions of particles

$$\{f(\mathbf{u} + \mathbf{a}dt, \mathbf{x} + \mathbf{u}dt, t + dt) - f(\mathbf{u}, \mathbf{x}, t)\}\, d\mathbf{u}\, d\mathbf{x} = \frac{\partial_e f}{\partial t}\, d\mathbf{u}\, d\mathbf{x}\, dt \tag{8.80}$$

where $\dfrac{\partial_e f}{\partial t}$ is the rate of change of f at a fixed point due to external effects and particle collisions. By dividing $d\mathbf{u}\, d\mathbf{x}\, dt$ and making dt tend to zero, Boltzmann's equation for f is obtained

$$\frac{\partial f}{\partial t} + \mathbf{u} \cdot \nabla f + \mathbf{a} \cdot \frac{\partial f}{\partial \mathbf{u}} = \frac{\partial_e f}{\partial t} \tag{8.81}$$

This equation is the fundamental equation of all PDF methods. The Boltzmann equation can be used to study the statistical behavior of any particle property. The governing equation for a property of the particle (defined as ϕ) is obtained by multiplying the Boltzmann equation with ϕ and integrating over all achievable possibilities. Although this is not difficult *per se*, there are numerous steps to be undertaken, and the details can be found in Chapman and Cowling (1990), Simonin (1996), or the review article by Reeks (2021). The result is the so-called Enskog equation

$$n\mathbb{C}(\phi) = n\frac{D\langle\phi\rangle}{Dt} + \frac{\partial n\langle\phi\mathbf{u}_p\rangle}{\partial\mathbf{x}} - n\left\langle\frac{D\phi}{Dt}\right\rangle - n\left\langle\mathbf{u}_p\frac{\partial\phi}{\partial\mathbf{x}}\right\rangle - n\left\langle\mathbf{a}_p\frac{\partial\phi}{\partial\mathbf{u}_p}\right\rangle$$
$$+ n\frac{D<\mathbf{u}_p>}{Dt}\cdot\left\langle\frac{\partial\phi}{\partial\mathbf{u}_p}\right\rangle + n\left\langle\frac{\partial\phi}{\partial\mathbf{u}_p}\mathbf{u}_p\right\rangle : \frac{\partial<\mathbf{u}_p>}{\partial\mathbf{x}}$$

$$(8.82)$$

Where n is the local number density, and $n\mathbb{C}(\phi)$ summarised the effect of collisions on property ϕ. The Enskog equation is a very powerful equation because it is the governing equation for the statistics of any particle property. It can be used to derive the continuity equation (by setting $\phi = 1$), the momentum equation (by setting $\phi = \mathbf{u}_p$), the particle fluctuating kinetic energy ($\phi = \mathbf{u}_p\mathbf{u}_p$), etc. The drawback of using the Enskog equation is that for some of the particle properties, several terms remain unknown. For instance, the governing equation for the particle fluctuating kinetic energy, $\phi = \mathbf{u}_p\mathbf{u}_p$, there is one unknown term that depends on the triple product $\mathbf{u}_p\mathbf{u}_p\mathbf{u}_p$, that represent the third moment of velocity fluctuations, or the transport of fluctuating kinetic energy by the velocity. For such terms, models need to be derived, and this is the main challenge in the Eulerian-Eulerian modelling framework.

8.6.5   THE EULERIAN-EULERIAN GOVERNING EQUATIONS

Using the Enskog equation, the continuity, momentum, and fluctuating energy equations for the dispersed phase can be derived (Lun *et al.*, 1984). The resulting equations are summarized as follows: First, there is a conservation of total volume, which is expressed by the equation

$$\alpha_p + \alpha_c = 1 \tag{8.83}$$

or

$$\frac{\partial}{\partial x}\left(\alpha_p\mathbf{u}_p + \alpha_c\mathbf{u}_c\right) = 0. \tag{8.84}$$

The first equation states that the sum of local volume fractions has to add up to unity. Each individual phase also has a continuity equation, expressed as

$$\frac{\partial\alpha_p}{\partial t} + \frac{\partial\alpha_p\mathbf{u}_p}{\partial x} = 0 \tag{8.85}$$

$$\frac{\partial\alpha_c}{\partial t} + \frac{\partial\alpha_c\mathbf{u}_c}{\partial x} = 0 \tag{8.86}$$

The final form of the solid phase momentum equation may be expressed as

$$\frac{\partial \rho_p \alpha_p \mathbf{u}_p}{\partial t} + \frac{\partial \rho_p \alpha_p \mathbf{u}_p \mathbf{u}_p}{\partial x} = \frac{\partial}{\partial x}\left[\alpha_p\left(\overline{\overline{R}}_p + \overline{\overline{T}}_p\right)\right] - \alpha_p \frac{\partial p}{\partial x} + \rho_p \alpha_p \mathbf{g} + \beta\left(\mathbf{u}_c - \mathbf{u}_p\right) \quad (8.87)$$

where the first term in brackets of the right-hand side, $\overline{\overline{R}}_p$, represents the kinetic stress tensor, which depends on the so-called granular temperature, Θ_p; and the second term, $\overline{\overline{T}}_p$, represents the solids collisional stress tensor of the particulate phase, also depending on the granular temperature (Lun et al., 1984). These closures of the solid phase momentum equation require a description for the solid phase stresses. When collisional interactions play an important role in the motion of the particles, concepts from kinetic gas theory (Chapman and Cowling, 1990; Lun et al., 1984; van Wachem et al., 2001a) can be used to describe the effective stresses in the solid phase resulting from particle streaming (kinetic contribution) and direct collisions (collisional contribution). These concepts depend on the local granular temperature of the particles.

The granular temperature is a measure of the fluctuating velocity of the particles, assuming that the fluctuating velocity stress tensor is isotropic. The granular temperature balance, which is equivalent to an energy balance, is expressed as

$$\frac{3}{2}\left[\frac{\partial \rho_p \alpha_p \Theta_p}{\partial t} + \frac{\partial \rho_p \alpha_p \mathbf{u}_p \Theta_p}{\partial x}\right] = -\frac{\partial \alpha_p \mathbf{J}_p}{\partial x} + \mathcal{P}_p - \alpha_p \chi_{energy} + \langle \mathbf{u}'_p \mathbf{f}_p \rangle \quad (8.88)$$

where $\mathbf{J}_p$ is the diffusion of fluctuating energy, usually modelled by a gradient assumption

$$\mathbf{J}_p = \kappa_p \nabla \Theta \quad (8.89)$$

with k_p being the so-called *solids thermal conductivity* (van Wachem et al., 2001a). The term $\mathcal{P}_p$ is the solids fluctuating energy production term. This production originates from the shear of the particle phase, which increases the fluctuating energy of the solid particles

$$\mathcal{P}_p = \left(\overline{\overline{R}}_p + \overline{\overline{T}}_p\right) : \nabla \mathbf{u}_p \quad (8.90)$$

The term χ_{energy} in Eq. (8.88) is the dissipation of fluctuating energy, originating from the collisions of the particles and their non-elastic nature. Finally, the term $\langle \mathbf{u}'_p \mathbf{f}_p \rangle$ is the rate of energy dissipation per unit volume resulting from the fluctuating forces on the particles, which are exerted by the continuous phase.

The momentum equations for the continuous phase of the Eulerian-Eulerian model can be expressed as

$$\frac{\partial \rho_c \alpha_c \mathbf{u}_c}{\partial t} + \frac{\partial \rho_c \alpha_c \mathbf{u}_c \mathbf{u}_c}{\partial x} = \frac{\partial}{\partial x}\left[\alpha_c\left(\overline{\overline{T}}_c + \overline{\overline{R}}_c\right)\right] - \alpha_c \frac{\partial p}{\partial x} - \beta\left(\mathbf{u}_c - \mathbf{u}_p\right) \quad (8.91)$$

where $\overline{\overline{T}}_c$ is the mean viscous stress tensor, typically closed by assuming a Newtonian rheology, and $\overline{\overline{R}}_c$ is the fluids Reynolds stress tensor. For calculating the fluid Reynolds stress tensor, the value of the fluid turbulent kinetic energy is required. The balance for the continuous phase fluid turbulent energy, k_c, in the Eulerian-Eulerian formulation is given as (Benavides and van Wachem, 2009)

$$\frac{\partial \rho_c \alpha_c k_c}{\partial t} + \frac{\partial \rho_c \alpha_c \mathbf{U}_c k_c}{\partial \mathbf{x}} = -\frac{\partial \alpha_c \mathbf{J}_c}{\partial \mathbf{x}} + \mathcal{P}_c - \alpha_c \varepsilon_c + \langle \mathbf{u}_c' \mathbf{f}_c \rangle \qquad (8.92)$$

where the term $\mathbf{J}_c$ represents the diffusion of turbulence kinetic energy, $\mathcal{P}_c$ is the production, and ε_c is the dissipation of turbulence energy. These terms are similar to the single phase equivalents of the $k - \varepsilon$ model. The last term on the right-hand side, $\langle \mathbf{u}_c' \mathbf{f}_c \rangle$ represents the rate of turbulent energy dissipation resulting from the acting of the fluctuating forces on the fluid exerted by the particles.

The Eulerian-Eulerian model has been used to successfully predict the behavior of many large-scale multiphase flows, such as of fluidized beds (Gidaspow, 1986; Neau et al., 2020; Pritchett et al., 1978; van Wachem et al., 1998), turbulent fluid-particle flows (Benavides and van Wachem, 2009; He and Simonin, 1993; Ozel et al., 2013; Simonin et al., 1993), bubbles in flows (Mudde and Simonin, 1999; Ma et al., 2015; Rzehak et al., 2015; Ziegenhein et al., 2015; Muniz and Sommerfeld, 2020), and droplets (Ma et al., 2016; Pai and Subramaniam, 2006).

8.6.6 MIXTURE MODELS

The complete Eulerian-Eulerian model framework can be complex to solve because it has a significant number of local unknowns: The common pressure, three velocity components per phase, and one volume fraction per phase. This means that for a two-phase flow, there are nine unknowns and, of course, just as many governing equations. The resulting system of equations is computationally expensive to solve. If there is a constitutive relation available between the relative velocity of the two phases, a more affordable option is available—the mixture model. The mixture model, which is based upon the Eulerian-Eulerian model, simplifies the set of governing equations, and it describes the complete multiphase flow with a single-fluid flow field, consisting of one continuity and one set of momentum equations. This is achieved by considering the *relative velocity between the phases* as given, so it should be determined through a constitutive equation or closure law, which typically depends on the type of flow and the flow regime. Considering two phases, a continuous phase and a particle phase, the expressions for the two velocities are then given as

$$\mathbf{u}_p = \mathbf{u}_m + \alpha_p \mathbf{u}_r$$

and

$$\mathbf{u}_c = \mathbf{u}_m - \alpha_c \mathbf{u}_r$$

where $\mathbf{u}_m$ is the mixture velocity, denoting the average velocity of the two phases weighted by their volume fraction, and $\mathbf{u}_r$ is the relative velocity, for which a closure expression or an assumption is required.

Inserting the equations for the particle and fluid velocity, expressed relative to the slip velocity, into the Eulerian-Eulerian model and adding the results, the continuity equation becomes

$$\frac{\partial \alpha_p}{\partial t} + \nabla \cdot \left(\alpha_p \mathbf{u}_m \right) + \nabla \cdot \left(\alpha_c \alpha_p \mathbf{u}_r \right) = 0 \tag{8.93}$$

And the momentum equation is

$$\frac{D \rho_m \mathbf{u}_m}{Dt} = -\nabla p \left(\frac{\alpha_c}{\rho_c} + \frac{\alpha_p}{\rho_p} \right) + \nabla \cdot \left(\overline{\overline{\tau}}_m \right) + \rho_m \mathbf{g} \tag{8.94}$$

where $\overline{\overline{\tau}}_m$ indicates the local stresses of the mixture. When this relative velocity is assumed non-zero, this mixture model is commonly referred to as the algebraic slip model (ASM). The success of the mixture model depends very much on the accuracy of the closure model for the relative velocity, $\mathbf{u}_r$. The ASM is often applied for a dispersed multiphase flow, where a small slip between the continuous phase and the dispersed phase is prescribed.

A second type of mixture model also exists, the homogeneous model (HM). In the HM, the relative velocity is assumed to be zero. This means that both phases move with the same velocity and are in mechancial equilibrium. Such an assumption may be applicable in two situations. In the first situation, the dispersed phase is very finely dispersed in the carrier fluid and has a very low Stokes number. This justifies the assumption that both phases move with the same velocity. In the second situation, both phases are very clearly separated, and a fully resolved simulation (i.e. DNS) is carried out. The resolution of the numerical method in this second situation must be so fine that each computational cell is occupied by one phase and one phase only (not a mixture) at any single instant in time. Interfacial forces, such as surface tension, must be added separately to the resulting mixture equations. The resulting momentum equations of the HM are given by

$$\frac{D \rho_m \mathbf{u}_m}{Dt} = -\nabla p + \nabla \cdot \left(\overline{\overline{\tau}}_m \right) + \rho_m \mathbf{g} + \mathbf{F}_s \tag{8.95}$$

where $\mathbf{F}_s$ represents the forces between the phases, and the mixture density is given by

$$\rho_m = \sum_{i=phases} \alpha_i \rho_i$$

and the other mixture properties, such as the viscosity, are determined similarly.

The two-mixture models are easier to solve than the full Eulerian-Eulerian model, and the computational cost of solving the mixture model is typically like solving a single-phase fluid. Their success depends on the validity of the stringent assumption of the relative velocity between the phases, which is either zero (HM) or is provided with a closure expression (ASM).

8.7 HYBRID EULER-LAGRANGE APPROACHES

The hybrid Euler-Lagrange approaches allow for more refined modelling of a dispersed phase compared to two-fluid methods, but their application is limited to dispersed multiphase flows. In describing the dispersed phase, the discrete nature of individual particles is maintained. Particles are considered as point-masses in order to allow computations of industrial large-scale processes. The point-mass approximation and the mostly higher values of particle Reynolds numbers necessitates the knowledge of resistance coefficients for all relevant interfacial (fluid dynamic) forces acting on the particles. In addition, other particle-scale transport processes (i.e. not included in the equation of motion), such as particle-wall collisions and droplet/bubble breakup and coalescence, require additional modelling, which may be descriptive. Naturally, point-particle Lagrangian tracking may be combined with the unsteady approaches as there are DNS (direct numerical simulation) and LES (large eddy simulations). The application to large-scale industrial processes, however, still requires the application of RANS (Reynolds-averaged Navier-Stokes) methods in connection with an appropriate turbulence model. The dispersed phase is modelled by tracking a large number of particles through the beforehand computed carrier fluid flow field and solving their equations of motion while accounting for all relevant forces on the particles as well as heat and mass transfer, if pertinent, as presented in Chapters 4 and 5 (see also Sommerfeld *et al.*, 1993a; Michaelides, 2003, 2006; Sommerfeld *et al.*, 2008). Since the particles are considered as point-particles, and the flow around individual particles is not resolved, an essential consequence of this approach is that the particles need to be significantly smaller than the dimensions of the numerical grid or control volumes as described in Section 8.2.

There is no limitation in the maximum particle volume fraction to be considered when applying the Euler-Lagrange approach, as long as all relevant elementary processes are accounted for. Hence, the applications of this method range from the dilute to the dense regime, as for example are found in fluidized beds. However, there are different forms of the Lagrangian particle tracking methods, suggesting a classification, which is summarized in Figure 8.28 (Sommerfeld, 2017). For dilute two-phase flows normally the classical Lagrangian approach, which considers parcels, is adopted, as shown in Figure 8.28, at left; whereas in a denser regime, both the discrete particle method, or DPM (e.g. Li and Kuipers, 2003, 2007), as well as the DEM (discrete element method) are being applied (e.g. Zhong *et al.*, 2006). Both methods, DPM and DEM, rely on tracking all real particles in the system, for which reason the system size is limited. The difference between these two methods is the treatment of inter-particle collisions (Deen *et al.*, 2007): In the DPM, a hard sphere collision model is used, as depicted in the middle part of Figure 8.28. This approach is physically more realistic, but also imposes limitations on the tracking time step in order to guarantee only binary collisions. For this reason, the method is also called event driven. Binary collisions only last for very small time periods and are described by the impulse equations in connection with Coulomb's law of friction, as described in Section 6.1.3, so that the only model parameters are normal restitution ratio and friction coefficient. In the DEM (Figure 8.28, right), multiple and

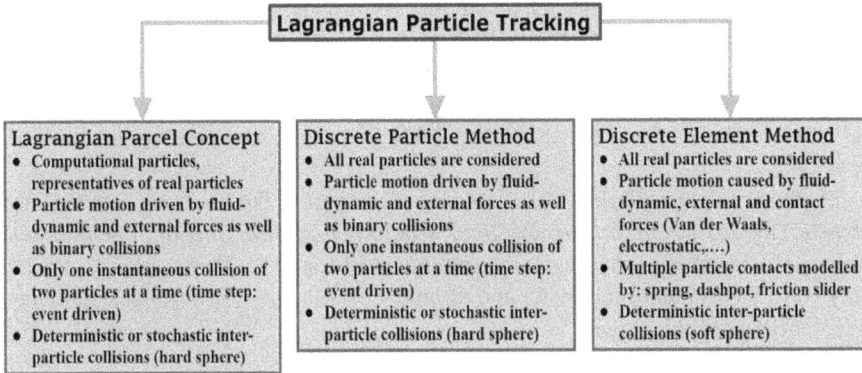

FIGURE 8.28 The classification of Lagrangian particle tracking methods (Sommerfeld, 2017).

overlapping particle contacts (called soft-sphere collisions and described in Section 6.2) are allowed, often modelled by a "spring, dashpot and friction slider element" according to Tsuji (1993) and Tsuji et al. (1993) and the pioneering work of Cundall and Strack (1979). This approach uses constant time steps (time-driven), which have to be properly chosen as it determines any possible particle overlap. In DEM, all real particles need to be tracked, unless a coarse-graining model is used. Currently, with the available computational resources, the number of the particles in the computations and the equipment size has strong limitations. When considering larger facilities, bigger particles must be used because the particle number is limited. Due to the modelling of multiple contacts, the DEM allows the consideration of much denser systems than DPM.

In the case of fine particles, for example in a spray, the tracking of all real particles in the system is computationally not affordable at present. Therefore, in the classical Lagrangian approach, the parcel concept (LPC, shown in Figure 8.28 left) is utilized, where the computational particles, mostly called *parcels*, represent a certain number of real particles with the same properties (i.e. size, velocity, and temperature). This procedure ensures that the correct particle mass flux is represented in the system. In stationary flows, a sequential or simultaneous tracking of the parcels may be adopted, while in unsteady flows, all parcels need to be tracked simultaneously. The parcel concept was introduced in connection with spray combustion modelling (Dukowicz, 1980), where after atomization, a very large number of droplets are carried by the flow field. Each computational particle represents a certain number of real droplets with the same properties (Dukowicz, 1980). This initiation was followed by numerous studies at Los Alamos National Laboratory, eventually ending up in the step-by-step extended and well-known KIVA-code, which was developed for the simulation of spray combustion (O'Rourke and Amsden, 1987; Amsden et al., 1989). Several other research groups were involved in the initial developments of the Lagrangian approaches for spray simulations (Rüger et al., 2000). In particular, the thesis of Bauman (2001) gives a detailed overview of the modelling approaches for the relevant elementary processes.

Local average properties such as the dispersed phase concentration, the number density, and velocities are obtained by ensemble averaging for each control volume in the computational domain, coupled with time averaging in a stationary flow. For three-dimensional flows, statistically reliable results for each computational cell require the tracking of 100,000 to 1,000,000 parcels, depending on the flow geometry.

Essential for the hybrid Euler-Lagrange approach is the coupling between the phases. Depending on the particle concentration or volume fraction, one may distinguish between one-way, two-way, and four-way coupling. This coupling may not only consider momentum transfer and turbulence modulation but also heat and mass transfer between the phases.

8.7.1 RANS CONTINUOUS-PHASE EQUATIONS

The Euler-Lagrange approach described in the following is based on the three-dimensional Reynolds-averaged conservation equations (RANS) in connection with the well-established k-ε turbulence model (Launder and Spalding, 1974). Although this turbulence model has some limitations, it allows the efficient numerical solution to most engineering problems. The general form of the conservation equations for an incompressible, unsteady, and three-dimensional flow with dispersed phase heat and mass transfer (Sommerfeld *et al.*, 1993a) is given by

$$\frac{\partial}{\partial t}\left(\rho\,\phi\right) + \frac{\partial}{\partial x_i}\left(\rho\,U_i\,\phi\right) = \frac{\partial}{\partial x_i}\left(\Gamma_\phi\,\frac{\partial\varphi}{\partial x_i}\right) + S_\phi + S_{\phi,P,m} + S_{\phi,P,ev} \qquad (8.96)$$

The continuity equation is obtained from this general equation with $\phi = 1$ and the momentum equations with $\phi = U_i$, where U_i ($i \in [1, 2, 3]$) are the three velocity components. The fluid temperature equation is obtained with $\phi = T$, and also different species conservation equations may be considered (e.g. gas phase and vapor phase). The conservation equations for the turbulent kinetic energy, k, and the dissipation rate, ε, are obtained with $\phi = k$ and $\phi = \varepsilon$, respectively. The diffusion coefficients for all the conservation equations are given in Table 8.2 together with the fluid phase source terms, S_ϕ. The particle phase source terms, $S_{\phi,P,m}$, account for the effect of the particles on mean flow and turbulence due to interfacial momentum transfer, and $S_{\phi,P,ev}$ is the transfer caused by particle/droplet evaporation or condensation. The source terms for the dispersed phase are summarised in Table 8.3 for the different flow variables. The left column provides the source terms due to interfacial momentum and heat transfer and the right column, those resulting from droplet evaporation, that is, mass transfer (Sommerfeld *et al.*, 1993a, 2008). In this case, two gaseous species are considered, for example, air and water vapor. It should be noted that momentum transfer due to evaporation occurs with the particle velocity, and in the case of condensation, the instantaneous fluid velocity seen by the particle has to be used. The summation of the source terms must be performed along all particle trajectories (over the index k) passing through a control volume. The gathering procedure for obtaining these source terms from the particle trajectory calculations will be described in the next section in more detail.

TABLE 8.2
Summary of the Variables, φ, the Fluid Source Terms, S_ϕ, and the Effective Transport Tensor, Γ, in Eq. (8.96) and Constants for the k-ε Turbulence Model

φ	S_ϕ	Γ
1	—	—
U_i	$\dfrac{\partial}{\partial x_j}\left(\Gamma_{U_i}\dfrac{\partial U_j}{\partial x_i}\right)-\dfrac{\partial p}{\partial x_i}+\rho\,g_i$	$\mu+\mu_t$
T	0	$\dfrac{\mu}{Pr}+\dfrac{\mu_t}{Pr_t}$
Y_k	0	$\dfrac{\mu}{Sc}+\dfrac{\mu_t}{Sc_t}$
k	$G_k-\rho\varepsilon$	$\mu+\dfrac{\mu_t}{\sigma_k}$
ε	$\dfrac{\varepsilon}{k}\left(C_1 G_k - C_2\rho\varepsilon\right)$	$\mu+\dfrac{\mu_t}{\sigma_\varepsilon}$
	$G_k = \mu_t\left(\dfrac{\partial U_i}{\partial x_j}+\dfrac{\partial U_j}{\partial x_i}\right)\dfrac{\partial U_i}{\partial x_j}$	$\mu_t = C_\mu\,\rho\,\dfrac{k^2}{\varepsilon}$

$C_\mu = 0.09$; $C_1 = 1.44$; $C_2 = 1.92$; $\sigma_k = 1.0$; $\sigma_\varepsilon = 1.3$

TABLE 8.3
Summary of Particle Phase Source Terms for the Different Fluid Flow Conservation Equations. See Eq. (8.96) (Left Column: Interfacial Momentum and Heat Transfer, $S_{\phi,P,m}$; Right Column: Interfacial Mass Transfer, $S_{\phi,P,ev}$)

φ	$S_{\phi,P,m}$	$S_{\phi,P,ev}$
1	0	$\displaystyle\sum_k \left(N_k\,\dot m_{k,ev}\right)/V_{CV}$
U_i	$\displaystyle-\sum_k \frac{\dot m_k N_k}{V_{CV}}\cdot\left[\left(u_{k,i}^{t+\Delta t}-u_{k,i}'\right)-g_i\left(1-\frac{\rho_F}{\rho_P}\right)\Delta t\right]$	$\displaystyle\sum_k \frac{N_k\,\dot m_{k,ev}\,u_a}{V_{CV}}$
T	$\displaystyle-\sum_k \frac{N_k}{V_{CV}}\left(H_{Lat}\,\dot m_k+\dot Q_L\right)$	$\displaystyle\sum_k \frac{N_k\,\dot m_{k,ev}}{V_{CV}}C_{pL}\left(T_k\right)T_k$
Y_1	0	0
Y_2	0	$\displaystyle\sum_k \frac{N_k\,\dot m_{k,ev}}{V_{CV}}$
k	$\displaystyle\sum_i\left(\overline{U_i\,S_{U_i,P,m}}-\overline{U_i}\;\overline{S_{U_i,P,m}}\right)$	$\overline{U_i\,S_{U_i,P,ev}}-\overline{U_i}\;\overline{S_{U_i,P,ev}}+$ $\dfrac{1}{2}\left(\overline{U_i\,U_i}\,S_{\rho,P,ev}-U_iU_iS_{\rho,P,ev}\right)$
ε	$C_3\dfrac{\varepsilon}{k}S_{k,P,m}$	$C_3\dfrac{\varepsilon}{k}S_{k,P,m}$

for evaporation: $u_{i,a}=u_{i,k}$
for condensation: $u_{i,a}=u_i$
Y_1: air; Y_2: water

8.7.2 PARTICLE TRACKING CONCEPTS

In the Lagrangian approach, the trajectories of the parcels moving through the flow field are calculated by solving the ordinary differential equations for the parcel location and the linear as well as angular velocity components. The change of particle linear velocity components is obtained by considering all relevant forces acting on the particle, and this depends on the flow system (Sommerfeld *et al.*, 2008; Sommerfeld, 2010). The change of the angular velocity along the particle trajectory results from the viscous interaction with the fluid (i.e. the torque). The equations of motion for spherical particles in vector form are summarized by the following system of equations

$$\frac{d\vec{x}_p}{dt} = \vec{u}_p \qquad (8.97)$$

$$m_p \frac{d\vec{u}_p}{dt} = \sum \vec{F}_n \qquad (8.98)$$

$$I_p \frac{d\vec{\omega}_p}{dt} = \vec{T} \qquad (8.99)$$

where $m_p = (\pi/6)\,\rho_P\,D_P^3$ is the particle mass, and $I_p = 0.1\,m_P\,D_P^2$ is its moment of inertia of a sphere. The different forces acting on the particle are fluid dynamic forces and external forces, for example, gravity, buoyancy, and electrical forces. The fluid dynamic forces acting on the particle surface are drag, added mass, pressure force, Basset/history force, and the transverse lift forces due to shear and particle rotation (Michaelides, 2003). In most industrial applications, the particle Reynolds number is larger than one so that appropriate correlations for the resistance coefficients (e.g. drag and lift coefficients) are used, as described in Chapter 4.

The consideration of the rotational motion of particles is of importance in the case of wall-bounded flows, where wall collisions induce high angular velocity, or it is modified due to wall collisions. Also, in cases with strong changes in fluid vorticity, the particle angular velocity should be considered, where the viscous torque will reduce the angular slip. In the non-equilibrium situation (i.e. high angular slip), the slip-rotation lift force, or often called Magnus force, should be considered and becomes of importance in gas-particle flows for particle diameters larger than about 70 μm (Sommerfeld, 1996).

It should be emphasized that the different forces and the torque in Eqs. (8.98) and (8.99) include the instantaneous fluid velocity experienced by the particles. In the case when LES or RANS are used for the calculation of the fluid velocity field, only the Reynolds-averaged velocities or filtered velocities are available. Therefore, the respective fluid fluctuating velocities have to be obtained from the turbulent kinetic energy or from the sub-grid-scale (SGS) turbulence. The respective models should describe the effect of turbulence on the particle motion (turbulent dispersion of particles) and should also account for the crossing trajectory effect induced by external forces.

If dispersed multiphase flows with heat and mass transfer between the phases are considered, two additional partial differential equations for the change of particle diameter and temperature have to be solved (Sommerfeld et al. 1993 a; Rüger et al. 2000)

$$\frac{dD_P}{dt} = -\frac{\dot{m}_{ev}}{\pi \ \rho_L \ D_P^2} \tag{8.100}$$

$$\frac{dT_P}{dt} = \frac{6 \ \dot{Q}_L}{\pi \ \rho_L \ C_{pL} \ D_P^3} \tag{8.101}$$

In order to consider the particle evaporation process, models for the determination of the evaporation rate $\dot{m}_{ev}$ and the heat flux $\dot{Q}_L$ are needed. For the determination of the limiting gas-side heat and mass transfer, a number of assumptions need to be made, such as spherical symmetry and film thickness around the droplet based on film theory (Sommerfeld *et al.*, 2008; Abramzon and Sirignano, 1989). There are mainly two approaches regarding the droplet temperature distribution inside the droplet: (a) The infinite conductivity model, which implies zero Biot number and a uniform temperature inside the droplet, and (b) the finite thermal conductivity model, which numerically resolves the temperature distribution inside the droplet based on spherical symmetry.

The time step for the particle tracking calculations should be automatically adjusted along the trajectory by considering all relevant timescales, which also are changing throughout the flow field. These timescales are as follows:

- The time required for a particle to cross a control volume t_{CV}
- The particle viscous response timescale τ_v
- The integral timescale of turbulence T_L
- The average time between binary particle collisions τ_c

The limitation of the time step by the inter-particle collision timescale is required in order to ensure that only one binary collision may occur during a Lagrangian tracking time step (Sommerfeld, 2001). Hence, the time step Δt must be a fraction of the minimum of these timescales for accurate calculations, for example 5–20%, depending on the degree of coupling between the phases

$$\Delta t = 0.05.....0.2 \cdot \min \left(t_{CV}, \tau_v, T_L, \tau_c \right). \tag{8.102}$$

In order to account for the influence of the particle phase on the fluid flow (two-way coupling), a successive solution of the Eulerian and Lagrangian part is required. The calculation starts with the solution of the fluid flow by not accounting for the source terms of the dispersed phase. After having reached convergence for the single-phase flow (inner fluid iterations), the particle trajectories are computed, and the particle phase properties (i.e. concentration and particle velocities) and the source terms are sampled for each control volume. In case a particle size distribution is considered, also the local particle size distributions and the size-velocity correlations should be determined. A more detailed description of possible two-way coupling procedures is given in Section 8.7.4.

8.7.3 GENERATION OF FLUID TURBULENT VELOCITIES

During the tracking procedure, the instantaneous fluid velocity along the particle trajectory must be determined since this information has been lost as a result of the Reynolds-averaging process applied to the conservation equations for the fluid flow (RANS methods). The generation of the instantaneous fluid velocity along the particle trajectory can be based on several stochastic approaches, as for example the simple *"eddy-lifetime model"* (Gosman and Ioannides, 1981), the single-step correlated model based on the Langevin equation (Zhou and Leschziner, 1991; MacInnes and Braco, 1992; Sommerfeld *et al.*, 1993b), or the multistep correlated model (Berlemont *et al.*, 1990). The principles of these methods were summarized by Gouesbet and Berlemont (1999) and in the handbook by Michaelides *et al.* (2017). Essential for all these schemes is the estimation of the relevant time and length scales of turbulence, which is not a simple task in complex anisotropic flows. A thorough comparison and validation of a modified eddy-lifetime and the single-step Langevin model was conducted by Sommerfeld (1996), and a brief summary of the models is also provided in Sommerfeld *et al.* (2008). Also, in the frame of LES for particle-laden flows, the developed sub-grid-scale turbulent transport model was based on the single-step Langevin model (Lipowsky and Sommerfeld, 2007).

One of the first stochastic approaches to model the turbulent particle dispersion (i.e. the instantaneous fluid velocity seen by the discrete particle) was the so-called *eddy lifetime or discrete eddy concept* proposed by Gosman and Ioannides (1981). This model assumes that the particle interacts with a sequence of turbulent eddies of given sizes and lifetimes. After the particles traversed the eddy or after the eddy is dissipated, it was assumed that the particle enters a new eddy with randomly sampled fluctuation intensity and, hence, a new size and dissipation timescale. The mean fluid velocity seen by the particle is interpolated from the neighboring grid points according to the particle position. The instantaneous fluid fluctuation velocity is sampled from a Gaussian distribution function with zero mean and a standard deviation corresponding to the local fluid mean velocity fluctuation (rms-value) independently for the three velocity components. Especially in connection with the k-ε-turbulence model, isotropic turbulence is assumed, and the rms-value is obtained from

$$\sigma_f = \sqrt{\frac{2}{3} k} \qquad (8.103)$$

The instantaneous fluid velocity fluctuation is assumed to influence the particle motion for a certain period of time before a new instantaneous velocity is sampled. In this simple model, the successively sampled fluid velocity fluctuations are assumed to be uncorrelated.

The interaction time of the particles with the individual eddies is limited by two criteria, namely, the eddy lifetime (or dissipation timescale), which is assumed to correspond to the integral timescale of the most energetic eddies and the time required for a particle to traverse the eddy. The latter considers the so-called crossing trajectory effect. The integral timescale can be estimated with the turbulent kinetic energy and dissipation rate as follows

$$T_E = C_T \frac{k}{\varepsilon} \quad \text{with} \quad C_T = 0.3 \qquad (8.104)$$

The constant C_T was correlated using experimental results on particle dispersion in grid turbulence (Snyder and Lumley, 1971; Wells and Stock, 1983) and/or developed pipe flow (Calabrese and Middleman, 1979). The "crossing trajectory effect" is accounted for by summing the particle travel distance within one eddy and comparing it with the eddy length scale given by the expression

$$L_E = \sigma_f T_E \qquad (8.105)$$

When the travel distance of the particle in the eddy becomes larger than the eddy size, a new eddy and, hence, a new fluid fluctuation is sampled, independent of the conditions regarding the interaction time. In the original model by Gosman and Ioannides (1981), the instantaneous fluid velocity fluctuation and the eddy time and length scales were kept constant during one eddy-particle interaction period.

In order to account for the inhomogeneity of turbulence along the particle trajectory, this model was modified by Milojevic (1990) and Sommerfeld et al. (1993b): The instantaneous fluid velocities experienced by the particle were continuously updated according to the particle position in the flow, within an eddy-particle interaction period. This implies the update of the mean fluid velocities and the fluid fluctuation values according to the following equations

$$u'_{n+1} = u'_n \frac{\sigma_{n+1}}{\sigma_n}; \quad v'_{n+1} = v'_n \frac{\sigma_{n+1}}{\sigma_n} \qquad (8.106)$$

using the new local value of the turbulent kinetic energy to determine σ_{n+1}. Similarly, the eddy time and length scales were updated according to the new local values of k and ε (Eqn. 8.104 and 8.105) at the new particle position. With these values, the eddy-particle interaction time and the particle position in the eddy are updated as follows

$$\Delta t_{int,n+1} = \Delta t_{int,n} \cdot \frac{T_{E,n+1}}{T_{E,n}}$$

$$x_{int,n+1} = \Delta x_{int,n} \cdot \frac{L_{E,n+1}}{L_{E,n}} \qquad (8.107)$$

A single-step Langevin equation model was developed by Sommerfeld et al. (1993b) and is detailed in Sommerfeld (1996). The mean fluid velocity is interpolated from the neighboring fluid velocity values, which are known at each control volume. The fluctuation velocity acting on the particle is composed of a random contribution of the local fluid fluctuation velocity, which appears in the second term on the right-hand side of Eq. (8.108), and a part which accounts for the correlation of the fluid fluctuation at the actual particle position with that of the particle position one time step earlier, the first term on the right-hand side of Eq. (8.108)

$$u_{i,n+1}^P = R_{P,i}\left(\Delta t, \Delta r\right) u_{i,n}^P + \sigma_i \sqrt{1 - R_{P,i}^2\left(\Delta t, \Delta r\right)} \ \chi_i \tag{8.108}$$

where χ_i is a vector of three independent Gaussian random numbers with mean value of zero and standard deviation of one. With these values, the local fluid root mean square (rms) values σ_i, and the correlation term $R_{P,i}(\Delta t, \Delta r)$, the random contributions for the three velocity directions (i = 1, 2, 3) are reconstructed. Further, the correlation term consists of a Lagrangian and a Eulerian part and defines the degree of correlation in the turbulence as experienced by the particle as

$$R_{P,i}\left(\Delta t, \Delta r\right) = R_{L,i}\left(\Delta t\right) \cdot R_{E,i}\left(\Delta r\right) \tag{8.109}$$

In this approach, the Lagrangian term is defined using the Lagrangian timescale T_L and the actual time step of the tracking procedure

$$R_{L,i}\left(\Delta t\right) = exp\left(-\frac{\Delta t}{T_{L,i}}\right) \tag{8.110}$$

The Lagrangian integral timescale can be estimated from the turbulent kinetic energy and the dissipation rate as follows

$$T_{L,i} = C_T \frac{\sigma_i^2}{\varepsilon} \tag{8.111}$$

$$C_T = 0.24; \quad \sigma_i^2 := \overline{u'^2}, \overline{v'^2}, \overline{w'^2}$$

It should be noted that in connection with a k-ε turbulence model, the three components of σ_i are identical; hence, the anisotropy of turbulence can be only accounted for by using a Reynolds-stress turbulence model (Sommerfeld et al., 2021). The constant $C_T = 0.24$ was obtained by calibration with experimental data (Sommerfeld et al., 1993b). The Eulerian part of the correlation term $R_{P,i}(\Delta t, \Delta r)$ can be calculated using the longitudinal and transversal correlation coefficients $f(\Delta r)$ and $g(\Delta r)$ (Von Karman and Horwarth, 1938)

$$R_{E,i,j}\left(\Delta r\right) = \left\{f\left(\Delta r\right) - g\left(\Delta r\right)\right\} \cdot \frac{r_i\, r_j}{r^2} + g\left(\Delta r\right) \cdot \delta_{i,j} \tag{8.112}$$

From the Eulerian correlation tensor, $R_{E,i,j}(\Delta r)$, only the three main components are considered, and the correlation coefficients, $f(\Delta r)$ and $g(\Delta r)$, are given by the expressions

$$f\left(\Delta r\right) = exp\left(-\frac{\Delta r}{L_{E,i}}\right) \quad \text{and} \quad g\left(\Delta r\right) = \left(1 - \frac{\Delta r}{2 \cdot L_{E,i}}\right) \cdot exp\left(-\frac{\Delta r}{L_{E,i}}\right) \tag{8.113}$$

The integral length scales $L_{E,i}$ are defined as (Sommerfeld, 2001)

$$L_{E,x} = C_L \, T_L \, \sigma_i \quad \text{and} \quad L_{E,y} = L_{E,z} = 0.5 \cdot L_{E,x} \tag{8.114}$$

where $L_{E,x}$ is the length scale in flow direction, the other two are the lateral components. The constant $C_L = 3.0$ was determined by comparison with experiments (Sommerfeld, 1996).

More sophisticated dispersion models were developed that are based on the general form of the Langevin equation for generating the instantaneous fluid velocity experienced by the particle (e.g. Minier and Peirano, 2001; Minier et al., 2004). These models can also account for an anisotropic turbulence structure in more complex flows in combination with full Reynolds-stress turbulence models, such as swirling flows (Lipowsky and Sommerfeld, 2005) or in narrow airway passages (Sommerfeld et al., 2021). An extensive review on modelling dispersion of heavy particles in homogeneous isotropic turbulence, including aspects of LES applications, is given by Huilier (2021).

8.7.4 POINT-MASS COUPLING APPROACHES

Because of the hybrid nature of the Euler-Lagrange approach (i.e. continuum assumption for the carrier phase and discrete particle method for the dispersed phase), special attention has to be given to the two-way coupling procedure. This means that during the particle tracking in the beforehand calculated flow field, the particle phase properties and the source terms for the continuous phase equations have to be evaluated by statistical averaging. Depending on the requirements for the temporal resolution of the flow and the coupling between the two phases, different approaches may be used.

For stationary particle-laden flows, the *steady-state coupling* is numerically very efficient since only the terminal coupled solution of the two-phase system at t → ∞ is of interest. Therefore, particle tracking may be done simultaneously or sequentially (i.e. one particle after the other) for a sufficiently large number of parcels. The solution procedure is illustrated in Figure 8.29 (Sommerfeld et al., 2008, Laín and Sommerfeld, 2012) and may be summarized as follows:

1. Solve the fluid flow (Eulerian) without particles until convergence is reached.
2. Track a large number of parcels for a sufficient statistical averaging of particle properties and source terms in all computational cells.
3. In this calculation mode, source term evaluation is based on temporal and ensemble averaging as depicted in Figure 8.30a.
4. The momentum coupling may be based on the change of particle velocity within a numerical cell, subtracting, however, external forces. Thus, the influence of all the interfacial forces acting on the particle is accounted for. This approach is numerically much more efficient compared to the summation of all the hydrodynamic forces acting on all particles (see Eq. 8.115).

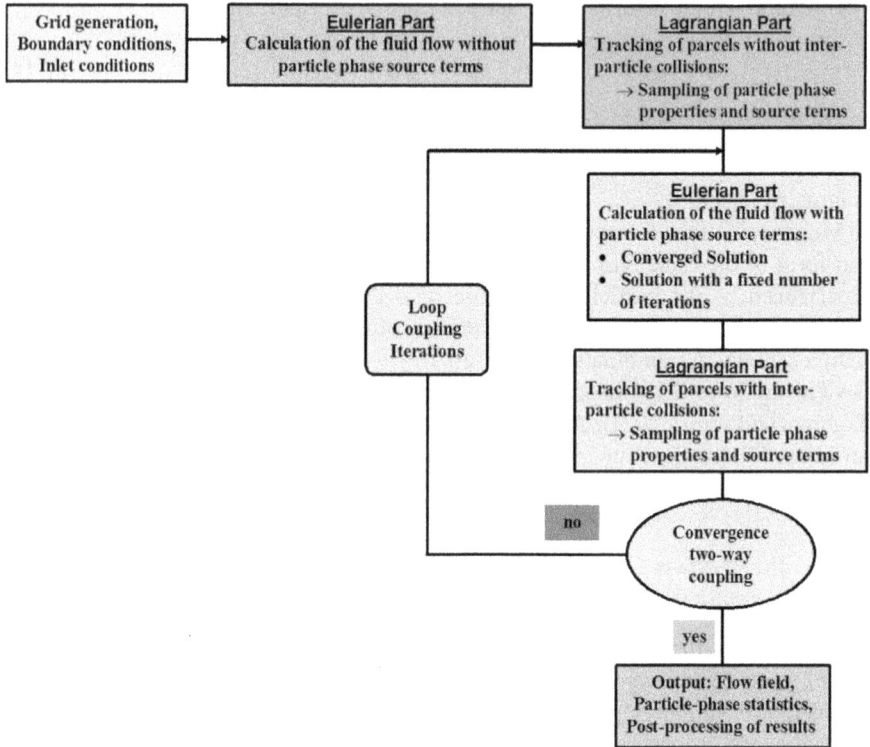

FIGURE 8.29 Flow chart of coupling iteration procedure for the hybrid Euler-Lagrange approach in the case of stationary flows.

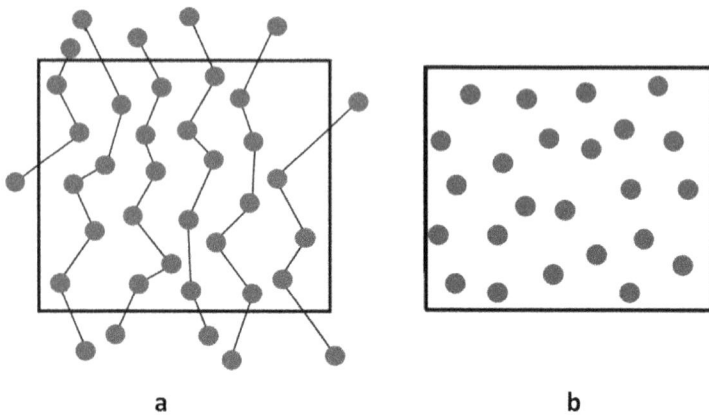

FIGURE 8.30 Illustration of the averaging procedure for different two-way coupling methods; (a) steady approach with temporal and ensemble averaging or semi-unsteady approach; (b) full unsteady calculations based on instantaneous fluid dynamic forces on all particles in a numerical cell.

5. Under-relaxation of source terms is suggested for smoothly approaching a converged solution for the coupled system (Kohnen et al., 1994; Laín and Sommerfeld, 2012).
6. Recalculate the flow field with the new source terms being obtained from Eq. (8.118) with a preset under-relaxation factor. The under-relaxation factor depends on the degree of coupling between the phases (e.g. particle concentration).
7. Recalculate the Lagrangian part by tracking a large number of parcels since the flow field has changed and perform averaging.
8. This sequential calculation of Euler and Lagrange program modules must be repeated until the coupled system has converged. This should preferably be decided based on the evolution of some fluid property in the computational domain (Laín and Sommerfeld, 2012; Sommerfeld, 1996).

This procedure also applies for thermal coupling and species transfer.

Using the change of particle velocities in a computational cell, the momentum source terms are determined by the expression (Kohnen et al., 1994)

$$\overline{S_{U_i,P}} = -\frac{1}{V_{CV}\,\Delta t_{ref}}\sum_{k=1}^{k_{tot}} m_k N_k \sum_{n=1}^{n_s}\left\{\left(u_{P,k,i}^{n+1}-u_{P,k,i}^{n}\right)-g_i\left(1-\frac{\rho_F}{\rho_P}\right)\Delta t\right\}$$
(8.115)

In this equation, the sum over n implies a time-space averaging along the particle trajectory, and the sum over all parcels k_{tot} passing the control volume is an ensemble average. To obtain statistically reliable source terms, a sufficient number of parcels need to pass the control volumes. This may be critical in flow fields with boundaries between a dispersed phase regions and fluid flow areas, for example, in a particle-laden free jet. In such a situation, the coupling and source term calculation may be improved by a spatial interpolation method over a certain number of neighboring computational cells. The source terms in the conservation equation of the turbulent kinetic energy k are obtained from the Reynolds-averaging procedure

$$S_{k,P} = \sum_{i=1}^{3}\overline{u_i\,S_{U_i,P}} - \overline{u_i}\,\overline{S_{U_i,P}}$$
(8.116)

The source term in the ε-equation is modelled in the following way

$$S_{\varepsilon,P} = C_{\varepsilon 3}\frac{\varepsilon}{k}S_{k,P}$$
(8.117)

As demonstrated by Squires and Eaton (1991) the constant $C_{\varepsilon 3}$ is not universal but depends on the particle response time and concentration. For many situations, a value of $C_{\varepsilon 3} = 1.1$ may be used.

Subsequently, the continuous phase is recalculated by accounting for the particle phase source terms, as illustrated in the flow chart of Figure 8.29. In order to avoid convergence problems, an under-relaxation procedure should be used (Kohnen et al., 1994)

$$S_{\varphi,P}^{new} = (1-\gamma)S_{\varphi,P}^{old} + \gamma\,S_{\varphi,P}^{samp}$$
(8.118)

where $S_{\varphi,P}^{new}$ are the source terms used to calculate the new flow field, $S_{\varphi,P}^{old}$ are the source terms used in the previous Eulerian calculation, and $S_{\varphi,P}^{samp}$ are the new source terms collected in the Lagrangian calculation. The under-relaxation factor depends on the degree of coupling (i.e. on the particle concentration and the particle size) and is selected accordingly in the range between zero and one (see for example: Kohnen *et al.*, 1994; Laín and Sommerfeld, 2012). After a prescribed number of Eulerian iterations or after a certain degree of convergence is reached for the fluid flow, the particle tracking is performed again, since the flow field has changed due to the two-way coupling (i.e. the influence of the particles on the flow field). With the sampled new source terms, the continuous phase is solved again. The process continues until convergence is reached for the coupled system, as shown in Figure 8.29. The overall convergence is decided based on the evolution of a certain reference value (Kohnen *et al.*, 1994), for example, the fluid velocity at a monitoring location, described in Section 8.7.6.

The next coupling mode should be used if the temporal evolution of the flow field needs to be fully resolved. For such fully *unsteady multiphase flows*, the timescales of flow and particle tracking may be in the same range, namely, $\Delta t_E \approx \Delta t_L$. This solution procedure applies mainly for DNS, but also for LES in some cases, and implies that the Eulerian and Lagrangian modules are sequentially solved with the same time steps. The minimum of all the relevant timescales should be selected. The main features of this solution approach are as follows:

- All particles or parcels need to be tracked simultaneously.
- The averaging of particle phase properties and the source terms is based on all parcels residing in the computational cells at one instant of time, that is, ensemble averaging, as shown in Figure 8.30b.
- It requires a very large number of particles to be tracked for obtaining reliable particle phase properties and source terms.
- Moreover, the momentum source terms need to be determined from the instantaneous values of all fluid dynamic forces acting on the particles within a cell, as shown in Figure 8.30b.
- Under-relaxation is not suitable for the full-unsteady approach.

In addition, there are many unsteady situations where the required Eulerian time step, Δt_E, is much larger than the expected Lagrangian time steps, Δt_L, which is often dictated by the particle response time. A typical Eulerian time step for RANS or LES simulations is between $\Delta t_E = [10^{-4}$ s, 10^{-2} s$]$ and Lagrangian time steps are typically in the range $\Delta t_L = [10^{-8}$ s, 10^{-4} s$]$, depending on the type of particles dispersed in the flow and other constraints, for example, the inter-particle collision time and the turbulence timescale. Especially for tiny particles, their very small response time mostly limits the Lagrangian time step. For such a situation, a so-called *semi-unsteady approach* may be used. This concept was introduced by Sommerfeld *et al.* (1997) and may be summarized as follows:

1. The flow field is calculated with Δt_E starting from a certain initial condition, where Δt_E determines the temporal resolution of the flow field and may be predetermined.

2. Then simultaneous particle tracking is conducted in this "frozen" flow field until the actual time level of the flow is reached for all particles. The particle tracking time step is dynamically adapted for each parcel according to the locally relevant timescales, as in Eq. (8.102). During each Eulerian time step, several thousands of computational particles are randomly injected.
3. The particle phase source terms and the particle properties are averaged over the time sequence Δt_E, and additionally, ensemble averaging is done (see Figure 8.30a), similar to the steady-state approach.
4. Under-relaxation of source terms shall not be applied in this situation.
5. The new source terms are introduced into the fluid flow calculations, which are then solved for the next Eulerian time step.

This semi-unsteady approach is illustrated in Figure 8.31, where the upper line shows the fluid timeline and the lower one the particle tracking timeline for several particles exemplarily. This approach is numerically very efficient compared to a full unsteady simulation, since mostly the particle timescales are very small and, hence, would dictate the Eulerian time step. Such a method was proposed by Sommerfeld *et al.* (1997) for URANS simulations of a locally aerated bubble column and was later used by Lipowsky and Sommerfeld (2005) for simulating small particle separation in a gas cyclone.

8.7.5 MESH SIZE REQUIREMENTS IN TWO-WAY COUPLED EULER-LAGRANGE SIMULATIONS

The two-way coupled Euler-Lagrange method was first introduced by Crowe *et al.* (1977) as the *particle-source-in-cell* (PSIC) method. In this approach, the

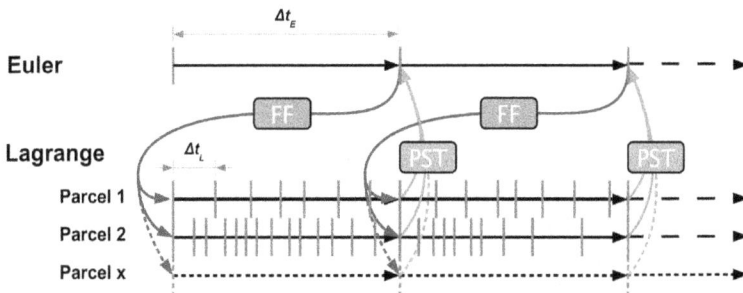

FIGURE 8.31 The timelines for a semi-unsteady Euler-Lagrange simulation (Lipowsky and Sommerfeld, 2007) with $\Delta t_E > \Delta t_L$. At the end of the Eulerian calculation, the "frozen" flow field (FF) is used for particle tracking, and the particle-phase source terms are coupled back to the fluid flow for the next Eulerian simulation steps (PST).

Lagrangian particles are sources or sinks of momentum for the carrier fluid as follows

$$M = \frac{1}{V_{cell}} \sum_{p \in cell} -F_{p,fluid} \tag{8.119}$$

This corresponds to the momentum contribution of all the particles present in each computational cell. The fluid force acting on each particle, $F_{p,fluid}$, is calculated using a closure model using the flow variables at a *larger scale*, also referred to as reduced unsteady models, such as those introduced by Gatignol (1983) and Maxey and Riley (1983) and presented in detail in the review by Michaelides (2003). The PSIC method applies only for tracking particles that are much smaller than the computational cells, that is, when $d_p \ll \Delta x$, with d_p the particle diameter, and Δx the computational mesh spacing. This limitation is due to the reduced models used for estimating the fluid forces acting on a particle and require knowledge of the *undisturbed fluid velocity* at the location of the particle. This is the conceptual flow velocity as though the particle under consideration had been taken out of the flow. The velocity available on the Eulerian grid, however, corresponds to a *disturbed fluid velocity* due to the disturbance caused by the momentum, which is fed back to the fluid by the particle under consideration. When interpolating the fluid velocity to the locations of the individual particles, an error is made, the magnitude of which depends on the ratio of the particle size over that of the fluid cell. The magnitude of the velocity disturbance induced by a particle, that is, of the difference between the undisturbed and disturbed fluid velocities, can be estimated by identifying the self-induced particle flow disturbance with a singular fundamental flow solution: the *Oseenlet* (Oseen, 1927). Integrating this singular solution over a volume similar to that of a computational cell, one may estimate the self-induced particle velocity disturbance and, hence, the error (Evrard *et al.*, 2021). When drag is the dominant component of the fluid force acting on each particle, the maximum relative error made in estimating the drag force can be shown to be approximately

$$Error \approx \frac{6}{5} \frac{d_p}{\Delta x} \psi_{Oseen}\left(Re_p, \frac{d_p}{\Delta x}\right) \tag{8.120}$$

This approximation is valid for $d_p < 2\Delta x$ and $Re_p < 500$. The function ψ_{Oseen} tends to 1 when Re_p tends to 0 and decreases proportionally to $1/Re_p$ as Re_p increases. A contour map of the maximum relative error made in estimating of the drag force acting on a particle in the PSIC framework, as a function of $d_p/\Delta x$ and Re_p, is given in Figure 8.31. As an example, when a particle is ten times smaller than a computational cell and its Reynolds number is very small, an error of up to 12% can be made when estimating the drag force imposed by the fluid on the particle. Figure 8.32 can be very easily used to make a good estimate of the error made when applying the PSIC method in a point-particle approach.

New approaches have been proposed to mitigate the errors associated with the two-way coupling of Euler-Lagrange simulations, often referred to as *volume-filtered*

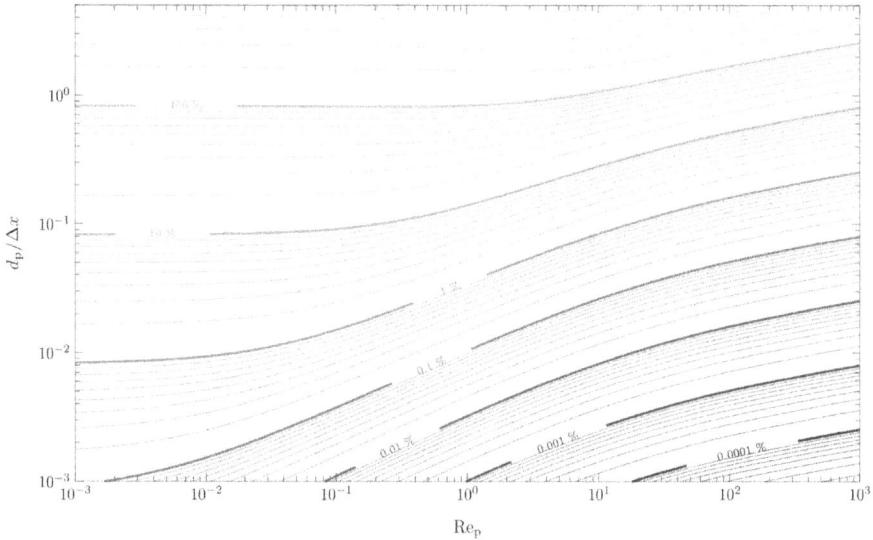

FIGURE 8.32 Contour plot of the errors made in the interpolation of the undisturbed flow velocity in the *PSIC* method as a function of the ratio between the particle diameter and mesh spacing and of the particle Reynolds number.

approaches (Balachandar *et al.*, 2019; Capecelatro and Desjardins, 2013). They rely upon two numerical components, which can be used separately or together:

- At first, the scale of the filter or regularization kernel for transferring momentum from the particles to the fluid is made proportional to the particle size, as shown in Figure 8.34. This is in contrast to being proportional to the mesh spacing, as with the PSIC model of Figure 8.33. The spreading of the momentum contribution over the support of the filter can be achieved by the solution of a diffusion equation (Capecelatro and Desjardins, 2013; Poustis *et al.*, 2019) or using an adaptive numerical quadrature (Evrard *et al.*, 2019). The choice of a regularization length-scale that is proportional to that of the particle means that, as the ratio $d_p/\Delta x$ increases, errors will reach a plateau instead of monotonically growing. The asymptotical value reached by these errors is directly related to the ratio of the kernel length-scale to that of the particle. The larger this ratio, the smaller the error— however, increasing this ratio also means that the momentum contribution associated with a particle is spread over an increasingly large region.
- Secondly, the velocity interpolated from the Eulerian mesh is corrected as to approximate the undisturbed fluid velocity. This can either be done by semi-empirical models. The coefficients of the models are fitted using the results of numerical simulations (Horwitz and Mani, 2018, 2019), by estimating the velocity disturbance induced by a particle with the solutions to the Stokes and Oseen equations (Balachandar *et al.*, 2019; Evrard *et al.*, 2020; Ireland and Desjardins, 2017).

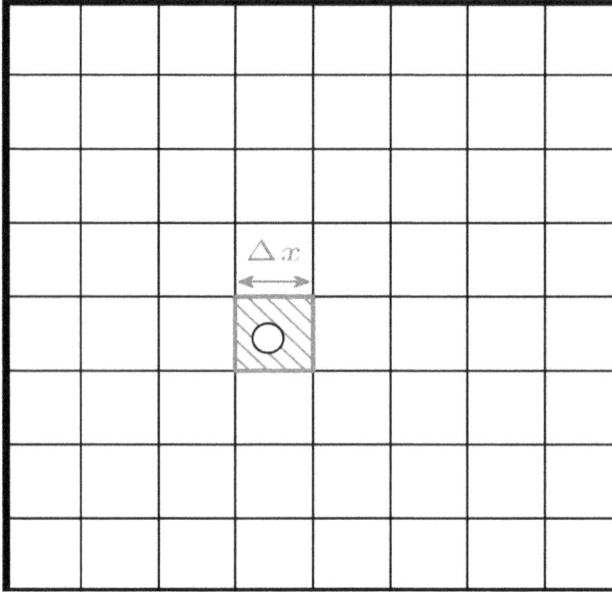

FIGURE 8.33 In the *PSIC* method, the momentum contribution associated with a particle is spread in the computational cell where the particle is located.

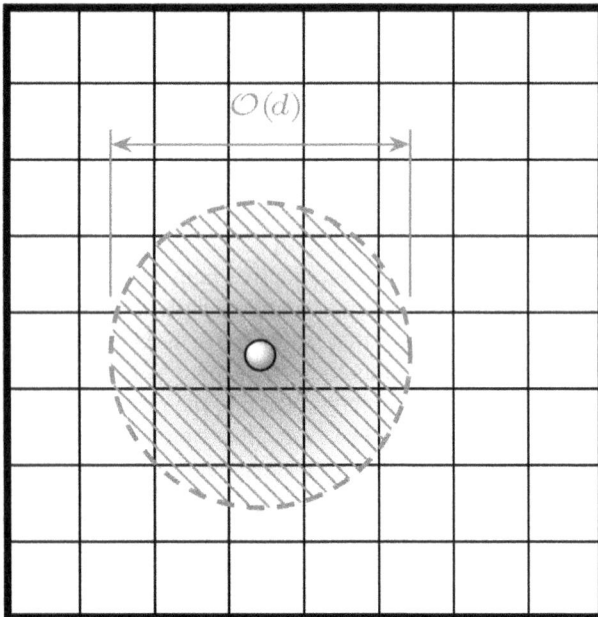

FIGURE 8.34 In the volume-filtered approach, the momentum contribution associated with a particle is spread in a region surrounding the particle, whose length-scale is proportional to that of the particle.

8.7.6 EXAMPLE EULER-LAGRANGE SIMULATIONS: PNEUMATIC CONVEYING

The pneumatic conveying of solid particles in channel or pipe flows is of industrial importance and met in numerous areas of process engineering, such as chemical, pharmaceutical and food industries, power generation (conveying of pulverized coal), and air-conditioning technology. The size of the conveyed particles ranges typically from a few micro-meters to several millimeters. The wide application of pneumatic conveying is connected to numerous advantages for flow-induced powder transport: dust-free transportation, high flexibility, and low cost as well as maintenance. Depending on the application, different conveying regimes are used, such as dense- and dilute-phase conveying (Siegel, 1991). Essential for the design of such systems is the pressure drop as a function of the superficial gas velocity. This pressure drop is composed of the gas phase pressure drop, which is quite well-known already for different pipe elements, and the additional pressure drop due to particle transport (Michaelides, 1987; Michaelides and Roy, 1987). The single-phase pressure loss depends on the wall friction coefficient (i.e. the wall roughness) and on the flow Reynolds number. For the additional pressure loss of the particles, different contributions may be identified, namely, pressure loss caused by particle-wall friction (i.e. resulting from the momentum loss due to particle-wall collisions), pressure loss due to particle lifting in vertical conveying, and pressure loss due to particle acceleration upon injection or after a pipe fitting (Westman *et al.*, 1987). Hence, the particle contribution is strongly governed by the pipe geometry considered. In addition, the particle phase pressure drop remarkably depends on pipe material and diameter, particle size, shape and material, wall roughness (since it is strongly correlated with the particle-wall collision frequency, see Sommerfeld, 2003), as well as particle phase mass loading η, defined as the ratio between the particle mass flow rate and the gas mass flow rate. As a result of this complexity, universal (for all gas-solid systems) and highly accurate correlations for the pressure drop as a function of conveying velocity are not available. Normally, experiments with particular systems and parameters are required to develop such correlations (Siegel, 1991), resulting in the empirical design of pneumatic conveying systems. Quite often, phase diagrams are used to correlate the pressure drop with the superficial gas velocity and the particle mass loading as a parameter. These experimental phase diagrams depend on specific parameters, such as pipe diameter, particle size, and size distribution, implying that conveying another type of particles requires new measurements. Other important design issues for dilute-phase pneumatic conveying systems are the temporal and spatial homogeneous delivery of gas particle mixtures and possible particle erosion in bends, branches, or other fittings and inserts.

An experimental design and layout of pneumatic conveying systems is rather tedious and costly. Therefore, during the past two decades, numerical methods based on CFD were developed for facilitating the understanding and design of such systems. The continuous phase computations are mainly based on Reynolds-averaged conservation equations (RANS) in connection with an appropriate turbulence model (e.g. k-ε turbulence model or a Reynolds-stress model). For describing the dispersed phase in the dilute regime, both the Euler/Euler method and the Euler-Lagrange method are commonly used.

As an application example, pneumatic conveying in a channel and a pipe are compared. The numerical scheme adopted to simulate the dispersed two-phase flow along channel and pipe is the hybrid, fully coupled, stationary, and three-dimensional Euler-Lagrange approach (Laín and Sommerfeld, 2012). The fluid flow was calculated based on the Euler approach by solving the Reynolds-averaged conservation equations in connection with the k-ε turbulence model, which was extended in order to account for modulation by the dispersed phase, that is, two-way coupling (Kohnen and Sommerfeld, 1997). The horizontal channel has a length of 6 m, a height of 35 mm, and a width of 350 mm. For allowing the comparison, the pipe has the same hydraulic diameter yielding a value of 63 mm. A block-structured mesh was used with 240,000 hexahedral computational cells for the channel and 260,000 for the pipe. The coupling between flow and particle phase is done applying the stationary approach (see Section 8.7.4). In this case, 240,000 parcels were tracked during each coupling iteration, considering drag, gravity/buoyancy, and transverse lift due to shear and rotation, as well as the stochastic inter-particle collision model and the wall collision model with wall roughness, both based on the hard-sphere concept.

For illustrating the convergence behavior of a three-dimensional coupled Euler-Lagrange computation, the evolution of the normalized residuals and the profiles of the stream-wise velocity and turbulent kinetic energy of the gas phase are shown in Figure 8.35 for the pipe flow with low roughness (i.e. R0 = 2.2 μm, Δγ = 1.4°) laden with 130 μm diameter glass particles at a loading ratio of 1.0. A source term under-relaxation of 0.2 was applied.

After each introduction of the new particle source terms at the beginning of a Eulerian calculation, all residuals jump up more than an order of magnitude followed by a more or less continuous decrease and then reach again a converged solution for the fluid flow, which is depicted in Figure 8.35a. During the first coupling iterations without inter-particle collisions (i.e. two-way coupling only), the number of inner fluid iterations between successive Eulerian calculations with updated source terms is significantly reduced after a few coupling iterations. After starting to account for inter-particle collisions (i.e. after seven iterations), the residuals jump up again, and a higher number of fluid iterations is needed within a coupling step to reach a converged fluid flow. The coupling iterations required to reach final convergence is much longer with inter-particle collisions than for two-way coupling only. Due to this behavior, one may conclude that for Euler-Lagrange calculations, the residuals are not a very useful criterion to judge the convergence behavior of the coupled system.

Therefore, the evolution of the gas phase properties, such as gas velocity and turbulent kinetic energy, at a reference position is recommended as a convergence criterion (Kohnen et al., 1994). Figures 8.35b and 8.35c demonstrate the deformation process of fluid property profiles as a result of the coupling between the phases. In these figures, it is clearly seen that in the case of low roughness, the profiles for the stream-wise air velocity and the turbulent kinetic energy become strongly skewed when inter-particle collisions are neglected (i.e. two-way coupling for the first seven coupling iterations). This evolving skewness is due to the high particle concentration near the pipe bottom and the associated high particle-wall collision frequency and resulting momentum loss, which slows down the gas phase (Figure 8.36b) and

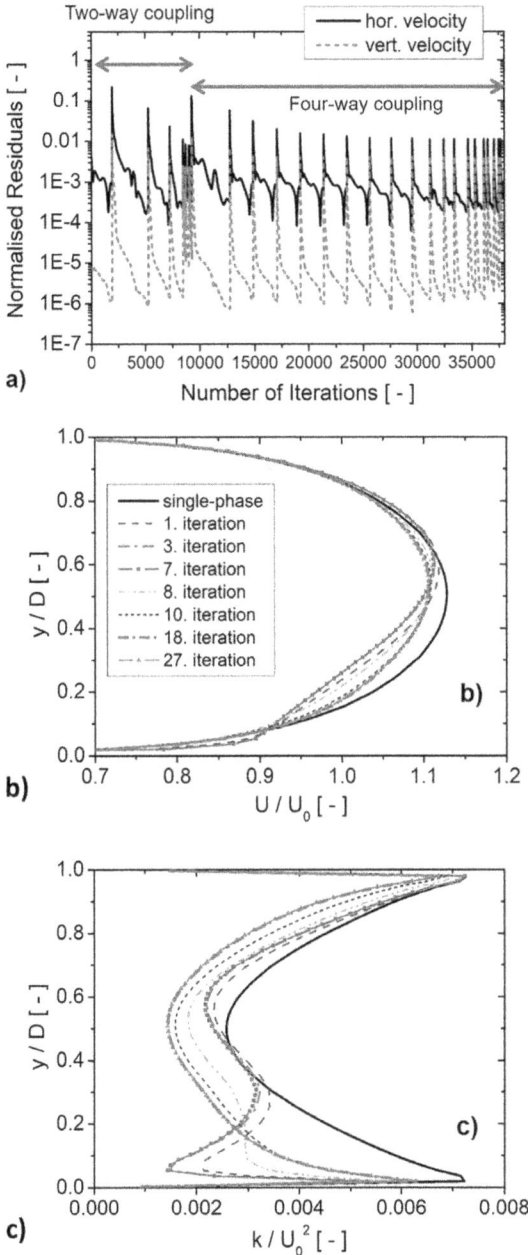

FIGURE 8.35 Convergence behavior for a particle-laden pipe flow with 130 μm diameter glass particles and low wall roughness (R0 = 2.2 μm, Δγ = 1.4°); (a) development of the residuals for the stream-wise and vertical gas velocity fields, (b) development of the stream-wise gas velocity profile at the end of the pipe, (c) development of gas turbulent kinetic energy profile at the end of the pipe (k-ε turbulence model, conveying velocity U_0 =, 20 m/s, 130 μm spherical particles, mass loading 1.0; from Laín and Sommerfeld, 2012).

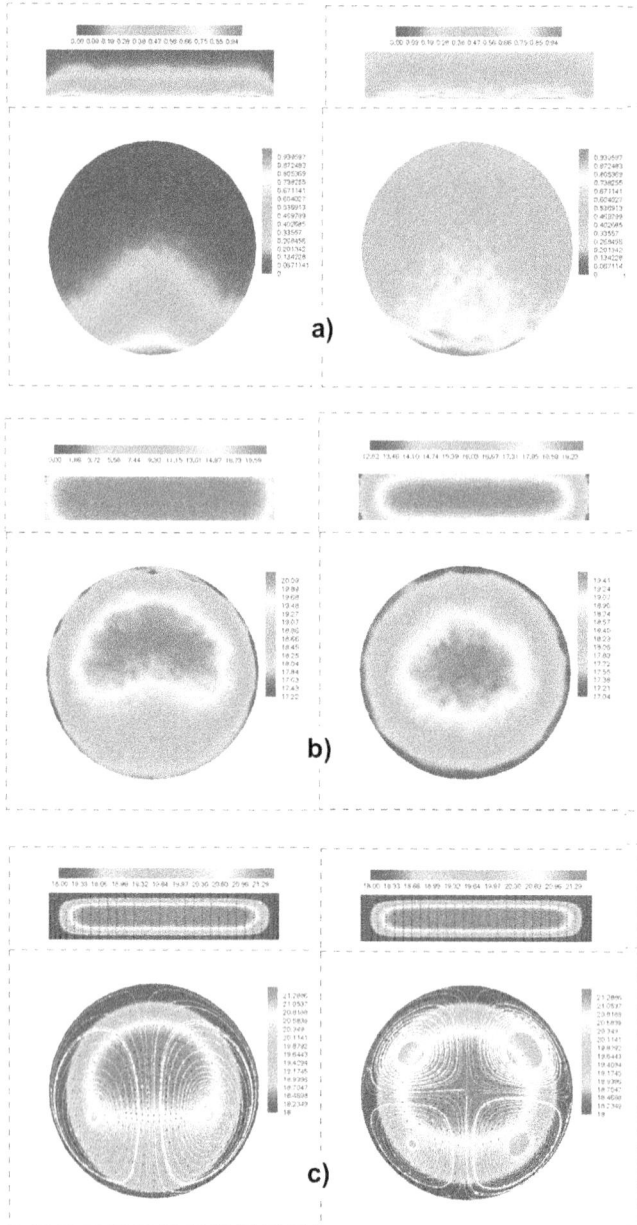

FIGURE 8.36 Comparison of calculated flow structure in a particle-laden developed channel and pipe flow for low roughness, $\Delta\gamma = 0.8°$. Left column is two-way coupling, right column is four-way coupling; (a) particle concentration distribution; (b) distribution of stream-wise particle velocity; (c) distribution of stream-wise gas velocity; and white lines streamlines of gas phase cross-sectional velocity, taken from Laín and Sommerfeld (2012) (k-ε turbulence model, average conveying velocity $U_0 =$, 20 m/s; particle diameter 130 μm; particle mass loading 1.0).

generates strongly asymmetric k-profiles, that is, pronounced turbulence dissipation due to the presence of the particles, for the first seven coupling iterations (Figure 8.36c). When inter-particle collisions are considered, the particles are redistributed within the pipe cross-section (Sommerfeld, 2003), and the profiles of the gas phase variables become more symmetric according to what is observed experimentally, as shown in the curve for 27 iterations in Figures 8.36b and 8.36c.

A comparison of cross-sectional distributions for gas and particle phase properties at the end of the channel and the pipe are shown in Figure 8.36, demonstrating the influence of inter-particle collisions for low wall roughness, with $\Delta\gamma = 0.8°$ and at constant mass loading of $\eta = 1.0$. It is apparent that neglecting inter-particle collisions results in an unrealistic accumulation of the particles near the bottom of pipe or channel due to gravity Figure 8.36a. Collisions between particles yield a somewhat better dispersion of the particles (Section 6.1.4), which are however still concentrated in the lower half of the cross-section.

The stream-wise mean velocity of the particles has a maximum in the core of the pipe, but the region of the highest velocity has a different shape for both calculations (i.e. without and with inter-particle collisions). Due to the strong gravitational settling of the particles, when neglecting inter-particle collisions, the highest particle velocities are found above the particle "rope" where the flow resistance due to particles is lower. Hence, the maximum velocity region has a kind of "kidney shape." When inter-particle collisions are considered, the particle velocity distribution is almost symmetric with respect to the pipe axis, as shown in Figure 8.36b. Similar cross-sectional distributions are found for the stream-wise mean air velocity (Figure 8.36c) since the particle velocity distribution coupled with the air-phase velocity distribution. With respect to the vertical direction, the particle velocity field in the channel flow are symmetric for both modelling assumptions.

The streamlines of the cross-sectional component of the gas velocity are also shown in Figure 8.36c. In both cases, a secondary flow develops in the cross-section of the pipe. Without inter-particle collisions and for the low roughness considered, two circulation cells are visible, whereas the other case yields four circulation cells. This phenomenon is only observed in circular pipes and not in channels. The secondary flow is originating from a so-called *focusing effect* (Sommerfeld and Lain, 2009), which may be explained as follows: In horizontal conveying, inertial particles are essentially moving in the horizontal direction ($u_p \gg v_p$) and settling under the influence of gravity. As a consequence, they will collide with the curved bottom section of the pipe wall and, therefore, rebound back toward the center of the pipe (focusing of trajectories). This is associated with a concentrated momentum transfer from the particle phase to the fluid, if the mass loading is high enough, pushing the fluid flow upward and inducing such a secondary flow. In the case of strong gravitational settling (i.e. without inter-particle collisions), the particles exclusively collide with the bottom wall. Almost no particles are colliding with the top wall in this case. Hence, for the low roughness, the cross-sectional flow is pushed upwards by the particle focusing forming only two circulation cells.

If the vertical dispersion of the particles is enhanced (i.e. due to inter-particle collisions), the collision frequency with the upper pipe wall increases. Consequently, these particles are also rebounding back toward the center of the pipe, and due to the momentum transfer to the fluid, four counter-rotating circulation cells develop in the pipe cross-section, as depicted in Figure 8.36c. It should be noted that the intensity of the secondary flow is very low for both two- and four-way coupling and accounts only about 0.2% of the average conveying velocity.

In the case of the horizontal channel, the particle-laden flow may be regarded as being almost two-dimensional in the vertical middle plane of the channel. With respect to the particle concentration distribution, inter-particle collisions also yield a slightly better vertical dispersion of the particles. The distributions of stream-wise gas and particle velocity fields are almost symmetric with respect to the horizontal center-line plane.

A comparison of the pressure drop along channel and pipe is shown in Figure 8.37 for the pure gas flow and the particle-laden flow with low and high roughness. The single-phase pressure drop is almost identical for channel and pipe since their hydraulic diameter was chosen to be the same. In the particle-laden flow, an additional pressure drop arises, mainly due to particle-wall collisions. For a high roughness case, shown in Figure 8.37b, this additional pressure drop is considerably larger than for the low roughness case, which is shown in Figure 8.37a, and is caused by the enhanced wall collision frequency (Sommerfeld, 2003), for both channel and pipe flow. The pressure drop in the pipe is significantly higher compared to the channel, again the result of the larger wall collision frequency for the particles in a pipe because the geometry of the channel cross-section implies that the probability of particle collisions with the side walls is much lower than with the upper or lower wall. However, in the pipe, the curvature of the walls increases the probability of particle collisions with any part of the wall. In the channel flow, the particle-laden flow is developing faster since the wall clearance is smaller than in the pipe. From the finding described in the previous section, one may conclude that a channel used for pneumatic conveying requires less energy than a pipe, as indicated in Figure 8.37b.

8.8   APPLICATIONS OF NUMERICAL METHODS TO FLUIDIZED BED REACTORS

Fluidized beds are a key element in many industrial processes. Strong interactions between the particulate and gas phase allow for efficient heat transfer and particle mixing and provide a favorable environment for chemical reactions (Kunii and Levenspiel, 1991). One of the most common applications is in fluidized bed reactors, used for large-scale reactions such as the Fluidic Catalytic Cracking (FCC) of fossil fuels, in which reactants are fed through a granular catalyst. Improving the performance of such reactors allows for faster and cheaper processes, and designing the system to optimize the reaction process is an ongoing area of work. This is a difficult task because of the complex nature of particle fluidization. The behavior of a fluidized bed is strongly dependent on the physical properties of both the solid and gas phase. For example, the dynamics are heavily influenced by the size of particles being fluidized. Powders were classified by Geldart (1986) according to their fluidization characteristics. The behavior of fluidized beds has been thoroughly researched in the Eulerian-Eulerian as well as the Eulerian-Lagrangian frameworks.

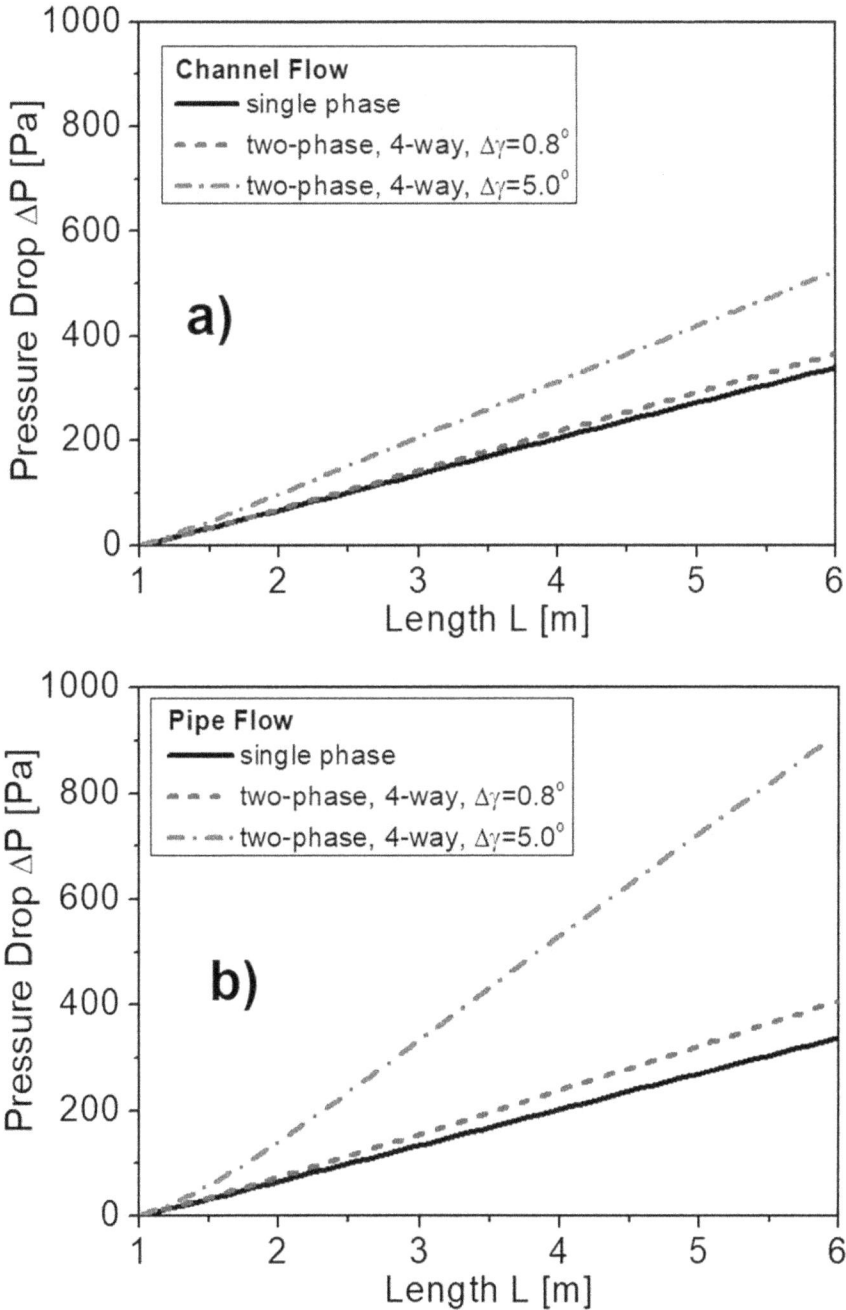

FIGURE 8.37 Comparison of the pressure drop for single phase flow and particle-laden flow with different wall roughness ($\Delta\gamma = 0.8°$ and $5.0°$) in the case of the channel (a) and the pipe (b); (k-ϵ turbulence model, four-way coupling, average conveying velocity $U_0 = 20$ m/s; particle diameter 130 µm; particle mass loading 1.0; from Laín and Sommerfeld (2012)

8.8.1 EULERIAN-EULERIAN PREDICTION OF FLUIDIZED BEDS

The predictions of CFD simulations of bubbling fluidized beds, slugging fluidized beds, and bubble injection into fluidized beds in the Eulerian-Eulerian framework, incorporating various closure models, have been frequently used and have shown a great deal of success. Before the "standard" Eulerian-Eulerian models with the kinetic theory of granular flows can be applied, a number of additional closure models are required, such as the fluid-particle drag and the frictional stress model.

Frictional Stress

At high solid volume fraction, sustained contacts between particles occur, and the stresses predicted by kinetic theory of granular flow (Lun et al., 1984) are insufficient. Hence, the additional frictional stresses must be accounted for in the description of the solid-phase stress. Several approaches have been presented in the literature to model the frictional stress (Johnson and Jackson, 1987; Srivastava and Sundaresan, 2003). Typically, the frictional stress is written in a Newtonian form and has a deviatoric stress-like contribution and a normal stress-like contribution. The frictional stress is added to the stress predicted by kinetic theory for $\alpha_s > \alpha_{s,min}$, where the subscript *min* stands for a threshold value, leading to

$$P_s = P_{kinetic} + P_{frictional} \tag{8.121}$$

$$\mu_s = \mu_{kinetic} + \mu_{frictional} \tag{8.122}$$

There are a number of expressions for the semi-empirical equation for the normal frictional stress, such as (Johnson and Jackson, 1987)

$$P_{frictional} = Fr \frac{(\alpha_s - \alpha_{s,min})^{nn}}{(\alpha_{s,max} - \alpha_s)^{pp}} \tag{8.123}$$

where Fr, nn, and pp are empirical material constants and $\alpha_s > \alpha_{s,min}$, $\alpha_{s,min}$ being the solid volume fraction when frictional stresses become important. The frictional shear viscosity is then defined by the frictional normal stress and a linear relationship (Schaeffer, 1987)

$$\mu_{frictional} = \frac{P_{frictional} \cdot \sin\phi}{\alpha_s \sqrt{\frac{1}{6}\left(\left(\frac{\partial u_s}{\partial x} - \frac{\partial v_s}{\partial y}\right)^2 + \left(\frac{\partial v_s}{\partial y}\right)^2 + \left(\frac{\partial u_s}{\partial x}\right)^2\right) + \frac{1}{4}\left(\frac{\partial u_s}{\partial y} + \frac{\partial v_s}{\partial x}\right)^2}} \tag{8.124}$$

where ϕ is the angle of internal friction of the particulate material. Values of $\alpha_{s,min}$ are typically in the range 0.55–0.6. Values for the empirical parameters are dependent of the material properties, with some examples given in Table 8.4.

TABLE 8.4
Values for the Empirical Parameters in Eq. (8.123), as Suggested by Various Researchers

Fr[N/m^2]	nn	pp	$\alpha_{s,min}$	ϕ	d_s [μm]	ρ_s [kg/m^3]	material	reference
0.05	2	3	0.5	28°	150	2,500	—	Ocone et al., 1993
$3.65 \cdot 10^{-32}$	0	40	—	25.0°	1,800	2,980	glass	Johnson and Jackson, 1987
$4.0 \cdot 10^{-32}$	0	40	—	25.0°	1,000	1,095	polystyrene	Johnson and Jackson, 1987
0.05	2	5	0.5	28.5°	1,000	2,900	glass	Johnson et al., 1990

Solving the Eulerian-Eulerian Equations

There are various numerical frameworks to solve the Eulerian-Eulerian governing equations, such as the modified SIMPLE (Patankar, 1980) and IPSA (Spalding, 1983) algorithms, to fully coupled strategies (van Wachem and Denner, 2014). When discretizing the equations, it is important that second order accurate schemes are used, and for the advection terms, a higher-order total variation diminishing scheme is highly recommended. Compressibility is an important effect in fluidized beds, as the gas density varies with 10%–30% over a typical fluidized bed due to the pressure drop over the fluidized bed. The grid spacing should be determined by refining the grid until average properties change by less than a few percent. Due to the deterministic chaotic nature of the system, the dynamic behavior always changes with the grid, but the statistics should converge.

Boundary Conditions

Simulations can be carried out in two-dimensional and three-dimensional geometries. The walls of the fluidized bed should be treated as no-slip velocity boundary conditions for the fluid phase and slip velocity boundary conditions should be employed for the particle phase. A possible boundary condition for the granular temperature follows from Johnson and Jackson (1987):

$$n \cdot (\kappa \nabla \Theta) = \frac{\pi \rho_s \varepsilon_s \sqrt{3\Theta}}{6\varepsilon_{s,max} \left[1 - \left(\frac{\varepsilon_s}{\varepsilon_{s,max}}\right)^{1/3}\right]} \left[\varphi' | v_{slip} |^2 - \frac{3\Theta}{2}\left(1 - e_w^2\right)\right] \tag{8.125}$$

where the left-hand side represents the conduction of granular energy to the wall, the first term on the right-hand side represents the generation of granular energy, due to particle slip at the wall, and the second term on the right-hand side represents the dissipation of granular energy due to inelastic collisions.

Another possibility for the boundary condition for the granular temperature is proposed by Jenkins (1992)

$$n \cdot (\kappa \nabla \Theta) = -v_{slip} \cdot M - D \tag{8.126}$$

where the exact formulations of M and D depend upon the amount of friction and sliding occurring at the wall region. Simulations that are done with an adiabatic boundary condition for the granular temperature at the wall ($\nabla\Theta = 0$) show very similar results.

The boundary condition at the outlet of the fluidized bed is a so-called pressure boundary. The pressure at this boundary is then fixed to a reference value. Neumann boundary conditions are applied to the gas, assuming a fully developed gas flow. For this, the freeboard of the fluidized bed needs to be of sufficient height; this should be validated through the simulations. In the freeboard, the solid volume fraction is very close to zero and this can lead to unrealistic values for the particle velocity field and poor convergence. For this reason, a solid volume fraction of 10^{-6} is typically set at the top of the freeboard. This way the whole freeboard is seeded with a very small number of particles, which gives more realistic results for the particle phase velocity in the freeboard, but does not influence the behavior of the fluidized bed itself.

The bottom of the fluidized bed should be made impenetrable for the solid phase by setting the solid-phase axial velocity to zero. For the freely bubbling fluidized bed and the slugging fluidized bed, Dirichlet boundary conditions are employed at the bottom with a uniform gas inlet velocity. In the case of the bubble injection, a Dirichlet boundary condition is employed at the bottom of the fluidized bed. The gas inlet velocity is kept at the minimum fluidization velocity, except for a small orifice in the center of the bed, at which a high inlet velocity is specified. Finally, the solids-phase stress, as well as the granular temperature, at the top of the fluidized bed are set to zero.

Initial Conditions

Initially, the bottom part of the fluidized bed should be filled with particles at rest, typically with a uniform solid volume fraction. The gas flow in the bed is set to its minimum fluidization velocity. In the freeboard, a solid volume fraction of 10^{-6} is set, as explained previously. The granular temperature is initially set to a low value, such as $10^{-10} \cdot m^2\, s^{-2}$.

Example Simulations

The test cases described in this section are a freely bubbling fluidized bed, a slugging fluidized bed, and a single bubble injection into a fluidized bed (van Wachem et al., 2001a). Schematic diagrams of the flow configurations for the three cases are given in Figure 8.38.

The particles in a fluidized bed move due to the action of the fluid through the drag force, and bubbles and complex solids mixing patterns result. Typically, the average solid volume fraction in the bed is fairly large, averaging about 40%, whereas in the freeboard of the fluidized bed (the top), there are almost no particles.

With increasing gas velocity above the minimum fluidization velocity, U_{mf}, bubbles are formed because of the inherent instability of the gas-solid system. The behavior of the bubbles significantly affects the flow phenomena in the fluidized bed, for example, solids mixing, entrainment, and heat and mass transfer. The test cases in this comparative study are used to investigate the capabilities of the closure

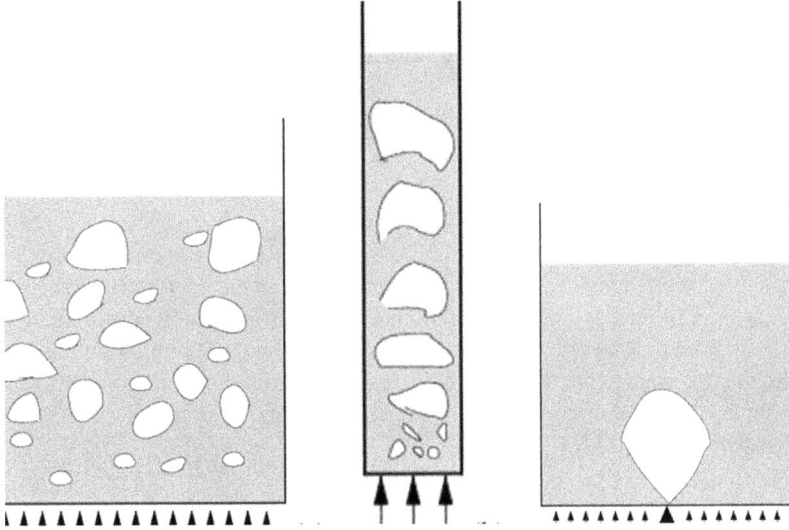

FIGURE 8.38   Setup for (a) bubbling, (b) slugging, and (c) bubble injection fluidized beds.

TABLE 8.5
System Properties and Computational Parameters of the Presented Test Cases

Parameter	Description	Freely Bubbling Fluidized Bed	Slugging Fluidized Bed	Bubble Injection into Fluidized Bed (Kuipers, 1990)
ρ_s [kg/m^3]	solid density	2640	2640	2660
ρ_g [kg/m^3]	gas density	1.28	1.28	1.28
μ_g [$Pa\,s$]	gas viscosity	$1.7 \cdot 10^{-5}$	$1.7 \cdot 10^{-5}$	$1.7 \cdot 10^{-5}$
d_s [μm]	particle diameter	480	480	500
e [-]	coefficient of restitution	0.9	0.9	0.9
ϵ_{max} [-]	max. solid volume fraction	0.65	0.65	0.65
U_{mf} [m/s]	minimum fluidization velocity	0.21	0.21	0.25
D [m]	inner column diameter	0.5	0.1	0.57
H [m]	column height	1.3	1.3	0.75
H_{mf} [m]	height at minimum fluidization	0.97	0.97	0.5
$\epsilon_{s,mf}$ [-]	solids volume fraction at minimum fluidization	0.42	0.42	0.402
Δx [m]	x mesh spacing	$7.14 \cdot 10^{-3}$	$6.67 \cdot 10^{-3}$	$7.50 \cdot 10^{-3}$
Δy [m]	y mesh spacing	$7.56 \cdot 10^{-3}$	$7.43 \cdot 10^{-3}$	$1.25 \cdot 10^{-2}$

models and governing equations to predict fluidization behavior, for example, bubble behavior and bed expansion. Simulation results of each test case are compared to generally accepted experimental data and (semi) empirical models. The system properties and computational parameters for each of the test cases are given in Table 8.5.

Results: Slugging Fluidized Beds

Figure 8.39 shows the predicted maximum bed expansion with increasing gas velocity during the slug flow as well as the slug rise velocity and the correlations of Kehoe and Davidson (1970). The results show that the prediction capability of the Eulerian-Eulerian model is good.

Results: Bubbling Fluidized Beds

Figure 8.40 shows the predicted bubble rise velocity employing different drag models in a freely bubbling fluidized bed compared to the empirical correlation of (Hilligardt and Werther, 1986). All the investigated drag models are in fairly good agreement with the empirical correlation. Moreover, the CFD simulations are able to predict the dynamic behavior in the fluidized bed, leading to a range of bubble sizes and bubble rise velocities, rather than just one statistical average, as is the case with empirical correlations.

Results: Bubble Injection

Figure 8.41 shows the quantitative bubble size prediction for a single jet entering a fluidized bed operating at the minimum fluidization velocity on the drag models of Wen and Yu (1966) and Syamlal *et al.* (1993), which are compared to the experimental data of Kuipers (1990). Frictional stresses can increase the total solid-phase stress by orders of magnitude and is an important contributing force in dense gassolid modelling, although the size of the bubble is not significantly influenced by the frictional stress, as shown in Figure 8.41. The Wen and Yu (1966) drag model yields better agreement with Kuipers' (1990) findings for both the bubble shape and size than the Syamlal *et al.* (1993) drag model. The Syamlal *et al.* (1993) drag model underpredicts the bubble size and produces a bubble that is more circular in shape than in the experiments of Kuipers (1990) and in the simulations with the Wen and Yu (1966) drag model.

The previous test cases show that the intrinsic instabilities and resulting bubbling behavior can be well described using the Eulerian-Eulerian framework. Nowadays, large simulations using this framework are conducted to elucidate the large-scale behavior of industrial fluidized beds, such as Neau *et al.* (2020). However, the success of these models requires applying the correct closures, which typically come with several assumptions.

8.8.2   EULERIAN-LAGRANGIAN PREDICTIONS FOR FLUIDIZED BEDS

The Eulerian-Lagrangian prediction of fluidized beds was proposed in the early 1990s (Tsuji, 1993; Tsuji *et al.*, 1993) and has become an increasingly popular methodology for studying fluidized beds. For the particle-particle collision models, both the hard-sphere model (Hoomans *et al.*, 1996; van Wachem *et al.*, 2001b) and the soft-sphere models (Liu and van Wachem, 2019; Tsuji *et al.*, 1993) are used, although the former is seldom applied because of the required assumptions of binary and instantaneous collisions, which are not applied to fluidized beds. The application of the Eulerian-Lagrangian model in conjunction with the soft-sphere model is also

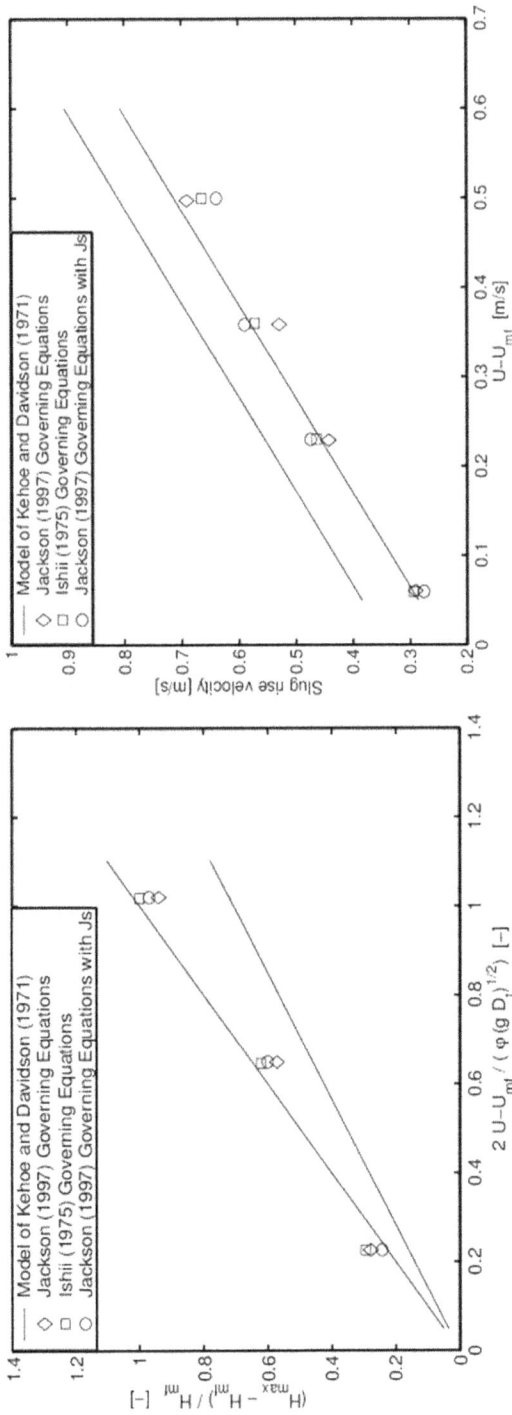

FIGURE 8.39 The predicted maximum bed expansion (left) and the slug rise velocity (right) as a function of the increase in gas velocity, compared to measurements (Kehoe and Davidson, 1970).

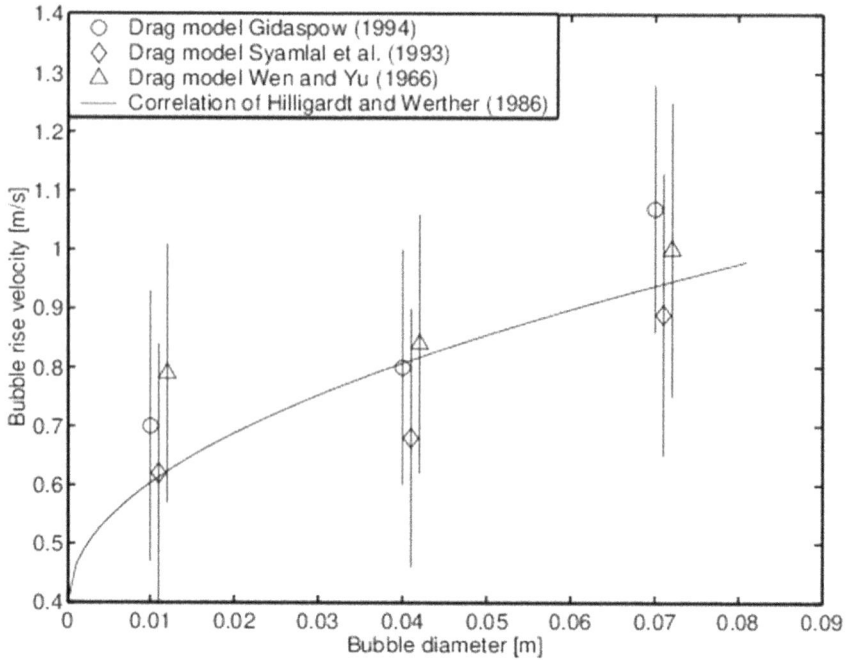

FIGURE 8.40 The bubble rise velocity in a freely bubbling fluidized bed simulation compared with the empirical correlation by Hilligardt and Werther (1986).

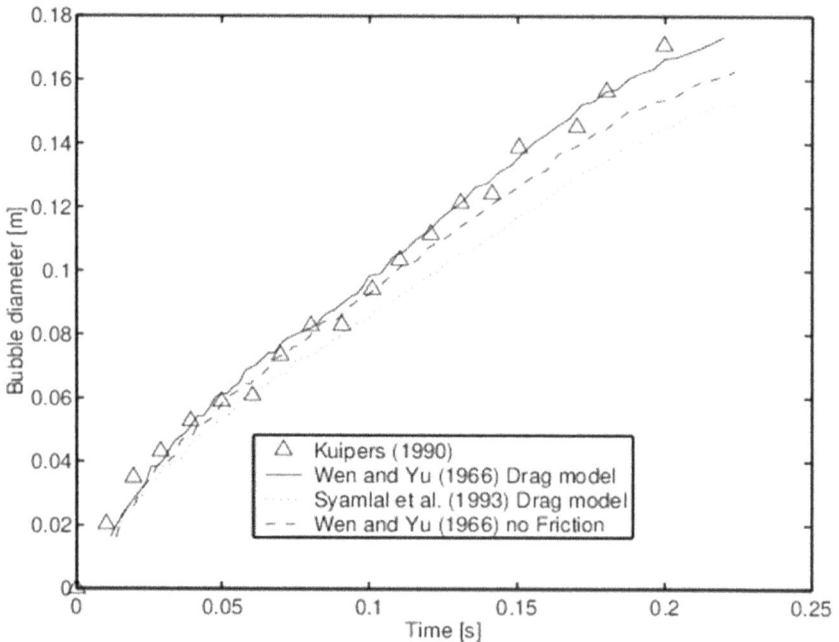

FIGURE 8.41 The bubble diameter as a function of time for the bubble injection case, comparing Eulerian-Eulerian simulations using various drag models with experimental data.

commonly referred to as *CFD-DEM*. The current advance of CFD-DEM methods continues at an accelerating pace. Simulations of particle-fluid flow in industrial systems with complex geometries are reported, and significant efforts are being made on the parallelization or efficient algorithms of CFD-DEM models, making it a promising technique for a large-scale simulation (Amritkar *et al.*, 2014; Dufresne *et al.*, 2019; Kafui *et al.*, 2011). It has become possible to use CFD-DEM models to study some industrial problems (Deen *et al.*, 2007).

Validation is still an important development for the further development of CFD-DEM (Garg *et al.*, 2010; Li *et al.*, 2012). It is important to validate simulations, as issues regarding mesh dependency, material property effects, and the importance of turbulence still exists (Liu and van Wachem, 2019).

CFD-DEM Model

In a CFD-DEM simulation, the modified Navier-Stokes equations are solved to capture the behavior of the fluid phase, which the motion of each individual particle is tracked with Newton's second law. Although it has become possible to perform simulations of fluidized beds with fully resolved methods, such as the immersed boundary method, this approach is still extremely costly and can only be performed for fairly small fluidized beds (Esteghamatian *et al.*, 2017; Ozel *et al.*, 2017). Mostly, a volume averaged representation of the fluid phase is used.

The mass and momentum conservation of the volume averaged fluid phase are given by the continuity equation

$$\frac{\partial \varepsilon_f \rho_f}{\partial t} + \nabla \cdot \left(\varepsilon_f \rho_f \vec{u}_f \right) = 0 \qquad (8.126)$$

and the momentum equation

$$\frac{\partial \varepsilon_f \rho_f \vec{u}_f}{\partial t} + \nabla \cdot \left(\varepsilon_f \rho_f \vec{u}_f \vec{u}_f \right) = \nabla \cdot \varepsilon_f \overline{\overline{\tau}}_f - \varepsilon_a \nabla p + \beta \left(\vec{u}_f - \vec{v}_p \right) \qquad (8.127)$$

where $\vec{u}_f$ is the fluid velocity, ρ_f the fluid density, p is the fluid pressure, β is the interphase momentum transfer coefficient, $\vec{v}_p$ is the particle velocity, and ϵ_f is the local fluid volume fraction.

When discretizing the mass and momentum conservation equations, it is important to consider on what volume the volume fraction is determined and to what volume the interphase momentum transfer is applied. In the original particle-source-in-cell (PSIC) method (Crowe *et al.*, 1977), these are determined in the discretized fluid cells in which the particle is currently located. Although this methodology is still frequently used, its application may lead to significant errors, and mesh converging results may not be possible (Evrard *et al.*, 2021). Instead, a filtering method will give predictions on which scale the volume fraction should be determined and the interphase momentum transfer be applied (Balachandar *et al.*, 2019; Capecelatro and Desjardins, 2013; Evrard *et al.*, 2020).

For each of the particles in the fluidized bed, Newton's second law is solved

$$m_p \frac{D\vec{v_p}}{Dt} = F_{\text{Drag}} + F_{\text{Press}} + F_{\text{Add}} + F_{\text{History}} + F_{\text{Bouyancy}} + F_{\text{Faxen}} + F_{\text{Collisions}}$$

This is essentially the so-called BBO equation of Chapter 4, with an additional term due to the forces arising from the collisions on the right-hand side (for a soft-sphere model only). Rotation can also be accounted for, as previously discussed in Chapters 4 and 6, although rotation is not thought to be of importance for fluidized beds with spherical particles (van Wachem *et al.*, 2001b).

8.8.3 EXAMPLE OF SIMULATIONS

This test case emanates from the experiments of van Wachem *et al.*, 2001b. The fluidized bed is a semi-two-dimensional fluidized bed of 90 mm wide, 500 mm high, and 8 mm deep. The particles are polystyrene spheres with a density of 1,150 kg m^{-3} and diameter of 1.54 mm. The filling of the bed was done with 39 g of particles. The minimum fluidization velocity is approximately 0.72 m/s, using air with a density of r = 1.28 kg m^{-3} and a viscosity of m = 1.7×10^{-5} Pa s.

The simulations in the next section are carried out with a standard CFD-DEM particle-source-in-cell approach (Crowe *et al.*, 1977). The boundary conditions are as follows: For the side, back, and front walls, a no-slip boundary condition was used. For the outlet, a constant gradient for the velocities is assumed, and the pressure is set to the ambient pressure. For the inlet, the superficial gas velocity is specified, and a constant gradient for pressure is assumed. Initially, the fluidized bed is at rest and is filled with 39 g of particles. A sequence of snapshots for the particle positions is shown in Figure 8.42.

With the standard CFD-DEM methods, a reasonable prediction of the fluidized bed behavior is obtained: Results for bubble sizes, bubble velocities, bed expansion, and pressure drop are all in fairly good agreement. However, to obtain fully mesh-independent results, a filtering method for the coupling between the particles and the fluid (Balachandar *et al.*, 2019; Capecelatro and Desjardins, 2013; Evrard *et al.*, 2020) needs to be employed. It is also straightforward to add physical phenomena

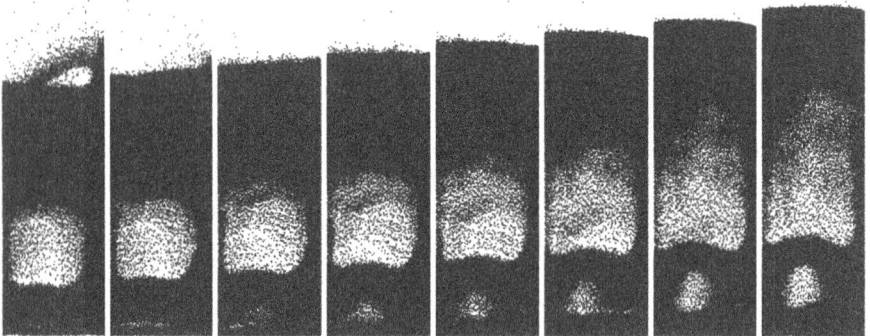

FIGURE 8.42 Sequential snapshots of particle locations in the fluidized bed with a velocity of U = 1.4 U$_{mf}$. The time spacing between the snapshots is Dt = 2.0×10^{-2} s.

pertinent to particle-particle and fluid-particle processes, such as combustion, heat transfer, thermophoresis, comminution, and agglomeration.

REFERENCES

Abramzon, B. and Sirignano, W., 1989, Droplet vaporisation model for spray combustion calculations, *Int. J. of Heat Mass Transfer*, **32**, 1605–1618.

Aidun, C.K. and Clausen, R., 2010, Lattice-Boltzmann method for complex flows, *Ann. Rev. Fluid Mech.*, **42**, 439–472.

Amritkar, A., Deb, S. and Tafti, D., 2014, Efficient parallel CFD-DEM simulations using Open MP, *J. of Computational Physics*, **256**, 501–519.

Amsden, A.A., O'Rourke, P.J. and Butler, T.D., 1989, KIVA-II: A computer program for chemically reactive flows with sprays, Los Alamos Scientific Laboratory Report, LA-11560-MS.

Anderson, T.B. and Jackson, R., 1967, A fluid mechanical description of fluidized beds, *I and EC Fundamentals*, **6**, 524–539.

Bakker, A., 2002, Applied computational fluid dynamics: Lecture 10: Turbulence models. www.bakker.org.

Balachandar, S., Liu, K. and Lakhote, M., 2019, Self-induced velocity correction for improved drag estimation in Euler—Lagrange point-particle simulations, *J. Comput. Phys.*, **376**, 160–185.

Bauman, S.D., 2001, *A Spray Model for an Adaptive Mesh Refinement Code*. Ph.D. Thesis, University of Wisconsin, Madison.

Bec, J., Celani, A., Cencini, M. and Musacchio, S., 2005, Clustering and collisions of heavy particle in random smooth flows, *Phys. Fluids*, **17**, 073301.

Benavides, A. and van Wachem, B.G.M., 2009, Eulerian-Eulerian prediction of dilute turbulent gas-particle flow in a backward-facing step, *Int. J. of Heat and Fluid Flow*, **30**, 452–461.

Benzi, R., Succi, S. and Vergassola, M., 1992, The lattice Boltzmann equation: Theory and applications. *Physics Reports*, **222**, 145–197.

Berlemont, A., Desjonqueres, P. and Gouesbet, G., 1990, Particle Lagrangian simulation in turbulent flows, *Int. J. Multiphase Flow*, **16**, 19–34.

Bhatnagar, P.L., Gross, E.P. and Krook, M., 1954, A model for collision processes in gases. I. Small amplitude processes in charged and neutral one-component systems, *Physical Review*, **94**, 511–525.

Bini, M. and Jones, W.P., 2008, Large-eddy simulation of particle-laden turbulent flows, *J. Fluid Mechanics*, **614**, 207.

Bouzidi, M., Firdaouss, M. and Lallemand, P., 2001, Momentum transfer of a Boltzmann-lattice fluid with boundaries, *Phys Fluids*, **13**, 3452–3459.

Brändle de Motta, J.C., Costa, P., Derksen, J.J., Peng, C., Wang, L.-P., Breugem, W.-P., Estivalezes, J.L., Vincent, S., Climent, E., Fede, P., Barbaresco, P. and Renon, N., 2019, Assessment of numerical methods for fully resolved simulations of particle-laden turbulent flows, *Computers & Fluids*, **179**, 1–14.

Burton, T.M. and Eaton, J.K., 2005, Fully resolved simulations of particle-turbulence interaction. *J. Fluid Mech.*, **545**, 67–111.

Caiazzo, A., 2008, Analysis of lattice Boltzmann nodes initialisation in moving boundary problems, *Progr. Comput. Fluid Dynam. Int. J.*, **8**, 3–10.

Calabrese, R.V. and Middleman, S., 1979, The dispersion of discrete particles in a turbulent fluid field, *AIChE J.*, **25**, 1025–1035.

Capecelatro, J. and Desjardins, O., 2013, An Euler—Lagrange strategy for simulating particle-laden flows, *J. Comput. Phys.*, **238**, 1–31.

Chapman, S. and Cowling, T.G., 1990, *The Mathematical Theory of Non-Uniform Gases: An Account of the Kinetic Theory of Viscosity, ThermalC, and Diffusion in Gases*, 3rd edition, Cambridge mathematical library, Cambridge University Press, Cambridge; New York.

Chen, S. and Doolen, G.D., 1998, Lattice Boltzmann method for fluid flows, *Ann. Rev. Fluid Mech.*, **30**, 329–364.

Chouippe, A. and Uhlmann, M., 2015, Forcing homogeneous turbulence in direct numerical simulation of particulate flow with interface resolution and gravity, *Physics of Fluids*, **27**, 123301.

Climent, E. and Maxey, M.R., 2009, The force coupling method: A flexible approach for the simulation of particulate flows. *Theoretical Methods for Micro Scale Viscous Flows*, 173–193.

Costa, P., Picano, F., Brandt, L. and Breugem, W.-P., 2016, Universal scaling laws for dense particle suspensions in turbulent wall-bounded flows, *Physical Review Letters*, **117**.

Crouse, B., 2003, *Lattice-Boltzmann Strömungssimulationen auf Baumdatenstrukturen*. Dissertation, Technische Universität München.

Crowe, C.T., Sharma, M.P. and Stock, D.E., 1977, The particle-source-in-cell (PSI-cell) model for gas-droplet flows, *ASME J. of Fluids Eng.*, **99**, 325–332.

Cui, Y., Schmalfuß, S., Zellnitz, S., Sommerfeld, M. and Urbanetz, N., 2014, Towards the optimisation and adaptation of dry powder inhalers, *Int. J. of Pharmaceutics*, **470**, 120–132.

Cui, Y. and Sommerfeld, M., 2015, Forces on micron-sized particles randomly distributed on the surface of larger particles and possibility of detachment, *Int. J. Multiphase Flow*, **72**, 39–52.

Cundall, P.A. and Strack, O.D.L., 1979, A discrete numerical model for granular assemblies. *Geotechnique*, **29**, 47–65.

Davis, A., Michaelides, E.E. and Feng, Z-G., 2012, Particle velocity near vertical boundaries—a source of uncertainty in two-fluid models, *Powder Technology*, **220**, 15–23.

Deen, N.G., van Sint Annaland, M., van der Hoef, M.A. and Kuipers, J.A.M., 2007, Review of discrete particle modeling of fluidized beds. *Chemical Engineering Science*, **62**, 28–44.

Derksen, J.J., 2012, Direct numerical simulations of aggregation of monosized spherical particles in homogeneous isotropic turbulence, *AIChE J.*, 58, 2589–2600.

Dhotre, M.T., Deen, N.G., Niceno, B., Khan, Z. and Joshi, J.B., 2013, Large Eddy Simulation for dispersed bubbly flows: A review, *Int. J. of Chemical Engineering*, **2013**, Article ID 343276.

Dietzel, M., Ernst, M. and Sommerfeld, M., 2016, Application of the Lattice-Boltzmann method for particle-laden flows: Point-particles and fully resolved particles. *Flow, Turbulence and Combustion*, **97**, 539–570.

Dietzel, M. and Sommerfeld, M., 2013, Numerical calculation of flow resistance for agglomerates with different morphology by the Lattice-Boltzmann method. *Powder Technology*, **250**, 122–137.

Dufresne, Y., Moureau, V., Lartigue, G. and Simonin, O., 2019, A massively parallel CFD/DEM approach for reactive gas-solid flows in complex geometries using unstructured meshes, *Computers & Fluids*, 104402.

Dukowicz, J.K., 1980, A particle-fluid numerical model for liquid sprays, *J. Computational Physics*, **35**, 229–253.

Eaton, J.K. and Fessler, J.R., 1994, Preferential concentration of particles by turbulence, *Int. J. Multiphase Flow*, **20**, 169–209.

Elghobashi, S., 1994, On predicting particle-laden turbulent flows, *Applied Scientific Research*, **52**, 309–329.

Ergun, S., 1952, Fluid flow through packed columns, *Chem. Eng. Prog.*, **48**, 89–94.

Ernst, M., 2016, *Analyse des Clustering-, Kollisions- und Agglomerationsverhaltens von Partikeln in laminaren und turbulenten Strömungen.* Dissertation Zentrum für Ingenieurwissenschaften, Martin-Luther Universität Halle-Wittenberg.

Ernst, M., Dietzel, M. and Sommerfeld, M., 2013, A lattice Boltzmann method for simulating transport and agglomeration of resolved particles, *Acta Mechanica*, **224**, 2425–2449.

Ernst, M. and Sommerfeld, M., 2012, On the volume fraction effects on inertial colliding particles in homogeneous isotropic turbulence, *J. of Fluids Engineering*, **134** (031302).

Ernst, M. and Sommerfeld, M., 2015, Resolved numerical simulation of particle agglomeration, in *Colloid Process Engineering* (Eds. M. Kind, W. Peukert, H. Rehage and H.P. Schuchmann), pp. 45–71, Springer Int. Publishing, Switzerland.

Ernst, M., Sommerfeld, M. and Laín, S., 2019, Quantification of preferential concentration of colliding particles in a homogeneous isotropic turbulent flow. *Int. J. Multiphase Flow*, **117**, 163–181.

Esteghamatian, A., Hammouti, A., Lance, M. and Wachs, A., 2017, Particle resolved simulations of liquid/solid and gas/solid fluidized beds, *Physics of Fluids*, **29**, 033302.

Eswaran, V. and Pope, S.B., 1988, An examination of forcing in direct numerical simulations of turbulence, *Computers & Fluids*, **16**, 257–278.

Evrard, F., Denner, F. and van Wachem, B., 2019, A multi-scale approach to simulate atomisation processes, *Int. J. Multiph. Flow*, **119**, 194–216.

Evrard, F., Denner, F. and van Wachem, B., 2020, Euler-Lagrange modelling of dilute particle-laden flows with arbitrary particle-size to mesh-spacing ratio, *J. Comput. Phys.*, **X 8** (100078).

Evrard, F., Denner, F. and van Wachem, B., 2021, Quantifying the errors of the particle-source-in-cell Euler-Lagrange method, *Int. J. Multiph. Flow*, **135** (103535).

Falkovich, G., Fouxon, A. and Stepanov, M.G., 2002, Acceleration of rain initiation by cloud turbulence, *Nature*, **419**, 151–154.

Fede, P. and Simonin, O., 2006, Numerical study of the subgrid fluid turbulence effects on the statistics of heavy colliding particles, *Physics of Fluids*, **18** (045103).

Fede, P. and Simonin, O., 2010, Effect of particle-particle collisions on the spatial distribution of inertial particles suspended in homogeneous isotropic turbulent flows, in Turbulence and Interactions. *Notes on Numerical Fluid Mechanics and Multidisciplinary Design*, **110** (Springer), 119–125.

Feng, Z.G. and Michaelides, E.E., 2002, Inter-particle forces and lift on a particle attached to a solid boundary in suspension flow, *Physics of Fluids*, **14**, 49–60.

Feng, Z.G. and Michaelides, E.E., 2003, Equilibrium position for a particle in a horizontal shear flow, *Int. J. Multiphase Flow*, **29**, 943–957.

Feng, Z.G. and Michaelides, E.E., 2004, The immersed boundary—Lattice Boltzmann method for solving fluid-particles interaction problems, *J. Computational Physics*, **195**, 602–628.

Feng, Z.G. and Michaelides, E.E., 2005, Proteus: A direct forcing method in the simulations of particulate flows, *J. Comput. Phys.*, **202**, 20–51.

Feng, Z.G. and Michaelides, E.E., 2008a, Inclusion of heat transfer computations for particle laden flows, *Phys. Fluids*, **20** (040604).

Feng, Z.G. and Michaelides, E.E., 2008b, Heat transfer in particulate flows with Direct Numerical Simulation (DNS), *Int. J. Heat Mass Transf.*, **52**, 777–786.

Feng, Z.G. and Michaelides, E.E., 2009, Robust treatment of no-slip boundary condition and velocity updating for the Lattice-Boltzmann simulation of particulate flows, *Computers in Fluids*, **38**, 370–381.

Ferziger, J.H. and Peric, M., 2002, *Computational Methods for Fluid Dynamics*, 3rd edition, Springer-Verlag, Berlin.

Filippova, O. and Hänel, D., 1998, Grid refinement for lattice-BGK models, *J. Comput. Phys.*, **147**, 219–228.

Fornari, W., Formenti, A., Picano, F. and Brandt, L., 2016, The effect of particle density in turbulent channel flow laden with finite size particles in semi-dilute conditions, *Physics of Fluids*, **28**, 033301.

Fröhlich, J., 2006, *Large Eddy Simulation turbulenter Strömungen*. Teuber Verlag, Wiesbaden.

Gan, H., Chang, J.Z. and Howard, H.H., 2003, Direct numerical simulation of the sedimentation of solid particles with thermal convection, *J. Fluid Mech.*, **481**, 385–411.

Gao, H., Li, H. and Wang, L.-P., 2013, Lattice Boltzmann simulation of turbulent flow laden with finite-size particles, *Computers and Mathematics with Applications*, **65**, 194–210.

Garg, R., Galvin, J., Li, T. and Pannala, S., 2010, Open-source MFIX-DEM software for gas—solids flows: Part I—Verification studies. *Powder Technology*, Selected Papers from the, 2010 NETL Multiphase Flow Workshop 220, 122–137.

Gatignol, R., 1983, The Faxén formulas for a rigid particle in an unsteady non-uniform Stokes-flow, *J. Mécanique Théorique Appliquée*, **2**, 143–160.

Geldart, D., 1986, *Gas Fluidization Technology*, John Wiley & Sons, New York, NY.

Gidaspow, D., 1986, Hydrodynamics of fluidization and heat transfer: Supercomputer modeling, *Applied Mechanics Reviews*, **39**, 1.

Glowinski, R., Pan, T.W., Hesla, T.I., Joseph, D.D. and Périaux, J., 2001, A fictitious domain approach to the direct numerical simulation of incompressible viscous flow past moving rigid bodies: Application to particulate flow, *J. of Computational Physics*, **169**, 363–426.

Gosman, A.D. and Ioannides, I.E., 1981, Aspects of computer simulation of liquid-fuelled combustors. Aerospace Science Meeting, Paper No. AIAA-81-0323.

Gouesbet, G. and Berlemont, A., 1999, Eulerian and Lagrangian approaches for predicting the behaviour of discrete particles in turbulent flows, *Progress in Energy and Combustion Science*, **25**, 133–159.

Guo, Z., Zheng, C. and Shi, B., 2002, An extrapolation method for boundary conditions in lattice Boltzmann method, *Phys. Fluids*, **14**, 2007–2010.

Hardy, B., Simonin, O., De Wilde, J. and Winckelmans, G., 2022, Simulation of the flow past random arrays of spherical particles: Microstructure-based tensor quantities as a tool to predict fluid-particle forces, *Int. J. of Multiphase Flow*, (103970).

He, J. and Simonin, O., 1993, Non-equilibrium prediction of the particle-phase stress tensor in vertical pneumatic conveying, *ASME*, **166**, 253–253.

He, X. and Luo, L.-S., 1997, Theory of the lattice Boltzmann method: From the Boltzmann equation to the lattice Boltzmann equation, *Phys. Rev. E*, **56**, 6811–6817.

Higuera, F.J., Succi, S. and Benzi, R., 1989, Lattice gas dynamics with enhanced collisions. *Europhysics Letters*, **9**, 345–349.

Hilligardt, K. and Werther, J., 1986, Local bubble gas hold-up and expansion of gas/solid fluidized beds, *Ger. Chem. Eng*, **9**, 215.

Hölzer, A. and Sommerfeld, M., 2009, Lattice Boltzmann simulations to determine drag, lift and torque acting on non-spherical particles, *Computers and Fluids*, **38**, 572–589.

Hoomans, B., Kuipers, J., Briels, W. and Van Swaaij, W., 1996, Discrete particle simulation of bubble and slug formation in a two-dimensional gas-fluidised bed: A hard-sphere approach, *Chem. Eng. Sci.*, **51**, 99–118.

Horwitz, J.A.K. and Mani, A., 2018, Correction scheme for point-particle models applied to a nonlinear drag law in simulations of particle-fluid interaction, *Int. J. Multiph. Flow*, **101**, 74–84.

Horwitz, J.A.K. and Mani, A., 2019, Accurate calculation of Stokes drag for point—particle tracking in two-way coupled flows, *J. Comput. Phys.*, **318**, 85–109.

Hu, H.H., Patankar, N.A. and Zhu, M.Y., 2001, Direct numerical simulations of fluid-solid systems using the arbitrary Lagrangian-Eulerian technique, *J. Comput. Phys.*, **169**, 427–462.

Huilier, D.G.F., 2021, An overview of the lagrangian dispersion modeling of heavy particles in homogeneous isotropic turbulence and considerations on related LES simulations, *Fluids*, **6**, 145.

Ibrahim, A.H., Dunn, P.F. and Brach, R.M., 2003, Microparticle detachment from surfaces exposed to turbulent air flow: Controlled experiments and modeling, *J. of Aerosol Science*, **34**, 765–782.

Igci, Y., Andrews IV, A.T., Sundaresan, S., Pannala, S. and O'Brian, T., 2008, Filtered two-fluid models for fluidized gas-particle suspensions, *AIChE J.*, **54**, 1431–1448.

Ireland, P.J. and Desjardins, O., 2017, Improving particle drag predictions in Euler—Lagrange simulations with two-way coupling, *J. Comput. Phys.*, **338**, 405–430.

Ishii, M., 1975, *Thermo-Fluid Dynamic Theory of Two-Phase Flow*, Eyrolles, Paris.

Ishii, M. and Hibiki, T., 2006, *Thermo-Fluid Dynamics of Two-Phase Flow*. Springer, New York.

Ishii, M. and Mishima, K., 1984, Two-fluid model and hydrodynamic constitutive relations. *Nuclear Engineering and Design*, **82**, 107–126.

Issa, R., 2009, Simulation of intermittent flow in multiphase oil and gas pipelines, *Seventh Int. Conference on CFD in the Minerals and Process Industries*. CSIRO, Melbourne, Australia, 9–11 December.

Jackson, R., 1997, Locally averaged equations of motion for a mixture of identical spherical particles and a Newtonian fluid, *Chemical Engineering Science*, **52**, 2457–2469.

Jenkins, J., 1992, Boundary conditions for rapid granular flow: Flat, frictional walls, *J. of Applied Mechanics*, **59**, 120.

Johnson, P.C. and Jackson, R., 1987, Frictional—collisional constitutive relations for granular materials, with application to plane shearing, *J. of Fluid Mechanics*, **176**, 67–93.

Johnson, P.C., Nott, P., Jackson, R., 1990, Frictional-collisional equations of motion for particulate flows and their application to chutes, *J. of Fluid Mechanics*, **210**, 501–535.

Kafui, D.K., Johnson, S., Thornton, C. and Seville, J.P.K., 2011, Parallelization of a Lagrangian—Eulerian DEM/CFD code for application to fluidized beds. *Powder Technology*, **207**, 270–278.

Kehoe, P. and Davidson, J., 1970, Continuously slugging fluidized beds, in: Chemeca. pp. 97–116.

Kim, J., Moin, P. and Moser, R., 1987, Turbulence statistics in fully developed channel flow at low Reynolds number, *J. of Fluid Mechanics*, **177**, 133–166.

Kjeldby, T.K., Henkes, R. and Nydal, O.J., 2011, Slug tracking simulation of severe slugging experiments, *Int. J. of Mechanical, Aerospace, Industrial, Mechatronic and Manufacturing Engineering*, **5**, 1156–1161.

Kohnen, G., Rüger, M. and Sommerfeld, M., 1994, Convergence behaviour for numerical calculations by the Euler-Lagrange method for strongly coupled phases, in *Numerical Methods in Multiphase Flows 1994* (Eds. C.T., Crowe et al.), pp. 191–202, ASME, New York, *FED-Vol. 185*.

Kohnen, G. and Sommerfeld, M., 1997, The effect of turbulence modeling on turbulence modification in two-phase flows using the Euler-Lagrange approach. *Proceedings of the Eleventh Symposium on Turbulent Shear Flows*, Grenoble, France, Sept., **3** (P3), 23–28.

Kuerten, J.G.M., 2016, Point-particle DNS and LES of particle-laden turbulent flow—a state-of-the-art review. *Flow Turbulence Combust*, **97**, 689–713.

Kuipers, J.A.M., 1990, *A Two-Fluid Micro Balance Model of Fluidized Beds*. Ph.D. Thesis, University of Twente, The Netherlands.

Kuipers, J.B., 2002, *Quaternions and Rotation Sequences*, Princeton University Press, Princeton.

Kunii, D. and Levenspiel, O., 1991, *Fluidization Engineering*, Butterworth-Heinemann, Boston.

Kussin, J. and Sommerfeld, M., 2002, Experimental studies on particle behaviour and turbulence modification in horizontal channel flow with different wall roughness, *Experiments in Fluids*, **33**, 143–159.

Ladd, A.J.C., 1994, Numerical simulations of particulate suspensions via a discretized Boltzmann equation part I. theoretical foundation, *J. Fluid Mech.*, **271**, 285–310.

Ladd, A.J.C., 1997, Sedimentation of homogeneous suspensions of non-Brownian spheres, *Phys. Fluids*, **9**, 491–502.

Ladd, A.J.C. and Frenkel, D., 1990, Dissipative hydrodynamic interactions via lattice-gas cellular automata, *Phys. Fluids*, **A2**, 1921.

Laín, S. and Sommerfeld, M., 2012, Numerical calculation of pneumatic conveying in horizontal channels and pipes: Detailed analysis of conveying behaviour. i, **39**, 105–120.

Lallemand, P. and Luo, L.S., 2003, Lattice Boltzmann method for moving boundaries, *J. Comput. Phys.*, **184**, 406–421.

Launder, B.E. and Spalding. D.B., 1974, The numerical computation of turbulent flows, *Comp. Meth. Appl. Mech. and Eng.*, **3**, 269–289.

Lee, J., Kim, J., Choi, H. and Yang, K.-S., 2011, Sources of spurious force oscillations from an immersed boundary method for moving-body problems, *J. of Computational Physics*, **230**, 2677–2695.

Lesieur, M., Metais, O. and Comte, P., 2005, *Large Eddy Simulations of Turbulence*, Cambridge University Press, Cambridge.

Li, J. and Kuipers, J.A.M., 2003, Gas-particle interactions in dense gas-fluidized beds, *Chemical Engineering Science*, **58**, 711–718.

Li, J. and Kuipers, J.A.M., 2007, Effect of competition between particle-particle and gas-particle interactions on flow patterns in dense gas-fluidized bed, *Chemical Engineering Science*, **62**, 3429–3442.

Li, T., Garg, R., Galvin, J. and Pannala, S., 2012, Open-source MFIX-DEM software for gas-solids flows: Part II—Validation studies. Powder Technology, Selected Papers from the, 2010 NETL Multiphase Flow Workshop 220, 138–150.

Lilly, D.K., 1967, The representation of small-scale turbulence in numerical simulation experiments. *Proc. IBM Sci., Comput. Symp. Environ. Sci.*, Chapter 14, IBM, New York, NY.

Lipowsky, J. and Sommerfeld, M., 2005, Time dependent simulation of a swirling two-phase flow using an anisotropic turbulent dispersion model, *Proceedings of the ASME Fluids Engineering Summer Conference*. Houston, Texas, Paper No. FEDSM2005–77210.

Lipowsky, J. and Sommerfeld, M., 2007, LES-simulation of the formation of particle strands in swirling flows using an unsteady Euler-Lagrange approach, *Proceedings of the 6th Int. Conference on Multiphase Flow, ICMF2007*, Leipzig Germany, Paper No. S3_Thu_C_54.

Liu, D. and van Wachem, B., 2019, Comprehensive assessment of the accuracy of CFD-DEM simulations of bubbling fluidized beds, *Powder Technology*, **343**, 145–158.

Loha, C., Chattopadhyay, H. and Chatterjee, P.K., 2012, Assessment of drag models in simulating bubbling fluidized bed hydrodynamics, *Chemical Engineering Science*, **75**, 400–407.

Lomholt, S. and Maxey, M.R., 2003, Force-coupling method for particulate two-phase flow: Stokes flow, *J. of Computational Physics*, **184**, 381–405.

Lun, C.K.K., Savage, S., Jeffery, D.J. and Chepurniy, N., 1984, Kinetic theories for granular flow: Inelastic particles in Couette flow and slightly inelastic particles in a general flow field, *J. Fluid Mech*, **140**, 223–256.

Luo, K., Tan, J., Wang, Z. and Fan, J., 2016, Particle-resolved direct numerical simulation of gas-solid dynamics in experimental fluidized beds, *AIChE J.*, **62**, 1917–1932.

Lurie, M.V. and Sinaiski, E., 2008, *Modeling of Oil Product and Gas Pipeline Transportation*, WILEY-VCH Verlag GmbH & Co. KGaA, Weinheim.

Ma, T., Ziegenhein, T., Lucas, D., Krepper, E. and Fröhlich, J., 2015, Euler—Euler large eddy simulations for dispersed turbulent bubbly flows, *Int. J. of Heat and Fluid Flow*, **C**, 51–59.

Ma, P.C., Esclape, L., Carbajal, S., Ihme, M., Buschhagen, T., Naik, S.V., Gore, J.P. and Lucht, R.P., 2016, High-fidelity simulations of fuel injection and atomization of a hybrid air-blast atomizer, in: *54th AIAA Aerospace Sciences Meeting, AIAA SciTech Forum. American Institute of Aeronautics and Astronautics.*

MacInnes, J.M. and Braco, F.V., 1992, Stochastic particle dispersion and the tracer particle limit, *Phys. Fluids A*, **4**, 2809–2824.

Majumdar, S., Iaccarino, G. and Durbin, P., 2001, RANS solvers with adaptive structured boundary non-conforming grids. *Center for Turbulence Research Annual Research Briefs*, 353–366.

Mallouppas, G. and van Wachem, B.G.M., 2013, Large Eddy simulations of turbulent particle-laden channel flow, *Int. J. of Multiphase Flow*, **54**, 65–75.

Marchioli, C., Soldati, A., Kuerten, J.G.M., Arcen, B., Taniere, A., Goldensoph, G., Squires, K.D., Cargnelutti, M.F. and Portela, L.M., 2008, Statistics of particle dispersion in direct numerical simulations of wall-bounded turbulence: Results of an Int. collaborative benchmark test. *Int. J. of Multiphase Flow*, **34**, 879–893.

Mark, A. and van Wachem, B.G.M., 2008, Derivation and validation of a novel implicit second-order accurate immersed boundary method, *J. of Computational Physics*, **227**, 6660–6680.

Maxey, M.R. and Riley, J.R.J., 1983, Equation of motion for a small rigid sphere in a nonuniform flow, *Phys. Fluids*, **26**, 883.

Mei, R., Yu, D., Shyy, W. and Luo, L.-S., 2002, Force evaluation in the lattice Boltzmann method involving curved geometry, *Phys. Rev. E*, **65** (041203).

Michaelides, E.E., 1987, Motion of particles in gases: Average velocity and pressure loss, *J. Fluids Engineering*, **109**, 172.

Michaelides, E.E., 2003, Freeman scholar paper—Hydrodynamic force and heat/mass transfer from particles, bubbles and drops, *J. Fluids Eng.*, **125**, 209–238.

Michaelides, E.E., 2006, *Particles, Bubbles and Drops—Their Motion, Heat and Mass Transfer*, World Scientific Publishers, Singapore.

Michaelides, E.E., Crowe, C.T. and Schwarzkopf, J.D. (eds.), 2017, *Multiphase Flow Handbook*, 2nd edition, CRC Press, Boca Raton.

Michaelides, E.E. and Roy, I., 1987, An evaluation of several correlations used for the prediction of pressure drop in particulate flows. *Int. J. of Multiphase Flows*, **13**, 433.

Milojevic, D., 1990, Lagrangian stochastic-deterministic (LSD) prediction of particle dispersion in turbulence. *Part. and Part. Systems Characterization*, 7, 181–190.

Minier, J.-P. and Peirano, E., 2001, The PDF approach to turbulent polydispers two-phase flows, *Physics Reports*, **352**, 1–214.

Minier, J.-P., Peirano, E. and Chibbaro, S., 2004, PDF model based on the Langevin equation for polydispersed two-phase flows applied to a bluff-body gas-solid flow, *Physics of Fluids*, **16**, 2419–2431.

Mittal, R., Dong, H., Bozkurttas, M., Najjar, F.M., Vargas, A., von Loebbecke, A., 2008, A versatile sharp interface immersed boundary method for incompressible flows with complex boundaries, *J. of Computational Physics*, **227**, 4825–4852.

Mohd-Yusof, J., 1997, Combined immersed boundaries/B-splines methods for simulations of flows in complex geometries. *Annual Research Briefs*, Center for Turbulence Research, Stanford University.

Monchaux, R., Bourgoin, M. and Cartellier, A., 2012, Analyzing preferential concentration and clustering of inertial particles in turbulence, *Int. J. Multiphase Flow*, **40**, 1–18.

Mudde, R.F. and Simonin, O., 1999, Two- and three-dimensional simulations of a bubble plume using a two-fluid model, *Chemical Engineering Science*, **54**, 5061–5069.

Muniz, M. and Sommerfeld, M., 2020, On the force competition in bubble columns: A numerical study, *Int. J. of Multiphase Flow*, **128** (103256).

Nadim, A. and Stone, H.A., 1991, The motion of small particles and droplets in quadratic flows, *Studies in Applied Mathematics*, **85**, 53–73.

Neau, H., Pigou, M., Fede, P., Ansart, R., Baudry, C., Mérigoux, N., Laviéville, J., Fournier, Y., Renon, N. and Simonin, O., 2020, Massively parallel numerical simulation using up to 36,000 CPU cores of an industrial-scale polydispersed reactive pressurized fluidized bed with a mesh of one billion cells, *Powder Technology*, **366**, 906–924.

Nguyen, N.Q. and Ladd, A.J.C., 2002, Lubrication corrections for lattice Boltzmann simulations of particle suspensions, *Phys. Rev. E.*, **66**, 046708.

Ocone, R., Sundaresan, S. and Jackson, R., 1993, Gas-particle flow in a duct of arbitrary inclination with particle-particle interactions. i, **39**, 1261–1271.

O'Rourke, P.J. and Amsden, A.A., 1987, The TAB method for numerical calculation of spray droplet breakup. i 872089.

Oseen, C.W., 1927, *Neuere Methoden Und Ergebnisse In Der Hydrodynamik, Mathematik und ihre Anwendungen in Monographien und Lehrbüchern.* Akademische Verlagsgesellschaft, Leipzig.

Ozel, A., Brändle de Motta, J.C., Abbas, M., Fede, P., Masbernat, O., Vincent, S., Estivalezes, J.-L., Simonin, O., 2017, Particle resolved direct numerical simulation of a liquid—solid fluidized bed: Comparison with experimental data, *Int. J. of Multiphase Flow*, **89**, 228–240.

Ozel, A., Fede, P. and Simonin, O., 2013, Development of filtered Euler-Euler two-phase model for circulating fluidised bed: High resolution simulation, formulation and a priori analyses, *Int. J. of Multiphase Flow*, **55**, 43–63.

Pai, M.G. and Subramaniam, S., 2006, Modeling interphase turbulent kinetic energy ransfer in Lagrangian-Eulerian spray computations, *Atomization and Sprays*, **16**, 807–826.

Parmentier, J.-F. and Simonin, O., 2012, A functional subgrid drift velocity model for filtered drag prediction in dense fluidized bed. i., **58**, 1084–1098.

Patankar, S.V., 1980, *Numerical Heat Transfer and Fluid Flow*, Hemisphere Publishing Company, Washington, DC.

Peskin, C.S., 1972, Flow patterns around heart valves: A numerical method. *J. of Computational Physics*, **10**, 252–271.

Peskin, C.S., 1977, Numerical analysis of blood flow in the heart, *J. of Computational Physics*, **25**, 220–252.

Peskin, C.S., 2003, The immersed boundary method, *Acta Numerica*, **11**, 479–517.

Poustis, J.-F., Senoner, J.-M., Zuzio, D. and Villedieu, P., 2019, Regularization of the Lagrangian point force approximation for deterministic discrete particle simulations, *Int. J. Multiph. Flow*, **117**, 138–152.

Prahl, L., Hölzer, A., Arlov, D., Revstedt, J., Sommerfeld, M. and Fuchs, L., 2007, On the interaction between two fixed spherical particles, *Int. J. of Multiphase Flow*, **33**, 707–725.

Pritchett, J.W., Blake, T.R. and Garg, S.K., 1978, A numerical model of gas fluidized beds. *AIChE Symposium Series*, **74**.

Prosperetti, A. and Tryggvason, G., 2009, *Computational Methods for Multiphase Flow*, Cambridge University Press, Cambridge, UK.

Reeks, M.W., 1983, The transport of discrete particles in inhomogeneous turbulence, *J. Aerosol Sci.*, **14**, 729–739.

Reeks, M.W., 2021, The development and application of a kinetic theory for modeling dispersed particle flows, ASME, *J. Fluids Eng.*, **143** (8), 080803.

Rüger, M., Hohmann, S., Sommerfeld, M. and Kohnen, G., 2000, Euler-Lagrange calculations of turbulent sprays: The effect of droplet collisions and coalescence, *Atomization and Sprays*, **10**, 47–81.

Rzehak, R., Krepper, E., Liao, Y., Ziegenhein, T., Kriebitzsch, S. and Lucas, D., 2015, Baseline model for the simulation of bubbly flows, *Chemical Engineering & Technology*, **38**, 1972–1978.

Sanjeevi, S.K.P., Kuipers, J.A.M. and Padding, J.T., 2018, Drag, lift and torque correlations for non-spherical particles from stokes limit to high Reynolds numbers, *Int. J. of Multiphase Flow*, **106**, 325–337.

Schaeffer, D.G., 1987, Instability in the evolution equations describing incompressible granular flow, *J. of Differential Equations*, **66**, 19–50.

Schneiders, L., Fröhlich, K., Meinke, M. and Schröder, W., 2019, The decay of isotropic turbulence carrying non-spherical finite-size particles. i, **875**, 520–542.

Seo, J.H. and Mittal, R., 2011, A sharp-interface immersed boundary method with improved mass conservation and reduced spurious pressure oscillations, *J. of Computational Physics*, **230**, 7347–7363.

Sgrott Júnior, O.L. and Sommerfeld, M., 2019, Influence of inter-particle collisions and agglomeration on cyclone performance and collection efficiency. *Canadian J. Chemical Engineering*, **97**, 511–522.

Siegel, W., 1991, *Pneumatische Förderung: Grundlagen, Auslegung, Anlagenbau, Betrieb*. Vogel Verlag, Würzburg.

Simonin, O., 1996, Continuum modelling of dispersed two-phase flows, *Lecture series—von Karman Institute for Fluid Dynamics*, **2**, K1–K47.

Simonin, O., Deutsch, E. and Minier, J.P., 1993, Eulerian prediction of the fluid/particle correlated motion in turbulent two-phase flows, *Applied Scientific Research*, **51**, 275–283.

Snyder, W.H. and Lumley, J.L., 1971, Some measurements of particle velocity autocorrelation functions in a turbulent flow. i., **48**, 41–71.

Soldati, A., and Marchioli, C., 2009, Physics and modelling of turbulent particle deposition and entrainment: Review of a systematic study, *Int. J. Multiphase Flow*, **35**, 827–839.

Sommerfeld, M., 1996, *Modellierung und numerische Berechnung von partikelbeladenen turbulenten Strömungen mit Hilfe des Euler-Lagrange-Verfahrens*. Habilitationsschrift (habilitation thesis), Universität Erlangen-Nürnberg, Shaker Verlag, Aachen.

Sommerfeld, M., 2001, Validation of a stochastic Lagrangian modelling approach for inter-particle collisions in homogeneous isotropic turbulence, *Int. J. of Multiphase Flows*, **27**, 1828–1858.

Sommerfeld, M., 2003, Analysis of collision effects for turbulent gas-particle flow in a horizontal channel: Part I. Particle transport, *Int. J. Multiphase Flow*, **29**, 675–699.

Sommerfeld, M., 2010, *Particle motion in fluids*, VDI-Buch: VDI Heat Atlas, Springer Verlag Berlin/Heidelberg, Part 11, pp. 1181–1196.

Sommerfeld, M., 2017, Numerical methods for dispersed multiphase flows, in *Particles in Flows* (Eds. T. Bodnár, G.P. Galdi and Š. Nečasová), pp. 327–396, Series Advances in Mathematical Fluid Mechanics, Springer Int. Publishing, Cham, Switzerland.

Sommerfeld, M., Cui, Y. and Schmalfuß, S., 2019, Potentials and constraints for the application of CFD combined with Lagrangian particle tracking to dry powder inhalers, *European J. of Pharmaceutical Sciences*, **128**, 299–324.

Sommerfeld, M., Decker, S. and Kohnen, G., 1997, Time-dependent calculation of bubble columns based on the time-averaged Navier-Stokes equations with turbulence model, *Proceedings of the Japanese-German Symposium on Multi-Phase Flow*, Tokyo, Japan, 323–334.

Sommerfeld, M., Kohnen, G. and Qiu, H.-H., 1993a, Spray evaporation in turbulent flow: Numerical calculations and detailed experiments by phase-Doppler anemometry, *Revue de L'Institut Francais du Petrole*, **48**, 677–695.

Sommerfeld, M., Kohnen, G. and Rüger, M., 1993b, Some open questions and inconsistencies of Lagrangian particle dispersion models, *Proceedings Ninth Symposium on Turbulent Shear Flows*, Kyoto Japan, Paper No. 15–1.

Sommerfeld, M. and Qadir, Z., 2018, Fluid dynamic forces acting on irregular shaped particles: Simulations by the Lattice-Boltzmann method, *Int. J. Multiphase Flows*, **101**, 212–222.

Sommerfeld, M., Sgrott Jr., O.L., Taborda, M.A., Koullapis, P., Bauer, K. and Kassinos, S., 2021, Analysis of flow field and turbulence predictions in a lung model applying RANS and implications for particle deposition, *European J. of Pharmaceutical Sciences*, **166** (105959).

Sommerfeld, M., van Wachem, B. and Oliemans, R., 2008, Best practice guidelines for computational fluid dynamics of dispersed multiphase flows. *ERCOFTAC*, ISBN 978-91-633-3564-8.

Spalding, D.B., 1983, Developments in the IPSA procedure for numerical computation of multiphase-flow phenomena with interphase slip, unequal temperatures, etc, *Numerical Properties & Methodologies in Heat Transfer*, 421–436.

Squires, K.D. and Eaton, J.K., 1991, Preferential concentration of particles by turbulence, *Phys. Fluids A*, **3**, 1169–1178.

Srivastava, A. and Sundaresan, S., 2003, Analysis of a frictional-kinetic model for gas-particle flow, *Powder Technology*, **129**, 72–85.

Sundaram, S. and Collins, L.R., 1997, Collision statistics in an isotropic particle-laden turbulent suspension. Part 1. Direct numerical simulations. *J. Fluid Mech.*, **335**, 75–109.

Syamlal, M., Rogers, W. and O'Brien, T.J., 1993, *MFIX Documentation: Theory Guide.* DOE/METC-94/1004.

Tavassoli, H., Kriebitzsch, S., van der Hoef, M., Peters, E., and Kuipers, J., 2013, Direct numerical simulation of particulate flow with heat transfer, *Int. J. Multiphase Flow*, **57**, 29–37.

Tenneti, S., Garg, R., and Subramaniam, S., 2011, Drag law for monodisperse gas—solid systems using particle-resolved direct numerical simulation of flow past fixed assemblies of spheres, *Int. J. of Multiphase Flow*, **37**, 1072–1092.

Tenneti, S. and Subramaniam, S., 2014, Particle-resolved direct numerical simulation for gas-solid flow model development, *Ann. Rev. Fluid Mech.*, **46**, 199–230.

Tenneti, S., Sun, B., Garg, R. and Subramaniam, S., 2013, Role of fluid heating in dense gas-solid flow as revealed by particle-resolved direct numerical simulation, *Int. J. Heat Mass Transfer*, **58**, 471–479.

Thömmes, G., Becker, J., Junk, M., Vaikuntam, A.K., Kehrwald, D., Klar, A., Steiner, K. and Wiegmann, A., 2009, A lattice Boltzmann method for immiscible multiphase flow simulations using the level set method, *J. Comput. Phys.*, **228**, 1139–1156.

Tsuji, Y., 1993, Discrete particle simulation of gas-solid flows, *KONA Powder and Particle J.*, **11**, 57–68.

Tsuji, Y., Kawaguchi, T. and Tanaka, T., 1993, Discrete particle simulation of two-dimensional fluidized bed, *Powder Technology*, **77**, 79–87.

Uhlmann, M., 2005, An immersed boundary method with direct forcing for simulation of particulate flows, *J. Comput. Phys.*, **209**, 448–476.

Uhlmann, M. and Chouippe, A., 2017, Clustering and preferential concentration of finite-size particles in forced homogeneous-isotropic turbulence, *J. of Fluid Mechanics*, **812**, 991–1023.

van Wachem, B.G.M. and Denner, F., 2014, A fully coupled solver approach for multiphase flow problems, *2nd Int. Conference on Numerical Methods in Multiphase Flows*. Darmstadt, Germany, 30 June–2 July.

van Wachem, B.G.M., Schouten, J.C. and Krishna, R., 1998, Eulerian simulations of bubbling behaviour in gas-solid fluidised beds, *Computers & Chemical Engineering*, **22**, 299–307.

van Wachem, B.G.M., Schouten, J.C., van den Bleek, C.M., Krishna, R. and Sinclair, J.L., 2001a, Comparative analysis of CFD models of Dense Gas—solid systems, *AIChE J.*, **47**, 1035–1051.

van Wachem, B.G.M., Van der Schaaf, J., Schouten, J.C., Krishna, R. and van den Bleek, C.M., 2001b, Experimental validation of Lagrangian-Eulerian simulations of fluidized beds, *Powder Technology*, **116**, 155–165.

van Wachem, B.G.M., Zastawny, M., Zhao, F. and Mallouppas, G., 2015, Modelling of gas—solid turbulent channel flow with non-spherical particles with large Stokes numbers, *Int. J. of Multiphase Flow*, **68**, 80–92.

Von Karman, T. and Horwarth, L., 1938, On the statistical theory of isotropic turbulence, *Proc. Royal Society London*, **A164**, 192–215.

Vreman, A.W., 2016, Particle-resolved direct numerical simulation of homogeneous isotropic turbulence modified by small fixed spheres, *J. Fluid Mech.*, **796**, 40–85.

Wang, L.P. and Maxey, M.R., 1993, Settling velocity and concentration distribution of heavy particles in homogeneous isotropic turbulence, *J. Fluid Mech.*, **256**, 27–68.

Wells, M.R. and Stock, D.E., 1983, The effect of crossing trajectories on the dispersion of particles in a turbulent flow, *J. Fluid Mech.*, **136**, 31–62.

Wen, C.Y. and Yu, Y.H., 1966, A generalized method for predicting the minimum fluidization velocity, *AIChE J.*, **12**, 610–612.

Westman, M.A., Michaelides, E.E. and Thompson, F.A., 1987, Pressure losses due to bends in pneumatic conveying, *J. of Pipelines*, 7, 15–28.

Wilcox, D.C., 2006, *Turbulence Modelling for CFD*, 3rd edition, Publisher DCW Industries.

Xu, Z.-J. and Michaelides, E.E., 2003, The effect of particle interactions on the sedimentation process of non-cohesive particles, *Int. J. Multiphase Flow*, **29**, 959–982.

Yang, J. and Balaras, El., 2006, An embedded-boundary formulation for large-eddy simulation of turbulent flows interacting with moving boundaries, *J. of Computational Physics*, **215**, 12–40.

Yeung, P.K.K. and Pope, S.B.B., 1988, An algorithm for tracking fluid particles in numerical simulations of homogeneous turbulence, *J. of Computational Physics*, **79**, 373–416.

Yu, Z. and Shao, X., 2007, A direct-forcing fictitious domain method for particulate flows. *J. Comput. Phys.*, **227**, 292–314.

Zastawny, M., Mallouppas, G., Zhao, F. and van Wachem, B., 2012, Derivation of drag and lift force and torque coefficients for non-spherical particles in flows, *Int. J. of Multiphase Flow*, **39**, 227–239.

Zhao, F., George, W.K. and van Wachem, B.G.M., 2015, Four-way coupled simulations of small particles in turbulent channel flow: The effects of particle shape and Stokes number. *Physics of Fluids*, **27** (083301).

Zhao, F. and van Wachem, B.G.M., 2013a, A novel quaternion integration approach for describing the behaviour of non-spherical particles, *Acta Mechanica*, **224**, 3091–3109.

Zhao, F. and van Wachem, B.G.M., 2013b, Direct numerical simulation of ellipsoidal particles in turbulent channel flow, *Acta Mechanica*, **224**, 2331–2358.

Zhong, W., Xiong, Y.Q., Yuan, Z.L. and Zhang, M.Y., 2006, DEM simulation of gas-solid flow behaviours in Spout-fluid bed, *Chemical Engineering Science*, **61**, 1571–1584.

Zhou, Q. and Leschziner, M.A., 1991, A Lagrangian particle dispersion model based on a time-correlated stochastic approach, in *Gas-Solid Flows* (Eds. D.E. Stock, Y. Tsuji, J.T. Jurewicz, M.W. Reeks and M. Gautam), *ASME, FED*, **121**, 255–260.

Ziegenhein, T., Rzehak, R. and Lucas, D., 2015, Transient simulation for large scale flow in bubble columns, *Chemical Engineering Science*, **122**, 1–13.

9 Experimental Methods

Detailed measurements of multiphase flow processes and equipment are necessary for process and quality control, to characterize the global flow behavior and flow regimes, to determine the flow rates of the different phases, and to obtain local information about the characteristic properties of all involved phases, such as velocities and particle concentration. Moreover, experimental techniques are required to analyze the behavior and motion of individual particles or droplets in order to assess micro-processes occurring on the scale of the particles. In addition, there is the need to determine characteristic particle dimensions and other properties, including size distribution, shape, and surface area. Principally, particle characterization can be only done offline in a laboratory analysis.

Examples of process control are the monitoring of particle concentration at the outlet/exit pipe of a cyclone separator in order to ensure regulations compliance with respect to fine dust emissions. Such type of control should be done online and could be based on optical sensors providing a signal, which is a function of the particle concentration. Another example is the supervision of the performance of powder-grinding processes by frequent analysis of the produced particle size distribution. This may be done by online particle sizing methods. But typically, a characteristic sample of powder has to be extracted after the mill and brought to a laboratory analysis. For this purpose, numerous techniques are available. Similar to this process is the procedure after an agglomeration process, where fine powders generate agglomerates or granulates of defined size and shape. In this case, one has to collect a representative sample from the product stream. Consequently, the proper collection of representative samples in time and space (distributed over the entire production vessel) and the subsequent sample division for obtaining a typical analysis sample (i.e. typically a few hundred grams) becomes very important.

The requirements for measurement principles and techniques as well as the necessary devices are quite different for industry and research institutions. In industry, an instrument must be robust, easy to handle, and give clear and well-defined answers. In the research field, the dispersed phase properties, which are needed or shall be measured, makes the challenge of designing and developing instruments based on certain physical principles. For that purpose, quite often novel instruments are developed for extracting so far unknown properties of a two-phase system or particle-scale processes, as for example, solid particle rotational speed or temperature distribution inside spray droplets.

From the industrial perspective, one may distinguish between four measurement approaches depending on which property is needed, namely, inline, online, and at-line as well as offline methods. An inline measurement is conducted directly inside the process of consideration, whereas an online measurement is done within a bypass system of the continuous process. At-line implies that the measurement is done just next to the process but based on extracting samples from the process.

DOI: 10.1201/9781003089278-9

Finally, offline is a measurement done in a separate laboratory using more sophisticated measurement devices and requires taking characteristic samples from the process. In the context of multiphase flows, such offline (and also at-line) measurements are mainly related to the characterization of particulate matter from the process, namely, particle size, shape, and composition.

An accurate in situ determination of particle or droplet sizes is only possible in limited cases due to the complexity and dimension of the instrument needed and therefore is restricted to research facilities. In industrial applications, typically, a characteristic sample has to be collected from the process and delivered to the laboratory analysis. Different instruments are available using a multitude of measurement principles, for example, analytic sieving methods, sedimentation, optical counting or light scattering, and eventually imaging—some of which will be described later. Regarding particle size determination, one may distinguish between the following commonly used methods of sizing and definitions (Rhodes, 2008):

1. Direct geometrical dimensions like a diameter of a sphere and a length and diameter of a cylinder. These properties are mostly obtained from an imaging analysis or through a microscope.
2. Statistical lengths measured on planar projections of mostly non-spherical particles placed on a microscope specimen slide, for example, Martin's diameter, Feret's diameter, and maximum chord length.
3. Geometrical equivalent diameters. Here, the measured volume, surface area, or planar projection cross-section of any type of particle, also non-spherical particles, are used to calculate a sphere-related diameter, as explained in detail in Section 3.1.
4. Physical equivalent diameters are related to a particle size through a physical process. This diameter may not be identical to a geometrical dimension of the particle. Here, a physical property of the particles, such as sedimentation velocity or light scattering intensity, is measured. From these properties, a particle diameter is calculated through available correlations or often also through a calibration with defined particles. For example, the measured sedimentation velocity of a suspension is used to calculate a particle diameter based on the fundamental equations for a sphere, including the drag coefficient. Consequently, the diameter obtained for a non-spherical particle may be different from the real geometrical dimensions of the particles. Similarly, a theoretically obtained scattering intensity (e.g. Mie theory (Mie, 1908) or geometrical optics) may be used to determine a particle size from a measured single particle scattering intensity.

Technical systems, where a sample collection and a laboratory analysis of a process outcome is not possible, are spray systems with a wide area of applications, such as spray combustion, spray cooling of surfaces, spray dryers, and fire extinguishers. Therefore, in order to characterize spray droplets and their behavior, only in-line measurements directly at the spraying system are possible.

Numerous measuring techniques are available for experimental analysis of two-phase flow systems depending on the level of application. The first classification is based on the way the information is extracted from the two-phase system, that is, offline and online measurements as shown in Figure 9.1. Sampling methods are frequently applied in the process and powder technology industries to provide a detailed characterization of bulk solids at the different stages of the process for quality and process control (e.g. after milling or classification processes). Using mechanical sampling equipment, a number of characteristic samples of the bulk solids are collected and analyzed. The analysis of the samples may be performed using a microscope as well as mechanical methods, such as sieving, flow classification, cascade impaction, and sedimentation. Optical methods such as light scattering or light attenuation may also be applied. The result of these analyses could be the characteristic dimensions of the particles, particle shape factors, equivalent particle diameters, or particle surface area.

Online measuring techniques can be directly applied within the process or in a bypass line to analyze the properties of both the dispersed and continuous phases. Generally, these methods can only be used in dispersed two-phase systems with relatively low concentrations of the particle phase. Online measuring techniques are widely used in research programs to characterize the development and the properties of a two-phase flow system. A further classification of online measurement techniques may be based on the spatial resolution of the measurement, as shown in Figure 9.1. Integral methods provide time-resolved, spatially averaged properties of two-phase systems over an entire cross-section of the flow or along a light beam passing through the flow, depending on the method applied.

Local or pointwise measurement techniques allow the determination of local properties in a two-phase system with a spatial resolution depending on the measurement technique applied. These techniques have a spatial resolution of the dimension of the probe volume, which is on the order of $100 \, \mu m$ and are mostly based on optical principles, such as light scattering or laser-Doppler and phase-Doppler anemometry. These methods are also called single particle counting techniques since they require that only one particle be in the probe volume at a given time. Therefore, the size of the probe volume needs to be very small in order to fulfill this requirement. Since the particle phase properties and the respective distribution functions are obtained for all particles that pass through the probe volume during the measurement time, the result obtained by such methods is weighted by the particle number flux.

Imaging techniques with volume illumination (i.e. light sheet or back-lightning) freeze the two-phase system under consideration and provide information either

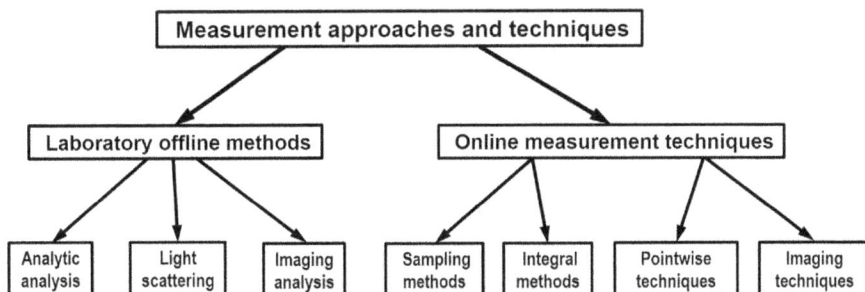

FIGURE 9.1 Measurement technique categories for the analysis of two-phase systems.

about the particle properties at a given instant of time, at multiple times, or during a time sequence within a finite but relatively large probe volume (e.g. field imaging techniques, such as particle image velocimetry (PIV) or holography). Distribution functions of particle phase properties are therefore obtained for all particles that happen to be in the probe volume at a given instant of time. Hence, the result is weighted by the particle number concentration.

For the characterization of two-phase flow systems, the following properties of the dispersed phase are of primary interest:

- Particle size (i.e. diameter of spherical particles, equivalent diameters of non-spherical particles and agglomerates)
- Particle shape and surface area
- Particle concentration, mass flux, or volume fraction
- Particle translational and rotational velocities
- Correlation between particle size and velocity

In many cases, the velocity of the continuous phase in the presence of the dispersed phase is also of interest in order to assess the influence of the particles onto the fluid flow—the two-way coupling. These properties are generally not obtained as single values but as distribution functions, which are characterized by mean values, standard deviations, and higher moments of the distribution functions.

A summary of the various measurement techniques and the dispersed phase properties, which can be obtained by the previously described measurement principles, is given in Table 9.1. There are several review papers on different types of measurement

TABLE 9.1

Categorization of Measurement Techniques with Respect to the Measurable Properties of a Two-Phase System

Particle Phase Property	Sampling or Offline Methods	Online Methods	
		Integral Methods	*Local Measurement Techniques*
Size	Sieving Coulter principle Sedimentation	Laser diffraction	Light scattering, Phase-Doppler anemometry
Concentration	Isokinetic sampling	Laser diffraction Light absorption	Isokinetic sampling, Fiber optic probes, Light scattering, Phase-Doppler anemometry
Velocity	Direct velocity measurements are not possible with sampling methods	Correlation technique	Fiber optic probes, Laser-Doppler anemometry, Phase-Doppler anemometry, Particle tracking velocimetry, Particle image velocimetry

techniques for multiphase flows, which are included in the references of this chapter. Techniques for measurements in two-phase flows laden with very small particles (i.e. nano- to micron-sized particles) were presented by Tu *et al.* (2017). In the following sections, the most common measurement techniques and their capabilities are introduced, and the limitations of their applications are summarized.

9.1 LIGHT SCATTERING FUNDAMENTALS

The basic theory of light scattering will be briefly summarized, and results for the dependence of scattering intensity on particle size will be introduced. The principles of light scattering are important for a number of online and offline measurement instruments and principles, such as light scattering intensity, laser diffraction, laser-Doppler (LDA), and phase-Doppler anemometry (PDA) and field imaging techniques, which will be introduced later.

The interaction of light with particles results in absorption and scattering of the light. The light scattered can be calculated using the theory of Mie (1908), which requires the solution of Maxwell's wave equations for the case of scattering of a plane electromagnetic wave by a homogeneous sphere. For more details about *Mie scattering theory*, the reader is referred to the relevant literature, for example, van de Hulst (1981). For the illumination of the measurement area or spot with visible light, mostly mono-chromatic light is used, which is essential when the measurement principle is based on interference, as for example in LDA and PDA. Typical light colors and wave lengths are blue, 450–495 nm; green, 495–570 nm; and red, 620–750 nm. For light scattering intensity measurements, quite often also white light sources are being used, which requires integration over the spectrum and yields a smoothing of the scattering intensity especially in the Mie scattering regime, as will be described later.

The absolute intensity of the light scattered by one particle, q, depends on the intensity of the incident light, I_0; the wavelength of the light, λ; the polarization angle of the light, γ; the particle diameter, D; the complex index of refraction, $\bar{n}$; and the scattering angle, φ

$$q = I_0 \frac{\lambda^2}{8\pi^2} I(\lambda, \gamma, \varphi, D, n, \kappa) \tag{9.1}$$

where I represents the so-called *Mie intensity*. The complex refractive index is given by

$$\bar{n} = n(1 - i\kappa) \tag{9.2}$$

Optical arrangements for scattering intensity measurements usually involve the integration of the scattered light over the receiving aperture for a selected mean scattering angle. Such a typical optical arrangement is illustrated in Figure 9.2, where φ is the scattering angle (0° forward scattering and 180° backward scattering), $\Delta\delta$ indicates the receiving solid aperture angle. For parallel and monochromatic light as the illuminating source, the resulting scattering intensity is obtained by integration

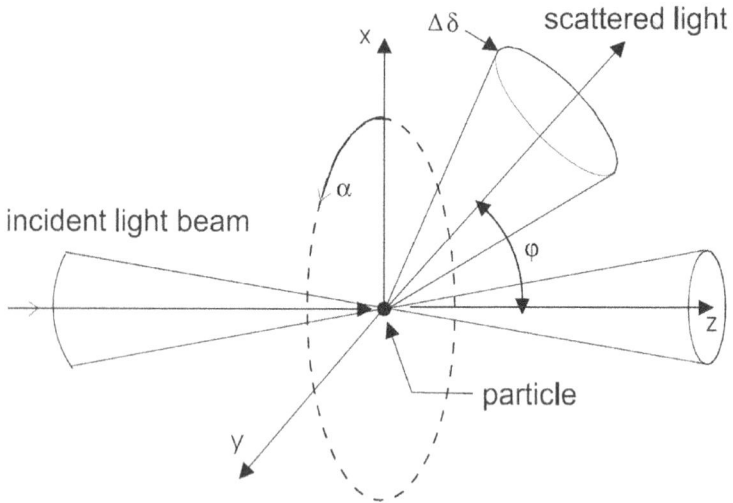

FIGURE 9.2 Optical arrangements for light scattering measurements with incident light beam and collected scattered light at a fixed scattering angle, φ, with a collection angle, $\Delta\delta$.

$$Q(\lambda, D, n, \kappa, \Delta\delta) = I_0 \frac{\lambda^2}{8\pi^2} \int_0^{2\pi} \int_{\Delta\delta} I(\lambda, \gamma, \varphi, D, n, \kappa) \sin\varphi\, d\varphi\, d\gamma \qquad (9.3)$$

When white light is used instead of monochromatic light, the scattering intensity is obtained by additionally integrating over the wavelength spectrum. This process has some advantages with regard to a smooth correlation between scattering intensity and particle size. In the following section, the principles of scattering intensity measurements will be introduced by using Mie calculations (DANTEC/Invent, 1994).

Typical examples of the scattering intensity around a particle for various particle diameters are shown in Figure 9.3 for parallel polarization. For the smallest particle (i.e. 0.1 μm), the angular intensity distribution has only two lobes for parallel polarization, which is characteristic of the Rayleigh scattering range. It is obvious that the absolute intensity of the scattered light increases remarkably with particle size. Moreover, the number of scattering lobes becomes larger when the particles become bigger. The highest scattering intensity is concentrated in the forward scattering direction (0 degree in the diagrams) and is mainly a result of diffraction. When the particles become bigger, this forward scattering diffraction lobe becomes narrower. For the largest particles, that is, 10 and 50 μm, the intensity distribution around the particles becomes noisy.

The influence of polarization on the scattering intensity around a 10 μm particle is illustrated in Figure 9.4. In the forward scattering direction, the intensity is comparable for both polarizations, but perpendicular yields some fluctuations. Parallel polarization gives larger fluctuations in the back-scatter region. Hence, for intensity measurements, the polarization should be selected according to the angular location of the receiver optics.

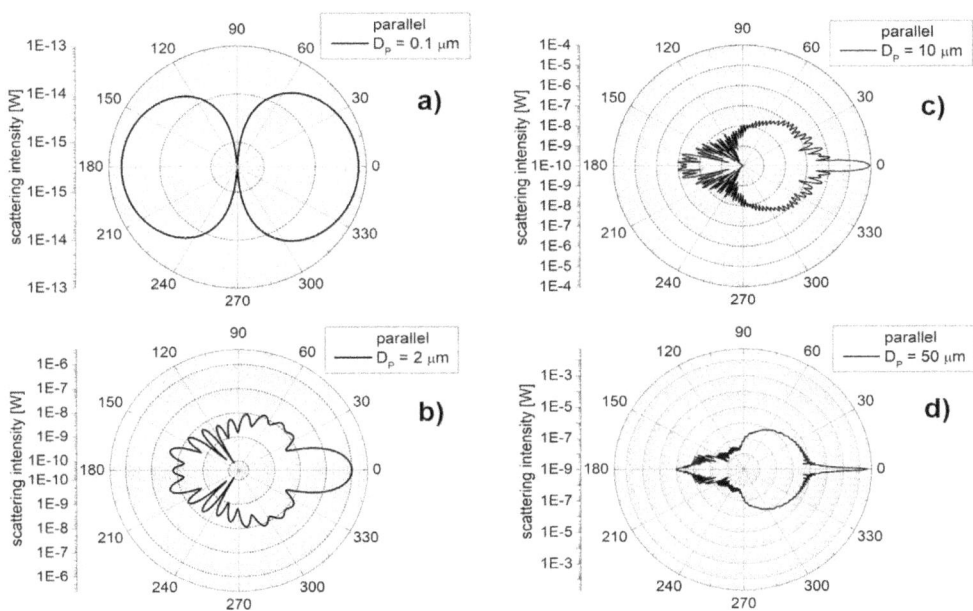

FIGURE 9.3 Angular distribution of scattering intensity around a particle ($I_0 = 10^7$ W/m², $\lambda = 632.8$ nm, $n = 1.5$; parallel polarization with respect to the incident light, forward scattering at 0 degree), particle diameters (a) 0.1 μm, (b) 2 μm, (c) 10 μm, and (d) 50 μm.

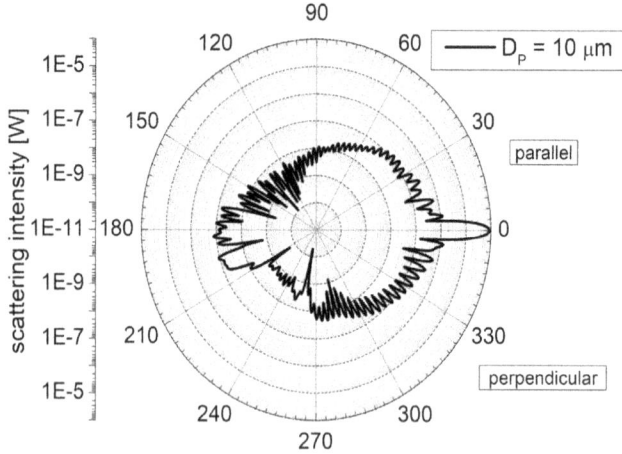

FIGURE 9.4 Angular distribution of scattering intensity around a particle; comparison of parallel and perpendicular polarization (upper and lower part of the graph) with respect to the polarization of the incident light ($I_0 = 10^7$ W/m^2, $\lambda = 632.8$ nm, $D = 10\,\mu$m, $n = 1.5$).

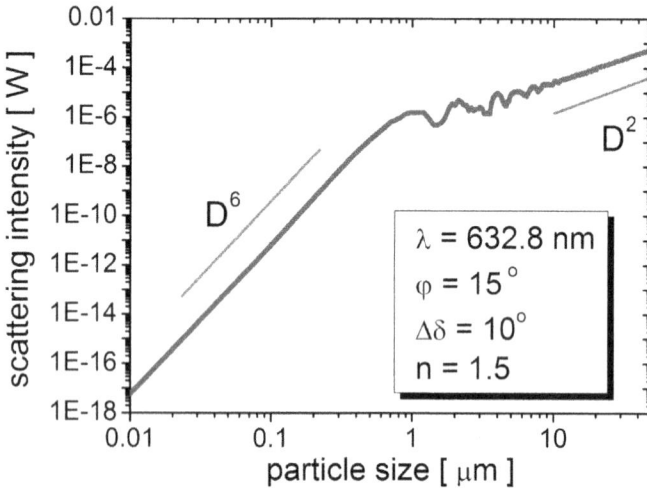

FIGURE 9.5 Dependence of scattering intensity on particle size obtained from a Mie computation ($I_0 = 10^7$ W/m^2, $\varphi = 15°$, $\Delta\delta = 10°$, $n = 1.5$).

The scattering intensity as a function of the particle diameter is analyzed in more detail for various properties of the particles and different configurations of the optical system with the parameters specified in Figure 9.5. Here the scattering angle is $\varphi = 15°$, the receiving aperture opening is $\Delta\delta = 10°$, and the particle consists of glass with $n = 1.5$. The results presented are again based on the Mie theory, and the calculations were performed with the software package "STREU" (DANTEC/Invent,

1994). In Figure 9.5, a typical dependence of the scattering intensity on the particle diameter is shown. From this correlation, three scattering regimes may be identified, which are often characterized by the so-called *Mie parameter*, $\alpha = \pi D/\lambda$.

1. The so-called *Rayleigh-scattering* applies to particles that are small compared with the wavelength of the incident light, that is, $\alpha \ll 1$ or $D < \lambda/10$. This regime is named after Lord Rayleigh, who first derived the basic scattering theory for such small particles (Strutt, 1871). Characteristic of this regime is a dependence of the scattering intensity on the fourth to sixth power of the particle diameter, hence, a region of high sensitivity.
2. For very large particles, that is, $\alpha \gg 1$ or $D > 4\,\lambda$, the laws of *geometrical optics* are applicable under certain conditions (van de Hulst, 1981). The light scattering intensity varies approximately with the square of the particle diameter.
3. The intermediate regime (i.e. $D \approx \lambda$) is called the Mie region, which is characterized by large oscillations in the scattering intensity, depending on the observation angle and the particle properties. These oscillations are also visible in the angular distributions (see Figure 9.3 c). Hence, the scattering intensity cannot be uniquely related to the particle size.

For large particles, that is, $D \gg \lambda$, the scattered light is composed of three components: namely, diffracted, externally reflected, and internally refracted light, as indicated in Figure 9.6. The refracted light may be separated in several modes depending on the number of internal reflections, that is, *P1, P2, P3 ... Pn*. Light diffraction is concentrated in the forward-scattering direction, that is, the so-called forward lobe, and is the dominant scattering phenomenon. This regime of geometrical optics is called the *Fraunhofer diffraction regime*. The diffraction pattern and the angular

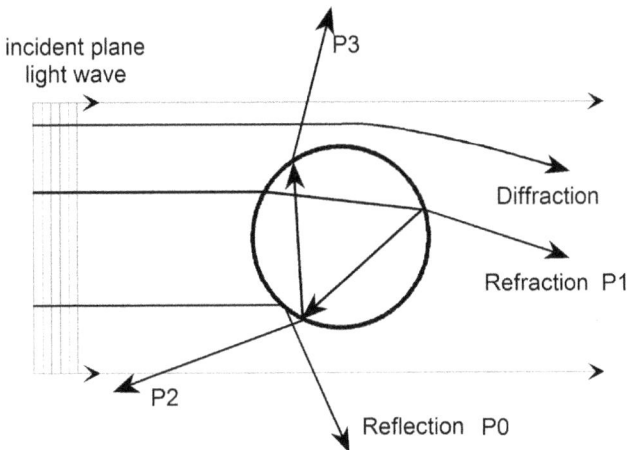

FIGURE 9.6 Different scattering modes for spherical particles in the geometric optics regime.

range of this forward lobe are dependent on the wavelength of the light and the particle diameter. As demonstrated in Figure 9.3, the angular range of diffracted light decreases with increasing particle diameter and is given by $\varphi < \pm\varphi_{\text{diff}}$ with

$$\sin\varphi_{\text{diff}} = \frac{5}{4}\frac{\lambda}{D} \tag{9.4}$$

It is important to note that the intensity of the diffracted light does not depend on the optical constants of the particle material. This is advantageous for sizing particles of different or unknown refractive indices.

Externally reflected light is scattered over the entire angular range, that is, $0 < \varphi < 180°$, whereas refracted light (i.e. $P1$) does not exceed an upper angular limit, $\varphi_{\text{refr.}}$, which is given by the expression

$$\cos\left(\frac{\varphi_{\text{refr}}}{2}\right) = \frac{n_m}{n_p} \tag{9.5}$$

for $n_p/n_m > 1.0$. Hence, this upper angular limit is determined by the relative refractive index. For a given fluid and for $n_p/n_m > 1.0$, φ_{refr} decreases, and more refracted light is concentrated in the forward direction with a decreasing refractive index of the particles. The properties of reflected and refracted light will be addressed in Section 9.2.5. The Mie response functions for three scattering angles, that is, $\varphi = 0$, 15, and 90°, are shown in Figure 9.7 for different refractive indices in the size range between 0.1 and 20 μm. The calculations were performed for a receiving aperture of $\Delta\delta = 10°$. The results may be summarized as follows:

- In the narrow forward-scattering range, the scattering intensity is almost independent of the refractive index of the particle (Figure 9.7 a). This is a result of the dominance of diffraction in this angular region. As described previously, diffraction is independent of the optical properties of the particles, and the scattering intensity varies approximately with the fourth power of the particle diameter.
- For off-axis locations, the scattering intensity shows large-scale fluctuations in the Mie range, that is, the scattering intensity cannot be uniquely related to the particle size. In the Rayleigh range, the scattering intensity varies with the sixth power of the particle diameter, and in the range of geometrical optics, the square law may be applied.
- At a 90° scattering angle, the largest differences in the scattering intensity are observed for the different refractive indices (Figure 9.7 c) and the absolute value of the scattering intensity is considerably lower than for the other scattering angles.

To reduce the intensity fluctuations, especially in the Mie range, and to establish a smoother size-intensity correlation, the aperture of the receiving optics may be increased so that the scattered light is received from a larger angular region, and the oscillations in the angular distribution of the scattering intensity are smoothed out over a larger region, as shown in Figure 9.8. In addition, the use of white light

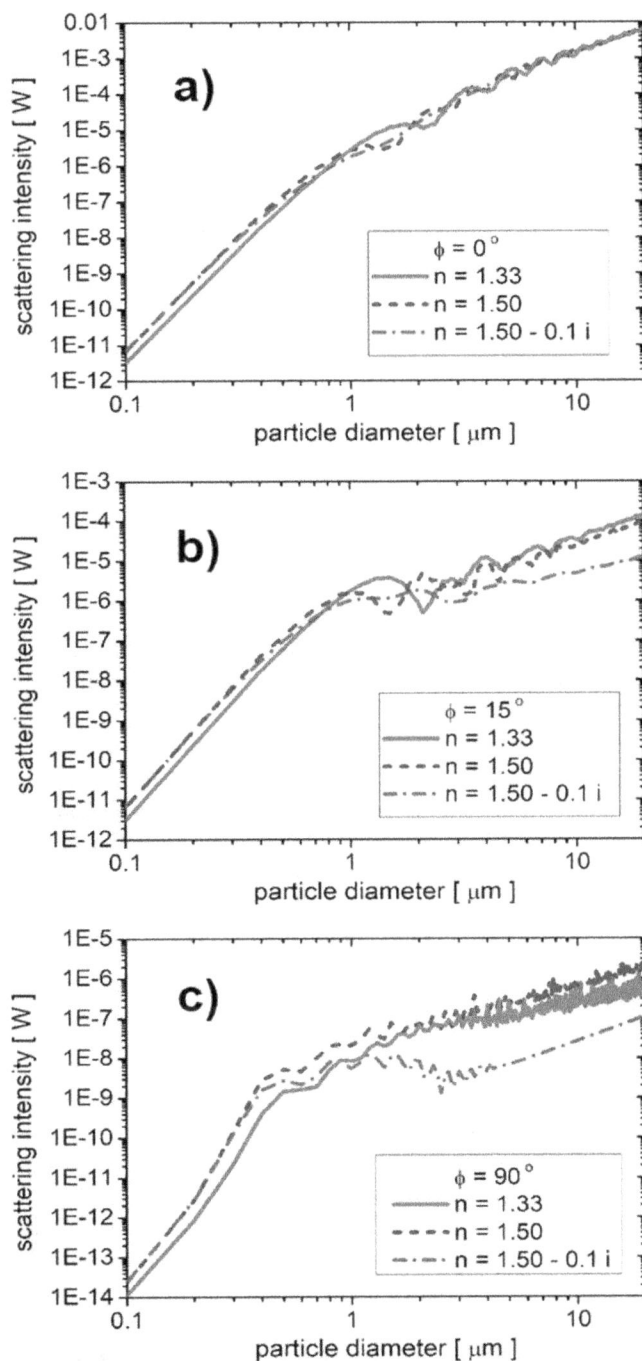

FIGURE 9.7 Mie response for three scattering angles: (a) $\varphi = 0°$, (b) $\varphi = 15°$, (c) $\varphi = 90°$, for spherical particles with different refractive indices ($I_0 = 10^7$ W/m^2, $\lambda = 632.8$ nm, $\Delta\delta = 10°$).

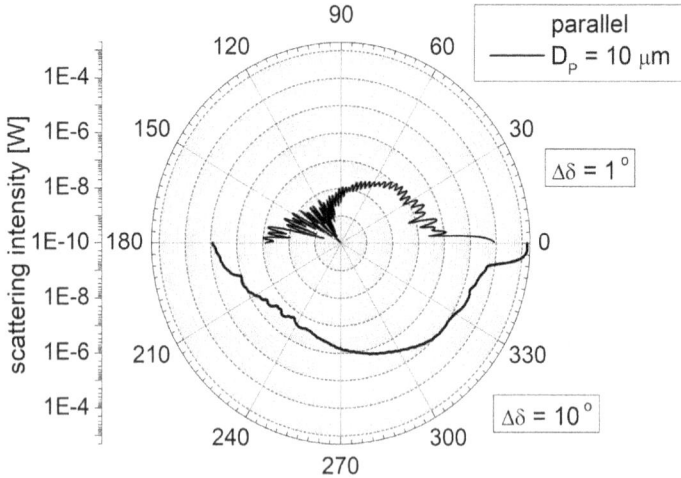

FIGURE 9.8 Influence of the receiving aperture opening on the Mie response function; upper part deldel = 1 degree; lower part deldel = 10 degree ($I_0 = 10^7$ W/m², $\lambda = 632.8$ nm, $\varphi = 15°$).

can improve the smoothness of the response curve since this is associated with an additional integration over the wavelength spectrum, as indicated by Eq. (9.3) (Broßmann, 1966).

Particle size measurements by the light intensity method require that only one particle be in the measurement volume at a time in order to enable the determination of the particle size from the measured signal amplitude. The simultaneous presence of more than one particle would result in an erroneous measurement, called *coincidence error*. Therefore, sizing methods based on scattering intensity measurements are also called *single particle counting techniques*. In order to fulfill this requirement, it is necessary to limit the measurement volume size, as will be presented in the next section.

9.2 SAMPLING AND OFFLINE METHODS

Sampling methods are mainly applied to characterize the shape, equivalent diameter, and surface areas of the solid particles. This implies that a number of representative samples of particles have to be collected from the flowing two-phase system or from the bulk solids conveyed on belts or stored in bins or containers. For this purpose, different sampling devices and methods are available, which enable collection from stationary or moving powders (Allen, 1990). The sampling should be performed in such a way that the collected samples closely represent the properties of the two-phase system under consideration. The process of sample collection and reduction of the sample size is illustrated in Figure 9.9. Usually, a number of samples have to be taken since the characteristics of the sample will vary over the process compartment, and it is important to ensure that the samples represent the "real" powder. A further important aspect in sampling is the reduction of the sample size

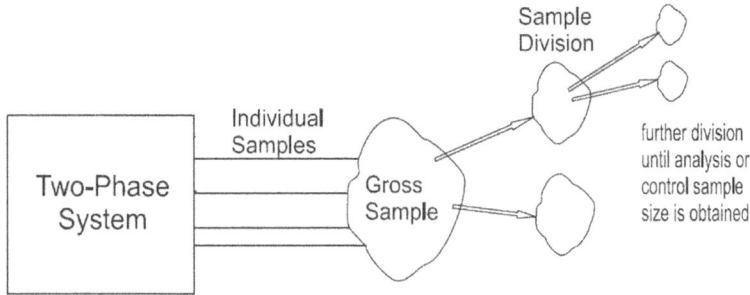

FIGURE 9.9   Process of sample collection and division.

to obtain a sample size for analysis, which typically is on the order of a few grams, depending on the method of analysis; for a sieving analysis, about 100 grams are needed. This dividing process needs to be performed one or more times in order to obtain a sample size, which can be analyzed as shown in the schematic of Figure 9.9.

For the analysis of the particle samples with regard to shape, linear dimension, and particle size, various methods, such as imaging methods, sieving, sedimentation, fluid classification, cascade impaction, laser light diffraction, or electrical sensing methods are available. Some of the most common methods are summarized here; others are described in detail by Allen (1990, 2003).

9.2.1   IMAGING METHODS, MICROSCOPY

Microscopy is a direct imaging method to determine particle diameters, characteristic dimensions, and shape factors of individual particles. This method can be used for the range of particle sizes between 1 and 200 μm. The lower limit is determined by diffraction (see Section 9.2.6 for more details), which may result in erroneous and blurred images of small particles. In order to determine the number distributions of linear dimensions, a relatively large number of individual particles must be analyzed, and this is time-consuming. In combination with digital image processing, however, such a direct imaging method becomes a powerful technique for particle characterization. The principle of conducting such a microscope imaging is illustrated in Figure 9.10. There are also instruments available, which allow automated particle size and shape analysis from falling powder samples using single or multiple CCD cameras (CCD stands for charged coupled device).

First, the sample has to be prepared in such a way that individual particles can be observed under a microscope, which implies a uniform dispersion of the particles on the sample slide. Then, images of a number of particles are taken using a CCD camera, which transforms the optical image into a digital image with a certain resolution of gray values for each pixel. Typically, 256 gray values are assigned to one pixel, which is one of the photo-sensitive elements of the CCD array. The number of pixels determines the spatial resolution of the CCD camera and is, for example, 1280 × 1024 for a standard CCD camera. After enhancement of the image—that is, background subtraction, thresholding, and removing noise objects—the image can

```
              ┌─────────────────────────────┐
              │   preparation of sample     │
              └─────────────────────────────┘
                           │  microscope
                           ▼
              ┌─────────────────────────────┐
              │      optical image          │
              └─────────────────────────────┘
                           │  CCD-camera
                           ▼
              ┌─────────────────────────────┐
              │     digitized image         │
              └─────────────────────────────┘
                           │  image enhancement
                           ▼
              ┌─────────────────────────────┐
              │  enhanced digitized image   │
              └─────────────────────────────┘
                           │  image processing
                           ▼
              ┌─────────────────────────────┐
              │  characteristic linear      │
              │  dimensions of particles    │
              └─────────────────────────────┘
```

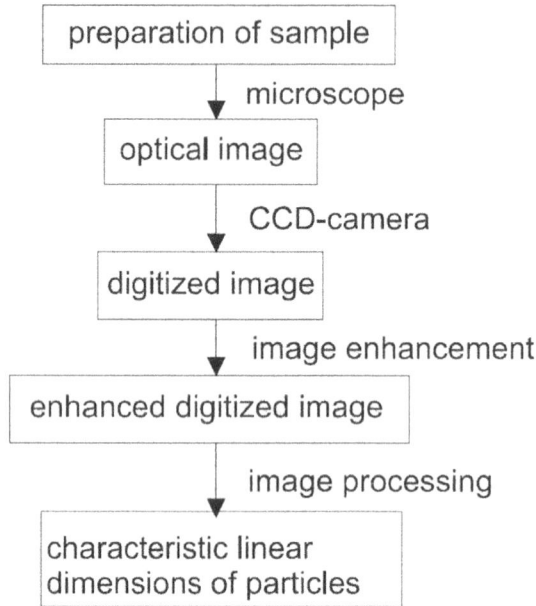

FIGURE 9.10   Principle of direct imaging technique, using digital processing.

be processed to determine the characteristic linear dimensions of the particle. Such linear dimensions may be the diameter of spherical particles or certain dimensions of non-spherical particles, which are determined from the particle's projection, since the image seen by a microscope or the camera is only two-dimensional. Therefore, the analysis of non-spherical particles by such an imaging method is biased by the fact that these particles will have a preferential orientation on the sample slide, that is, plate-like particles will lay flat on their slide.

Typical statistical linear dimensions of non-spherical particles are shown in Figure 9.11. The so-called *Feret-diameter, D_f,* is the distance between two tangents on opposite sides of the particle image, parallel to some fixed direction. The *Martin-diameter, D_m,* is the length of the line, which bisects the particle image and should be obtained in a fixed direction for one analysis. Other linear dimensions of the particle image are the *maximum chord $D_{c,max}$,* the largest linear dimension, and the perimeter diameter, which is the diameter of a circle having the same circumference as the perimeter of the image. A *projected area diameter* may be determined, which is the diameter of a circle having the same area as the projected image of the particle.

9.2.2   SIEVING ANALYSIS

Sieving is one of the oldest, simplest, and most widely used methods for particle size classification. It yields the mass fractions of particles in specific size intervals from which the resulting particle size distribution can be obtained. Sieving analysis can be performed with a wide range of mesh sizes, from approximately 5 μm to about

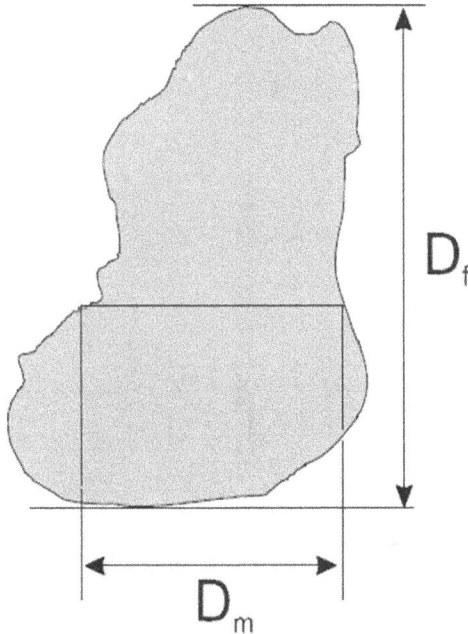

FIGURE 9.11   Typical linear dimensions of images of non-spherical particle.

TABLE 9.2
Sieving Methods with Respect to Particle Size Range and Particle Properties

Sieving Method	Particle Size Range	Particle Properties
Hand or machine sieving	63–125 μm	Dry powder
Air-jet sieving	10–500 μm	Cohesive and wet powder
Wet sieving with micro-mesh sieves	5–50 μm	Wet powder

---

125 mm. In order to cover this size range, different sieves and sieving methods have to be used. Also, the properties of the bulk solids determine the method of sieving. Woven metal sieves are quite often characterized on the basis of mesh size, which is the number of wires per linear measure (dimension). For example, the opening for a 400 mesh sieve is 37 μm, and the wire thickness is 26.5 μm (see the American ASTM). This corresponds to an open sieve area of 34%.

A classification of the sieving methods to be applied is based on the size of the particles, and the particle properties are given in Table 9.2. The sieves used for hand and machine sieving are usually woven from metal wires, which are soldered together and clamped to the bottom of a cylindrical container. For sieving analysis, a series of sieves are placed on top of one another with decreasing mesh size from top to bottom, as shown in Figure 9.12. This allows the determination of a complete size distribution with one measurement analysis.

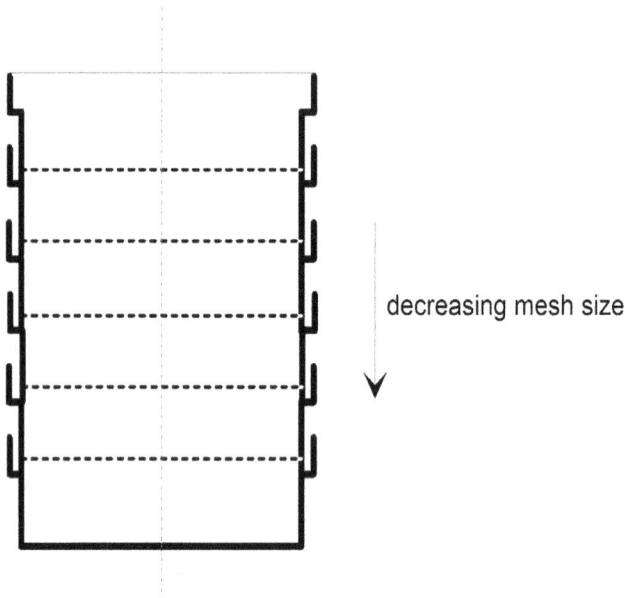

FIGURE 9.12   Arrangement of sieves for particle size analysis in the case of sieving.

The powder to be analyzed is put into the upper sieve bin. By shaking the apparatus for a certain period of time, particles that are smaller than the mesh size fall through the sieves and accumulate in the sieve bins according to the particle size. This suggests that for a series of sieves with mesh sizes of 1.6, 0.8, 0.4, 0.2, and 0.1 mm, particles in the size intervals > 1.6 mm, 0.8–1.6 mm, 0.4–0.8 mm, 0.2–0.4 mm, 0.1–0.2 mm, and < 0.1 mm are collected in the different sieve bins. Then the content in the different sieve bins is weighed, and a probability histogram of the mass fraction or probability density in each size interval is obtained as a function of the particle size. This corresponds to the discrete mass distribution discussed in Chapter 3. The mass fraction in each size interval is determined from $\Delta m_i/m$, where Δm_i is the mass of particles collected in the different sieve bins, and m is the total mass of the analysis sample. The mass probability density (frequency) distribution $f_m(D_i)$ is obtained from

$$f_m(D) = \lim_{\Delta D \to 0} \frac{\Delta m}{m\,\Delta D}$$

$$f_m(D_i) \simeq \frac{\Delta m_i}{m\,\Delta D_i} \tag{9.6}$$

where ΔD_i is the size interval $D_{i+1} - D_i$. A typical probability histogram of such an analysis is shown in Figure 9.13. The sample size for sieve analysis with sieves, which have a 200 mm bin diameter, usually range between 100 and 200 grams. The

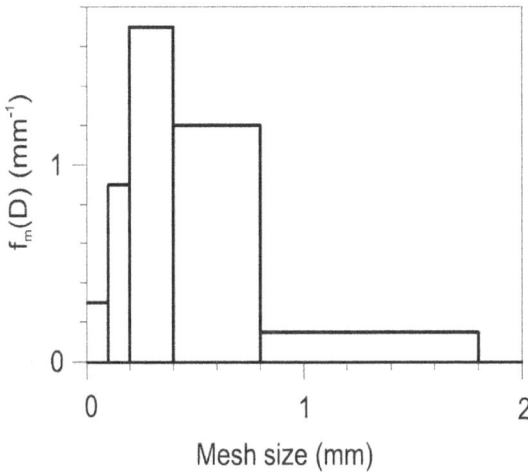

FIGURE 9.13 Probability histogram of a particle size distribution.

TABLE 9.3 Approximate Sieving Times for Different Mesh Sizes Using Hand or Machine Sieving

Mesh Size	Sieving Times
$> 160\ \mu m$	5–10 min
71–60 μm	10–20 min
40–71 μm	20–30 min

sieving time for dry powder depends on the mesh size and generally lies in the range specified in Table 9.3.

Machine sieving is usually performed by placing the set of sieves on a vibrating support. Hand or machine sieving of powders, which are cohesive or wet, may be enhanced by putting additional large granular materials on the sieves. This leads to de-agglomeration of the powder and avoids the blockage of the apertures; however, the sieving time is increased. Other sieving methods, such as air-jet sieving (see Table 9.2), should be used for powders, which are difficult to sieve. For air-jet sieving, only a single sieve is used, and only a single point on a cumulative mass distribution is obtained. For very fine and wet powders, roughly below 50 μm, wet sieving is recommended. Electroformed micromesh sieves are generally used for this method.

The characteristic particle diameter obtained by sieving analysis is a linear dimension and only corresponds to the diameter if spherical particles are analyzed. It should be noted that in case strongly non-spherical particles (having a small sphericity < 0.5) are measured, the throughput may include particles, which have a volume-equivalent diameter being larger than the mesh size. Therefore, such measurements suffer from significant uncertainty.

9.2.3 Sedimentation Methods

The principle of the *sedimentation method* is based on recording the sedimentation of powders dispersed in a liquid and deriving the particle size distribution from the measured temporal change of the particle concentration at a certain location (Fayed and Otten, 1997; Rhodes, 2008). Since the sedimentation process is governed by the free-fall velocity of the particles, a free-fall or Stokes diameter will be determined. Consequently, the obtained particle size is a physical equivalent diameter. Usually, small sedimentation columns are used to perform the measurement. Two methods may be applied to introduce the powder into the liquid. In the two-layer method, a thin layer of powder is introduced at the top of the liquid column; while in the second method, the powder is uniformly dispersed in the liquid by shaking the sedimentation column or by ultrasonic agitation before the measurement. In order to maintain a stabilized dispersion, wetting agents are quite often introduced into the liquid to reduce the attractive forces and to increase the repulsive forces between the particles in the suspension. Thereby, particle agglomeration is avoided, which eventually will yield larger and thereby wrong particle sizes.

Two approaches may be used to determine the particle size distribution using the sedimentation method. The first approach is the *incremental method*, in which the rate of change of particle number density or concentration is measured at a given location in the sedimentation column. Various methods, such as light attenuation or X-ray attenuation, can be applied for the measurement of the density or the concentration of the particles at the measurement location. In the second approach, the rate at which the particles accumulate at the bottom of the sedimentation column is measured. This method is called the *cumulative method* and also referred to as sedimentation balance. The remaining task in these measurements is to relate the measured properties, namely, particle concentration or sediment weight to a free-fall velocity and eventually a meaningful equivalent particle diameter. This is done mostly by using the drag coefficient of a spherical particle.

9.2.4 Cascade Impactor

For laboratory analysis of particle size distributions, a number of different flow-induced classification principles are used, such as aerosol centrifuges, elutriators, and different constructions of impactors. For all these principles, the particle classification or fractionation is based on the competition between a field force (i.e. gravity and centrifugal force) and a fluid force acting on the particle (typically the drag force). In a *cascade impactor*, the flow out of a small nozzle is deflected by a disc plate mounted a certain distance in front of the nozzle, and a stagnation point flow is established. The particles, depending on their size, are not able to follow this flow deflection and hit the plate where they should be deposited. Only smaller particles will follow the flow redirection and are carried to the next stage, as shown in Figure 9.14. Since the injected particles have normally a size distribution, each stage of an impactor has a distribution of the separation or collection efficiency (Hering, 2001), which should be as steep as possible to ensure a good quality classification. The collection efficiency is characterized by the cut size, where 50% of the mass are

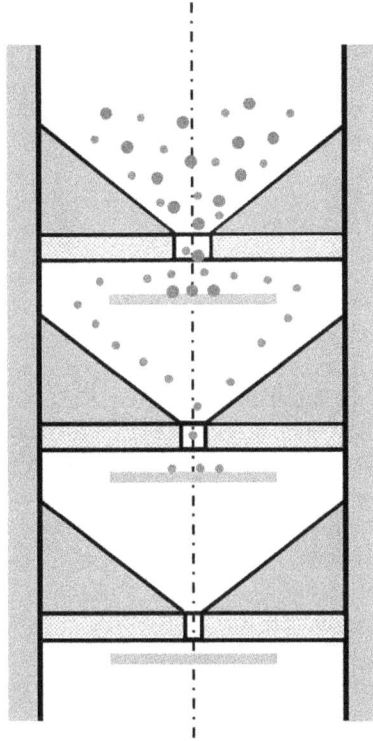

FIGURE 9.14   Principle of a single-orifice cascade impactor with three stages with flow from top to bottom for a binary particle mixture, that is, small and large particles.

collected on the plate. The recommended jet-to-plate spacing, normalized by the jet diameter, should be in the range between 1 and 5 in order to yield a cut size that is almost independent of the geometrical design (Hering, 2001). After a certain time, the plate is removed, and the collected particle weight is determined. This value divided by the injected powder mass is the separated particle fraction and yields one point on the particle size distribution. Since the particle motion through such a stagnation point flow is determined by the drag force, the obtained equivalent diameter is an aerodynamic diameter, which is determined by calibration with defined test aerosols.

Generally, cascade impactors have a number of stages, for example, about 5–9, in order to allow the determination of the entire size spectrum. Their design is based on the following: With increasing stage number (i.e. from top to bottom), the jet diameter is reduced, and the plate-to-nozzle distance is decreased. The flow through the impactor is induced by a suction pump, and the powder to be analyzed needs to be well dispersed. Then the particles entering the stages are accelerated by the converging nozzle and interact with the stagnation point flow. Heavy or large particles will hit the plate and stick to the surface, while smaller particles move into the next stage (Figure 9.14). Since their velocity becomes now larger and the distance to the collection plate reduces, the size of deposited particles is decreasing. Hence, the cut-size of each impactor element is continuously decreasing and should cover the entire aerosol particle size spectrum. The particle-laden flow through such a single-nozzle cascade impactor is determined by a Stokes number (Marple and Liu, 1974), defined as the ratio of particle response time to the characteristic flow timescale of the equipment

$$St = \frac{\tau_v}{\tau_c} = \frac{\rho_P D_P^2 Cu}{18\mu} \frac{u_j}{D_j} \tag{9.7}$$

Since very small particles are considered, the Cunningham correction factor for possible slip flow on the particle surface needs to be accounted for. The Stokes number decreases with reducing particle size. For the same particle size, the Stokes number increases when going down the stages of the impactor due to decreasing nozzle diameter and, hence, increasing jet exit velocity. Eventually, each stage gives a fractional deposition efficiency and, hence, a relative frequency distribution or by integration also a cumulative distribution based on the quantity particle mass or volume. The aerodynamic particle diameter has to be obtained from a calibration, for example, with spherical aerosols.

A problem in the operation of cascade impactors is the possible rebound of the particles from the collector discs. The sticking probability depends on the plate surface characteristics and the particle size (see Section 7.3). To avoid this common-in-practice problem, coating is applied to the collector plates (Hering, 2001). There is significant influence of the kind of coating on the deposition efficiency.

The nozzle Reynolds number ($Re = (\varrho_{air} u_j D_j)/\mu_{air}$) is an important criterion for the operation of a cascade impactor. The recommended range of Reynolds numbers is $500 < Re < 3000$. At lower Re-numbers, very broad collection efficiency curves were reported (Hering, 2001), most likely due to developing vortical structures. Beyond $Re = 3,000$ turbulence effects will modify very fine particle transport, and again, a broadening of the collection efficiency curve is observed (Hering, 2001). In the recommended range, the collection efficiency curves are rather steep, which is an indication of a very good classification.

Since the aerodynamic response of the particles is the governing physical effect transporting the particles, it is also possible to adapt the range of obtained cut sizes through the modification of the system pressure. With a reduction of the system pressure using partial vacuum, the cut size may be reduced to the nanometer range, for example, 50 nm as studied by Hering et al. (1978) for an eight-stage small impactor.

In the past, numerous other empirical principles of cascade impactors have been developed (Marple, 2004), depending on the type of application. A well-known impactor type is the Andersen cascade impactor, typically having six stages, where the particle-laden flow passes through a kind of sieve plates having multiple orifices (initially these were 400 orifices) with decreasing diameter and increasing number from top to bottom. Hence, multiple jets hit the collector plates, which are Petri dishes containing a culture medium in order to ensure that the hitting particles are also sticking.

In addition to the common use of cascade impactors for laboratory analysis of fine particle samples and environmental applications, another very important application is the characterization of aerosols released from medical inhalers via mimicked breathing processes (Mitchell and Nagel, 2003). For such aerosols' characterization, the so-called "next generation impactor" is used, where the different stages have horizontal planar layouts with seven stages, with an increasing number of nozzles and decreasing nozzle diameters. The cut sizes of these stages are between 0.5 μm and 5 μm, a range typical for inhalable aerosols.

9.2.5 ELECTRIC SENSING ZONE METHOD (COULTER PRINCIPLE)

The principle of electric sensing zone methods is based on the disturbance of an electrical field by a particle passing a defined probe volume. The disturbance can be related to the particle size by calibrating the system with mono-dispersed particles. A well-known instrument based on this principle is the Coulter counter, illustrated in Figure 9.15, which originally was used in medicine to count blood cells and subsequently modified to allow measurements of particle size and number.

The particles are suspended in an electrolyte and forced to pass through the small orifice by a pumping system. Electrodes are immersed in the fluid on both sides of the orifice and produce an electrical field. A particle passing the orifice will change the electrical impedance and generate a voltage pulse. The amplitude of the pulse is proportional to the volume of the particle, as shown in Figure 9.16. The Coulter counter principle belongs to the class of single particle counting techniques, together with other optical methods described in Section 9.4.3. Therefore, the pulse amplitude can be uniquely related to the particle size when only one particle is in the probe volume at one time. To avoid plugging the orifice, the diameter of the orifice needs to be larger than the largest particle in the sample.

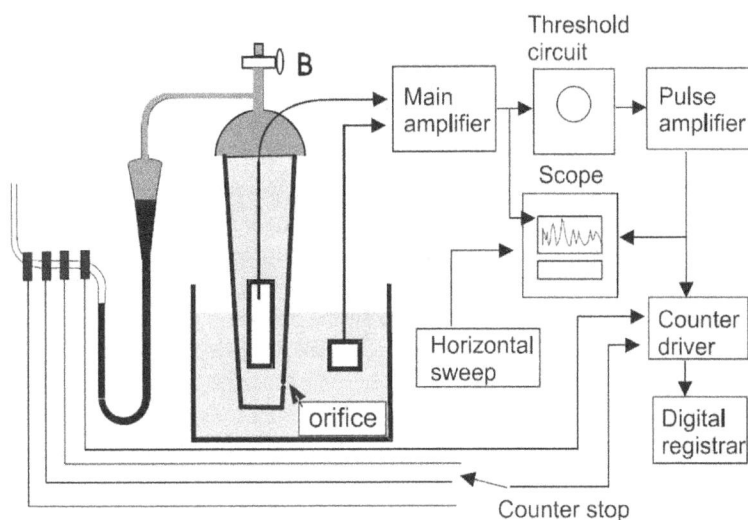

FIGURE 9.15   Principle of electrical sensing zone method, that is, Coulter Counter.

Source: (from *Particle Size Measurement, Vol. 1*, Allen, T., Chapter 9, p. 329, Figure 9.2, Springer and Chapman and Hall, 1997, with kind permission of Springer Science-Business Media B.V.)

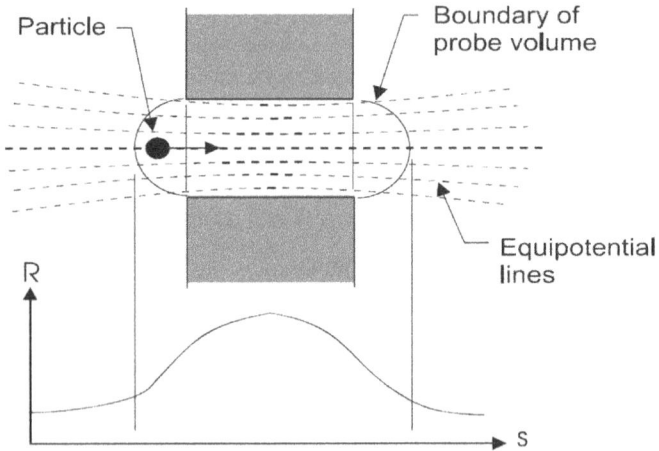

FIGURE 9.16 Probe volume and modification of resistance by particle in volume for Coulter principle.

The probe volume is the volume of the orifice and a certain region outside the orifice where the signal amplitude is above the noise level, as shown in Figure 9.16. The measurable particle size range is typically between 0.02 and 0.6 of the orifice diameter. This results in a dynamic range of 1:30. By using multiple measurement cells with different orifice diameters, a particle size range of 0.5 to 1,200 μm can be resolved. The highest measurable particle size is determined by the limitation due to keeping the particles in suspension. Large particles with density larger than that of the electrolyte tend to settle out of the suspension. A solution to this problem is to increase the viscosity of the electrolyte or to apply stirrers in the reservoirs. The lower limit of detectable particle size is determined by the electronic noise of the system.

Before the measurement, a vacuum is applied to the glass tube, shown in Figure 9.15, to produce an imbalance in the mercury siphon. After closing the valve (B), the flow through the orifice (A) is created as the mercury siphon is returned to equilibrium. The measurement is initiated and terminated by means of start and stop switches on the siphon. The resistance across the orifice is recorded by means of the electrodes mounted in the glass tube and in the reservoir, where the glass tube is immersed. The voltage pulses generated by the modification of the resistance across the orifice as particles pass through are amplified, sized, and counted. By setting an appropriate threshold level, only pulses above the noise level are detected and counted. This threshold level also determines the dimension of the effective probe volume as depicted in Figure 9.16.

The counts are collected in a number of pre-defined amplitude classes. By using a calibration curve, the measured amplitude is related to a particle size. Hence, the measurement yields an equivalent diameter of a particle, which gives the same voltage pulse as the calibration particles. More details about the theory of the Coulter principle and signal processing can be found in Allen (1990).

9.2.6 LASER-DIFFRACTION METHOD

The principle of laser light diffraction is mainly used for the determination of particle size distributions. Such instruments are quite complex and, therefore, used in the laboratory for an offline analysis. For simple situations also, an online measurement by laser diffraction is possible, for example, for an integral characterization of spraying systems.

The light scattering of particles considerably larger than the wavelength may be described by *geometrical optics* and *Fraunhofer diffraction theory*. It has been demonstrated for this regime that most of the light is scattered in the forward direction as a result of diffraction. The resulting forward scattering intensity pattern depends on particle size (as in Figure 9.3), so the analysis of the scattering pattern may be used to infer information about particle size distribution. The principle of laser diffraction has been used to analyze particle samples and two-phase flows, such as sprays (Swithenbank *et al.*, 1977; Hirleman *et al.*, 1984; Sijs *et al.*, 2021). A number of commercial instruments is based on this principle.

Fraunhofer diffraction is one limit of the basic *Fresnel-Kirchhoff theory* of diffraction, which describes the interaction of a monochromatic light beam with an aperture. In the *Fraunhofer limit*, the diffraction pattern of an aperture is the same for an opaque object with identical cross-section area and shape, except for the shadow produced by the object. In the far-field, however, the diffraction pattern is much larger than the geometrical image. Two requirements have to be met in order to obtain the Fraunhofer limit (Weiner, 1984): (1) The area of the object must be smaller than the product of the wavelength of light and the distance from the point source of the light to the diffracting object, and (2) the area of the object must be smaller than the product of the wavelength and the distance between the object and the observation plane. The first requirement is easily met for a parallel light beam since the point source can be considered to be at infinity. In order to fulfill the second requirement, the detection plane must be positioned far away from the object. Therefore, Fraunhofer diffraction is also known as far-field diffraction. Alternatively, a lens may be used to focus the diffracted light onto a photo-detector positioned in the focal plane of the lens. For such an optical configuration, the undeflected light is brought to a point focus on the axis (i.e. center of the detector), and the diffracted light is focused around this central spot, as shown in Figure 9.17. Therefore, the diffraction

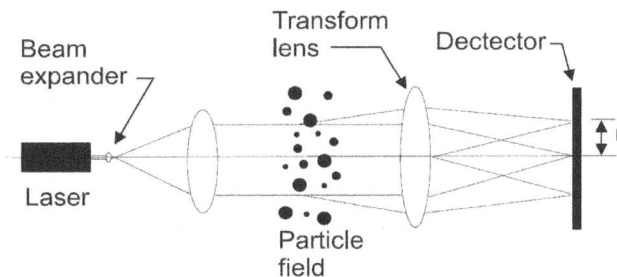

FIGURE 9.17 Optical arrangement of instruments for particle sizing by laser diffraction.

pattern will also be stationary for a moving object, and hence, the size measurement is not biased by the particle velocity, which is an important fact for in-process two-phase flow analysis. The intensity distribution of the diffraction pattern in the detector plane for a particle with diameter D may be obtained from the expression

$$I(x) = I_0 \left[\frac{2J_1(x)}{x} \right]^2 \tag{9.8}$$

where I_0 is the intensity at the center of the pattern, and J_1 is the first order spherical Bessel function. The parameter x is a function of the particle diameter and the optical configuration

$$x = \frac{\pi Dr}{\lambda \hat{f}} \tag{9.9}$$

where r is the radial distance from the center of the detection plane, $\hat{f}$ is the focal length of the transform lens, and λ is the wavelength of the incident light. By introducing

$$\sin \theta = \frac{r}{\hat{f}} \tag{9.10}$$

and using the fact that θ is usually very small, that is, $\sin \theta \approx \theta$, one obtains

$$x = \frac{\pi D\theta}{\lambda}. \tag{9.11}$$

The intensity at the center of the diffraction pattern is given by

$$I_o = cI_{inc} \frac{\pi^2 D^4}{16\lambda^2} \tag{9.12}$$

where c is a constant of proportionality and I_{inc} is the intensity of the incident light. This results in the following equation for the intensity pattern

$$I(D_p, \theta) = cI_{inc} \frac{\pi^2 D^4}{16\lambda^2} \left[\frac{2J_1\left(\frac{\pi D\theta}{\lambda}\right)}{\frac{\pi D\theta}{\lambda}} \right]^2 \tag{9.13}$$

This functional relationship is known as the *Airy function* and is plotted in Figure 9.18. The result shows that the diffraction pattern consists of a series of bright and dark concentric rings surrounding the central spot of non-diffracted light with a radius, which depends only on the dimensionless parameter x.

However, for online measurements of particle size distributions, the measurement of the radial intensity distribution is very cumbersome. Therefore, the photodetector consists of a number of concentric ring elements, as shown in Figure 9.19, and the light energy over only a finite area between r_i and r_{i+1} needs to be measured

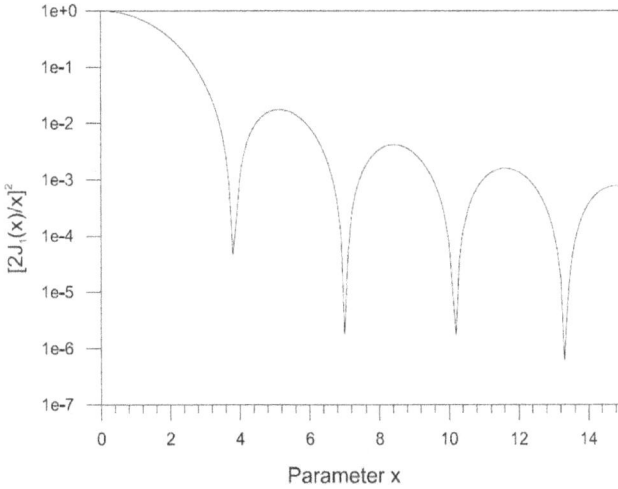

FIGURE 9.18 Fraunhofer diffraction pattern for a circular aperture or an opaque disk according to Eq. (9.13).

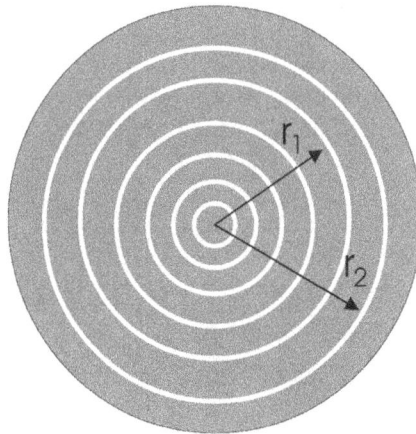

FIGURE 9.19 Illustration of ring detector element for the measurement of diffraction pattern.

and analyzed. By integrating Eq. (9.13) from r_i and r_{i+1}, one obtains the light energy received by the detector element

$$E_{i,i+1} = c\pi D^2 \left\{ \left[J_o^2(x) + J_1^2(x) \right]_i - \left[J_o^2(x) + J_1^2(x) \right]_{i+1} \right\} \qquad (9.14)$$

The constant c depends on the power of the light source and the sensitivity of the photodetector.

Since the diffraction analysis is done not only for one particle being in the laser beam but for a suspension (i.e. particles dispersed in air or liquid) with a certain particle size distribution, the light energy falling on one ring element has to be multiplied by the number of particles. For a particle size distribution with M size classes, each with N_j particles, one obtains by neglecting multiple scattering

$$E_{i,i+1} = c\pi \sum_{j=1}^{M} N_j D_j^2 \left\{ \left[J_o^2\left(x_j\right) + J_1^2\left(x_j\right) \right]_i - \left[J_o^2\left(x_j\right) + J_1^2\left(x_j\right) \right]_{i+1} \right\} \quad (9.15)$$

Where x_j corresponds to the value of x for the particle size D_j. This equation allows one to convert the measured light energy to the particle number distribution represented by M size classes each with N_j particles. The number of resolved particle size classes, M, is equal to the number of ring elements, L. Therefore, Eq. (9.15) has to be solved L-times, for each ring element. The lower and upper limits of the size classes depend on the magnitude of the smallest and largest ring radii.

The set of L equations is usually solved using the least-squares criteria. Initial values of N_i are either estimated from the raw data or calculated from an assumed functional form of the particle size distribution, e.g. Rosin-Rammler distribution function. Using Eq. (9.15), L light energy values are calculated and compared with the measured light energy values. The assumed N_j-values are then corrected, and the final result is obtained iteratively by minimizing the least-square error.

A laser light diffraction measurement requires careful preparation of the sample by dispersing the particles either in air or liquid. The dispersion must ensure that agglomerates in the sample break up into the primary particles in order to allow the measurement of their size distribution. The specified sizing range of laser diffraction instruments is mostly between about 20 nm to 3,500 μm. Such a wide sizing range is, however, only possible by combining small particle static light scattering measurements with laser light diffraction pattern recording for large particles describable by geometrical optics.

9.3   ONLINE INTEGRAL METHODS

Integral methods are characterized as those measurement techniques, which provide time-resolved but spatially integrated properties of dispersed flowing two-phase systems. Such methods are based on the distortion or attenuation of some energy source such as light, sound, or atomic radiation passing through a particle-laden mixture. Online integral methods are nonintrusive but provide only information integrated along the beam of the energy source. This is frequently an advantage for process control in industry, for example, in monitoring the performance of a cyclone separator or any other gas cleaning device with respect to particle concentration. In the following sections, the most common integral methods for applications to dispersed two-phase flows are summarized.

9.3.1   LIGHT ATTENUATION

The intensity of a light beam passing through a fluid-particle mixture will be attenuated due to scattering and absorption by all the particles within the beam. The intensity of the transmitted light is recorded by a photodetector, and the change of the photodetector resistance can be monitored by an appropriate electronic circuit. According to the Lambert-Beer law, the light attenuation by a suspension of mono-disperse particles is given by the expression

$$\frac{I}{I_o} = \exp\left(-k\frac{\pi}{4}D^2 nL\right) \tag{9.16}$$

where k is a constant of proportionality known as the extinction coefficient, D is the particle diameter, n is the particle number density (i.e. particles/m^3), and L is the optical path through the suspension, as shown in Figure 9.20. Eq. (9.16) shows that the light extinction depends on two properties of the particle phase: The particle number concentration and the particle diameter. Hence, one of these properties has to be known to enable the measurement of the other property. Introducing $\bar{\rho}_d = nm$ into Eq. (9.16) gives

$$\ln\left(\frac{I}{I_o}\right) = -k'\left(\frac{L}{D}\right)\left(\frac{\bar{\rho}_d}{\rho_d}\right) \tag{9.17}$$

The change in light intensity I for a change in particle bulk density $\bar{\rho}_d$ is given for constant values of L and D by the equation

$$\frac{dI}{I} = -k'\left(\frac{L}{D}\right)\frac{1}{\rho_d}d\bar{\rho}_d \tag{9.18}$$

Hence, for a given geometry, particle composition and particle diameter (i.e. particle diameter distribution), the change in light intensity is proportional to the change in particle cloud density so one has

$$\frac{dI}{I} = -K d\bar{\rho}_d \tag{9.19}$$

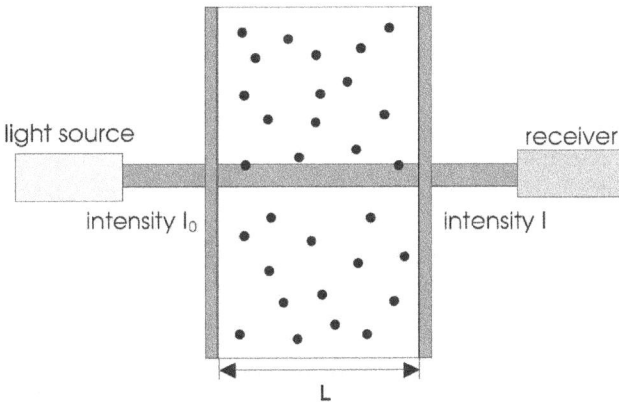

FIGURE 9.20   Principle of light attenuation method emitted light source intensity I_0 and received attenuated intensity I.

The constant of proportionality, K, depends on the geometry of the flow system and properties of the particle material, such as refractive index, material density, ρ_d, and diameter distribution. Therefore, the application of the light attenuation method requires the determination of the calibration constant K for each particle composition considered. For polydisperse particles, the dependence of the extinction coefficient on the particle size may result in larger measurement errors for the determination of the particle bulk density. Since the method requires optical access to the flow system, the buildup of particles at the windows must be either negligible or nonvarying during the measurement.

As mentioned previously, the particle number concentration obtained by the light attenuation method provides an integral value along the optical path through the flow system. In many situations, this concentration is not representative for the entire cross-section since there will always be a concentration distribution, as in a particle-laden jet. In this case, the particle concentration distribution along the optical path of a light beam through the centerline of the jet follows a normal distribution function. Hence, the value for the concentration obtained by light attenuation is some type of average value of this distribution. For the determination of the local particle concentration based on the principle of light attenuation, fiber optical probes may be used, as will be described in Section 9.4.2.

9.3.2 CROSS-CORRELATION METHOD

The cross-correlation technique allows the determination of the mean transit time of a flowing medium passing two sensors located at a fixed distance apart, as for example in a pipe or channel. An average velocity can be obtained as the ratio of the sensor separation L to the transit time T_t

$$V_0 = \frac{L}{T_t} \tag{9.20}$$

The averaging region depends on the way the relevant quantity is measured, for example, by a wall probe or a transmitting sensor. This principle has been used for the design of flow meters (Beck and Plaskowski, 1987). The cross-correlation technique may also be used to measure an average velocity in a flowing two-phase mixture, such as in pneumatic conveying (Kipphan, 1977; Williams *et al.*, 1991), as illustrated in Figure 9.21a. The technique requires monitoring the fluctuations of any property of the dispersed phase with two sensors. The type of sensor may be based on a variety of methods, such as ultrasound attenuation, light attenuation or scattering, electrostatic charge variation, and conductance or capacitance. In most of these situations, the resulting signals are proportional to the temporal variation of a particle concentration. If the sensors are not positioned too far apart, the two signals generated by the fluctuation of particle concentration are, to some extent, identical or strongly correlated, but time shifted as shown in Figure 9.21b. The time delay between the two signals $x(t)$ and $y(t)$ can be determined efficiently by computing the *cross-correlation function* of both signals over a certain measurement time period, T_m. The cross-correlation function for the time delay, τ, is obtained from the expression

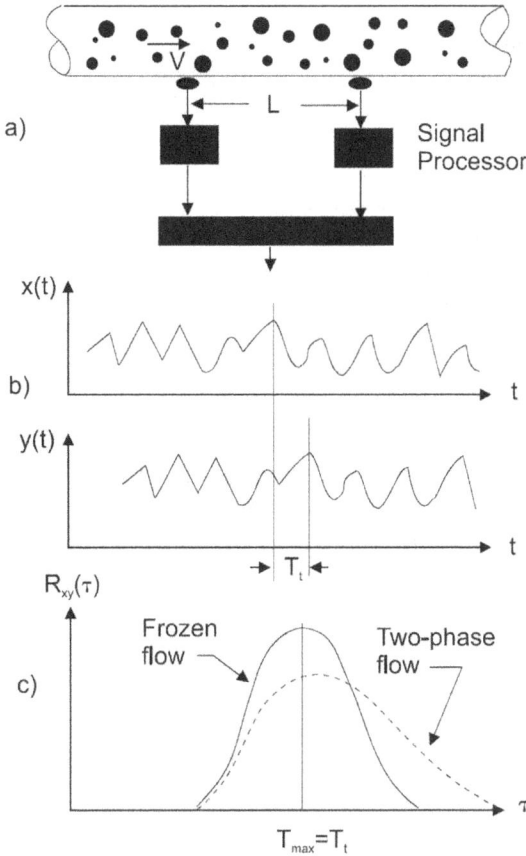

FIGURE 9.21 Cross-correlation method for particle velocity measurements: (a) Pneumatic conveying line and sensor locations, (b) signals from both sensors, and (c) cross-correlations functions.

$$R_{xy}(\tau) = \frac{1}{T_m} \int_0^{T_m} x(t-\tau) y(t) \, dt \tag{9.21}$$

In an ideal situation, that is, a frozen concentration pattern passing across the sensors, the signal $x(t)$ is identical to the signal $y(t)$, but shifted by the transit time T_t, that is, $y(t) = x(t - T_t)$. In this case, the shape of the cross-correlation function is identical to the auto-correlation function (Kipphan, 1977)

$$R_{xx}(\tau) = \frac{1}{T_m} \int_0^{T_m} x(t-\tau) x(t) \, dt \tag{9.22}$$

and shifted by the transit time, T_t, as illustrated in Figure 9.21c. In real situations, and especially in turbulent two-phase flows, the particle concentration signals are of stochastic nature, which results in a broadening of the cross-correlation function, as shown in Figure 9.21c. In these cases, the transit time is obtained from the location of the maximum in the cross-correlation function T_{max} as demonstrated by Kipphan (1977) using different models for the transport process. Kipphan (1977) also showed that the broadening of the cross-correlation function in comparison with the *autocorrelation function* may be used to estimate the average particle velocity fluctuations. It is also obvious that the degree of correlation of the signals depends on the separation of the sensors and the intensity of the concentration fluctuation or particle fluctuating motion. With an increase of the sensor spacing, the degree of correlation decreases. In order to achieve a maximum degree of correlation with a minimum scatter of the data, Kipphan (1977) suggests the following optimum sensor separation

$$L_{opt} \approx 0.35 \frac{b}{\sigma_v / V_0}$$ (9.23)

where b is the linear dimension or diameter of the sensor in the mainstream direction, and σ_v is the mean intensity of the particle velocity fluctuation.

Since most of the sensor principles mentioned previously enable the estimation of the particle concentration or bulk density, it is also possible to determine the particle mass flow rate from the expression

$$\dot{m}_p = \bar{\rho}_d V_0 A$$ (9.24)

where A is the cross-section of the sensing region perpendicular to the mainstream flow direction. To what extent the measured integral properties represent the real mean values in the flow system strongly depends on the type of sensor and the homogeneity of the dispersed phase. Considering horizontal pneumatic or hydraulic conveying, segregation effects due to gravitational settling may yield false measurements, as illustrated in Figure 9.22. A horizontal arrangement of the two opposed sensors (e.g. based on the light attenuation method) will completely underestimate

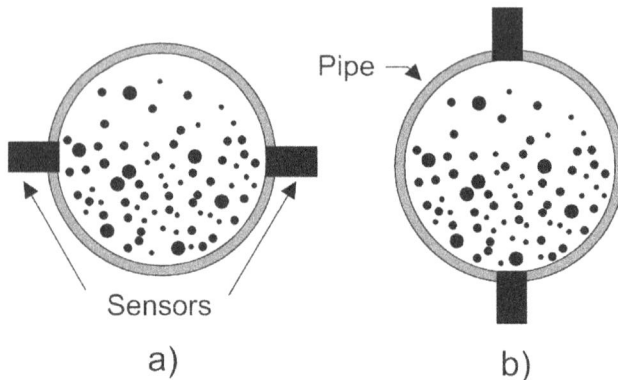

FIGURE 9.22   Sensor arrangements in horizontal pneumatic conveying: (a) Horizontal sensor arrangement, (b) vertical sensor arrangement.

the particle mass flow rate, while a vertical arrangement results in an overestimation. Hence, accurate measurements are only possible when the condition of the dispersed phase in the instrument probe volume is representative of the entire cross-section in the flow domain. Such a condition can only be realized in vertical conveying and when the sensors are placed far enough away from pipe bends or feeding systems to ensure that the mixture is homogeneous.

9.4   LOCAL MEASUREMENT TECHNIQUES

Local measurement techniques enable the determination of two-phase flow properties with a relatively high spatial resolution, depending on the method applied. According to Table 9.1, local measurement techniques may be divided into three groups: Probing methods (e.g. isokinetic sampling), single particle counting methods (e.g. light scattering, laser-Doppler and phase-Doppler anemometry), and field imaging techniques (e.g. particle image velocimetry). The latter method is grouped into local measurement techniques since the measured property can be exactly correlated with a location in space. Probing methods are intrusive and may disturb the flow considerably depending on the application and the dimension of the facility. They are, however, quite robust and widely used in industry for process control. The other methods described in this chapter are nonintrusive optical methods. These optical methods may be also divided according to the following two measurement principles:

1. Single particle counting methods: The particle phase properties and the respective distribution functions are obtained for those particles passing through a small finite probe volume within a given measurement time. The result is weighted by the particle number flux.
2. Field imaging techniques: The distribution functions of the particle phase properties are obtained for those particles, which happen to be in the probe volume (i.e. the probe volume is normally a relatively large light sheet) at a given instant of time. Hence, the result is weighted by the particle number concentration.

Therefore, these two measurement principles yield different results, which can be related, as demonstrated by Umhauer *et al.* (1990).

9.4.1   ISOKINETIC SAMPLING

Isokinetic sampling can be used to measure particle mass flux, concentration, and density in a flowing suspension. The collected particles may be further analyzed offline in the laboratory to obtain the particle size distribution or other properties, such as surface area and shape factor, using methods described in Section 9.2.

The basic requirement for measuring local particle flux is that the sample extracted must be representative of the suspension at the sample point in the two-phase stream, which is not an easy task when using isokinetic sampling. The principle of this method is based on inserting a sampling probe into the flowing two-phase system and gathering a representative sample of particles by a suction fan, as shown in Figure 9.23. The sampled particles are collected in a bag filter for a defined time interval, and then the particle mass is obtained by weighing the bag filter with and without particles. Hence, a particle mass flux is obtained, which, under certain conditions, may be related to the local particle flux, concentration, or density of the flowing suspension.

The first problem to be considered with isokinetic sampling is that the suction velocity u_s established in the sampling probe should be identical with the local gas velocity (i.e. isokinetic sampling condition). This may be achieved by adjusting the suction velocity in such a way that the static pressures outside and inside the sampling probe become identical, as shown in Figure 9.23.

If isokinetic conditions are not achieved, the region from where particles are collected is not identical with the cross-section of the sampling probe, as shown in Figure 9.24 (Soo et al., 1969). When the suction velocity is lower than the local fluid velocity, the region from where the particles are collected is smaller than the cross-section of the probe, as seen in Figure 9.24a. When the suction velocity is higher, the particles are collected from a larger area, as illustrated in Figure 9.24b. The boundary streamline of the flow and the boundary trajectory of the particles are not the same when the suction velocity does not match the local fluid velocity because of particle inertia, which is determined primarily by particle size and Stokes number. In this case, it is not possible to accurately determine the particle mass flux or concentration.

The determination of particle mass flux and bulk density by isokinetic sampling will be described in more detail. All the definitions used for this analysis are specified in Figure 9.25. The area A_d is the capture area for the particles and the area A_0 is the area of the continuous phase in the free stream that enters the probe. The mass flow rate of particles through the capture area in the free stream is equal to the mass flow rate in the probe.

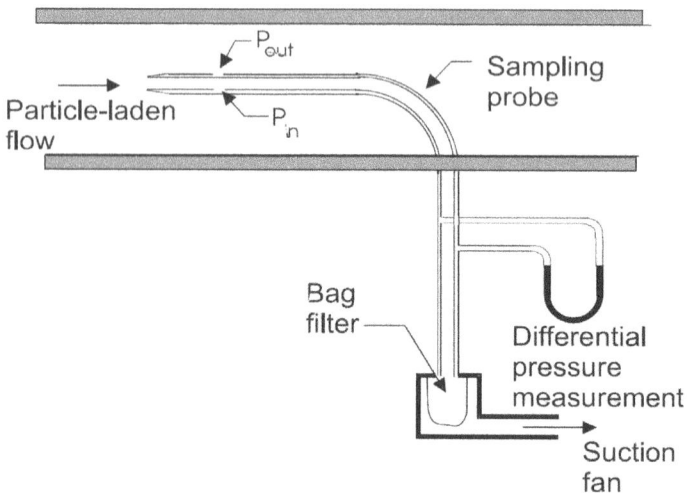

FIGURE 9.23 Schematic diagram for the operation of isokinetic sampling probes.

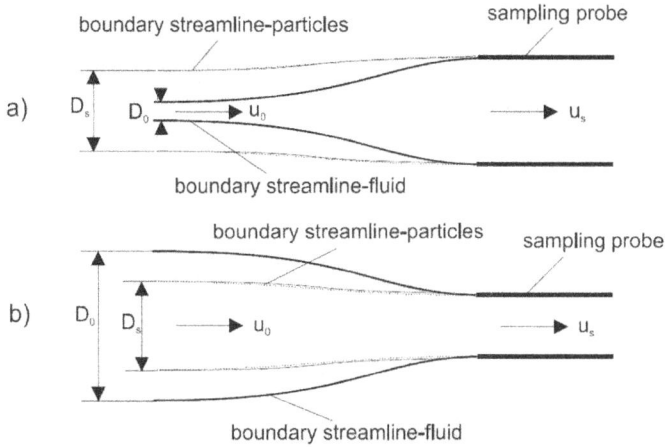

FIGURE 9.24   Streamline and particle trajectories for different sampling conditions: (a) $u_s < u_0$, (b) $u_s > u_0$.

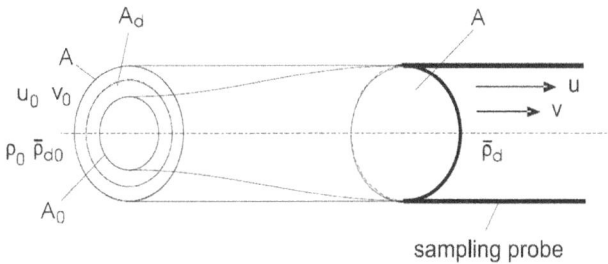

FIGURE 9.25   Definitions of particle-laden flow properties for an isokinetic sampling method.

$$\dot{m}_d = \bar{\rho}_{d,0} v_0 A_d = \bar{\rho}_d v A \tag{9.25}$$

Under isokinetic sampling conditions, $A_d = A$ and $v_0 = v$ yield

$$\frac{\bar{\rho}_{d,0}}{\bar{\rho}_d} = 1 \tag{9.26}$$

If ΔM_d is the mass of material collected on the bag filter during time Δt, then the particle mass flux is

$$\bar{\rho}_{d,0} v_0 = \frac{\dot{M}_d}{A} = \frac{\Delta M_d}{A \Delta t} \tag{9.27}$$

The free stream bulk density is

$$\bar{\rho}_{d,0} = \frac{\Delta M_d}{v_0 A \Delta t} \tag{9.28}$$

Thus, an additional measurement is needed for the particle velocity in the free stream.

Under non-isokinetic conditions, the ratio of the particle mass flux in the free stream to the mass flux in the sampling probe is

$$\frac{\bar{\rho}_{d,0} v_0}{\bar{\rho}_d v} = \frac{A}{A_d} \tag{9.29}$$

where the area ratio A/A_d is not necessarily known. However, if the particles have high inertia and travel in rectilinear trajectories, then $A_d = A$ and

$$\frac{\bar{\rho}_{d,0} v_0}{\bar{\rho}_d v} = 1 \tag{9.30}$$

The mass flux in the probe is measured directly from the accumulation in the filter using Eq. (9.27). The determination of the bulk density requires data on the free stream particle velocity.

If it is assumed that the particles are in velocity equilibrium with the conveying fluid ($u = v$) and that the conveying fluid is incompressible, then the continuity equation between the free stream and the probe can be written as

$$v_0 A_0 = v A \tag{9.31}$$

and the equation for the ratio bulk densities becomes

$$\frac{\bar{\rho}_{d,0}}{\bar{\rho}_d} = \frac{A_0}{A_d} \tag{9.32}$$

Bohnet (1973) performed an analysis based on the previous assumptions and developed correlations for the area ratio A_0/A_d. Since the particle trajectories around the sampling probe depend on particle inertia, Bohnet (1973) correlated his results with a non-dimensional Stokes number defined by

$$St_{Pr} = B = \frac{u_0 \tau_v}{D_{Pr}} = \frac{\rho_d D^2 u_0}{18 \mu D_{Pr}} \tag{9.33}$$

where D_{Pr} is the probe diameter. A diagram of the relative dust concentration plotted versus the ratio of sampling velocity to flow velocity is shown in Figure 9.26 with B as a parameter, which is the Stokes number of the particles moving around the probe head. It is obvious that the relative dust concentration increases with larger deviations from the isokinetic condition and larger values of the parameter B.

Isokinetic sampling cannot give accurate values for the particle cloud density in a carrier phase with a velocity gradient, that is, shear flow or wall boundary layer, since the flow around the sampling probe becomes asymmetric, and isokinetic conditions can hardly be established. In addition, wide particle size distributions will alter the motion of the different sized particles whereby it is also not possible to adapt accurate isokinetic conditions.

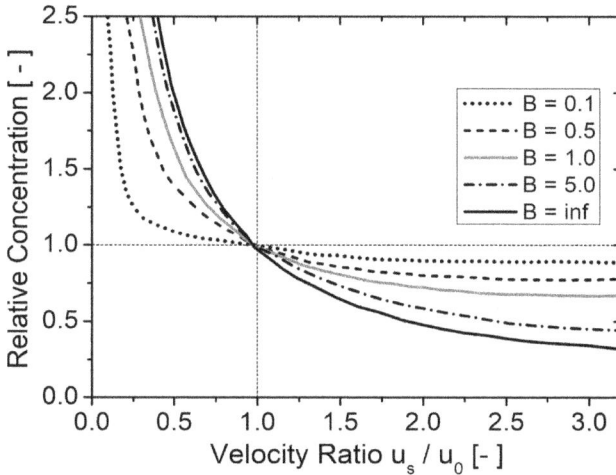

FIGURE 9.26 Measured dust concentration in the probe divided by the real concentration in the two-phase system (relative error) as a function of the velocity ratio (velocity in the probe/velocity in the flow system) for different values of the Stokes number of the probe.

Source: (from Bohnet, M., *1973*, Staubgehaltsbestimmung in strömenden Gasen, *Chemie-Ing.-Tech.*, 1973, **45**, 22. Copyright Wiley-VCH Verlag GmBH & Co. KGaA. Reproduced with permission.)

9.4.2 OPTICAL FIBER PROBES

Measurement systems based on optical fiber probes have been used in various configurations to perform local measurements of particle velocity, size, and concentration in two-phase flows. This includes particle-laden flows but also bubbly flows (Cartellier, 1992; Boyer *et al.*, 2002). An optical fiber probe system consists of the probe head, which is inserted into the flow; a light source; a photodetector; and the signal processing unit. The probe construction may differ in the arrangement of the light emitting fiber, which is connected to the light source, and the light receiving fiber, which transmits the light to the photodetector. Two different measurement principles may be identified:

1. Light attenuation method
2. Light scattering or reflection method

The principle of operation also determines the arrangement of the emitting and receiving fibers. In the light attenuation method, the optical fibers are arranged opposite to each other, and the light propagates a fixed distance through the two-phase mixture before reaching the receiving fiber, as shown in Figure 9.27a. Particle concentration measurements using this optical fiber arrangement may be based on the Lambert-Beer law for light attenuation (see also Section 9.3.1), or on counting individual particles passing through the probe volume between the two fibers. The method that is most suitable depends on the ratio of fiber diameter to particle diameter. The particle counting technique may be applied when the particle diameter is

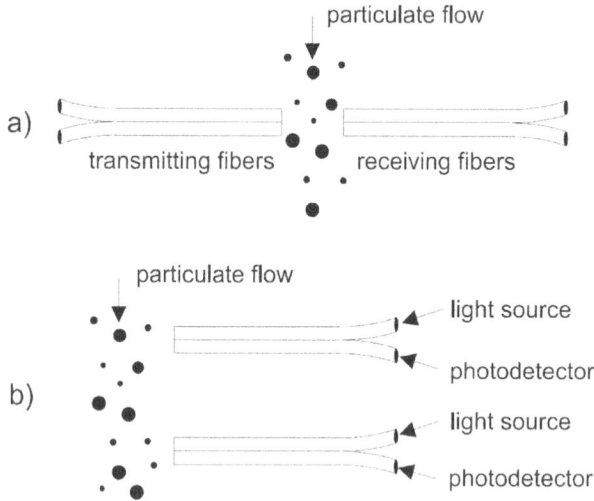

FIGURE 9.27 Typical arrangement of fiber optic probes: (a) light attenuation, (b) light scattering (reflection) method.

of the same order as the fiber diameter so that the instrument operates as a single particle counter.

Fiber probes based on the reflection from particles are constructed in such a way that the transmitting and receiving fibers are mounted parallel in the probe head, as shown in Figure 9.27b. Particles moving in front of the probe head are illuminated by the transmitting fiber and scatter light in the backward direction. This light is received by a second fiber, which is usually mounted in line with the emitting fiber.

Instead of separate transmitting and receiving fibers, a single fiber may also be used. In this case, the light from the light source is coupled into the sensor fiber by a fiber optical beam splitter, and the light scattered by the particles is received by the same sensor fiber, diverted by the beam splitter and transmitted to the photodetector. Such single fiber reflection probes have been used by Lischer and Louge (1992) and Rensner and Werther (1992) for measurements in dense particle-laden two-phase flows, such as fluidized beds. The advantages of the single fiber reflection probe are its small size (typically smaller than 1 mm in diameter) and the ability to withstand erosion, which may become a problem in gas-solid flows. On the other hand, it is only possible to measure particle concentration with this method. This requires an estimate of the effective probe volume dimensions, that is, the area from which the scattered light is received by the probe. A detailed analysis of the effective measuring volume of single fiber reflection probes was performed by Rensner and Werther (1992).

Particle velocity measurements by fiber optical probes are possible without calibration, based on the passage time of the particles between two successive sensors or the frequency method, which requires a special arrangement of a number of emitting and receiving fibers in the probe head. The *passage time method* is based on the cross-correlation of the signals received by two sets of emitting and receiving

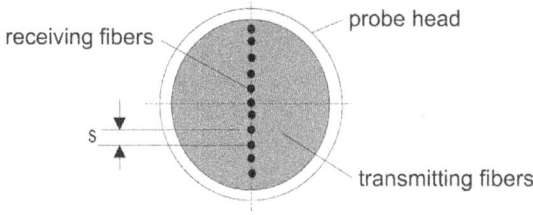

FIGURE 9.28   Cross-section of fiber optic probe used for the frequency method.

fibers and can be realized for both the light attenuation and the light scattering method, except for the single fiber reflection probe, which only allows for concentration measurements. The cross-correlation method requires that the two sensor pairs are aligned with the main flow direction in order to achieve a high degree of correlation. The required separation of the sensor pairs has to fulfill certain requirements in order to give a high degree of correlation, as recommended by Kipphan (1977). A more detailed description of the cross-correlation method has been given in Section 9.3.2.

The frequency method for measuring particle velocities (Petrak and Hoffmann, 1985) is based on using a number of illuminating fibers combined with a regular arrangement of receiving fibers at a given separation, as illustrated in Figure 9.28. All the receiving fibers are connected to one photodetector; therefore, the line array of the receiving fibers acts as a spatial grid whereby the scattered light is modulated so that the signal from the photodetector exhibits a sinusoidal shape with a dominant frequency f_0. This frequency can be related to the particle velocity by the expression

$$v = f_0 \, s \tag{9.34}$$

where s is the spacing between the receiving fibers. From the arrangement of the receiving fibers, it is obvious that the frequency method requires particle motion almost parallel to the line of fibers; otherwise, the signal will have only a few zero-crossings, which make an evaluation of the signal frequency difficult.

Comprehensive reviews on the measurement techniques applicable to fluidized bed analysis and process control were provided by Werther (1999) and van Ommen and Mudde (2008). The reviews summarize the operation of heat flux meters, sampling probes, optical probes for particle size, concentration and velocity measurements, as well as tomographic methods, which are capable of imaging an entire cross-section of the particulate two-phase flow field.

9.4.3   LIGHT SCATTERING INSTRUMENTS

The fundamentals of light scattering from spherical particles were presented in Section 9.1. This chapter is related to the technical details for accurate light scattering measurements and instruments. With light scattering instruments, it is possible to measure particle concentration, since the scattered light intensity is proportional to particle size and the number of particles within the established probe volume. It is possible to measure particles sizes based on light scattering intensity when the instrument operates as a single particle counting instrument, that is, only a single particle is in the probe volume at any time. Such a size measurement is accomplished using a calibration curve that measures a scattering intensity physical equivalent diameter. For both applications, it is essential to know exactly the dimensions of the probe volume.

In principle, the measuring volume dimensions can be fixed in two ways: The first method is based on directing a narrow particle-laden gas or liquid stream through the center portion of the measurement volume, as shown in Figure 9.29a. In this case, the diameter of the stream determines the measurement volume size, not the optical arrangement. This method is also called *aerodynamic* or *hydrodynamic focusing* and can only be applied by sampling the particles from the two-phase system under consideration, that is, by isokinetic sampling or a bypass system. The second method for defining the measurement volume is based on an appropriate optical design of the transmitting and receiving optics using imaging masks in order to allow the fixing of the probe volume, as shown in Figure 9.29b. It is also possible to apply the light scattering instrument for an online determination of particle size and concentration. Again, it should be remembered that particle sizing requires a single particle to pass though the probe volume.

The size of the measurement volume has to be selected in such a way that, for a given maximum particle number concentration, the coincidence error is significantly

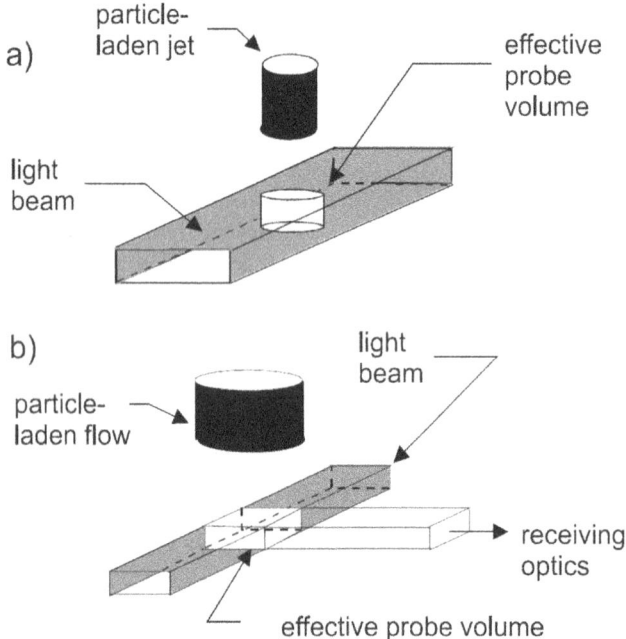

FIGURE 9.29 Methods to differentiate the size of the measurement volume: (a) Aerodynamic or hydrodynamic focusing, (b) optical demarcation of the probe volume by using light beam shaping masks.

reduced. The probability P_k for the presence of k-particles in the measurement volume follows a Poisson distribution

$$P_k = \frac{N^k}{k!} e^{-N} \tag{9.35}$$

where $N = n V_m$ is the number of particles in the probe volume, V_m. The relative probability for the presence of two particles is the following

$$P_2' = \frac{P_2}{P_1} = \frac{N}{2} \tag{9.36}$$

For a maximum allowable coincidence error of 5% (i.e. $P_2' = 0.05$), the limiting averaged particle number in the measurement volume must be $N = 0.1$. This results in a maximum particle number concentration of

$$n_{max} = \frac{0.1}{V_m} \tag{9.37}$$

This equation gives an estimate of the required size of the measurement volume for a given particle concentration. Especially for online measurements, this criterion is quite often a limiting factor, so rather small measurement volumes must be used, which are only possible by a specific optical configuration. Therefore, quite often, off-axis orientations of the receiving optics have to be selected in order to allow a better definition of the measurement volume. This yields lower scattering intensities compared to forward scattering arrangements (shown in Figure 9.3) and may limit the lower detectable particle size.

The measurement volume must be illuminated uniformly, and this can be achieved by properly shaping the light beam using masks in the transmitting optics. A more complex approach is the elimination of the boundary zone error through coincidence and amplitude discrimination, as illustrated in Figure 9.30. Particles passing through the edge of the illuminating light beam (i.e. particle 2 in Figure 9.30) will

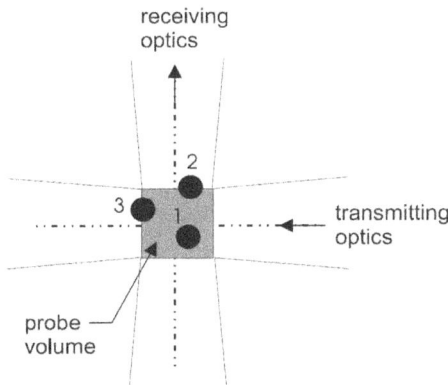

FIGURE 9.30 The boundary zone error occurring for particle sizing by light scattering intensity.

be only partly illuminated, and hence, their scattered light intensity will be too low and not proportional to their size. If the scattering intensity is still above the detection level (i.e. trigger level), such particles will be detected as smaller particles. For particles passing the edge of the beam, which is imaged into the photodetector (i.e. particle 3 in Figure 9.30), only a portion of the scattered light will be received, and again their size will be underestimated. The boundary zone error can be eliminated by extended optical systems with two collocated measurement volumes.

A typical optical setup for a particle size analyzer operating with a 90° scattering angle together with the signal processing system (Umhauer, 1983) is shown in Figure 9.31. Although this is an instrument not available in the market, it illustrates the required complexity in the optical and electronic system for allowing accurate particle size measurements. The transmitting optics consist of a white light source, a condenser, an imaging mask, and an imaging lens. The scattered light is focused into the photomultiplier using a lens system and an imaging mask, which limits the area from which scattered light is collected.

The photomultiplier provides an analog signal, which is first filtered to remove signal noise and then digitized to obtain the pulse height of the signal. A counter is used to determine the number of signals detected. The data are then acquired by a computer and further processed to determine the particle size distribution. The number frequency distribution is obtained by grouping the pulse height U into a number of classes of width ΔU_i as follows

$$f_n\left(U_i\right) = \frac{1}{N} \frac{\Delta N\left(U_i\right)}{\Delta U_i} \qquad (9.38)$$

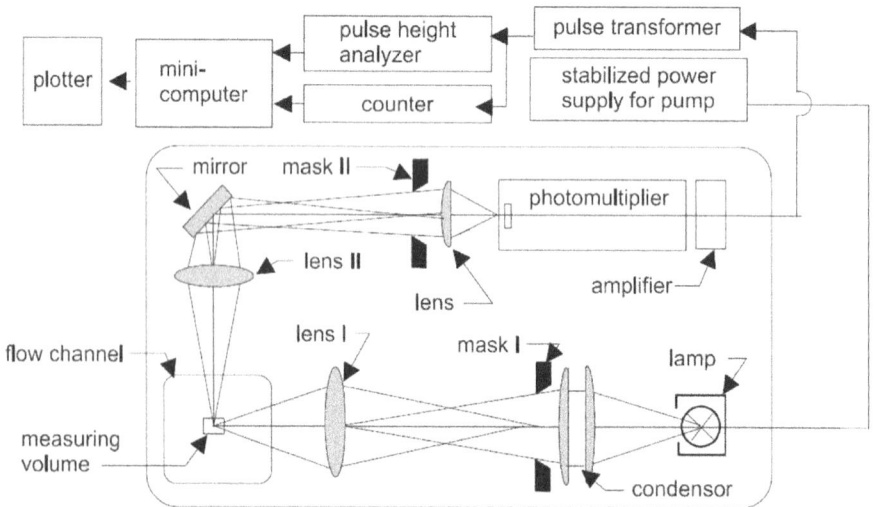

FIGURE 9.31   Optical configuration of a particle size analyzer operating at 90° scattering angle and signal processing.

Source: (from *J. Aerosol Sci.*, **14**, Umhauer, H., Particle size distribution analysis by scattered light measurements using an optically defined measuring volume, 765, 1983, with kind permission from Elsevier.)

where $\Delta N(U_i)$ is the number of samples acquired in the amplitude interval i, and N is the total number of samples acquired during the measuring time. In general, a calibration curve for U as a function of the particle size, D, must be used to transform the distribution with respect to pulse height into a number frequency size distribution.

It is possible to determine the particle number flux in the direction perpendicular to the probe volume cross-section from

$$\dot{n}_{tot} = \sum_{i=1}^{I} \dot{n}_i = \sum_{i=1}^{I} \frac{\Delta N(D_i)}{t_m A_m} = \frac{N}{t_m A_m} \tag{9.39}$$

where I is the number of size classes, t_m is the measuring time, and A_m is the cross-section of the measurement volume. By multiplying the number flux $\dot{n}_i$ by the particle mass $m_{d,i}$, the particle mass flux can also be obtained.

Generally, a calibration curve has to be used to relate the pulse height (i.e. scattering amplitude) to a characteristic particle size, such as the diameter of a spherical particle or an equivalent scattering diameter. A determination of the calibration curve using calculations based on the Mie theory is only possible for the rather limited case of homogeneous and smooth spherical particles.

The light scattering instrument introduced by Kiselev et al. (2005) allows for an in situ measurement of aerosol sizes using an optically defined dimension of the probe volume. This is achieved using two receiving optics in forward scattering direction. The use of a white light source yields the integration and smoothing of the scattering intensity as a function of the aerosol size (Eq. 9.3). A wide angle scattered light collection is used in order to integrate over the strong angular variations of scattering intensity. Hence, a smooth response curve is obtained for water droplets with sizes below about 1 μm (Kiselev et al. (2005).

9.4.4 LASER-DOPPLER ANEMOMETRY

Laser-Doppler and phase-Doppler anemometry,[1] LDA and PDA, respectively, are the most advanced and accurate nonintrusive measuring techniques to measure the velocities of the fluid and the particles. These measuring techniques enable one to determine instantaneous (i.e. time series) and time-averaged measurements of particle velocities with a high spatial resolution. In order to tag the fluid velocities, the flow needs to be seeded with fine tracer particle, which closely flow the fluid flow fluctuations, especially in turbulent flows. Using PDA, the size of spherical particles, the refractive index of the particle, and the particle concentration can be determined accurately as well.

The physical principle underlying LDA and PDA for velocity measurements is the *Doppler effect*, which relates the interaction of sound or light waves with a moving observer or the modulation of sound or light waves received by a stationary observer from a moving emitter. In this application, the moving emitter is the particle passing through the light beam. In LDA, this principle is used in such a way that a laser emits plane light waves, which are received and transmitted from the moving particles. Hence, the frequency or wavelength of the light received by the particle is already modulated. Since the moving particles scatter the light into space, an additional

Doppler shift occurs when the scattered light is received from a stationary observer, as shown in Figure 9.32. Hence, the frequency of light received at the photodetector can be determined from the following equation

$$f_r = f_e \frac{1 - \dfrac{\vec{v} \cdot \vec{l}}{c}}{1 - \dfrac{\vec{v} \cdot \vec{k}}{c}} \tag{9.40}$$

where f_e is the frequency of the laser source (emitter), $\vec{v}$ is the velocity of the moving particle, c is the speed of the light, and $\vec{k}$, $\vec{l}$ are unit vectors in the directions shown in Figure 9.32. The frequency of the scattered light f_r (at the receiver) is, however, too high to allow direct detection by a photodetector. Therefore, two different methods are used in LDA so that the frequency of light to be detected is considerably reduced: The *reference beam method* and the *Doppler frequency difference method.*

Frequency detection is achieved in the reference beam method by illuminating the particle with a strong light beam and interfere the resulting scattered light with a weak reference beam from the laser light source on the photodetector, as shown in Figure 9.33. Subtracting the frequency of the reference beam gives the Doppler frequency

$$f_D = f_r - f_e \tag{9.41}$$

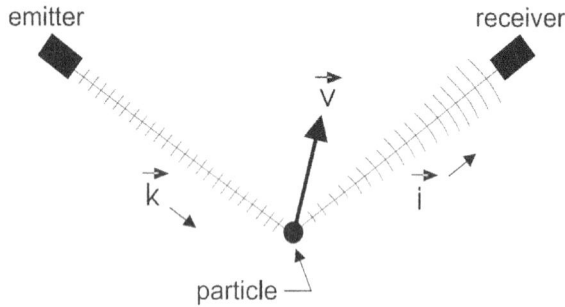

FIGURE 9.32   Doppler shift of scattered light from a moving particle.

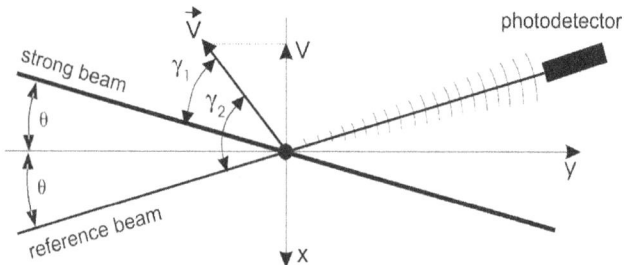

FIGURE 9.33   Configuration of reference beam LDA system.

Using Eq. (9.40) and introducing $\vec{v} \cdot \vec{k} = -|\vec{v}| \cos \gamma_1$ and $\vec{v} \cdot \vec{l} = -|\vec{v}| \cos \gamma_2$ the last expression yields

$$f_D = f_e \frac{1 + \dfrac{|\vec{v}|}{c} \cos \gamma_1}{1 + \dfrac{|\vec{v}|}{c} \cos \gamma_2} - f_e \tag{9.42}$$

and

$$f_D = \frac{1}{\lambda_e} \left| \frac{|\vec{v}| (\cos \gamma_1 - \cos \gamma_2)}{1 + \dfrac{|\vec{v}|}{c} \cos \gamma_2} \right| \tag{9.43}$$

The velocity component perpendicular to the bisector of the two incident beams is

$$v = -|\vec{v}| \sin \frac{\gamma_1 + \gamma_2}{2}$$

Using the trigonometric relationship,

$$\cos \gamma_1 - \cos \gamma_2 = -2 \sin \frac{\gamma_1 + \gamma_2}{2} \sin \frac{\gamma_1 - \gamma_2}{2}$$

one obtains

$$f_D = \frac{1}{\lambda_e} \left| \frac{2v \sin \theta}{1 + \dfrac{|\vec{v}|}{c} \cos \gamma_2} \right| \tag{9.44}$$

where $\theta = \dfrac{1}{2} (\gamma_1 - \gamma_2)$. Since in general, $|\vec{v}| \ll c$, Eq. (9.44) finally becomes

$$f_D = \left(\frac{2v \sin \theta}{\lambda_e} \right) \tag{9.45}$$

It should be noted that the reference beam mode can only be operated at a fixed observation angle, which coincides with the reference beam angle θ. Furthermore, the solid angle for light collection is limited to satisfy coherence requirements, that is, the amount of collected scattered light is restricted.

The Doppler frequency difference method is used for LDA measurements more frequently. The moving particle is illuminated by two laser beams from different directions, as shown in Figure 9.34. In this case, the frequency of the scattered light is obtained from the difference of the contributions from the two incident beams

$$f_D = f_{r1} - f_{r2} \tag{9.46}$$

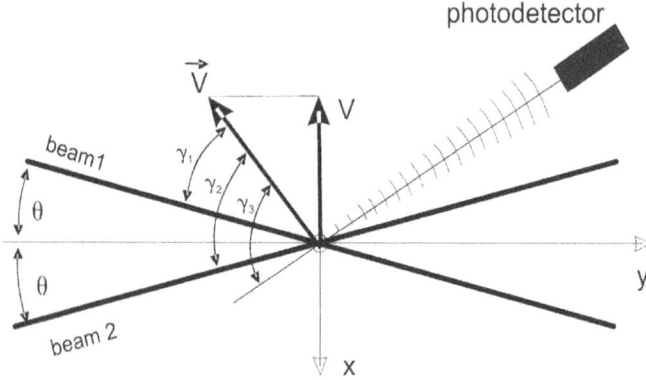

FIGURE 9.34 Configuration of Doppler difference frequency method (dual-beam LDA system).

Using once again Eq. (9.40) and the velocity components in the directions of the two beams, one obtains

$$f_D = f_e \left(\frac{1 + \left(\frac{|\vec{v}|}{c}\right) \cos \gamma_1 \quad 1 + \left(\frac{|\vec{v}|}{c}\right) \cos \gamma_2}{1 + \left(\frac{|\vec{v}|}{c}\right) \cos \gamma_3 \quad 1 + \left(\frac{|\vec{v}|}{c}\right) \cos \gamma_3} \right) \tag{9.47}$$

For the velocity component perpendicular to the bisector of the two incident beams and with $|\vec{v}| \ll c$, Eq. (9.47) yields

$$f_D = \frac{2v \sin \theta}{\lambda_e} \tag{9.48}$$

This expression is identical to the one obtained for the reference beam method. However, the observation angle can be arbitrarily selected in the Doppler difference method. This implies that the observation angle and solid angle of scattered light collection may be selected to fit a desired application.

The principle of the LDA may be also explained using the *fringe model* in the following way. If two coherent light beams cross, the interference of the light waves results in a fringe pattern parallel to the bisector plane, that is, the y-z plane, shown in Figure 9.35, which can be visualized on a screen when a lens of small focal length is placed at the intersection region of the beams. As the particle passes through the LDA probe volume, the scattering intensity detected by a photodetector is modulated in such a way that the Gaussian-shaped absolute scattering intensity (which results from the Gaussian intensity distribution in the probe volume) is superimposed with an alternating pattern produced by the particles passing through the bright and dark fringe pattern. However, as pointed out by Durst (1982), the fringe pattern does not exist for the particle and is the result of integration by the human eye and the photodetector, both of which have a response time much larger than the inverse of the

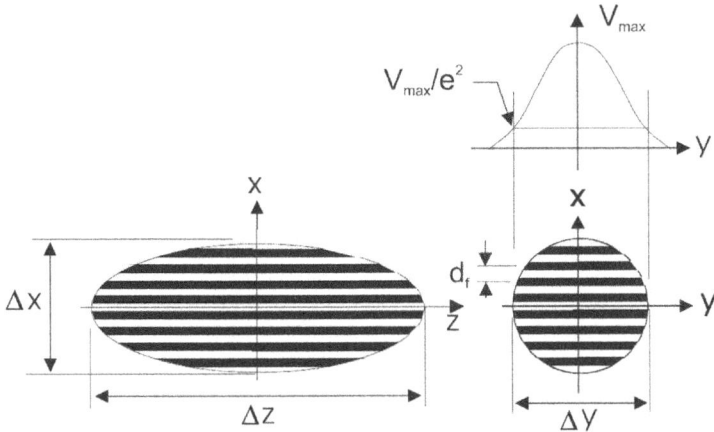

FIGURE 9.35   Dimensions of LDA probe volume and illustration of the fringe model.

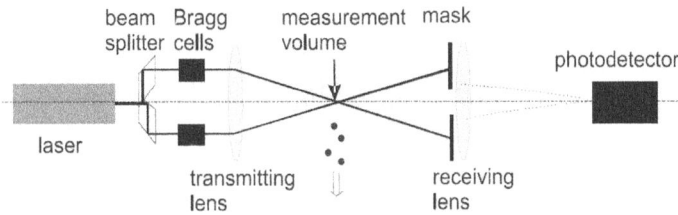

FIGURE 9.36   Typical optical setup of the dual beam LDA.

frequency of the light waves. The fringe spacing d_f is basically the conversion factor to determine the particle velocity from the measured Doppler difference frequency as follows

$$v = f_D \frac{\lambda_e}{2\sin\theta} = f_D d_f \qquad (9.49)$$

For particles smaller than the fringe spacing, a completely modulated Doppler signal will be generated, as shown in Figure 9.40, as high visibility. As the particle becomes larger than the fringe spacing, the signal modulation is reduced, and the scattering intensity received by the photodetector does not reduce to zero in the Doppler signal (low visibility region in Figure 9.40)

A typical optical setup of an LDA system operated in the forward-scattering mode is shown in Figure 9.36. The transmitting optics consist of the laser, a beam splitter, one or two Bragg cells, and a transmitting lens. The Bragg cells introduce a frequency difference between the two incident beams, and this makes it possible to detect the direction of particle motion in the measurement volume (Durst *et al.*, 1981). The receiving optics consists of an imaging lens with a mask in front of it and a photodetector with a pinhole.

The spatial resolution of the velocity measurement depends on the dimensions of the LDA probe volume, which is determined by the initial laser beam diameter, the beam crossing angle (determined by the initial beam spacing and focal length of the transmitting lens), the focal length of the receiving lens, and the observation angle of the receiving optics. Since the incident, focused laser beams have a Gaussian intensity distribution, as shown in Figure 9.35, their waist diameter at the focal plane is taken to be that value at which the light intensity has diminished to $1/e^2$ of the maximum value at the beam axis. Thus, the waist diameter is given by the expression

$$d_m = \frac{4\hat{f}_e \lambda_e}{\pi d_0} \tag{9.50}$$

where d_0 is the $1/e^2$ unfocused laser beam diameter, and $\hat{f}_e$ is the transmitting lens focal length. The probe volume established by the two crossing beams has an ellipsoidal shape, as shown in Figure 9.35, and the dimensions of the $1/e^2$ ellipsoid are as follows

$$\Delta x = \frac{d_m}{\cos\theta} \tag{9.51a}$$

$$\Delta y = d_m \tag{9.51b}$$

$$\Delta z = \frac{d_m}{\sin\theta} \tag{9.51c}$$

The number of fringes in the $1/e^2$ measurement volume can be determined from

$$N_f = \frac{\Delta x}{d_f} = \frac{d_m}{\cos\theta}\frac{2\sin\theta}{\lambda_e} = \frac{8}{\pi}\frac{\hat{f}_e}{d_0}\tan\theta = \frac{4}{\pi}\frac{\Delta b}{d_o} \tag{9.52}$$

where Δb is the initial spacing of the transmitting beams.

By using an off-axis orientation of the receiving optics, the length of the portion of the measurement volume imaged into the photodetector can be further reduced and the spatial resolution improved. However, due to the angular dependence of the light scattering intensity, any off-axis orientation of the receiving optics results in reduced scattering intensities, as mentioned in Section 9.1.

For more details about the principle of the LDA, frequency shifting, properties of photodetectors, and signal processing methods, the reader is referred to the relevant literature, such as Durst et al. (1981, 1987) and Tropea (2011). A review of signal processing methods is found in Tropea (1995). Several interesting applications of LDA for two-phase flow measurements and some further fundamental issues with measurements on dispersed particulate flows are described in the following paragraphs.

The basic ideas for LDA applications for two-phase flows were introduced by Farmer (1972, 1974), Durst and Zaré (1975), and Roberts (1977). They showed that LDA may also be used for velocity measurements of large reflecting and refracting particles. The light waves produced by the two incident laser beams either reflect or refract at large particles, as indicated in Figure 9.37, generate fringes projected

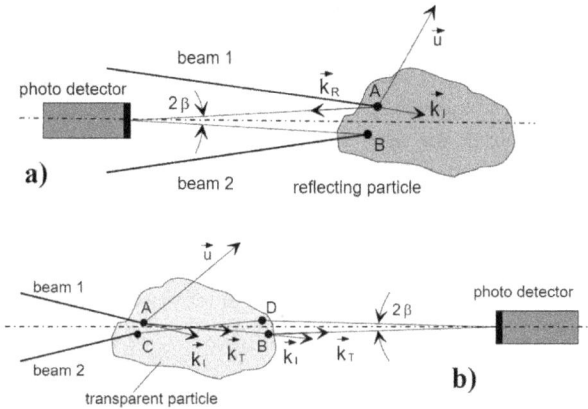

FIGURE 9.37 Interference of two laser beams for large (a) reflecting and (b) refracting particles. The intersection points of the incident and refracting beams with the particle surface are indicated by A, B, C, and D.

into space. The rate at which the fringes cross any point in space, that is, at the photodetector, is the same at all points in the surrounding space and is linearly related to the velocity component of large non-deformable particles perpendicular to the symmetry line between the two incident beams. The theoretical derivations of Durst and Zaré (1975) revealed that the relations for the Doppler difference frequency for large reflecting or refracting particles are identical to the universal equation for laser-Doppler anemometry (Eq. 9.48) when the intersection angle of the two incident beams is small and the photo-detector is placed at a distance much larger than the particle diameter from the measurement volume.

These findings are the basis for the application of LDA for dispersed particle velocity measurements in two-phase flows. Because LDA is a nonintrusive optical technique, it may be used for measurements in two-phase flow systems as long as optical access is possible and the two-phase system is dilute enough to allow the transmission of the laser beam and the scattered light. Numerous studies have been published in the past where LDA has been applied to various types of dispersed gas-solid two-phase flows, liquid sprays, and bubbly flows. In liquid systems with dispersed solid particles at higher concentration a refractive index matching approach may be also applied (Wiederseiner *et al.*, 2011; Poelma, 2020). In order to adapt the liquid refractive index accurately, matching that of the dispersed solid particles, a very accurate temperature control is required.

There also have been several attempts to apply laser-Doppler anemometry for the simultaneous measurement of particle velocity, size, and concentration (Farmer, 1972; Chigier *et al.*, 1979; Durst, 1982; Hess, 1984; Hess and Espinosa, 1984; Allano *et al.*, 1984; Negus and Drain, 1982). The sizing of particles by LDA is based on two principles:

1. The absolute value of the scattering intensity (i.e. pedestal of the Doppler signal, Figure 9.39 and 9.40)
2. The signal visibility, Eq. (9.64) (see Figure 9.40)

The *pedestal* of the Doppler signal is the low frequency component of the signal obtained by using a low pass filter, as shown in Figure 9.39. As introduced in Section 9.1, the intensity of the scattered light depends on the particle size. However, the size-amplitude relation shows strong fluctuations in the Mie region when the particle size is comparable to the laser wavelength. Particle sizing based on intensity measurements generally requires calibration. An additional problem with sizing particles by a standard LDA system is the effect of the non-uniform distribution of light intensity within the measurement volume. Laser beams normally have a Gaussian intensity distribution, as depicted in Figure 9.38. Particles passing through the edge of the measurement volume emit a lower scattering intensity and are detected as smaller

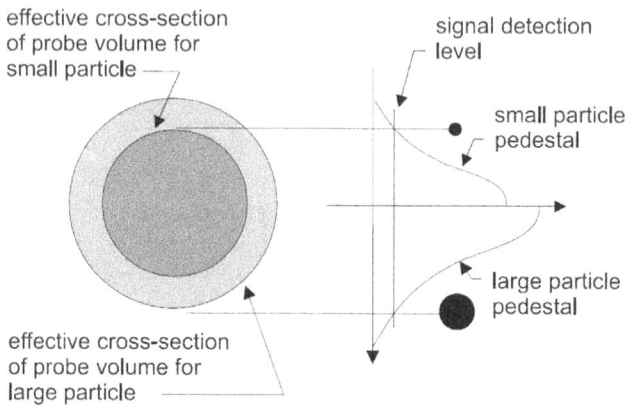

FIGURE 9.38 The Gaussian beam effect on intensity measurements by LDA and its influence on the effective cross-section of the measurement volume for small and large particles.

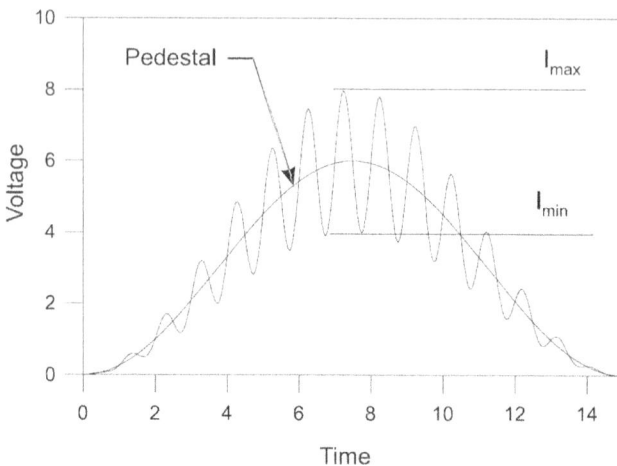

FIGURE 9.39 Shape of an ideal Doppler signal and definition of pedestal.

particles. This effect, which is also called *trajectory ambiguity*, or *Gaussian beam effect*, results in the effective dimensions of the probe volume being dependent on the particle size. A small particle passing the edge of the measurement volume might not be detected by the data acquisition system due to its low scattering intensity, while a large particle at the same location still produces a signal, which lies above the detection level, as shown in Figure 9.38. This effect also has consequences for an accurate determination of particle concentration. Therefore, measurements of particle size and concentration by LDA require extensions of the optical system or data acquisition procedures in order to reduce the errors associated with the Gaussian beam effect.

In addition, signal modulation may be used for particle sizing by LDA (Farmer, 1972). Compared to the scattering intensity measurements, the method has a number of advantages since the visibility does not depend on the scattering intensity and, hence, is neither biased by laser power nor detector sensitivity. The visibility is determined from the maximum and minimum amplitudes of the low-pass filtered Doppler signal as indicated in Figure 9.40 and obtained from the equation

$$V = \frac{I_{max} - I_{min}}{I_{max} + I_{min}} \tag{9.53}$$

The visibility of the Doppler signal decreases with increasing particle size, as illustrated in Figure 9.40. The first lobe in the visibility curve covers the measurable particle size range. With a further increase in particle size, secondary maxima appear in the visibility curve.

The visibility curve strongly depends on the optical configuration of the receiving optics, that is, the off-axis angle and the size and shape of the imaging mask in

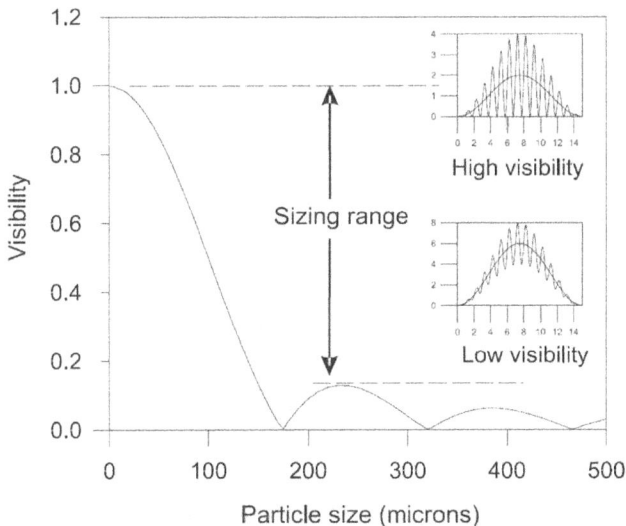

FIGURE 9.40 Variation of signal visibility with particle size (off-axis collection) and illustration of signals with low and high visibility.

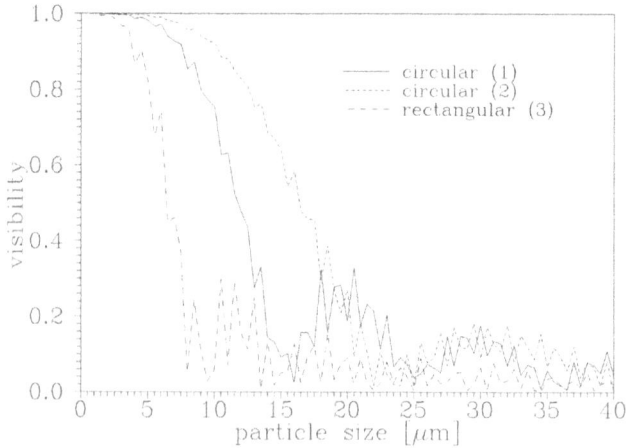

FIGURE 9.41 Visibility curves for different optical configurations of the receiving optics ($\lambda = 632.8$ nm, $\varphi = 0°$; (1) $d_f = 10.2$ μm circular mask, $\Delta\delta = 4°$; (2) $d_f = 18$ μm circular mask, $\Delta\delta = 4°$; (3) $d_f = 6.55$ μm rectangular mask, receiving aperture angle in horizontal and vertical direction, $\Delta\delta_h = 11°$, $\Delta\delta_v = 4°$).

the receiving optics. The latter effect was evaluated in detail by Negus and Drain (1982). Example Mie calculations for the visibility curves for different optical configurations are shown in Figure 9.41. It is obvious that the shape of the imaging mask significantly influences the measurable size range. The measurable particle size range is significantly increased by using an off-axis arrangement of the receiving optics (see also Chigier, 1991).

Extensive research has been performed on the suitability of the visibility method for particle sizing (Adrian and Orloff, 1977). It was found that this method appears to be very sensitive to the positioning of the aperture mask, the accuracy of the mask dimensions, and the particle trajectory through the LDA probe volume. Therefore, additional optical systems are needed to enable a combination of visibility and signal amplitude for particle size evaluation. A detailed review on the visibility method was contributed by Tayali and Bates (1990), where a number of other LDA-based methods are also described.

The following methods have been used in the past for particle sizing by LDA:

1. Limitation of the probe volume size by additional optical systems (i.e. gate photodetector or two-color systems)
2. Modification of the laser beam to produce a "top-hat" intensity distribution
3. Combined measurements of visibility and pedestal amplitude

There are several examples of particle size measurements using LDA, which were developed before the introduction of PDA. Most of the techniques, based on intensity and visibility measurements described in the next section, were shown not to be very reliable even though, in a few cases, they were used to develop commercial instruments. It should be emphasized, however, that there is still a need for reliable instruments for local, single-point size and velocity measurements in two-phase flows with non-spherical particles common to industrial processes.

In order to limit the region of the probe volume from where signals are received, Chigier *et al.* (1979) used an additional receiving optics placed at 90° off-axis to trigger the LDA receiving system mounted in forward-scattering mode. For further reduction of the trajectory ambiguity, a sophisticated signal processing system was developed. A comparison of particle size distribution measurements by LDA with results obtained by the slide impaction method gave only fair agreement.

By superimposing two probe volumes of different diameters and color, it is possible to trigger the data acquisition system only when the particles pass through the central part of the larger probe volume where the intensity is more uniform, as depicted in Figure 9.42. Such a co-axial arrangement of the two probe volumes may be realized by using a two-component LDA system with a different waist diameter for each color (Yeoman *et al.*, 1982; Modarress and Tan, 1983) or by overlapping a large diameter single beam with the LDA probe volume (Hess, 1984; Morikita *et al.*, 1994). When a particle passes through the LDA probe volume, the light-scattering intensity from the larger single beam is measured to determine the particle size, as shown in Figure 9.42.

The "top-hat" technique may be applied to generate laser light beams with uniform intensity distribution. Allano *et al.* (1984) used a holographic filter and related the scattering intensity with the particle diameter by using the Lorenz-Mie theory. Simultaneous measurements of particle size and velocity with such a system appear in the study by Grehan and Gouesbet (1986). Also, a combination of LDA with light scattering instruments has been applied for simultaneous particle size and velocity measurements (Durst, 1982).

Because of the particle size-dependent dimensions of the probe volume (see Figure 9.38) and the difficulties associated with particle size measurements using LDA, the measurement of particle concentration is generally based on a calibration

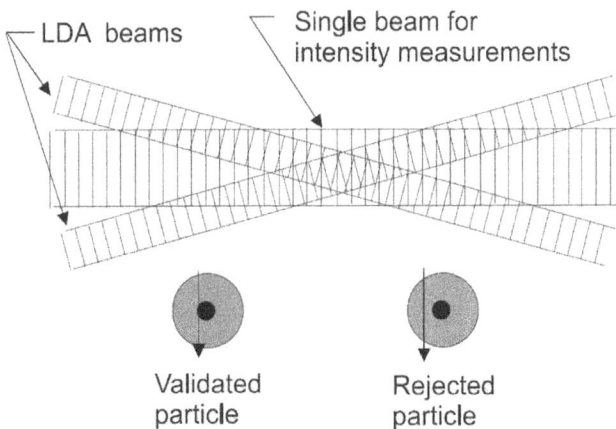

FIGURE 9.42 Co-axial arrangement of two probe volumes of different color for combined particle size and velocity measurements.

procedure using the information from a global mass balance. In practice, accurate particle concentration measurements by laser-Doppler anemometry are only possible for simple one-dimensional flows with mono-sized particles. In this case, the probe volume size may be determined by calibration. This, however, does not remove the problem related to a spatial distribution of particle concentration in the flow system.

The scattering intensity received by the photodetector is also influenced by the turbidity from the measurement location to the receiver. A scan of the probe volume through the two-phase flow for measuring a velocity profile yields different optical path lengths through the particle-laden flow and associated different light absorption (Kliafas *et al.*, 1990). Because of this, the use of the scattering intensity for determining particle phase properties is not recommendable.

For a simultaneous determination of fluid and particle velocity by LDA, the fluid flow has to be additionally seeded by small tracer particles, which are able to follow the turbulent fluctuations. The remaining task is the separation of the Doppler signals resulting from tracer particles and the dispersed phase particles. In most cases, this discrimination is based on the scattering intensity combined with some other method to reduce the error due to the Gaussian beam effect.

An improved amplitude discrimination procedure using two superimposed measurement volumes of different sizes and color was developed by Modarress and Tan (1983). The smaller or pointer probe volume was used only to trigger the measurements from the larger control volume. This arrangement ensures that the sampled signals were received only from the center part of the larger probe volume, where the spatial intensity distribution is relatively constant, and thus it is possible to classify the signal amplitude with respect to tracer and dispersed phase particles.

A combined amplitude-visibility discrimination method, which did not rely on additional optical components, was proposed by Börner *et al.* (1986). After first separating the signals based on the signal amplitude, the visibility of all signals was determined to ensure that no samples from large particles passing through the edge of the measurement volume were collected as tracer particles. This method required additional electronic equipment and complex software for signal processing.

A much simpler amplitude discrimination method was introduced by Hishida and Maeda (1990). To ensure that only particles traversing the center of the measurement volume were sampled, a minimum number of zero crossings in the Doppler signal were required for validation.

The discrimination procedures described here can be successfully applied only when the size distribution of the dispersed phase particles is well separated from the size distribution of the tracer particles.

LDA combined with a direct imaging technique, referred to as the *shadow-Doppler technique*, has been developed for combined size and velocity measurements in two-phase flows with arbitrary shaped particles as observed in coal combustion systems (Hardalupas *et al.*, 1994). A particle moving through the LDA probe volume produces a planar shadow of the non-spherical particle, which is recovered by a line array of photodiodes. The temporal signals due to particle motion are converted to a planar image of the particle. With this information, the particle shape and orientation can be inferred. When using a typical shape indicator, such as sphericity or eccentricity, one should keep in mind that these values are not necessarily a property of the particles but are biased by the particle orientation, which is affected by flow structure and turbulence. Since the image produced is not influenced by the optical properties of the particle, shadow imaging can be also applied to optically inhomogeneous particles (e.g. suspension droplets) to measure their size (Morikita and Taylor, 1998). The shadow Doppler technique has emerged in commercially available instruments.

9.4.5 PHASE-DOPPLER ANEMOMETRY

The principle of phase-Doppler anemometry (PDA) method is based on the Doppler difference method used for conventional laser-Doppler anemometry and was introduced by Durst and Zaré (1975). Using an extended receiving optical system with two or more photodetectors, it is possible to measure particle size (actually a local particle curvature is measured) and velocity simultaneously. The phase shift of the light scattered by a spherical particle through either refraction or reflection from the two intersecting laser beams is used to infer the particle size. It should be emphasized that the particle diameter can be obtained with a PDA only for spherical particles.

The operational principle of PDA can be explained using the simple fringe-type model assuming that the interference fringes in the intersection region of the two incident light beams of the LDA are parallel light rays (Saffman, 1987a). A spherical transparent particle placed into this fringe pattern will act as a kind of lens with focal length $\hat{f}$, which will project the light rays into space, as indicated in Figure 9.43. The separation of the projected fringes at a distance ℓ from the particle is given approximately by

$$\Delta s \approx \left(\ell - \hat{f} \right) \frac{d_f}{\hat{f}} \tag{9.54}$$

where d_f is the fringe spacing in the measurement volume. The focal length of the particle is given by

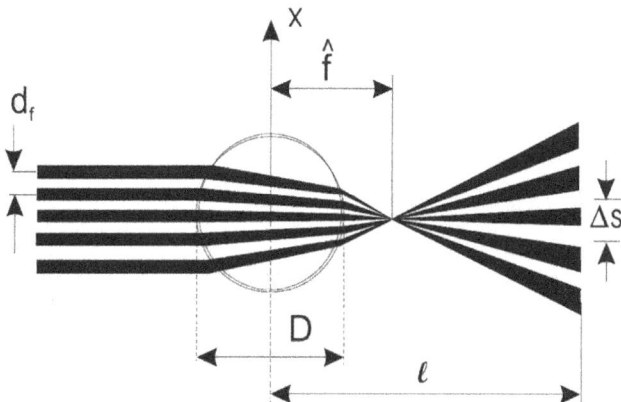

FIGURE 9.43 Fringe model for the phase-Doppler principle in the case of refraction.

$$\hat{f} = \frac{m}{m-1}\frac{D}{4}$$ (9.55)

where $m = n_d/n_m$ is the ratio of the refractive index of the particle to that of the surrounding medium. Since small particles are considered and ℓ is usually much larger than the particle diameter, one obtains

$$\Delta s \approx \ell\frac{d_f}{\hat{f}}$$ (9.56)

The separation of the projected fringes is obtained from

$$\Delta s \approx \frac{4\ell d_f}{D}\frac{m-1}{m}$$ (9.57)

Hence, the separation varies inversely as the particle diameter. A measurement of Δs can be used to determine particle size.

A simple schematic diagram illustrating the operation of the PDA is shown in Figure 9.44. A mask with two slits is placed at the distance ℓ from the particle. The two slits, separated by a distance $\Delta s'$, allow light to reach two photodetectors, A and B. As the particle passes through the probe volume, the fringe pattern sweeps across the mask and generates signals from the photodetectors. If Δs were equal to $\Delta s'$, the two signals would be in phase. However, because Δs does not equal $\Delta s'$, a phase shift occurs between the two high-pass signals, as shown in Figure 9.44. This phase shift can be related directly to Δs and used in Eq. (9.57) to determine the particle diameter. The frequency of the signal is the Doppler difference frequency corresponding to the particle velocity, v.

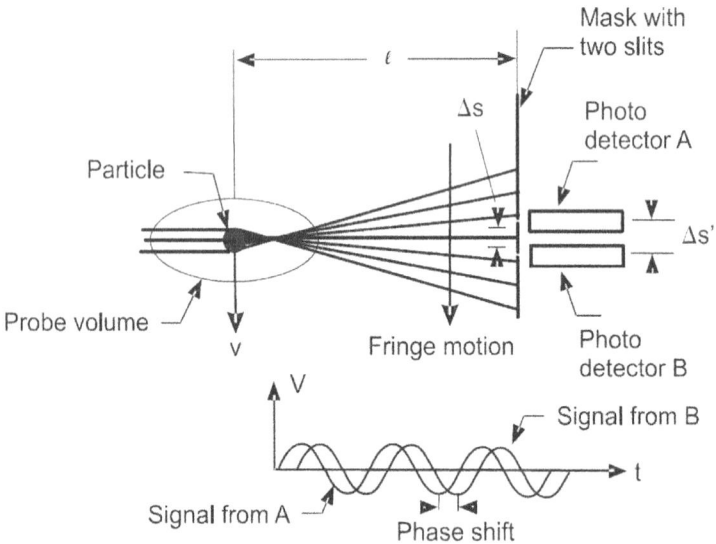

FIGURE 9.44   Sketch showing the basic concept of the PDA system and the measurement of the phase shift.

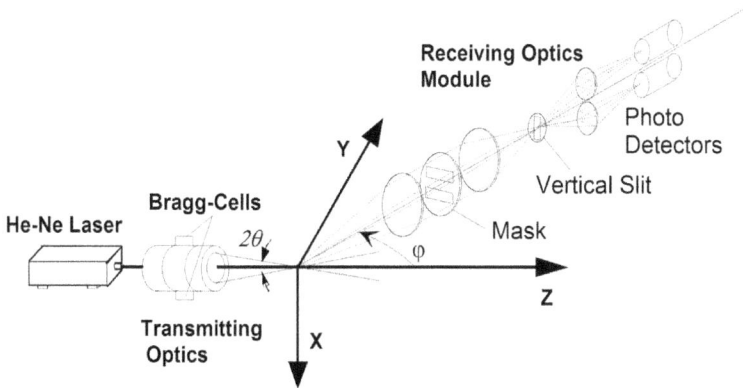

FIGURE 9.45 Optical configuration of a two-detector PDA.

A typical optical setup of a two-detector PDA-system is shown in Figure 9.45. The transmitting optics are the conventional dual-beam LDA optics, which, in this case, make use of two Bragg cells for frequency shifting. The PDA receiver is positioned at the off-axis angle φ (scattering angle), and a collection (or receiving) lens, lens 1, produces a parallel beam of scattered light. This parallel light beam passes through a mask, which defines the elevation angles of the two photodetectors. In this case, the mask has two rectangular slits in order to allow the light to pass through to the photodetectors. The slits are located parallel to the y-z plane at the elevation angles $\pm\psi$. The light is then focused by lens 2 on a spatial filter, that is, a vertical slit, which defines the effective length of the probe volume from where the scattered light can be received. The effective length of the probe volume is determined by the width of the spatial filter l_s (typically 100 μm) and the magnification of the receiving optics; that is, the focal length of the collecting lens to the second lens, $L_s = l_s \hat{f}_1 / \hat{f}_2$. Finally, the scattered light passing the two rectangular slits is focused onto the photodetectors using two additional lenses.

The signals seen by the two detectors will have a relative phase difference

$$\Delta\phi = 2\pi \frac{\Delta s'}{\Delta s} = 2\pi \frac{2\ell}{\Delta s} \sin\psi \tag{9.58}$$

where ψ is the elevation angle of one photodetector measured from the centerline in Figure 9.45, which is also the bisector plane of the two incident laser beams producing the measurement volume. The length, ℓ, now becomes the focal length of lens 1, $\hat{f}_1$. By using Eq. (9.57) for the fringe separation distance, the phase shift becomes

$$\Delta\phi = \pi \frac{D}{d_f} \frac{m}{m-1} \sin\psi \tag{9.59}$$

Finally, using the equation relating the laser wave length and fringe spacing,

$$d_f = \frac{\lambda_e}{2\sin\theta}$$

gives

$$\Delta\phi = \frac{2\pi D}{\lambda}\frac{m}{m-1}\sin\theta\sin\psi \qquad (9.60)$$

It should be emphasized that Eq. (9.60) is an approximation valid only for small scattering angles, ϕ, which represents the off-axis angle measured from the forward-scattering direction, to the optical axis of the receiver. The equation is very useful, however, to provide a rough estimate of the measurable particle size range for a given system or to perform a preliminary design of the optical configuration for small scattering angles.

To determine the particle size from the measured phase difference, the required correlations are derived from geometrical optics. This process is valid for particles that are large compared to the wavelength of light (van de Hulst, 1981). The phase due to the length of optical path is given by

$$\delta = \frac{2\pi Dm}{\lambda}\left(\sin\tau - pm\sin\tau'\right) \qquad (9.61)$$

The parameter p indicates the type of scattering, that is, $p = 0, 1, 2 \dots$ for reflection, first-order refraction, second-order refraction, and so on; τ and τ' are the angles between the incident ray and the surface tangent and the refracted ray and the surface tangent, respectively, as shown in Figure 9.46. The relationship between τ and τ' is given by Snell's law

$$\cos\tau' = \frac{1}{m}\cos\tau \qquad (9.62)$$

For a dual-beam LDA system, the phase difference of the light scattered from each of the two beams is given in a similar way

$$\Delta\phi = \frac{2\pi Dm}{\lambda}\left[\left(\sin\tau_1 - \sin\tau_2\right) - pm\left(\sin\tau_1' - \sin\tau_2'\right)\right] \qquad (9.63)$$

where the subscripts 1 and 2 are used to indicate the contributions from both incident beams. For two photodetectors placed at a known off-axis angle ϕ and placed

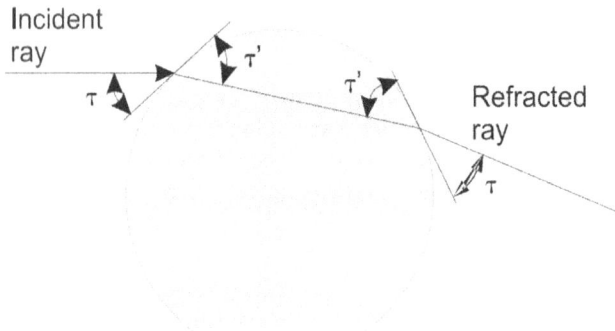

FIGURE 9.46 Phase difference of a light refracted at a spherical particle.

symmetrically with respect to the bisector plane at the elevation angles $\pm\psi$, one obtains the phase difference (Bauckhage, 1988)

$$\Delta\phi = \left(2\pi Dm / \lambda\right)\Phi \qquad (9.64)$$

The parameter Φ depends on the scattering mode. For reflection ($p = 0$)

$$\Phi = \sqrt{2}[\left(1 + \sin\theta\sin\psi - \cos\theta\cos\psi\cos\varphi\right)^{1/2}$$
$$-\left(1 - \sin\theta\sin\psi - \cos\theta\cos\psi\cos\varphi\right)^{1/2}] \qquad (9.65)$$

and for refraction ($p = 1$)

$$\Phi = 2\left\{ \begin{array}{l} \left[1 + m^2 - \sqrt{2}m\left(1 - \sin\theta\sin\psi + \cos\theta\cos\psi\cos\varphi\right)^{1/2}\right]^{1/2} \\ -\left[1 + m^2 - \sqrt{2}m\left(1 + \sin\theta\sin\psi + \cos\theta\cos\psi\cos\varphi\right)^{1/2}\right]^{1/2} \end{array} \right\} \qquad (9.66)$$

where 2θ represents the total angle between the two incident beams. Since the phase difference is a function of p, one expects a linear relation for the correlation between particle size and phase for only those scattering angles, where one scattering mode is dominant (i.e. reflection or refraction). Therefore, the values for Φ have been given for these two scattering modes only (i.e. Eqs. 9.65 and 9.66). Other scattering modes, that is, $p = 2$, may be also used for phase measurements, especially in the region of backscattering. Such a backscatter arrangement may be used for opaque particles and also might have advantages with regard to optical access since both the incident beams and the scattered light may be transmitted through one window.

By recording the band-pass filtered Doppler signals from the two photodetectors the phase difference $\Delta\phi$ is determined from the time lag between the two signals as indicated in Figure 9.47.

$$\Delta\phi = 2\pi\frac{\Delta t}{T} \qquad (9.67)$$

where T is the time of one cycle of the signal. It is now possible to determine the particle diameter for a given refractive index n_m and wavelength λ from the expression

$$D = \frac{\lambda}{2\pi m}\frac{1}{\Phi}\Delta\phi \qquad (9.68)$$

Obviously, a PDA system with only two detectors can only distinguish a phase shift between 0 and 2π, which limits the measurable particle size range for a given optical configuration. Therefore, three-detector systems are used in applications, where two phase differences are obtained from two detector pairs having different spacing, as shown in Figure 9.48. The measurement of a $\Delta\phi_{1-2}$ phase shift with a two-detector system would yield three possible particle diameters. However, the phase shift $\Delta\phi_{1-3}$ measured with the 1–3 detector removes this ambiguity and determines the actual diameter. This method enables one to extend the measurable particle size range while maintaining the resolution of the measurement. With the detector pair 1–2,

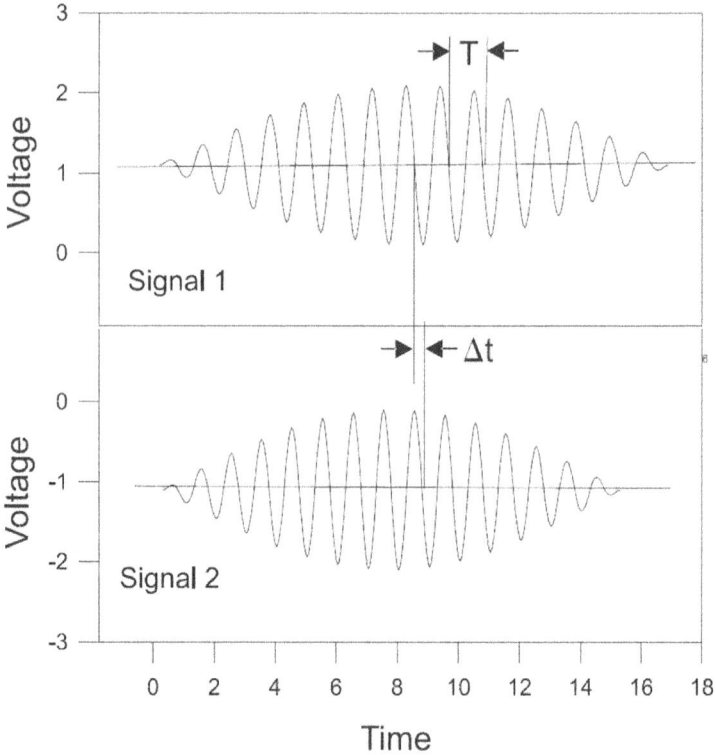

FIGURE 9.47   Determination of phase shift from two band-pass filtered Doppler signals.

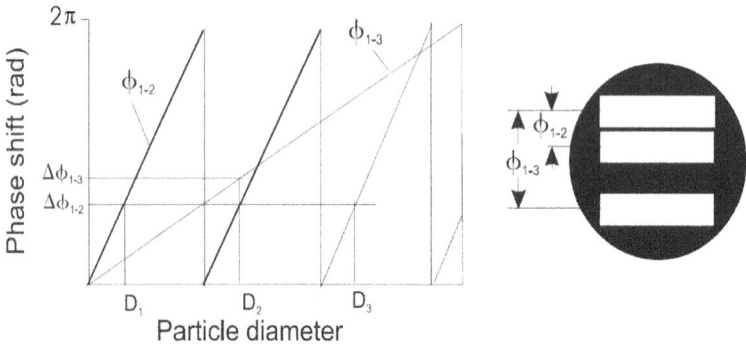

FIGURE 9.48   Phase size relations for a three-detector phase-Doppler system.

a high resolution of the size measurement is achieved. However, the upper measurable particle size is limited by the 2π ambiguity. The corresponding upper size limit is eventually given by the phase size correlation ϕ_{1-3}. The ratio of the two-phase measurements may be used for additional validation, such as checking the sphericity of deformable particles (liquid droplets or bubbles).

The principle of geometrical optics, introduced previously to calculate light scattering from particles, is limited to particle sizes larger than the wavelength of light and takes into account reflection and refraction. This implies that no interference of the scattering modes is considered (van de Hulst, 1981). In most cases, this theory is sufficient to support the optical layout of PDA systems. Also, this theory can be easily extended to account for the Gaussian light intensity distribution in the probe volume (Sankar and Bachalo, 1991).

For small particles, diffraction represents a special contribution to the light scattering, which may affect and disturb the phase measurement. Therefore, the more general Mie theory must be applied to determine the scattering characteristics for a particle of any given size. The Mie theory relies on the direct solution of Maxwell's equations for the case of the scattering of a plane light wave by a homogeneous spherical particle for arbitrary size and refractive index. In order to calculate the scattered field for a PDA system, it is necessary to add the contributions of the two incident beams and to average over the aperture of the receiving optics by taking into account the polarization direction and phase of each beam. Hence, it is possible to determine the intensity, visibility, and phase for arbitrary optical configurations. To allow for the influence of the Gaussian beam, the generalized Lorenz-Mie theory (GLMT) has also been applied to design and optimize PDA systems (Grehan *et al.*, 1992).

There are several issues related to the optimum selection of the optical systems, which can be explored for different types of particles (i.e. reflecting and transparent particles) based on calculations by geometrical optics and Mie theory (DANTEC/ Invent, 1994). The calculations using geometrical optics are performed for a point-like aperture, while the Mie calculations account for the integration over a rectangular aperture with given half-angles in the horizontal (δ_h) and vertical (δ_v)directions with respect to the y-z plane. The half-angles are defined as

$$\delta_h = \frac{\Delta W}{2} \hat{f}_1; \ \delta_v = \frac{\Delta H}{2} \hat{f}_1 \tag{9.69}$$

where ΔW and ΔH are the width and height of the aperture, respectively. It should be noted that the integration of the scattered light over the receiving aperture is important for obtaining a linear phase-size relation because depending on particle size, the scattering intensity field may have strong fluctuations (see Figure 9.3).

Transparent particles may be distinguished between those having a refractive index larger or smaller than the surrounding medium. Liquid droplets or glass beads in air have a relative refractive index (*i.e.* n_p/n_m), which is larger than one—typically in the range 1.3 to 1.5—and bubbles in liquid have a relative refractive index less than one, as shown in Table 9.4.

The selection of the optimum optical configuration should be based mainly on the relative importance of the scattering mode considered (i.e. reflection, refraction, or second-order refraction) with respect to the other modes and the resulting linearity of the phase-size relationship. The intensity of the considered scattering mode should be at least ten times larger than the sum of all the other scattering modes. The intensities of the different scattering modes are determined by using

TABLE 9.4
Characteristic Scattering Angles for Different Combinations of the Dispersed and Continuous Phase, That Is, Different Relative Refractive Indices

Two-Phase Flow System	$m = n_p/n_m$	ϕ_B	ϕ_C	ϕ_R
air bubbles in water	1.0/1.33	106.12	82.49	–
water droplets in oil	1.33/1.50	96.88	55.09	–
oil droplets in water	1.5/1.33	283.1	55.09	94.10
water droplets in air	1.33/1.0	83.12	82.49	137.48
diesel droplets in air	1.46/1.0	68.82	93.54	153.34
glass particles in air	1.52/1.0	66.68	97.72	158.92

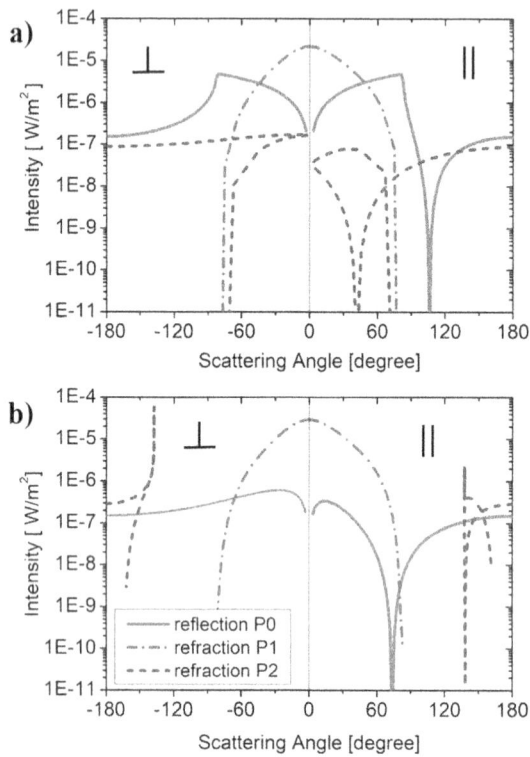

FIGURE 9.49   Angular intensity distribution of different scattering modes obtained by geometrical optics for a point receiving aperture and parallel (∥) as well as perpendicular (⊥) polarization ($\lambda = 632.8$ nm, half angle of the incident beams $\theta = 2.77°$, $D = 30\ \mu$m; (a) m = 0.75 (air bubbles in water), (b) m = 1.33 (water droplet in air).

calculations based on geometrical optics where both perpendicular (⊥) and parallel (∥) polarization are considered, as shown in Figure 9.49 (perpendicular is the left part and parallel the right part of each graph as indicated). Diffracted light (not shown in Figure 9.49) is concentrated in the forward-scattering direction and independent of polarization. This region should be avoided since this scattering mode is not suitable for PDA particle sizing. Reflected light covers the entire angular range for refractive index ratios below and above one. However, a distinct minimum is found for parallel polarization at the so-called *Brewster's angle*, which is given by the expression

$$\varphi_B = 2\tan^{-1}(1/m) \tag{9.70}$$

The Brewster angle decreases with an increasing refractive index ratio, as indicated in Table 9.4. First-order refraction (P1) is concentrated in the forward-scattering range and extends to the *critical angle*, which, for different relative refractive indices $m = n_p/n_m$, is given as

$$\phi_c = 2\cos^{-1}(m) \qquad m < 1 \tag{9.71}$$

$$\phi_c = 2\cos^{-1}(1/m) \qquad m > 1 \tag{9.72}$$

The critical angle increases with an increasing relative refractive index for $m > 1$, and the first-order refraction becomes dominant over reflection for a wider angular range. For $m < 1$, the critical angle increases with decreasing refractive index ratio. Second-order refraction (P2) again covers the entire angular range for a relative refractive index less than one. For m larger than one, the second-order refraction is concentrated in the backward-scattering range and is limited by the *rainbow angle*, given by the expression

$$\phi_R = \cos^{-1}\left[\frac{2}{m^4}\left(\frac{4-m^2}{3}\right)^3 - 1\right] \tag{9.73}$$

The angular range of second-order refraction is reduced with an increasing relative refractive index and the rainbow angle increases. These characteristic scattering angles are summarized in Table 9.4 for the different typical refractive indices. A map of the presence of the different scattering modes as a function of scattering angles and relative refractive indices was introduced by Naqwi and Durst (1991) to support the layout of the optical configuration of PDA systems. In the book by Albrecht *et al.* (2003), detailed calculations of the intensity of different scattering modes based on the Lorentz-Mie theory and Debye series decomposition are presented for parallel and perpendicular polarization.

Mie calculations are now presented for a range of the optimum scattering angles suggested by the relative intensity distributions. For bubbles in water, the optimum scattering angle seems to be rather limited, that is, between 70° and about 85°, where reflection is dominant for either polarization, as shown in Figure 9.49a. The phase-size relations show reasonable linearity in this range, and a scattering angle of 55° also gives a linear response function, as shown in Figure 9.50. Strong interference with refracted light exists in the forward-scattering mode, and the phase-size relation becomes nonlinear (i.e. at a scattering angle of $\varphi = 30°$).

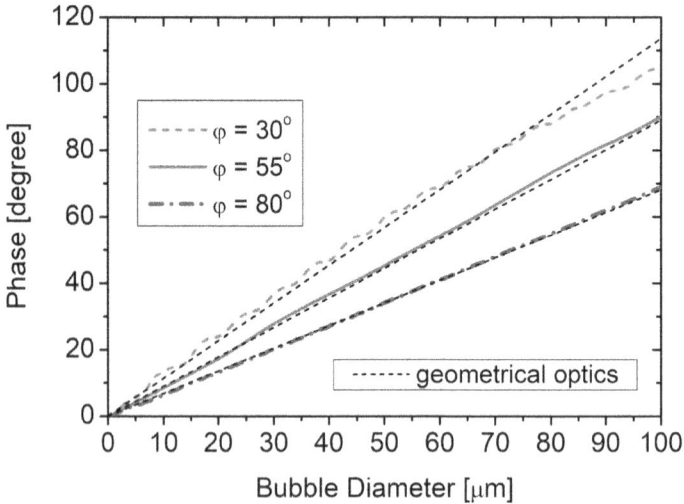

FIGURE 9.50 Mie calculation of the phase-size relations for different scattering angles between 30° and 80° and comparison with geometrical optics. (λ = 632.8 nm, p-polarization, m = 0.75 (i.e. air bubble in water), θ = 2.77°, Ψ = 1.85°, δ_h = 5.53°, δ_v = 1.85°).

For two-phase systems with relative refractive indices larger than one, refraction is dominant for both polarization directions in the forward-scattering range up to about 70° to 80°, depending on the value of the refractive index ratio, as depicted in Figure 9.49b. Since below 30° the diffraction interferes with the refracted light (especially for small particles), the lower limit of the optimum scattering angle is bounded by this value. This is also observed from the angular distribution of the phase for different particle diameters shown in Figure 9.51. It is obvious in this figure that the forward-scattering direction (i.e. between 30° and 80°) is dominated by the first-order refraction (P1, negative phase). Between 100° and 125° reflection (positive phase) appears to be dominant, and between 140° and 160°, the second-order refraction (P2) is the main contributor. This does not imply that the phase-size relations are everywhere linear. The phase-size relation for water droplets in air, which are illustrated in Figure 9.52, show that a reasonable linearity is obtained in the range between 30° and 80°.

The intensity distributions shown in Figure 9.49b also suggest that for $m > 1$, the reflected light is dominant between the critical angle ϕ_c and the rainbow angle ϕ_R. However, interference with third-order refraction exists in this range, which is not shown in Figure 9.49. This angular range can be only recommended for perpendicular polarization, where a reasonable linearity of the phase response curve is obtained for water droplets in air only around 100°, as shown in Figure 9.53. Some oscillations of the phase-size relation are observed up to droplet sizes of 60 μm, and this will result in errors of size measurements.

In the backscatter region, the intensity of secondary refraction (P2) is dominant only within a narrow range above the rainbow angle ϕ_R for perpendicular polarization.

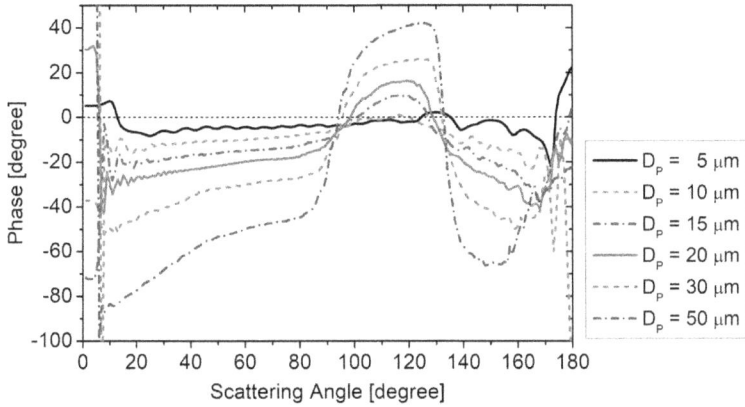

FIGURE 9.51   Angular distribution of phase for different particle diameters. ($\lambda = 632.8$ nm, p-polarization, $m = 1.33$ (i.e. water droplets in air), $\theta = 2.77°$, $\Psi = 1.85°$, $\delta_h = 5.53°$, $\delta_v = 1.85°$).

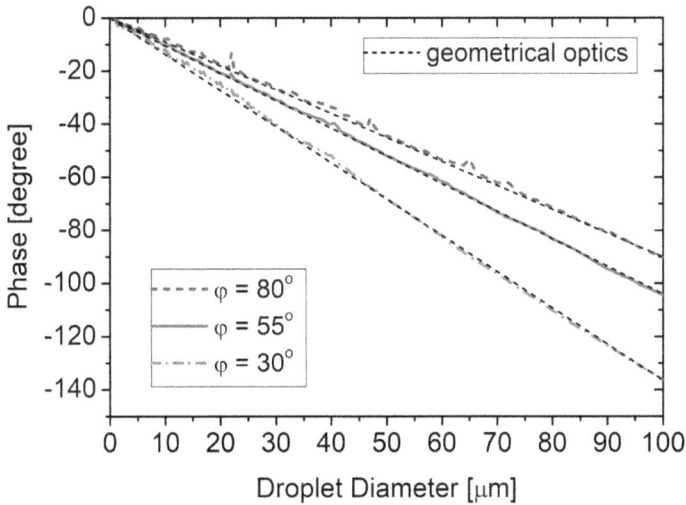

FIGURE 9.52   Mie calculation of phase-size relations for different scattering angles between 30° and 80° and comparison with geometrical optics. ($\lambda = 632.8$ nm, p-polarization, $m = 1.33$ (i.e. water droplets in air), $\theta = 2.77°$, $\Psi = 1.85°$, $\delta_h = 5.53°$, $\delta_v = 1.85°$).

The optimum location of the receiving optics, however, strongly depends on the value of the relative refractive index, as shown in Figure 9.54. This is critical in fuel spray applications, where the refractive index varies with droplet temperature, and hence, the location of the rainbow angle is not constant. It is apparent in Figure 9.54 that just above the rainbow angle, that is, for $\phi = 140°$, a linear phase-size relation is also obtained. At an angle of 160°, strong fluctuations and deviations from linearity are observed, and hence, this scattering angle is not recommended.

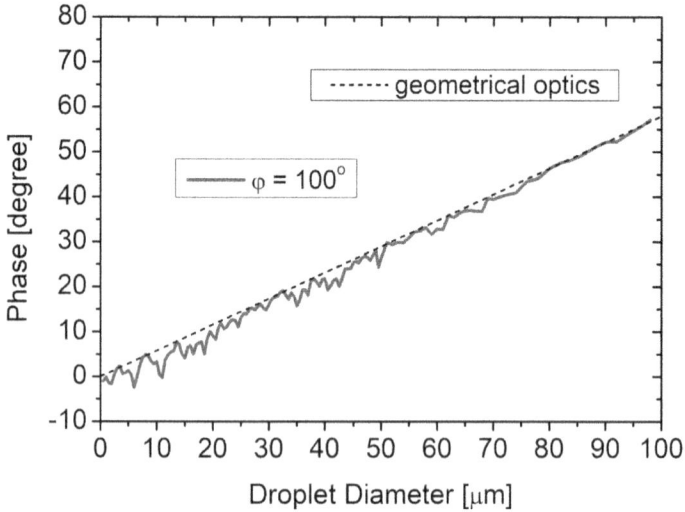

FIGURE 9.53 Mie calculation of phase-size relations for scattering angles of 100° and 120°. ($\lambda = 632.8$ nm, s-polarization, $m = 1.33$ (i.e. water droplets in air), $\theta = 2.77°$, $\Psi = 1.85°$, $\delta_h = 5.53°$, $\delta_v = 1.85°$).

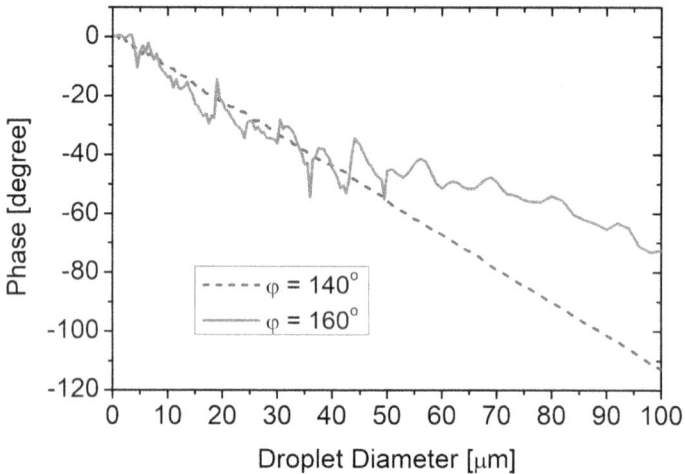

FIGURE 9.54 Mie calculations of phase-size relations for scattering angles of 140° and 160°. ($\lambda = 632.8$ nm, s-polarization, $m = 1.33$ (i.e. water droplets in air), $\theta = 2.77°$, $\Psi = 1.85°$, $\delta_h = 5.53°$, $\delta_v = 1.85°$).

The correct application of PDA requires that one scattering mode is dominant, and the appropriate correlation (i.e. Eq. 9.65 or Eq. 9.66) has to be used to determine the size of the particle from the measured phase. However, on certain trajectories of the particles through the focused Gaussian beam, the wrong scattering mechanism might become dominant and lead to erroneous size measurements (Sankar and Bachalo, 1991; Grehan *et al.*, 1992). This error is called *trajectory ambiguity* and is illustrated in Figure 9.55, where a transparent solid particle moving in air is considered, and the desired scattering mode is refraction, which is dominant for collection angles between 30° and 80°. When the particle passes through the part of the measurement volume located away from the detector (i.e. negative y-axis), it is not illuminated homogeneously. The refracted light is coming from the outer portion of the measurement volume, where the light intensity is relatively low, while the reflected light comes from a region close to the center of the probe volume, where the illuminating light intensity is considerably higher due to the Gaussian profile. In this situation, the reflected light might become dominant, resulting in an incorrect size measurement because the particle diameter is determined from the correlation for refraction. The opposite effect occurs for an air bubble in water, where the desired scattering mechanism is reflection. When the bubble moves through the probe volume facing the detector, the reflected scattering intensity is significantly lower than the intensity due to refraction. When using the correlation for reflection in this case, a wrong size measurement is the result.

It is obvious from Figure 9.55 that the trajectory ambiguity is potentially very important for large particles with sizes comparable to the dimensions of the probe

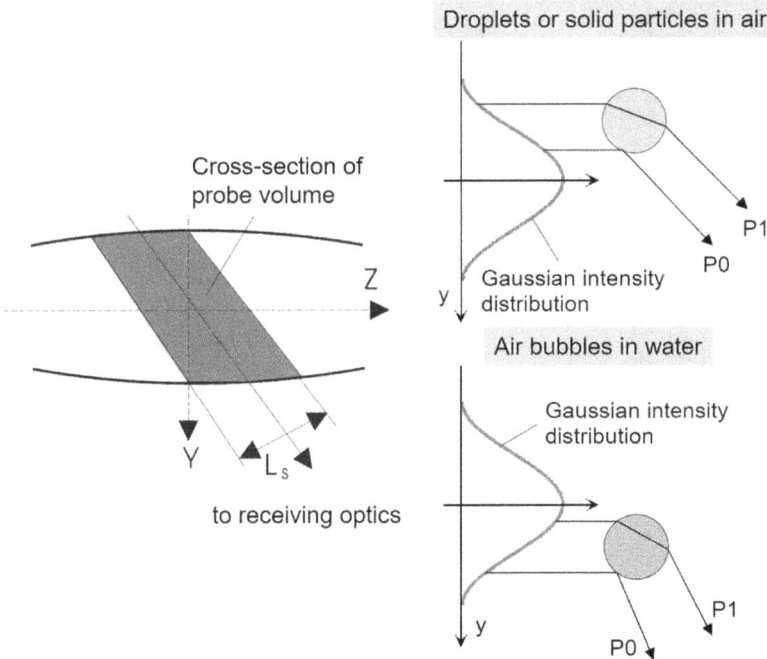

FIGURE 9.55 The Gaussian beam effect, trajectory ambiguity, and detection of the wrong scattering mechanism. Left: Top view of probe volume, cut out by the receiver geometry. Upper right: Trajectory effect for droplets or solid particles in air. Lower right: Trajectory effect for air bubbles in water.

volume. A similar ambiguity occurs due to the so-called slit effect since a vertical slit (spatial filter) is used to demarcate the length of the probe volume (Xu and Tropea, 1994).

The phase error and scattering amplitude along the y-axis for different particle sizes as they pass through a probe volume with a 100 μm probe volume diameter is illustrated in Figure 9.56 (Qiu et al., 2000). The phase and amplitudes are calculated using GLMT (Grehan et al., 1992). The phase error may become negative or positive depending on the particle diameter. The smallest errors are observed, however, for smaller particles, and this leads to the recommendation that the probe volume diameter should be about five times larger than the largest particles in the size spectrum considered. This requirement has restrictions for applications in dense particle-laden flows, where the probe volume must be small enough to ensure that

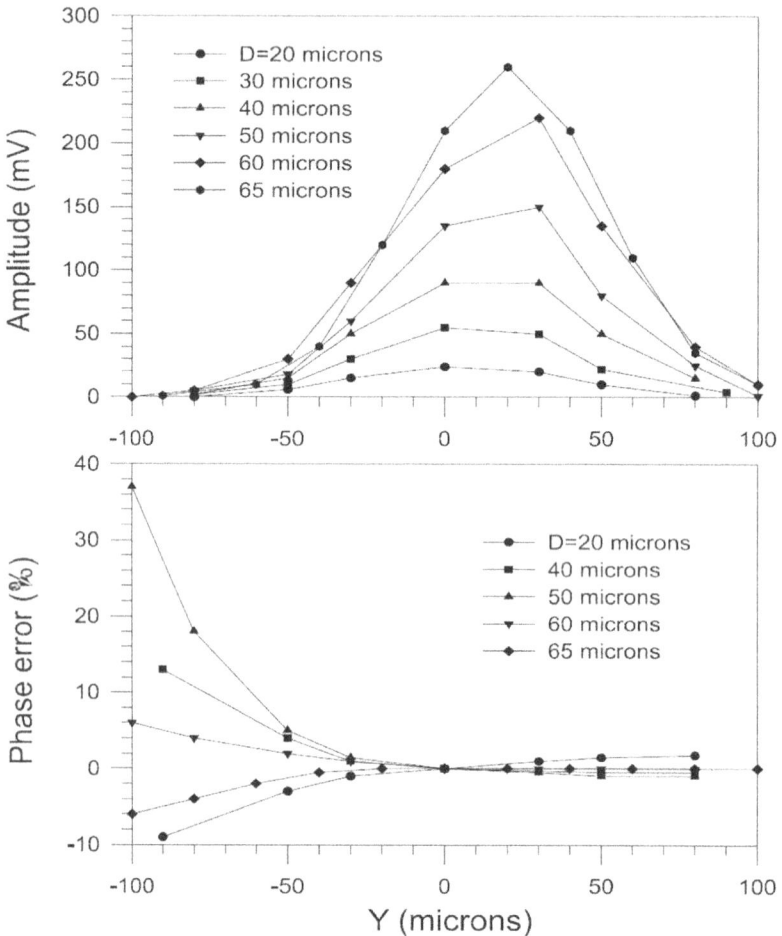

FIGURE 9.56   Phase error and scattering amplitude along the y-axis for different particle diameters and for probe volume diameter of 100 μm (Qiu et al., 2000).

the probability of two particles being in the probe volume at the same time is very small, as explained in Section 9.4.3.

In order to eliminate or reduce the errors due to the Gaussian beam and slit effects (Xu and Tropea, 1994), a dual-mode PDA system was developed (Tropea *et al.*, 1996). Both errors can be grouped into the general category of trajectory ambiguity and can lead, as the result of detecting the wrong scattering mode, to significant sizing errors as illustrated in Figure 9.57. The dual-mode PDA is a combination of the standard PDA (SPDA) and the planar PDA (PPDA), which supposedly has no Gaussian beam effect (Aizu *et al.*, 1993) but suffers from strong oscillations in the phase-size relation. The optical system for the dual-mode PDA is based on standard two-component transmitting optics and a receiver with four detectors where two of them are aligned vertically (SPDA) and the other two horizontally (PPDA), as illustrated in Figure 9.57. Typically, the four detectors are mounted in a single fiber-based probe head with an exchangeable mask having four rectangular openings (Sommerfeld and Tropea, 1999). With such a dual-mode PDA, the SPDA is basically used for the size measurement and the PPDA, with lower sensitivity, is used to resolve the 2π ambiguity. The correlation of the SPDA and PPDA phase differences are used to validate the measurement (Tropea *et al.*, 1996). Hence, this method essentially eliminates the errors due to trajectory effects (Gaussian beam and slit effects) and is also sensitive to droplet sphericity. Finally, the higher accuracy of the dual-mode PDA enables more reliable particle mass flux measurements.

Since PDA allows the measurement of particle size and velocity, it is also possible to estimate the particle number or mass concentration and the particle mass flux. The particle number concentration is defined as the number of particles per unit volume. This quantity, however, cannot be measured directly since the PDA is a single

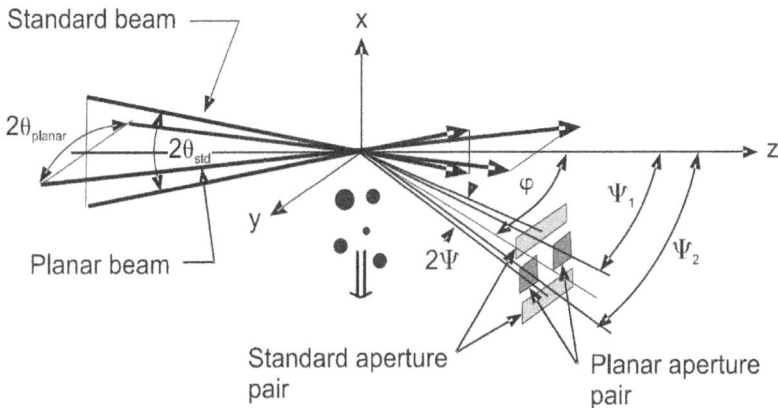

FIGURE 9.57 Optical configuration of a dual-mode PDA system with four receiving apertures.

Source: From Tropea, C., Xu, T.-H., Onofri, F., Grèhan, G., Haugen, P. and Stieglmeier, D., Dual mode phase-Doppler anemometer, *Particle and Particle System Characterization*, 1996, **13**, 165. Copyright Wiley-VCH Verlag GmbH & Co., KGaA. Reproduced with permission.

particle counting instrument, which requires that, at most, only one particle at a time be in the probe volume. The particle concentration has to be obtained from the number of particles moving through the probe volume during a given measurement period. For each particle, one has to determine the volume of fluid, which passes through the probe volume cross-section together with the particle during the measurement time Δt_s. The volume depends on the instantaneous particle velocity $\mathbf{v}$ and the probe volume cross-section perpendicular to the velocity vector, that is, $Vol = A'$ $|\mathbf{v}| \Delta t_s$, as shown in Figure 9.58. In addition, the effective cross-section of the probe volume is a function of the particle size and, therefore, $A = A(\alpha_k, D_i)$, where α_k is the particle trajectory angle for each individual sample k, and D_i is the particle diameter of size class i. Hence, the concentration associated with one particle is

$$n = \frac{1}{Vol} = \frac{1}{|\mathbf{v}| A'(\alpha_k, D_i) \Delta t_s} \tag{9.74}$$

This implies that for accurate particle concentration measurements, one has to know the instantaneous particle velocity and the effective probe volume cross-section. The dependence of the probe volume cross-section on the particle size is a result of the Gaussian intensity distribution in the probe volume and the finite signal noise. As illustrated in Figure 9.38, a large particle passing the edge of the probe volume will scatter enough light to produce a signal above the detection level. A small particle will generate a detectable signal for only a small displacement from the probe volume center. Therefore, the probe volume cross-section decreases with particle size and approaches zero for $D_p \to 0$.

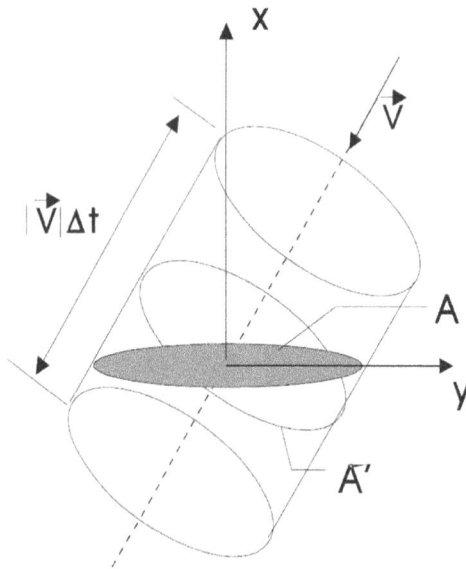

FIGURE 9.58 Probe volume associated with one particle moving across the detection region during the measurement time Δt.

The effective probe volume cross-section is determined by the off-axis position of the receiving optics and the width of the spatial filter used to limit the length of the probe volume imaged onto the photodetectors. For a one-dimensional flow along the x-axis, as shown in Figure 9.59, the effective size-dependent cross-section of the measuring volume is determined from the equation:

$$A(D_i)_x = 2r(D_i)L_s / \sin\varphi \qquad (9.75)$$

where L_s is the width of the image of the spatial filter in the receiving optics, as shown in Figure 9.59, which depends on the slit width and the magnification of the optics; D_i is the diameter of size class i; and $r(D_i)$ is the particle size dependent radius of the probe volume. For any other particle trajectory through the probe volume, the effective cross-section obtained with the particle trajectory angle α_k is

$$A'(\alpha_k, D_i) = \frac{2r(D_i)L_s}{\sin\varphi}\cos\alpha_k \qquad (9.76)$$

The particle trajectory angle can be determined from the different instantaneous particle velocity components:

$$\cos\alpha_k = \frac{1}{\sqrt{1+\left(\dfrac{w}{u}\right)^2+\left(\dfrac{v}{u}\right)^2}} \qquad (9.77)$$

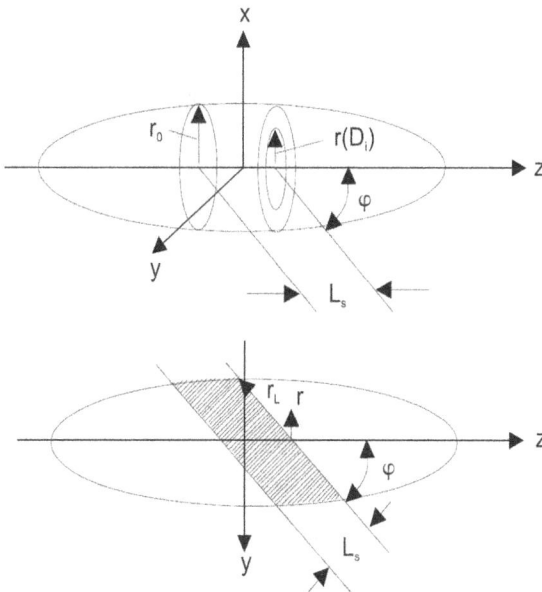

FIGURE 9.59 Geometry of a PDA measurement volume showing the diameter and the width as a result of the image slit width.

The particle size-dependent radius of the probe volume $r(D_i)$ may be determined in situ by using the burst length method (Saffman, 1987b) or the so-called logarithmic mean amplitude method (Qiu and Sommerfeld, 1992). The latter is more reliable for noisy signals, that is, low signal-to-noise ratio as demonstrated by Qiu and Sommerfeld (1992). The reader is referred to the previous publications for more details.

These findings suggest that in complex two-phase flows with random particle trajectories through the probe volume, a three-component PDA system is required for accurate concentration measurements. For a spectrum of particle sizes, the local particle number density is evaluated from the expression

$$n = \frac{1}{\Delta t_s} \sum_{k=1}^{M} \left[\sum_{i=1}^{N_k} \frac{1}{A'(\alpha_k, D_i)} \sum_{j=1}^{N_j} \frac{1}{|\mathbf{v}_{k,j}|} \right] \qquad (9.78)$$

The sums in Eq. (9.78) involve the summation over individual realizations of particle velocities (index j), in predefined directional class (index k) and size class (index i). The summation over particle size classes includes the appropriate particle-size dependent cross-section of the measurement volume for each size class, as in Eq. (9.76), and finally the summation over all directional classes.

The particle bulk density can be obtained by multiplying Eq. (9.78) with the mass of the particles. The particle mass flux, which is a vector quantity, becomes a useful quantity to characterize types of multiphase flow. The mass flux in the direction n is obtained from the expression

$$\dot{M}_d = \frac{1}{\Delta t_s} \sum_{k=1}^{M} \left[\sum_{i=1}^{N_k} \frac{m_p}{A'(\alpha_k, D_i)} \sum_{i=1}^{N_j} \frac{v_n}{|\mathbf{v}_{k,j}|} \right] \qquad (9.79)$$

Where v_n is the particle velocity component in the direction n, which is to be determined, and m_p is the mass of one particle. For a directed two-phase flow, that is, when the temporal variation of the particle trajectory through the probe volume is relatively small (such as in a spray), the particle trajectory angle may be determined from independent measurements of the individual velocity components using Eq. (9.77) (Qiu and Sommerfeld, 1992).

Accurate particle concentration or mass flux measurements, even in complex flows using one-component PDA systems, were also possible using the integral value method (Sommerfeld and Qiu, 1995). This integral value was determined under the envelope of the band-pass filtered Doppler signal in order to estimate the instantaneous particle velocity and its trajectory through the probe volume. It should be noted that dual-mode PDA systems allow for very accurate measurements of particle mass flux, because they provide more information on the effective size of the measurement volume (Tropea et al., 1996).

9.5   IMAGING TECHNIQUES AND PTV/PIV

Imaging techniques have a wide area of application and are, as a first step, very often used for a qualitative visualization of the multiphase flow system. Examples for applications range from characterization of separated two-phase flows (i.e. flow regimes of liquid-gas flows in pipelines), to liquid atomization through spraying systems, to dispersed particle-laden flows in powder handling devices. The present description of the fundamental principles is limited to dispersed multiphase flows, which are accessible by conventional optical techniques (mostly using visible light) in a nonintrusive way. Imaging methods for opaque or very dense dispersed multiphase flows were reviewed by Williams and Beck (1995), Powell (2008), and Poelma (2020), who considered electrical capacitance tomography, magnetic resonance imaging and x-ray imaging, and tomography.

Imaging techniques applying visible light for illumination may be used to determine the instantaneous spatial distribution of particles over a finite region of interest in the flow, but also properties of individual particles such as particle velocities, particle size and shape (the latter with appropriate magnification). Such instantaneous images can be obtained by either using pulsed light sources, which can deliver a high light energy during a very short time period combined with a slow image recording system or by using a continuous light source in combination with high-speed photography or other high-speed recording systems, for example, high-speed CCD cameras. With these two procedures, the recorded images yield a frozen pattern of the particle distributions. A continuous illumination combined with long-exposure photography can yield information about the time-averaged distribution of the dispersed phase within a flow field.

In most applications, the illumination of the flow system by the light source defines the "measurement volume," which may be a light sheet or beam and a volume illumination with a certain rectangular cross-section. In addition, the entire system may be illuminated by diffuse front-light or backlighting, used mainly for shadow imaging. The following light sources may be used for imaging techniques:

- Continuous white light sources, such as halogen lamps
- Spark discharges or flashlights (e.g. Xenon flash lamp)
- Continuous or pulsed laser light sources (e.g. Argon-Ion and Nd-Yag laser)
- High-power LEDs or LED arrays emitting different light colors.

The required power of the light sources is determined by the illumination region and the particle sizes under consideration and the scattering properties of the particle (i.e. reflecting, transparent, or absorbing particles). In the range of geometrical optics, that is, for particles larger than the wavelength of the light, the scattering intensity is approximately proportional to the square of the particle diameter. It must be noted that the sensitivity of the image recording system affects the required laser power.

The image recording system determines the possible spatial resolution of the images. The highest spatial resolution is obtained by conventional film photography or high-speed motion picture cameras, which also use photographic films. Such recording systems, however, have the disadvantage that the image evaluation is rather cumbersome and cannot be easily automated. The evaluation of the photographs requires post-processing by scanning and digitizing, which generally results in a reduction of the spatial resolution. More convenient for image analysis are methods using direct electronic digitization, such as CCD (charged coupled device) cameras, which, however, have a limited spatial resolution. Typically, CCD cameras have 1024×768, 1280×1024, or 1048×2048 photo diodes, which are called pixels. High-resolution CCD cameras have up to 4096×4096 pixels. On the other hand, digitized images can be more easily enhanced by various software-based algorithms and filter operations. Also, statistical information on the properties of the dispersed phase can be evaluated more easily when recording multiple images.

Imaging techniques for two-phase flows may be categorized according to the applied illuminating and recording systems, and the information can be extracted from the images in the following way:

- Direct imaging and visualization techniques for particle-scale processes
- Whole field visualization to determine the particle phase spatial distribution
- Particle tracking velocimetry (PTV) or streak line technique
- The classical particle image velocimetry (PIV)

Laser-speckle velocimetry and holographic methods will not be considered here. More information about laser-speckle velocimetry can be found in Adrian (1986, 1991) and for studies of sprays by holographic methods the reader is referred to the papers of Chigier (1991) and Chavez and Mayinger (1990).

Direct imaging techniques are mostly based on continuous illumination with high-speed recording and are used to visualize the time-dependent motion and behavior of individual particles, droplets, and bubbles. Typical illuminations are diffuse front lightning and background illumination. Such a technique allows one to study the shape of bubbles or droplets under different flow conditions. Common applications are droplet-scale transport processes, such as collisions between droplets and droplet-wall collisions (Frohn and Roth, 2000). Since there are numerous examples for applications of such techniques in two-phase flows, the reader is also referred to the book by Van Dyke (1982) to appreciate the potential of imaging techniques. Some research on the application of imaging techniques and shadow imaging for the analysis of various types of dispersed two-phase flows, that is, sprays and particulate, as well as bubbly flows were summarized by Sommerfeld (2007).

For analyzing the outcome of binary droplet collisions, a high-resolution direct imaging technique was developed by Sommerfeld and Pasternak (2019) applying a stereoscopic observation of such collision events. The main objective was the quantification of the collision outcome using a so-called *collision map* that summarizes bouncing, coalescence, stretching, and reflexive separation. The collisions were induced by two piezo-electrically generated droplet chains directed toward each other. The imaging was based on shadow graphs using two LED arrays providing a homogeneous background illumination. The recording of the collision event was based on two high-speed CCD cameras (Sommerfeld and Pasternak, 2019).

Multidimensional particle sizing techniques have been developed based on interferometric particle imaging (IPI) and the phase Doppler principle, called global phase-Doppler (GPD). The combination of both methods yields more information. A detailed description of these methods and the underlying principles is provided by Damaschke *et al.* (2005). In both techniques, a light sheet illumination is used.

In the IPI, a single light sheet illuminates a particle cloud such as a spray. A defo-cused image (far-field image) of the droplets is recorded, for example, by a CCD camera. Due to the interference of the glare points from different scattering modes (e.g. reflection and refraction) on the droplet surface, the droplet interference pat-tern appears as a circular object with fringes. The spatial separation of the fringes is proportional to droplet size. Similar to the PDA, the GPD uses two illuminating light sheets, which are adjusted to intersect within the measurement region. The far-field interference pattern resulting from glare points of the same scattering mode is recorded. In this case, the droplets appear as circles with fringe patterns whose spacing can be related to the droplets' sizes. Both techniques provide field images with size information for all droplets captured in the light sheet. However, their application is limited to very dilute systems due to the rather large images of the interference pattern recorded.

The visualization of the instantaneous distribution of the dispersed phase par-ticles in a two-phase flow is generally performed by producing a thin laser light sheet, which illuminates a specific plane of the two-phase flow field. Images are taken by a recording system, oriented perpendicular to the light sheet, as shown in Figure 9.60. To obtain instantaneous images of the dispersed phase distribution, either the light source has to be operated in a single pulse mode (i.e. a single frozen image is obtained), or the recording systems must allow for short time exposure, for example, high-speed video camera. The exposure time or the pulse duration of the laser determines the temporal resolution and has to be adjusted according to the flow velocity in order to yield sharp images of the dispersed-phase objects. The image recording system must be well-focused. Such a visualization technique has been applied by Longmire and Eaton (1992) to evaluate the particle concentration

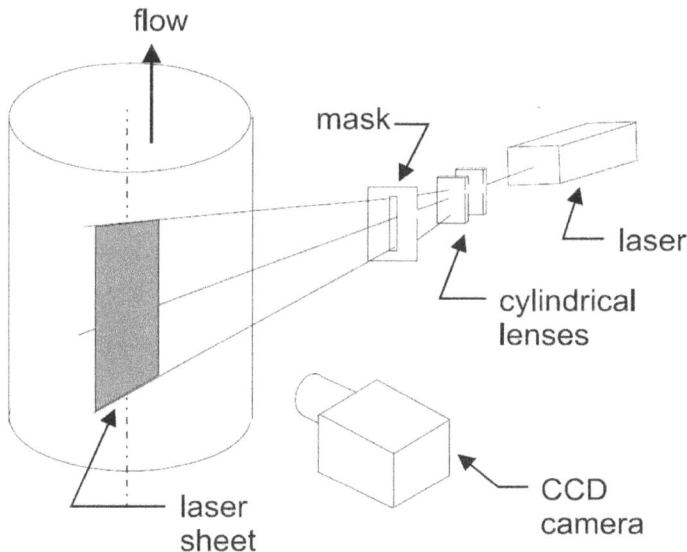

FIGURE 9.60 Optical arrangement of a light sheet flow visualization technique.

distribution and, in turn, the particle response characteristics in a forced particle-laden free jet. A pulsed copper vapor laser was used as a light source, and the light sheet was produced by a cylindrical-spherical lens combination. The light sheet was transmitted through the center region of the jet and aligned parallel to the flow direction. Images of the particles were taken with a 35 mm camera activated by an electronic trigger. The single-pulse photographs were printed on a 8 in. × 10 in. photographic paper so that the individual particle images could be identified. A scanner was used to produce a binary image matrix, which was transferred to a PC for further processing. The images were enhanced, and all individual particle images were identified. The particle concentration field was evaluated by dividing the images into a number of small cells and then counting the individual particles in each of the cells. In order to correlate the particle concentration distribution with the flow structure, independent images were recorded when the jet was only seeded with smoke particles for visualizing the gas flow field structure.

A similar visualization technique was applied by Wen *et al.* (1992) for studies of the particle response to the vortical structures evolving in a plane shear layer. Either a diffuse flashlight unit or a continuous Argon-Ion laser light sheet were used as light sources. The images were recorded by a large format camera (6 × 4 cm) using exposure times between 1/250 and 1/10,000 seconds. The photographs were processed by a commercial image processing system. The airflow and the vortical structures in the shear layer were visualized using the smoke wire technique. From the instantaneous images of the particle distribution in the shear layer, the response characteristics of different-sized particles were analyzed.

In a study by Huber and Sommerfeld (1994), a continuous Argon-Ion laser light sheet was used together with a CCD camera for recording images of the cross-sectional particle concentration distribution in pneumatic conveying systems. Using digital image processing, the images were enhanced and instantaneous particle concentration distributions in the pipe cross-section were evaluated. The mean particle concentration field was determined by averaging a number of individual images.

Particle tracking and particle image velocimetry (PTV and PIV) enable the determination of the instantaneous particle velocity distribution over a finite domain of the flow field, or specifically in the case of PTV along a particle trajectory. Both techniques have been extensively used in single-phase flows, especially for the analysis of unsteady and turbulent flows. In single-phase flows, small tracer particles are added as light scattering centers in order to trace the fluid elements. By observing the locations of the images of these marker particles over two- or multiple-time steps, the velocity can be determined from the displacement $\Delta \vec{x}(\vec{x},t)$ and the time period Δt between two successive light pulses or recordings from

$$\vec{u}(\vec{x},t) = \frac{\Delta \vec{x}(\vec{x},t)}{\Delta t} \tag{9.80}$$

The response time of the marker particles must be sufficiently small that the particles follow the turbulent fluctuations in the flow, that is, the tracer particle Stokes number should be much smaller than 1.

In dispersed two-phase flows, the velocity distribution of the dispersed phase can easily be determined by PTV or PIV since the particles are usually rather large and, therefore, scatter the light very well. Challenging in two-phase flow applications of PIV and PTV is the simultaneous measurement of the velocity fields of the continuous phase as well as the dispersed phase particles. In the following paragraphs, the basic principles of PTV and PIV are explained, and some examples of applications to two-phase flow studies are given.

According to the classification by Adrian (1991), both methods, that is, PIV and PTV, may be grouped under the name pulse-light velocimetry (PLV), which indicates that pulsed light sources are used for illumination of the flow region of interest. For PLV, powerful laser light sources are used most of the time where the emitted light beam is expanded by a system of lenses to form a narrow light sheet typically up to a few millimeters in thickness. It is possible to sweep a laser beam through the flow field using a rotating mirror (e.g. polygon mirror), whereby the effective intensity in the light sheet is larger. The time separation Δt between subsequent laser pulses determines the measurable particle velocity range, and the pulse duration establishes the degree to which the image of the particle is frozen on the recording system. Lasers typically used for PLV are the following:

- Continuous lasers (e.g. Argon-Ion laser), which are chopped using acoustic-optical modulators
- Continuously pulsed metal-vapor lasers (e.g. copper-vapor laser)
- Q-switched ruby lasers
- Frequency double-pulsed Nd:Yag lasers, which can produce trains of double pulses

Several combinations of laser pulse coding (i.e. single, double, or multiple pulse) and image recording (e.g. single frame recording and multi-framing) may be used to determine the velocity of (tracer) particles as well as their direction of motion (Adrian, 1991).

The differences between PTV and PIV are the way the images are processed, and the displacement is evaluated. With PTV, the tracks of individual particles are reconstructed from single or multiple images; hence, this technique is only applicable for relatively dilute two-phase flows where the length of the recorded particle tracks is smaller than the mean inter-particle spacing. A sophisticated point-by-point analysis of the images is generally required. PIV requires higher particle concentrations since the displacement of a group of particles located in an interrogation area (a small subregion of one image) is determined by special processing techniques such as auto- and cross-correlation methods. Hence, an average displacement of the light scattering pattern and an average velocity of all particles located in the interrogation area is determined. The number of particles in each interrogation area should typically be from 5 to 10 in order to yield reasonable averages of the particle velocity.

The resolution of velocity measurements by PIV and PTV depends on the accuracy of the displacement measurement, Δx, and the uncertainty in the time difference between the light pulses, Δt. Normalized by the maximum velocity in the flow field, u_{max}, the root mean square (rms) uncertainty of the velocity is determined from (Adrian, 1986)

$$\frac{\sigma_u}{u_{max}} = \frac{\sigma_{\Delta x}}{\Delta x} + \frac{\sigma_{\Delta t}}{\Delta t} = \frac{\sigma_{\Delta x}}{u_{max}\Delta t} + \frac{\sigma_{\Delta t}}{\Delta t} \tag{9.81}$$

where $\sigma_{\Delta x}$ is the root mean square uncertainty in the displacement between two images, and $\sigma_{\Delta t}$ is the uncertainty in the time between two successive light pulses or recordings. In the determination of the displacement, the major source of error is related to estimating the centers of the particle images, which is determined by the applied processing algorithm and the image size. With the assumption that this uncertainty is related to the image diameter through a number c, which is established by the accuracy of the processing method used for the particle location, one obtains

$$\sigma_{\Delta x} = c\, d_I \tag{9.82}$$

The image size, d_I, depends on the resolution of the recording system, d_r (i.e. photographic film or electronic recording), and the magnification of the imaging system, M. It can be estimated from the following relation (Adrian, 1986)

$$d_I = \left(M^2 D^2 + d_s^2 + d_r^2\right)^{1/2} \tag{9.83}$$

where D is the particle diameter, and d_s is the diffraction-limited spot diameter of the optical system given by (Adrian, 1986)

$$d_s = 2.44\left(M+1\right) f^{\#} \lambda \tag{9.84}$$

where $f^{\#}$ is the f-number of the lens, and λ is the wavelength of light. The relative importance of the first two terms in Eq. (9.83) depends on particle size. For an optical system with $M = 1$, $f^{\#} = 8$, and $\lambda = 514.5$ nm, the image diameter becomes independent of the particle size for diameters much less than 20 μm. On the other hand, for large particles (i.e. much larger than 20 μm, which is mostly the case in two-phase flows), the image diameter is given by

$$d_I = \left(M^2 D^2 + d_r^2\right)^{1/2} \tag{9.85}$$

In addition, the image size may be increased by particle movement during the exposure time or pulse duration, Δt_{exp}. Hence, the maximum velocity in the flow field times the pulse duration should be smaller than the image diameter to avoid this blur effect

$$u_{max}\Delta t_{exp} \leq d_I \tag{9.86}$$

With the assumption that the pulse frequency of the light source can be adjusted with high accuracy, that is, the time jitter may be neglected, the uncertainty of velocity measurement becomes

$$\sigma_u = c\frac{d_I}{\Delta t} \tag{9.87}$$

This equation reveals that high accuracy for velocity determination is achieved with small image diameters.

Two different methodologies may be used for PTV, the single-frame/multi-pulse technique and the multi-frame/single-pulse method, where the particle trajectories are reconstructed by overlapping several single exposure images. The second method corresponds to conventional cinematography and has the advantage that the particles may be tracked for a long time (i.e. to evaluate Lagrangian trajectories) and that the direction of particle motion is fully determined from the sequence of the recordings. The disadvantage is that the frequency of the image recording system must be equal to the laser pulse frequency or a known integer multiple smaller. This limits the measurable velocity range when CCD cameras with a standard framing rate of 50 Hz are used. Otherwise, high speed videos or motion picture cameras are required. Such a method was used by Perkins and Hunt (1989) to experimentally study particle motion in the wake of a cylinder. In order to reconstruct the particle trajectories, a cross-correlation method was used to avoid a time-consuming search algorithm to identify images belonging to one particle track.

In the other mode of PTV, the single-frame multi-pulse method, the time sequence of the particle images is recorded on one frame for a certain time interval. This method has the advantage that slow image recording systems such as standard CCD cameras may be used. However, the direction of particle motion cannot be resolved by this method and its applicability is limited to simple flows with a known direction of particle motion. This limitation can be overcome with pulse coding or image shifting, which is described later. In combination with low-frequency laser light sources, this method produces a sequence of particle streak lines, where the particle velocity is determined from the displacement of the centers of individual streak lines. For this mode of operation, the pulse duration needs to be longer than the image diameter divided by the particle velocity, that is, $\Delta t > d_I / |\vec{u}|$ and the time between pulses needs to be longer than the pulse duration.

The analysis of PTV images usually consists of the following steps (see Hassan *et al.*, 1992):

1. Thresholding of the image, that is, removal of noise objects with intensity below a certain preselected gray level scale
2. Analysis of the image to determine the location of the particle images
3. Point-by-point matching of the particle images from one frame to the next (if the multi-frame technique is used)
4. Determination of the displacement and particle velocity for each individual particle
5. Post-processing in order to erase unreasonable velocity vectors resulting from errors in the previous analysis steps

Some interesting examples of the application of PTV techniques to two-phase flows are presented in the following. The particle streak-line technique was used by Ciccone *et al.* (1990) to study the saltating motion of particles over a deposited particle layer. A section of the flow field was illuminated with a laser light sheet, which was chopped by a rotating solid grid with an equal number of open and closed sections.

The images were recorded by conventional film photography. Subsequently, the photographs were digitized and enhanced using fast Fourier transforms (FFT) to enable reconstruction of the trajectories of individual particles from a number of imaged streak lines.

A similar streak-line technique was used by Sommerfeld *et al.* (1993c) and Sommerfeld and Huber (1999) to obtain statistical information on the collision of individual particles with rough walls. The laser light sheet was produced in such a way that it propagated in the streamwise direction along a horizontal channel and illuminated thereby a thin sheet perpendicular to the wall of the channel. The light beam from a continuous Argon-Ion laser was chopped using a Bragg-cell. The images of particle trajectories in the near-wall region were recorded by a CCD camera, which was mounted perpendicular to the light sheet and was equipped with a macro-zoom lens and an extension bellow. A special processing software package was developed to determine the streak lines belonging to an individual particle, which collided with the wall. The processing routines involved the enhancement of the images, the identification of streak lines with a minimum of 50 pixels and the grouping of two streak lines before and after wall impact for each wall collision event. The particle velocity components and the trajectory angles before and after the impact were determined from the separation of the geometric centers of the two respective streak lines.

The two most commonly used PIV methodologies are the single-frame/double-pulse (or less common multi-pulse) technique and the multi-frame (i.e. mostly two-frame)/single-pulse method. The first technique is usually combined with a spatial autocorrelation algorithm in order to evaluate the average displacement of image pairs in each interrogation area. This technique has the advantage of not requiring sophisticated recording systems with high-speed shutters. When used with a double-pulse laser, it cannot resolve the particle direction of motion. Hence, additional measurements are required to determine the particle motion direction in complex or turbulent flows. Two such methods are most common:

1. The recording system (e.g. CCD or photographic camera) is placed on a linear rail and moves at a constant and defined speed during the exposure.
2. A rotational mirror is mounted between the observed flow field and the recording system.

Both methods result in an additional known shift of the second image with respect to the first one. By subtracting this shift from the recorded image displacement, the effective displacement is determined, which may be either positive or negative, depending on the sense of particles velocity vector.

The second methodology of PIV is the multi-frame/single-pulse method. In this case, subsequent images are recorded on separate frames, and hence, similar to the corresponding PTV method, a high-speed recording system is required with a speed identical to the laser pulse frequency or a frequency, which is an integer multiple smaller than the laser pulse frequency. The shutter of the camera has to be synchronized with the laser pulses. Since the displacement is evaluated from at least two subsequent images, the direction of particle motion is fully determined, and no additional equipment is needed. Cross-correlation methods are most favorable for the determination of the average displacement of the particles in the interrogation area.

Examples of the application of PIV in two-phase flows are numerous. In many cases, however, the velocity fields of both phases were not recorded simultaneously. Therefore, it was not possible to assess the effect of the dispersed phase on the fluid flow, the so-called two-way coupling. For example, the influence of vortices on the particle motion in a forced jet impinging onto a circular plate was investigated by Longmire and Anderson (1995). The velocity fields of both phases in the vicinity of the wall were evaluated independently using the single-frame/double-pulse PIV method. In order to resolve directional ambiguity, the image shift technique, based on a rotating mirror placed between the flow field and the camera, was used. For analyzing the vortex structure in the single-phase jet, both the jet and the outer flow were seeded with smoke particles. For the determination of the particle phase velocity field, it was required that a sufficient number of particles be present in each interrogation area in order to ensure a good autocorrelation. Since the particle loading was very low (i.e. dilute system), a complete velocity vector field could not be obtained for the dispersed phase, although ensemble averages of ten realizations were evaluated. Problems in resolving the velocity of particles moving toward the wall and those rebounding from the wall using PIV became obvious. In those interrogation areas, where both classes of particles were present, an average velocity depending on the number of particles in each class was determined. This case reveals one disadvantage of the PIV method.

In order to assess the influence of the particle phase onto the fluid flow, the so-called two-way coupling, simultaneous imaging of both phases, is required. Hence, the fluid flow has to be seeded with appropriate tracer particles, which are able to follow the velocity fluctuations of the flow. For such a purpose, the tracer particle Stokes number—calculated with the relevant smallest timescale of turbulence—has to be much smaller than 1. Very often, the particle phase velocities are obtained by the PTV approach due to their mostly low number concentration, and the tracer particle velocity field is evaluated by PIV algorithms (Elhimer *et al.*, 2016). In order to discriminate the images from tracer and dispersed phase particles, which is the most important step in multiphase flow imaging, two approaches are most common:

1. Discrimination based on the size of the particle images. This requires a considerable difference in the particle size of the dispersed phase particles and the tracer particles as, for example, in a bubbly and droplet flows.
2. Introduction of fluorescent tracer particles, which are excited by the incident laser light and emit light with a different wavelength, also called optical discrimination. This method requires two recording systems, each equipped with an appropriate color filter.

The size discrimination method is reliable in case the particulate-phase images are resolved by quite a number of pixels. The tracer particle images need to cover at least one pixel in order to be detectable. The success of such discrimination relies on a number of image filtering and manipulation steps. The initial image includes both tracer and dispersed phase particles. First, such an image must be conditioned by erasing background noise yielding image *A*. Afterwards, the edges of the object boundaries, especially those of the dispersed phase particles, need to be enhanced through edge detection filters obtaining image *B*. Using now a median filter, all small object images, namely those of the tracer particles, can be removed giving image *C*. This image contains only the dispersed phase objects and is used to evaluate particle properties, such as size and velocity by PTV. The subtraction of image *C*, only containing the particle objects, from image *B* yields an image that contains only tracer particles named image *D*. This tracer image can be further processed using a PIV algorithm to obtain the fluid velocity field. However, since the initial location of the larger dispersed phase elements leaves empty holes, velocities near the particle boundary can be not determined easily (Elhimer *et al.*, 2016). Such a rather sophisticated but accurate method was applied by Khalitov and Longmire (2002) for measurements in a turbulent particle-laden channel flow. The solid particles were differently sized glass beads with diameters around 50 µm, and 1 µm oil droplets were used as tracer particles. Consequently, for allowing such measurements, where the solid particles cover several pixels, high resolution imaging with a large magnification is required. The same approach was applied by Bröder and Sommerfeld (2007) for measurements in a laboratory bubble column, where in addition to the velocities of both phases, also bubble size and shape were determined. It should be emphasized, that in the case of quite large dispersed phase elements, a backlighting (i.e. yielding shadow images) should be preferred compared to a light-sheet illumination. In such cases, only the glare-points of the particles are appearing on the images, and it is not possible to determine their size.

The second discrimination approach is based on laser-induced fluorescence (LIF). In this case, either the seed particles or the dispersed phase particles (i.e. solid particles or droplets) are treated with a fluorescent dye solution. Solid particles must be impregnated with the dye solution, while in the case of liquid droplets, the dye solution may be mixed with the liquid prior to atomization. When the dyed particles are illuminated by laser light with the appropriate wavelength, they emit light at a different wavelength. For example, rhodamine 6G may be excited by green light (i.e. 514.5 nm), and the emitted fluorescent light has a wavelength around 580.5 nm. By using two CCD cameras (or any other cameras) with appropriate color filters, one camera may receive only the scattered light around the laser wavelength, and the other camera responds to the fluorescent light. Hence, separate images of the seed particles and the dispersed phase particles are produced.

A similar technique combined with shadow imaging was used by Tokuhiro *et al.* (1998) to measure the fluid velocity distribution around a large, stationary bubble, especially within the wake. In this case, a laser light sheet for PIV and an infrared LED array for background illumination, providing a planar image of the bubble, were combined. Two opposed mounted CCD cameras were used in order to image the bubble shape using a shadow technique and for recording the fluorescing tracer images for doing a PIV.

A similar LIF-PIV system based on a light sheet illumination was developed by Bröder and Sommerfeld (2002) for a bubble column (i.e. diameter 140 mm). The velocity field of the fluid was obtained, adding 50 µm, almost neutrally buoyant tracer particles doped with Rhodamine 6G dye. Also, the bubbles (i.e. volume fractions up to about 5%) were illuminated by the light sheet, and hence, only the scattering from the glare points was received by the CCD camera. Therefore, the bubbles were not resolved by the imaging system. For recoding the fluorescent light from the tracer and the scattered light from the bubbles, two CCD cameras equipped with appropriate band pass filters were used, as shown in Figure 9.61. In addition, another optical discrimination between the phases was applied. The bubble-CCD camera was placed at a scattering angle of 80° with the dominant reflected light. The tracer-CCD camera was placed at a scattering angle of 106°, close to the Brewster's angle for the bubbles where the reflection from the bubbles is very low, as shown on Figure 7.49a (with parallel polarization). In order to compensate the slightly oblique view of the CCD cameras, they were equipped with Scheimpflug optical systems, and two double images are obtained for tracer (i.e. fluid) and point bubbles (only glair points are detected). Since the bubble number concentration was sufficiently high, both velocity fields were obtained by PIV using an accurate and fast MQD (minimum quadratic difference) approach proposed by Gui *et al.* (1997). In order to map the entire flow field along the column (height 650 mm), the light sheet with about 3 mm thickness and 100 mm height was scanned along the column with a computer-controlled system. In each plane about 2,000 double images were collected and averaged to obtain mean velocity fields of both phase as well as the associated mean fluctuations (Bröder and Sommerfeld, 2002). These few examples presented on imaging techniques show the wide range of applications and the potential of imaging techniques for analyzing two-phase flows. Further examples may be found in relevant journals and conference proceedings.

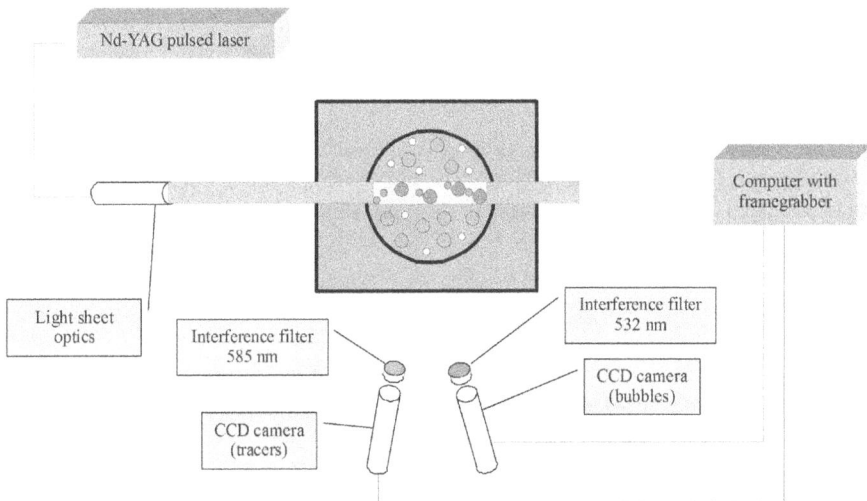

FIGURE 9.61 Schematic diagram of the bubble column test facility. The circular bubble column is placed in a square tank also filled with water in order to partly match the refractive indices. Also shown are the optical arrangement of the LIF-PLV-system, light sheet illumination with a pulsed Nd-YAG laser, two CCD cameras with Scheimpflug optics and band pass filters for recording fluorescent light from the trace particles, and scattered light from the bubbles.

9.6   SUMMARY

In this chapter, some of the most common measurement techniques for particle characterization (i.e. sampling methods) and online analysis of two-phase systems are introduced. Most of the techniques for the analysis of particle samples with regard to size, shape, and surface area are well established.

Local or pointwise online methods such as light scattering, laser-Doppler, and phase-Doppler anemometry for particle characterization and for the reliable and accurate measurement of flow properties in industrial processes or in research laboratories are well established, and commercial instruments are available. Further developments are required to measure additional properties, such as the refractive index and temperatures of droplets. Due to the powerfulness of imaging techniques, LDA and PDA systems are currently less often used in research.

All types of imaging techniques introduced are developed for very specific purposes, mostly in the research area. One may globally distinguish between high-resolution particle-scale imaging and larger scale field imaging for obtaining information on flow regimes and particle phase distributions. Velocity field imaging techniques, like PIV and PTV, are well established, and commercial instruments are available. There are still numerous developments for specific applications and combining these methods with other techniques. An important limitation in applications is the requirement of the transparency of the two-phase flow system.

Industrial applications of single particle counting instruments for process control are, at present, very few. More robust and easy-to-use instruments need to be developed for such applications. Miniaturized optical systems based on semiconductor lasers, which have been developed lately, may give rise to compact and robust systems in the future.

However, one should keep in mind that optical techniques are limited to applications in dilute two-phase systems only. For online measurements and control in dense systems, radiation sources, such as X-rays and gamma rays, which are highly penetrative due to their extremely short wavelengths, have to be used. Tomographic methods, based on these radiation sources, have received considerable attention for imaging the phase distribution in dense multiphase systems.

Example: The optical configuration of a two-detector PDA system for measurements in a particle-laden channel flow (air with $\rho_c = 1.18$ kg/m^3) with a mass concentration of two is to be designed. The particles (spherical glass beads with $\rho_d = 2,500$ kg/m^3, $n_p = 1.52$) have a size range from 40 to 150 μm with a number averaged diameter of 60 μm. The relative refractive index is 1.52. Transmitting optics with a focal length of 500 mm and a beam spacing Δb of 30 mm together with a He-Ne laser ($\lambda = 632.8$ nm) with a beam diameter d_o of 1 mm are available. The receiving optics should be positioned at an off-axis angle of 30°, where refraction is the dominant scattering mode. The spatial filter in the receiving optics should have a width of 100 μm and a magnification of about 2 so that a length of 200 μm of the probe volume is imaged onto the receiver.

- Determine a suitable elevation angle for the rectangular slits to ensure that particles up to 200 μm can be measured.

- Estimate the coincidence error for the presence of two particles in the probe volume. Assume that the gas and particles have the same mean velocity and carry out the estimate assuming that the particles are monodisperse with a size corresponding to the most probable diameter in the size distribution, that is, 60 μm.
- Redesign your optical system to reduce the coincidence error to 5% by considering changing the focal length transmitting lens.

Solution:

The half angle of the incident beam, θ, is

$$\tan\theta = \frac{\Delta b}{2f_e}, \quad \theta = 1.72^0$$

Using Eq. (9.60), one can solve for ψ in the form

$$\sin\psi = \Delta\phi \frac{\lambda_e}{2\pi D} \frac{m-1}{m} \frac{1}{\sin\theta}$$

With $\Delta\phi = 2\pi$, $D = 200$ μm, the value for ψ is 2.06°. From Eq. (9.64) with $\Delta\phi = 2\pi$, the maximum diameter is

$$D_{max} = \frac{\lambda}{m\Phi} \tag{9e.1}$$

Using Eq. (9.66) with $\Psi = 2°$ gives $\Phi = 0.0027$ and

$$D_{max} = 154\mu m \tag{9e.2}$$

which is within the size range. However, extending the range to a 200 μm particle requires a smaller angle. For $\Psi = 1.5°$, $\Phi = 0.00202$, which would allow measurement of a 206 μm particle. With a 500 mm lens, this would require a 2.6 cm spacing. The final choice will depend on the slit separation and receiver focal lengths that are available.

From Eq. (9.50), the waist diameter is

$$d_m = \frac{4\hat{f}_e\lambda_e}{\pi d_o} = 402\mu m$$

and the probe volume is

$$V_m = \frac{\pi}{4}d_m^2 L_s = 0.0254\text{mm}^3$$

The relative probability of having two particles in the probe volume is given by Eq. (9.36) as

$$P_2' = \frac{N_m}{2} = \frac{nV_m}{2}$$

For a mass concentration of 2,

$$C = \frac{\alpha_d \rho_d}{\alpha_c \rho_c} = 2 \tag{9e.3}$$

Taking $\alpha_c \simeq 1$

$$\alpha_d = 2\frac{1.18}{2500} = 9.4 \times 10^{-4} \tag{9e.4}$$

The volume fraction and number density are related by

$$\alpha_d = n\frac{\pi}{6}D^3 \tag{9e.5}$$

For 60 μm particles

$$n = 8.3/\text{mm}^3 \tag{9e.6}$$

Thus

$$\frac{nV_m}{2} = \frac{8.3 \times 0.254}{2} = 0.1 \tag{9.94}$$

Hence, the coincidence error is 0.1 or 10%.

   For a coincidence error of 5%, the measuring volume has to be halved, which can be done by decreasing the probe diameter by a factor $\sqrt{2}$. This requires the focal length to be 350 mm. A transmitting lens with a focal length of 350 mm or less would be adequate.

NOTE

1 The terms *anemometry* and *velocimetry* are interchangeably used in this field.

REFERENCES

Adrian, R.J., 1986, Multi-point optical measurements of simultaneous vectors in unsteady flow: A review, *Int. J. Heat and Fluid Flow*, **7**, 127–145.

Adrian, R.J., 1991, Particle-imaging techniques for experimental fluid mechanics, *Ann. Rev. Fluid Mech.*, **23**, 261–304.

Adrian, R.J. and Orloff, K.L., 1977, Laser anemometer signals: Visibility characteristics and application to particle sizing, *Applied Optics*, **16**, 677–684.

Aizu, Y., Durst, F., Gréhan, G., Onofri, F. and Xu, T.-H., 1993, PDA-system without Gaussian beam defects, *The Third Int. Congress on Optical Particle Sizing*, Yokohama, Japan, 23–26 August.

Albrecht, H.-E., Damaschke, N., Borys, M. and Tropea, C., 2003, *Laser Doppler and Phase Doppler Measurement Techniques*, Series: Experimental Fluid Mechanics, Springer-Verlag, Berlin.

Allano, D., Gouesbet, G., Grehan, G. and Lisiecki, D., 1984, Droplet sizing using a top-hat laser beam technique, *J. of Physics D: Applied Physics*, **17**, 43–58.

Allen, T., 1990, *Particle Size Measurements*, 4th edition, Chapman and Hall, London.

Allen, T., 2003, *Powder Sampling and Particle Size Determination*, 1st edition, Elsevier, Amsterdam, Holland.

Bauckhage, K., 1988, The phase-Doppler-difference-method, a new laser-Doppler technique for simultaneous size and velocity measurements, *Part. Part. Syst. Charact.*, **5**, 16–22.

Beck, M.S. and Plaskowski, A., 1987, *Cross Correlation Flowmeters: Their Design and Application*, Adam Hilger, Bristol.

Bohnet, M., 1973, Staubgehaltsbestimmung in strömenden Gasen, *Chemie-Ing.-Techn.*, **45**, 18–24.

Börner, T., Durst, F. and Manero, E., 1986, LDV measurements of gas-particle confined jet flow and digital data processing, *Proceedings, 3rd Int. Symposium on Applications of Laser Anemometry to Fluid Mechanics*, Paper 4.5.

Boyer, C., Duquenne, A.-M. and Wild, G., 2002, Measuring techniques in gas-liquid and gas-liquid-solid reactors, *Chemical Engineering Science*, **57**, 3185–3215.

Bröder, D. and Sommerfeld, M., 2002, An advanced LIF-PLV system for analysing the hydrodynamics in a laboratory bubble column at higher void fraction, *Experiments in Fluids*, **33**, 826–837.

Bröder, D. and Sommerfeld, M., 2007, Planar shadow image velocimetry for the analysis of the hydrodynamics in bubbly flows, *Measurement Science and Technology*, **18**, 2513–2528.

Broßmann, R., 1966, *Die Lichtstreuung an kleinen Teilchen als Grundlage einer Teilchengrößenbestimmung*. Doctoral Thesis, University of Karlsruhe, Faculty of Mechanical and Process Engineering.

Cartellier, A., 1992, Simultaneous void fraction measurement, bubble velocity, and size estimate using a single optical probe in gas-liquid two-phase flows, *Rev. Sci. Instrum.*, **63**, 5442–5453.

Chavez, A. and Mayinger, F., 1990, Evaluation of pulsed laser hologramms of spray droplets using digital image processing, *Proc. 2nd Int. Congress on Optical Particle Sizing*, Arizona, 462–471.

Chigier, N., 1991, Optical imaging of sprays, *Prog. Energy Combust. Sci.*, **17**, 211–262.

Chigier, N.A., Ungut, A. and Yule, A.J., 1979, Particle size and velocity measurements in planes by laser anemometer, *Proc. 17th Symp. (Int.) on Combustion*, 315–324.

Ciccone, A.D., Kawall, J.G. and Keffer, J.F., 1990, Flow visualization/digital image analysis of saltating particle motions, *Experiments in Fluids*, **9**, 65–73.

Damaschke, N., Nobach, H., Nonn, T.I., Semidetnov, N. and Tropea, C., 2005, Multi-dimensional particle sizing techniques, *Experiments in Fluids*, **39**, 336–350.

DANTEC/Invent, STREU, 1994, A computational code for the light scattering properties of spherical particles, *Instruction Manual*.

Durst, F., 1982, Review-combined measurements of particle velocities, size distribution and concentration: Transactions of the ASME, *J. of Fluids Engineering*, **104**, 284–296.

Durst, F., Melling, A. and Whitelaw, J.H., 1981, *Principles and Practice of Laser-Doppler Anemometry*, 2nd edition, Academic Press, London.

Durst, F., Melling, A. and Whitelaw, J.H., 1987, *Theorie und Praxis der Laser-Doppler Anemometrie*, G. Braun Verlag, Karlsruhe.

Durst, F. and Zaré, M., 1975, Laser-Doppler measurements in two-phase flows, *Proceedings of the LDA-Symposium*, University of Denmark.

Elhimer, M., Praud, O., Marchal, M., Cazin, S. and Bazile, R., 2016, Simultaneous PIV/PTV velocimetry technique in a turbulent particle-laden flow, *J. of Visualization*, **20**, 1–16.

Farmer, W.M., 1972, Measurement of particle size, number density and velocity using a laser interferometer, *Applied Optics*, **11**, 2603–2612.

Farmer, W.M., 1974, Observation of large particles with a laser interferometer, *Applied Optics*, **13**, 610–622.

Fayed, M.E. and Otten, L. (eds.), 1997, *Handbook of Powder Science & Technology*, 2nd edition, Chapman & Hall, New York, USA.

Frohn, A. and Roth, N., 2000, *Dynamics of Droplets*, Springer Verlag, Berlin.

Grehan, G. and Gouesbet, G., 1986, Simultaneous measurements of velocities and size of particles in flows using a combined system incorporating a top-hat beam technique, *App. Opt.*, **25**, 3527–3538.

Grehan, G., Gouesbet, G., Nagwi, A. and Durst, F., 1992, On elimination of the trajectory effects in phase-Doppler systems, *Proc. 5th European Symp. Particle Characterization (PARTEC 92)*, 309–318.

Gui, L., Lindken, R. and Merzkirch, W., 1997, Phase-separated PIV measurements of the flow around systems of bubbles rising in water, *Proceedings of the ASME Fluids Engineering Summer Meeting*, Paper No. FEDSM97–3103.

Hardalupas, Y., Hishida, K., Maeda, M., Morikita, H., Taylor, A.M.K.P. and Whitelaw, J.H., 1994, Shadow Doppler technique for sizing particles of arbitrary shape, *Applied Optics*, **33**, 8417–8426.

Hassan, Y.A., Blanchat, T.K., Seeley, C.H. and Canaan, R.E., 1992, Simultaneous velocity measurements of both components of a two-phase flow using particle image velocimetry, *Int. J. Multiphase Flow*, **18**, 371–395.

Hering, S.V., 2001, Impactors, cyclones, and other inertial and gravitational collectors, in *Air Sampling Instruments for Evaluation of Atmospheric Contaminants*, 9th edition (Eds. B.S. Cohen and C. McCammon), pp. 316–376, ACGIH, Cincinnati, USA.

Hering, S.V., Plagan, R.C. and Friedlander, S.K., 1978, Design and evaluation of new low-pressure impactor, *Environmental Science & Technology*, **12**, 667–673.

Hess, C.F., 1984, Non-intrusive optical single-particle counter for measuring the size and velocity of droplets in a spray, *Applied Optics*, **23**, 4375–4382.

Hess, C.F. and Espinosa, V.E., 1984, Spray characterization with a nonitrusive technique using absolute scattered light, *Optical Engineering*, **23**, 604–609.

Hirleman, D.E., Oechsle, V. and Chigier, N.A., 1984, Response characteristics of laser diffraction particle size analyzers: Optical sample volume extent and lens effects, *Optical Engineering*, **23**, 610–619.

Hishida, K. and Maeda, M., 1990, Application of laser/phase Doppler anemometry to dispersed two-phase flow, *Part. Part. Syst. Charact.*, 7, 152–159.

Huber, N. and Sommerfeld, M., 1994, Characterization of the cross-sectional particle concentration distribution in pneumatic conveying systems, *Powder Technology*, **79**, 191–210.

Khalitov, D.A. and Longmire, E.K., 2002, Simultaneous two-phase PIV by two-parameter phase discrimination, *Experiments in Fluids*, **32**, 252–268.

Kipphan, H., 1977, Bestimmung von Transportkenngrößen bei Mehrphasenströmungen mit Hilfe der Korrelationstechnik, *Chem.-Ing.-Techn., Jahrg.*, **49**, 695–707.

Kiselev, A., Wex, H., Stratmann, F., Nadeev, A. and Karpushenko, D., 2005, White-light optical particle spectrometer for *in situ* measurements of condensational growth of aerosol particles, *Applied Optics*, **44**, 4693–4701.

Kliafas, Y., Taylor, A.M.K.P. and Whitelaw, J.H., 1990, Errors due to turbidity in particle sizing using laser-Doppler anemometry. Trans. of the ASME, *J. Fluid Engineering*, **112**, 142–148.

Lischer, J. and Louge, M.Y., 1992, Optical fiber measurements of particle concentration in dense suspensions: Calibration and simulation, *Applied Optics*, 31, 5106–5113.

Longmire, E.K. and Anderson, S.L., 1995, Effects of vortices on particle motion in a stagnation zone, in *Gas-Particle Flows* (Eds. D.E. Stock, Y. Tsuji, M.W. Reeks, E.E. Michaelides and M. Gautam), Vol. 228, pp. 89–101, ASME FED.

Longmire, E.K. and Eaton, J.K., 1992, Structure of a particle-laden round jet, *J. Fluid Mechanics*, **236**, 217–257.

Marple, V.A., 2004, History of impactors: The first 110 years, *Aerosol Science and Technology*, **38**, 247–292.

Marple, V.A. and Liu, B.Y.H., 1974, Characteristics of laminar jet impactors, *Environmental Science & Technology*, **8**, 648–654.

Mie, G., 1908, Beiträge zur Optik trüber Medien, speziell kolloidaler Metallösungen, *Ann. der Physik, Jahrg.*, **25**, 377–422.

Mitchell, J.P. and Nagel, M.W., 2003, Cascade impactors for the size characterization of aerosols from medical inhalers: Their uses and limitations, *J. Aerosol Med.*, **16**, 341–377.

Modarress, D. and Tan, H., 1983, LDA signal discrimination in two-phase flows, *Experiments in Fluids*, **1**, 129–134.

Morikita, H., Hishida, K. and Maeda, M., 1994, Simultaneous measurement of velocity and equivalent diameter of non-spherical particles, *Part. Part. Syst. Charact.*, **11**, 227–234.

Morikita, H. and Taylor, A.M.K.P., 1998, Application of shadow Doppler velocimetry to paint spray: Potential and limitations in sizing optically inhomogeneous droplets. *Meas. Sci. Tech.*, **9**, 221–231.

Naqwi, A.A. and Durst, F., 1991, Light scattering applied to LDA and PDA measurements. Part 1: Theory and numerical treatments, *Part. Part. Syst. Charact.*, **8**, 245–258.

Negus, C.R. and Drain, L.E., 1982, Mie calculations of scattered light from a spherical particle traversing a fringe pattern produced by two intersecting laser beams, *J. Phys. D: Applied Physics*, **15**, 375–402.

Perkins, R.J. and Hunt, J.C.R., 1989, Particle tracking in turbulent flows, in *Advances in Turbulence*, Vol. 2, pp. 286–291, Springer Verlag, Berlin.

Petrak, D. and Hoffmann, A., 1985, The properties of a new fiberoptic measuring technique and its application to fluid-solid flows. *Advan. Mech.*, **8**, 59.

Poelma, C., 2020, Measurement in opaque flows: A review of measurement techniques for dispersed multiphase flows, *Acta Mech.*, **231**, 2089–2111.

Powell, R.L., 2008, Experimental techniques for multiphase flows, *Physics of Fluids*, **20** (040605).

Qiu, H.-H., Jia, W., Hsu, C.T. and Sommerfeld, M., 2000, High accuracy optical particle sizing in phase-Doppler anemometry, *Measurement Science Technology*, **11**, 142–151.

Qiu, H.-H. and Sommerfeld, M., 1992, A reliable method for determining the measurement volume size and particle mass fluxes using phase-Doppler anemometry, *Experiments in Fluids*, **13**, 393–404.

Rensner, D. and Werther, J., 1992, Estimation of the effective measuring volume of single fiber reflection probes for solids concentration measurements, *Preprints of the 5th European Symposium Particle Characterization (PARTEC 92)*, 107–118.

Rhodes, M., 2008, *Introduction to Particle Technology*, 2nd edition, John Wiley & Sons, New York, NY.

Roberts, D.W., 1977, Particle sizing using laser interferometry, *Applied Optics*, **16**, 1861–1868.

Saffman, M., 1987a, Automatic calibration of LDA measurement volume size, *Applied Optics*, **26**, 2592–2597.

Saffman, M., 1987b, Optical particle sizing using the phase of LDA signals, *Dantec Information*, No. 5, 8–13.

Sankar, S.V. and Bachalo, W.D., 1991, Response characteristics of the phase-Doppler particle analyzer for sizing spherical particles larger than the wavelength, *Applied Optics*, **30**, 1487–1496.

Sijs, R., Kooij, S., Holterman, H.J., van de Zande, J. and D. Bonn, D., 2021, Drop size measurement techniques for sprays: Comparison of image analysis, phase Doppler particle analysis, and laser diffraction, *AIP Advances*, **11** (015315).

Sommerfeld, M., 2007, Application of extended imaging techniques for analyzing elementary processes in multiphase flow, in *Multiphase Flow: The Ultimate Measurement Challenge*, (Eds. X. Cai, Y. Wu, Z. Huang, S. Wang and M. Wang), Vol. 914, pp. 20–30, AIP Conference Proceedings, Meville, New York.

Sommerfeld, M. and Huber, N., 1999, Experimental analysis and modelling of particle-wall collisions, *Int. J. of Multiphase Flow*, **25**, 1457–1489.

Sommerfeld, M., Huber, N. and Wächter, P., 1993c, Particle-wall collisions: Experimental studies and numerical models, in *Gas-Solid Flows* (Eds. D.E. Stock, M.W. Reeks, Y. Tsuji, M. Gautam, E.E. Michaelides and J.T. Jurewicz), FED-Vol. 166, 183–191, ASME Fluids Engineering Conference, Washington, DC.

Sommerfeld, M. and Pasternak, L., 2019, Advances in modelling of binary droplet collision outcomes: A review of available results, *Int. J. of Multiphase Flow*, **117**, 182–205.

Sommerfeld, M. and Qiu, H.-H., 1995, Particle concentration measurements by phase-Doppler anemometry in complex dispersed two-phase flows, *Experiments in Fluids*, **18**, 187–198.

Sommerfeld, M. and Tropea, C., 1999, Single-point laser measurement, Chapter 7 in *Instrumentation for Fluid-Particle Flow* (Ed. S.L. Soo), pp. 252–317, Noyes Publications, Park Ridge, NJ.

Soo, S.L., Stukel, J.J. and Hughes, J.M., 1969, Measurement of mass flow and density of aerosols in transport, *Environmental Science and Technology*, **3**, 386–393.

Strutt, J.W., 1871, On the light from the sky, its polarization and colour, *Philos. Mag.*, **41**, 107–120.

Swithenbank, J., Beer, J.U., Taylor, D.S., Abbot, D. and McCreath, G.C., 1977, A laser diagnostic technique for the measurement of droplet and particle size distribution, *Progress in Astronautics and Aeronautics, AIAA*, 421–447.

Tayali, N.E. and Bates, C.J., 1990, Particle sizing techniques in multiphase flows: A review, *Flow Meas. Instrum.*, **1**, 77–105.

Tokuhiro, A., Maekawa, M., Iizuka, K., Hishida, K. and Maeda, M., 1998, Turbulent flow past a bubble and an ellipsoid using shadow-image and PIV techniques, *Int. J. of Multiphase Flow*, **24**, 1383–1406.

Tropea, C., 1995, Laser Doppler anemometry: Recent developments and future challenges, *Meas. Sci. Techn.*, **6**, 605–619.

Tropea, C., 2011, Optical particle characterization in flows, *Annu. Rev. Fluid Mech.*, **43**, 399–426.

Tropea, C., Xu, T.-H., Onofri, F., Gréhan, G., Haugen, P. and Stieglmeier, M., 1996, Dual-mode phase-Doppler anemometer, *Part. Part. Syst. Charact.*, **13**, 165–170.

Tu, C., Yin, Z., Lin, J. and Bao, F., 2017, A review of experimental techniques for measuring micro- to nano-particle-laden gas flows, *MDPI, Appl. Sci.*, **7**, 120.

Umhauer, H., 1983, Particle size distribution analysis by scattered light measurements using an optically defined measuring volume, *J. Aerosol Science*, **14**, 765–770.

Umhauer, H., Löffler-Mang, M., Neumann, P. and Leukel, W., 1990, Pulse holography and phase-Doppler technique: A comparison when applied to swirl pressure-jet atomizers, *Particle and Particle Systems Characterization*, 7, 226–238.

Van de Hulst, H.C., 1981, *Light Scattering by Small Particles*, Dover Publications, Inc., New York.

Van Dyke, M., 1982, *An Album of Fluid Motion*, The Parabolic Press, Stanford, CA.

van Ommen, R. and Mudde, R.F., 2008, Measuring the gas-solids distribution in fluidized beds: A review, *Int. J. of Chemical Reactor Engineering*, **6**.

Weiner, B.B., 1984, Particle and droplet sizing using Fraunhofer diffraction, in *Modern Methods of Particle Size Analysis* (Ed. H.G. Barth), pp. 135–172, John Wiley & Sons, New York.

Wen, F., Kamalu, N., Chung, J.N., Crowe, C.T. and Troutt, T.R., 1992, Particle dispersion by vortex structures in plane mixing layers, Transactions of the ASME, *J. of Fluids Engineering*, **114**, 657–666.

Werther, J., 1999, Measurement techniques in fluidized beds, *Powder Technol.*, **102**, 15–36.

Wiederseiner, S., Nicolas Andreini, N., Epely-Chauvin, G. and Ancey, C., 2011, Refractive-index and density matching in concentrated particle suspensions: A review, *Exp Fluids*, **50**, 1183–1206.

Williams, R.A. and Beck, M.S. (eds.), 1995, *Process Tomography*, Butterworth-Heinemann Ltd., Oxford.

Williams, R.A., Xie, C.G., Dickin, F.J., Simons, S.J.R. and Beck, M.S., 1991, Review: Multi-phase flow measurement in powder processing, *Powder Technology*, **66**, 203–224.

Xu, T.-H. and Tropea, C., 1994, Improving the performance of two-component phase Doppler anemometers, *Meas. Sci. Technol.*, **5**, 969–975.

Yeoman, M.L., Azzopardi, B.J., White, H.J., Bates, C.J. and Roberts, P.J., 1982, Optical development and application of a two-color LDA system for the simultaneous measurement of particle size and particle velocity, *ASME Winter Annual Meeting*, Arizona, 127–135.

PROBLEMS

9.1   The particle concentration in a particle-laden pipe flow is to be measured using a simple sampling probe. A homogeneous particle concentration in the pipe cross-section may be assumed. The suction velocity is adjusted to the average gas velocity in the pipe, that is, 20 m/s. An accumulation of 60 grams of particles is measured on the filter over a one-minute interval. Assume that the gas is incompressible and the gas velocity profile is given by

$$\frac{u}{U_o} = \left(\frac{y}{R}\right)^{1/n} \quad (\text{with } n = 7)$$

$$\frac{\bar{u}}{U_o} = \frac{2n^2}{(n+1)(2n+1)}$$

where U_o is the centerline gas velocity, u is the average velocity, and y is the distance from the wall, and R is the pipe radius.

Determine the particle concentration measured on the pipe centerline and 50 mm away from the wall using the result given in Figure 9.26.

The following properties are given:

pipe diameter	200 mm
sampling probe diameter	10 mm
sampling time	20 s
particle size	200 μm
particle material density	2500 kg/m^3
density of gas	1.18 kg/m^3
kinematic viscosity	18.4×10^{-6} Ns/m^2

9.2   Determine the required dimensions of the probe volume for a light scattering instrument allowing a coincidence error of 5%. The measurements will
be carried out in a gas-solid flow with a mass loading ratio of 2. Consider
two particle sizes, 15 and 60 μm. The particle material density is 2,500 kg/m³.
Assume the gas density is 1.2 kg/m³.

9.3   Design an LDA system for measurements in a high-speed, gas-particle
flow (initial velocity 500 m/s) through a normal shock wave. You have
available a signal processor, which is able to resolve a signal frequency
of 50 MHz. No frequency shifting is to be used. The laser has a beam
diameter of 2 mm. Determine the beam-crossing angle and select a suitable combination of initial beam spacing (typically: 10, 20, 30, 40, 50
mm) and focal length (typical focal lengths of lenses are 160, 300, 500,
700, and 1,000 mm). Also calculate the diameter of the probe volume and
the number of fringes. Ensure that the number of fringes is larger than 10
to allow for reliable signal processing. Assume the laser wave length is
514.5 nm.

9.4   Select an appropriate configuration of a PDA system for measurements in
a spray of water droplets (n$_p$ = 1.33) with a size spectrum from 50 to 200
μm. The spray is issuing into an airflow. A three-detector system with the
following configuration is available:

laser wavelength	514.5 nm
laser beam diameter	1.35 mm
focal length of transmitting lens	500 and 1,000 mm
initial beam spacing	variable between 20 and 40 mm
receiving lens focal length	variable between 160 and 400 mm

The receiving optics have interchangeable masks with the smallest detector
spacing (i.e. detectors 1–2 in Figure 9.48) being 5, 10, and 15 mm.
    Select a scattering angle of 30 degrees and try to minimize the Gaussian
beam effect by choosing the diameter of the probe volume to be larger than
the largest drop.

9.5   Estimate the errors for a PTV measurement in a flow with particles of 80 μm
and a maximum velocity of 10 m/s. A copper-vapor laser is available with
a pulse duration of 30 ns and a pulse-to-pulse jitter (variability of time
between pulses) of 2 ns. Assume that the image can be located within an
accuracy of 20%; that is, c = 0.2 in Eq. (9.82).
    The following properties of the optical system are given:

magnification	1
laser wavelength	510.6 nm
f-number of lens	8
resolution of recording system	20 μm

Determine the following parameters:

a) Image diameter
b) A suitable pulse frequency to allow for an image separation of 5 d_I
c) Error due to blur in relation to the image diameter uncertainty
d) Uncertainty in velocity measurement

10 Nanoparticles and Nanofluids

The term *nanofluid* was coined in the 1990s to denote heterogeneous suspensions of tiny particles—nanoparticles, with typically less than 200 nm equivalent diameters—in common fluids, which are called *base fluids*. Because of the presence and Brownian movement of the nanoparticles, the heat and mass transfer characteristics of nanofluids are markedly different than those of the base fluids and have become the subject of investigations of numerous research projects. Most of the experimental data indicate that the addition of small quantities of nanoparticles increases the effective thermal conductivity of the carrier fluid, increases the convective heat transfer coefficients, increases the mass transfer coefficient, increases the viscosity and friction factor, increases the convective boiling coefficient of the fluid, increases the critical heat flux (CHF) of the fluid, and in general, decreases the pool boiling coefficient of the fluid.

10.1 CHARACTERISTICS OF NANOPARTICLES AND NANOFLUIDS

For all practical purposes, nanofluids are heterogeneous mixtures of solid nanoparticles, of sizes less than 200 nm, and the carrier fluids. All the definitions, length scales, and timescales cited in Chapter 2 of this book also apply to the nanofluids. Salient characteristics of the nanofluids are as follows:

1. Because of their very small size, the particles have almost vanishing response times, τ_v and τ_T. As a result, momentum and thermal equilibrium with the base fluid are achieved very fast.
2. The tiny nanoparticles follow the fluid streamlines and, largely, do not contribute to the erosion of channels.
3. Because the surface to volume ratio of a heterogeneous suspension is $6/D$, the nanoparticles exhibit much higher surface-to-volume ratios than fine and coarse particles. Processes that depend on the interfacial area of the particles, such as chemical reactions, catalysis, mass absorption, and interfacial heat transfer take place faster in nanofluids.
4. The Brownian movement of the particles becomes very important. Since thermophoresis is a consequence of the Brownian movement, thermophoretic velocities are faster in nanofluids, especially when the nanoparticles are very small (Michaelides, 2015).
5. Because of the very large surface-to-volume ratio, the electrical forces and the *electrical double layer* around the nanoparticles are very important.

The *zeta potential* characterizes the stability of the nanofluid, with nanofluids that have low zeta potentials being unstable and their particles subject to aggregation.

6. Since particle aggregation is possible, the addition of surfactants and dispersants become necessary to keep the nanoparticles separate and the nanofluid stable. If nanoparticles aggregate, the fluid-solid suspension becomes one of larger particles, and its special characteristics fade.

10.2 EFFECTIVE TRANSPORT PROPERTIES OF NANOFLUIDS

The transport properties—viscosity, thermal conductivity, diffusivity—of homogeneous fluids at any point (x,y,z) of the suspension are defined as ratios of conjugate forces—shear stress, temperature, and concentration gradients—and the corresponding conjugate fluxes—velocity gradient, energy flux, and mass flux (Onsager, 1931)

$$\mu \equiv \frac{\tau}{\left(\dfrac{\partial u}{\partial y}\right)_{x,y,z}} \quad , \quad k \equiv \frac{q''}{\left(\dfrac{\partial T}{\partial y}\right)_{x,y,z}} \quad , \quad D_i \equiv \frac{m_i''}{\left(\dfrac{\partial \rho_i}{\partial y}\right)_{x,y,z}} \tag{10.1}$$

For anisotropic homogeneous materials, the transport properties are defined by the appropriate vectors of conjugate fluxes and forces. The spatial derivatives in the denominators are defined according to the continuum hypothesis (section 2.1). Since nanofluids are heterogeneous materials, their transport properties must be defined with respect to a larger volume that contains a sufficiently large number of nanoparticles. The length scale of this volume must necessarily be significantly larger than the radius of the nanoparticles, that is, $\delta y \gg D$ (Michaelides, 2014). Consequently, the *effective transport properties* of nanofluids are defined in analogy to Eq. (10.1) as

$$\mu_e \equiv \frac{\tau}{\left(\dfrac{\delta u}{\delta y}\right)_{x,y,z}} \quad , \quad k_e \equiv \frac{q''}{\left(\dfrac{\delta T}{\delta y}\right)_{x,y,z}} \quad , \quad D_{ie} \equiv \frac{m_i''}{\left(\dfrac{\delta \rho_i}{\delta y}\right)_{x,y,z}} \quad \text{with } \delta y \gg D. \tag{10.2}$$

The average velocity, temperature, and density gradients may be defined at any point (x,y,z) in the heterogeneous mixture of the nanofluid using the larger length scale, δy. Hence, the effective transport properties of nanofluids are well-defined in the entire domain of these heterogeneous mixtures.

This definition of the effective transport properties of heterogeneous media is implicit in all the analytical derivations of their transport properties, as for example in the viscosity derivations for the liquid-solid heterogeneous mixtures (Einstein, 1906, 1911; Brinkman, 1952; Batchelor, 1977). All the experimental studies prove that the effective transport properties are uniform in suspensions where the particles are uniformly distributed and are independent of the size of the length-scale δy, provided that $\delta y > 10D$. When the nanoparticles are not uniformly distributed—for example, in centrifugal separators, or during thermophoretic deposition—the effective transport properties of the nanofluids are non-uniform too, and in general, they are monotonic functions of the volumetric fraction of the nanoparticles.

10.3 EFFECTIVE VISCOSITY

10.3.1 EXPERIMENTAL DATA AND CORRELATIONS

The addition of solid particles in a homogeneous fluid, either gas or liquid, always has an enhancing effect on the viscosity of the fluid because of the following two mechanisms:

1. Particles move as rigid bodies with uniform velocity and disrupt, locally, the velocity profile of the fluid. The velocity gradients in the fluid part of the suspension become sharper to accommodate the entire fluid velocity gradient. The increased strain in the interstitial fluid, between neighboring particles, increases the shear stress, and this shows as a higher effective viscosity.
2. There is an additional momentum transfer in the direction perpendicular to the velocity gradient due to the Brownian movement of the solid particles. This momentum transfer always increases the effective viscosity of the suspension.

Eq. (2.55) applies to the viscosity of dilute heterogeneous suspensions of spheres, including nanofluids with spherical particles. However, most nanoparticles are of oblong or irregular shapes, and the analytically derived Eq. (2.52) underpredicts the experimental data. Several experimental studies determined the effective viscosity of classes of nanofluids, among which are the following:

1. The data by Pak and Cho (1998) as correlated by Buongiorno (2006) for alumina (Al_2O_3)-water nanofluids

$$\mu_e = \mu_c(1+39.11\alpha_d +534\alpha_d^2). \tag{10.3}$$

and for the titania (TiO_2)-water nanofluids

$$\mu_e = \mu_c(1+5.45\alpha_d +108\alpha_d^2). \tag{10 4}$$

2. The experimental data by Wang et al. (1999) as correlated by Maiga et al. (2005) for alumina (Al_2O_3)-water nanofluids

$$\mu_e = \mu_c(1+7.3\alpha_d +123\alpha_d^2). \tag{10.5}$$

3. A group of authors, including Nguyen et al. (2007) and Tseng and Lin (2003), correlated their data using the functional form of the viscosity for concentrated suspensions

$$\mu_e = A\mu_f \exp(B\varphi). \tag{10.6}$$

However, the values of the coefficients A and B vary significantly in these studies: the coefficient pairs (A, B) by Nguyen *et al.* (2007) for a water-Al$_2$O$_3$ nanofluid are (0.904, 14.8) while by Tseng and Lin (2003) for water-TiO$_2$ nanofluids are (13.47, 35.98). Simple calculations show that the last set of coefficients gives very high, unrealistic values for the effective Newtonian fluid viscosity. Perhaps the high viscosity values were observed because the solid-fluid suspension became non-Newtonian (Michaelides, 2014).

Of the available correlations, the polynomial functional form of the effective viscosity rather than the exponential form is recommended to be used because of the following: (a) Nanofluids are dilute mixtures, (b) the polynomial form is indicated by all analytical studies, and (c) most of the experimental studies favor polynomial expressions (Michaelides, 2013; Michaelides, 2014). However, there is a great deal of uncertainty on the actual values and no clear recommendation on the polynomial coefficients to be used.

10.3.2 NON-NEWTONIAN BEHAVIOR

When nanofluids are at rest, the particles tend to aggregate, and the heterogeneous suspensions of nanoparticles with high surface energy make up particulate structures and weak gels. The formation of such structures and particle networks, which strongly depends on the electrostatic forces between nanoparticles and the pH of the base fluid, may cause the transition of the suspensions to non-Newtonian fluids. The viscosity of non-Newtonian fluids depends on the other thermodynamic properties, for example, T and P, and also on the local value of the velocity shear, γ. This signifies that the viscosity is not an equilibrium property of the non-Newtonian fluid. For this reason, the non-Newtonian fluid viscosity is often called an "effective viscosity" and is denoted by a different symbol, η, with $\eta = \eta(T,P,\gamma)$.[1]

Krieger and Dougherty (1959) studied the rheology of fluid-solid suspensions, composed of micro- and nanoparticles in liquids. They determined that the local strain is best represented by a dimensionless shear Peclet number: $Pe_\gamma = \mu_f a^3 \gamma / k_B T$, which is a measure of the ratio of the shear force to the Brownian force. Figure 10.1

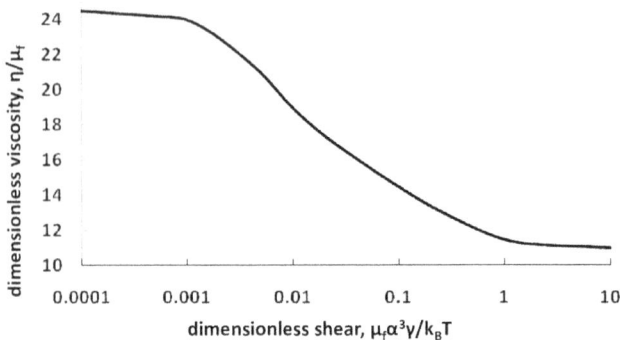

FIGURE 10.1   Shear dependence of the effective viscosity of a non-Newtonian nanofluid.

---

[1] As with the Newtonian fluids and the dynamic viscosity, μ, the effective viscosity, η, primarily depends on the temperature and is a weak function of the pressure, P.

shows the typical dependence of the dimensionless effective viscosity, η/μ, of a nanofluid on the rate of shear, γ, at constant T and P. It is apparent that the function $\eta(\gamma)$ has two asymptotic limits: One at the low value, and one at the high value of the shear, η_0, and η_∞ respectively, and that both asymptotic limits are significantly higher than the viscosity of the base fluid. Also, that the effective viscosity at the higher shear rates is by orders of magnitude lower than the effective viscosity at the lower shear rates. This phenomenon is called *shear thinning* and has been observed in most experimental studies with nanofluids (Kim *et al.*, 2011).

10.4 EFFECTIVE THERMAL CONDUCTIVITY

Early experimental work with carbon nanotube (CNT) suspensions reported that the addition of a small fraction of CNT in engine oil increased the thermal conductivity of the fluid by a factor of 2 to 3. Since this increase also promises significant convective heat transfer enhancement, nanofluids have been touted as "the heat transfer media of the future," their thermal behavior was characterized as "anomalous" (Choi *et al.*, 2001), and this sparked an immense scientific interest and a multitude of experimental studies with nanofluids during the first decade of the twenty-first century.

10.4.1 EXPERIMENTAL STUDIES

Figure 10.2 shows typical results from experiments on the effective conductivity of nanofluids with several types of nanoparticles (data from Michaelides, 2013). Based

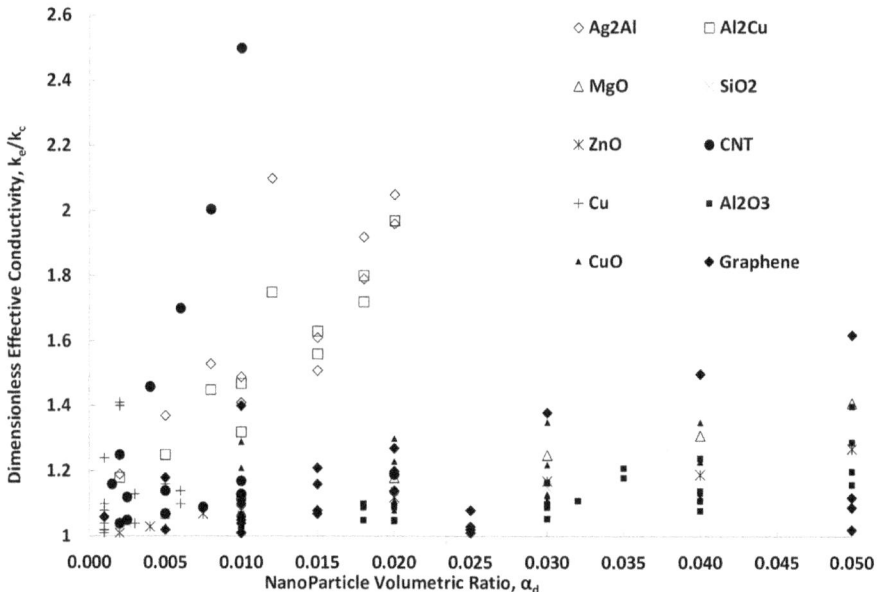

FIGURE 10.2 Effective thermal conductivity for several types of nanofluids.

on results from the several experimental studies, it becomes apparent that the most important parameters, which affect the thermal conductivity of nanofluids are as follows:

1. The type and properties of the nanoparticles
2. The nominal "size" of the nanoparticles
3. The shape of the nanoparticles
4. The type of the base fluid
5. The volumetric fraction of the nanoparticles
6. The pH and the type of surfactants used to stabilize the nanofluid and prevent aggregation

The highest increases of thermal conductivity have been observed with suspensions of carbon nanotubes (CNT)—both single-walled and multi-walled—and with metallic nanoparticle suspensions. The lowest enhancements were observed with oxides of nanoparticles, such as TiO_2 and Al_2O_3. These are solid particles with lower thermal conductivities. While almost all the experimental studies agree on the conductivity enhancement, there is a large discrepancy on the actual enhancement among sets of data emanating from different studies. Figure 10.3, which depicts the relative conductivity of CNT nanofluids (data from Choi et al., 2001; Liu et al., 2005; Ding et al., 2006; Xie et al., 2003; Biercuk et al., 2002), illustrates this point. From the apparent discrepancy among the different data sets, one may conclude that, while the material of the nanoparticles and their concentration are determinants of thermal conductivity enhancement, other variables also play equally important roles.

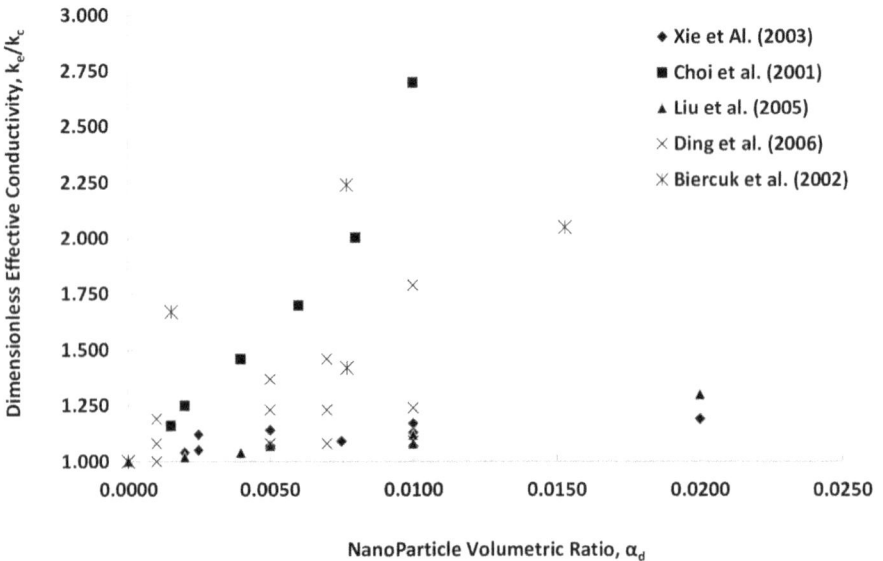

FIGURE 10.3 Effective thermal conductivity data of CNT nanofluids from several experiments.

10.4.2 ANALYTICAL EXPRESSIONS

Maxwell (1881) derived an analytical expression for the effective electrical conductivity of a heterogeneous mixture composed of spherical particles. Because of the analogy in thermal and electrical conduction processes, this expression is also valid for the effective thermal conductivity of heterogeneous mixtures composed of spherical particles

$$\frac{k_e}{k_c} = 1 + \frac{3(k_d - k_c)\alpha_d}{(k_d + 2k_c) - (k_d - k_c)\alpha_d} \Rightarrow \frac{k_e}{k_c} \to \frac{3\alpha_d}{1-\alpha_d} \quad \text{when } k_d \gg k_c, \quad (10.7)$$

where k_c is the conductivity of the carrier fluid, k_e the effective conductivity of the suspension, and α_d is the volumetric fraction of the nanoparticles. While Maxwell's expression is valid for spherical particles, of particular interest for nanofluids, which are invariably composed of non-spherical particles, is the analytical study by Nan et al. (2003), which pertains to nanofluids with high-aspect ratio particles (e.g. CNT and other elongated nanoparticles) at random orientations and low concentrations

$$\frac{k_e}{k_c} = \frac{3 + \alpha_d (k_d / k_c)}{3 - 2\alpha_d} \approx 1 + \frac{\alpha_d (k_d / k_c)}{3} \quad \text{when } \alpha_d \ll 1. \quad (10.8)$$

Because for most nanoparticles $k_d \gg k_c$, the addition of even very small amounts of highly conducting fibrous nanoparticles significantly enhances the effective conductivity of the suspension. Eq. (10.8) largely explains the experimentally observed conductivity enhancements with CNT and metallic nanoparticles and proves that the thermal conductivity of nanofluids is not "anomalous."

10.4.3 MECHANISMS OF THERMAL CONDUCTIVITY ENHANCEMENT

The plethora of experimental data on the enhanced effective conductivity of nanofluids generated several analytical studies on the mechanisms of the enhanced conductivity. The most important mechanisms that affect the effective conductivity of nanofluids are the following:

1. The significantly higher thermal conductivity of the nanoparticles.
2. The shape (spheres, fibers, irregular) of the particles. Even if randomly oriented, fibrous and elongated nanoparticles contribute significantly more to the effective conductivity of the suspension.
3. The development of aggregates and chains that form highly conductive paths within the nanofluid or contribute to the development of such paths.

Other mechanisms that influence the higher effective thermal conductivity of nanofluids are as follows:

1. The Brownian movement of the particles.
2. The added mass of the fluid that follows the random Brownian movement (micro-convection).

3. The formation of "solid-layers" composed of liquid molecular layers at the solid-fluid interface that locally changes the thermodynamic properties of the carrier fluid.
4. The electric charge on the surface of the particles.
5. Transient local heat transfer effects (history term).
7. Nanofluid preparation methods and the addition of surfactants, which modify the carrier fluid properties.
8. Magnetic fields on magnetic nanoparticles, which affect the orientation of elongated nanoparticles and assist the formation of highly conducting aggregates and structures.
9. The amount of fluid "trapped" within the volume defined by the boundary of irregular particles.
10. Velocity discontinuity (slip) and thermal discontinuity at the fluid-particle interface as elucidated in Section 5.5.2.4.

10.5   FORCED CONVECTION

If nanofluids are to become actual heat transfer media, they will be used in channels of all sizes to remove heat from other materials (e.g. from electronic components). In these cases, the main parameter of interest is the convective heat transfer coefficient, h, which is typically given by a closure equation in terms of the Nusselt number

$$\dot{Q} = hA\Delta T \quad with \quad h = Nu\frac{k_e}{L_{ch}}, \tag{10.9}$$

where L_{ch} is the characteristic dimension of the heat transfer conduit, typically the diameter or the hydraulic diameter of the channel. The Nusselt number is given by an empirical equation, $Nu = f(Re, Pr)$, which may be easily found in monographs on heat transfer (e.g. Incropera and DeWitt, 2000). It is apparent from Eq. (10.9) that if the effective thermal conductivity of a heat transfer medium increases, the convective coefficient increases proportionately. However, several experimental studies that measured independently the thermal conductivity and the convective heat transfer coefficients of nanofluids in laminar flows concluded that the increase of h is significantly higher than what the relationship of Eq. (10.9) indicates (Michaelides, 1986). There is an additional augmentation of the convective heat transfer coefficient, h, beyond and above what follows by the increased effective thermal conductivity of the nanofluid. The experiments also indicate that the additional augmentation of h increases with the volumetric fraction of the dispersed phase, α_d, but the relationship is not linear.

Figure 10.4 shows representative experimental data for the convective heat transfer coefficient augmentation for nanofluids in laminar channel flows (data from Li and Xuan, 2002; Wen and Ding, 2004; Ding et al., 2006; Hojjat et al., 2010). In this figure, Nu is the experimentally measured Nusselt number and Nu_e is the Nusselt number analytically calculated using the effective conductivity of the nanofluid ($Nu_e = hL_{ch}/k_e$) using the analytical expressions for the Nusselt number, $Nu_e = f(Re, Pr)$, for laminar flows in channels. It is observed that the data consistently demonstrate that this ratio is higher than one, sometimes by as much as 60%. This general

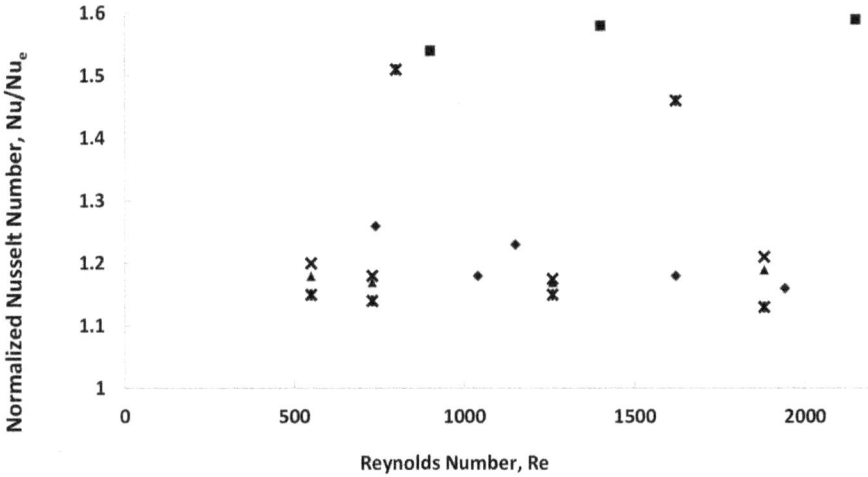

FIGURE 10.4 Augmentation of Nusselt numbers for several types of nanofluids.

observation implies that, in addition to the enhanced conductivity of the nanofluids, there is an additional mechanism that augments the coefficient h.

The experimental evidence for turbulent flows with nanofluids is not clear: A few experimental studies observed a modest augmentation, in the range 10–20%, while most of the other experiments did not observe any augmentation, within the limits of the experimental uncertainty (Michaelides, 2014).

The most likely explanation for the observed augmentation of the convective heat transfer coefficient—above and beyond the values calculated by the higher thermal conductivity of the nanofluid—is the *fluid micro-convection*, which is induced by the Brownian movement of the particles.

Figure 10.5 is a schematic diagram that illustrates the micro-convection process in a carrier fluid. As the ellipsoidal particle is advected by the flow, it experiences Brownian movement and translates from position 1 to position 2. The carrier fluid rushes to fill the vacuum created in position 1 by the volume of the particle as well as that of the added (virtual) mass of the fluid, which roughly follows the particle. Simultaneously, the fluid rushes out of position 2 to create space for the solid particle and for the added mass of the fluid that follows the particle. This process creates a local fluid agitation in the vicinity of particles, which augments the overall laminar convective heat transfer coefficient.

The apparent absence of such augmentation in the turbulent heat transfer regime corroborates this mechanism: the magnitude of turbulent velocity fluctuations and the agitation of the carrier fluid by turbulence are by far greater than the agitation induced by the micro-convection process, which is caused by the Brownian movement. In turbulent flows, the contribution of micro-convection to the value of the convective heat transfer coefficient would be negligible and almost vanishing. On the contrary, in laminar flows, any systematic lateral velocity fluctuations significantly contribute to the overall heat transfer coefficient.

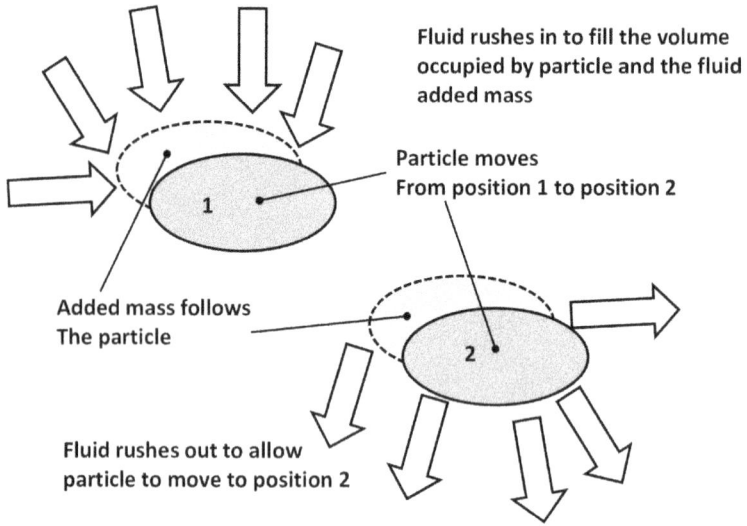

FIGURE 10.5   Fluid micro-convection due to particle movement.

10.6   NATURAL CONVECTION

The addition of nanoparticles in a carrier fluid has several effects, some of which promote and others that inhibit the natural (free) convection:

1. Effective fluid viscosity enhancement (inhibiting).
2. Damping of instabilities on the solid surface of the particles (inhibiting).
3. Interactions with the heating surface. Depending on the nature of particles and the interactions, this may either promote or inhibit natural convection.
4. Particle sedimentation. This induces long-range, upward fluid currents that assist the formation of instabilities (promoting).
5. Enhancement of the effective fluid conductivity (promoting).
6. Chemically active nanoparticles generate locally thermal and flow instabilities around the particles (promoting).

It appears that there is not a single and clear mechanism for the enhancement or reduction of natural convection when nanoparticles are added to the base fluid, and this is reflected in the sets of experimental data, which are divided on the overall effect on the natural convection: Almost all the analytical studies (e.g. Tzou, 2008; Shukla and Dhir, 2008) and a few experimental studies (e.g. Okada and Suzuki, 1997) advocate the enhancement of natural convection heat transfer in nanofluids. On the other hand, several experimental studies (e.g. Putra *et al.*, 2003; Wen and Ding, 2005) show that there is a clear and systematic reduction of the natural heat transfer coefficients. More accurate and better documented experiments, where the size and distribution of nanoparticles are also measured, are needed to improve our understanding of the effect of nanoparticle addition on the natural convection process.

10.7   BOILING

10.7.1   Pool Boiling

The addition of nanoparticles has the following promoting and inhibiting effects (Michaelides, 2014):

1. Nanoparticles provide additional nucleation sites (promoting).
2. Nanoparticles may deposit and deactivate a fraction of the nucleation sites on the heated surface (inhibiting).
3. Deposited nanoparticles after evaporation change the characteristics and structure of the heating surface (either promoting or inhibiting, depending on the type of the deposited nanoparticles).
4. Surfactants added for the stability of nanoparticles in suspensions modify the surface tension of the fluid, thus affecting bubble formation and liftoff (usually inhibiting).
5. Nanoparticles enhance the micro-layer evaporation on the heated surfaces (promoting).
6. Nanoparticles modify the thermodynamic and transport properties of the base fluid (either promoting or inhibiting, depending on the type of the deposited nanoparticles).

Most of the experimental studies with nanofluids indicate that the addition of metal oxide nanoparticles in a fluid reduces the pool boiling heat transfer coefficient (Das et al., 2003; Kim et al., 2007). However, some studies with metallic and CNT nanoparticles indicate an increase of the pool boiling heat transfer coefficient (Park and Jung, 2007). Typical experimental results with the addition of metal oxide nanoparticles are shown in Figure 10.6, where the Nukiyama curve for

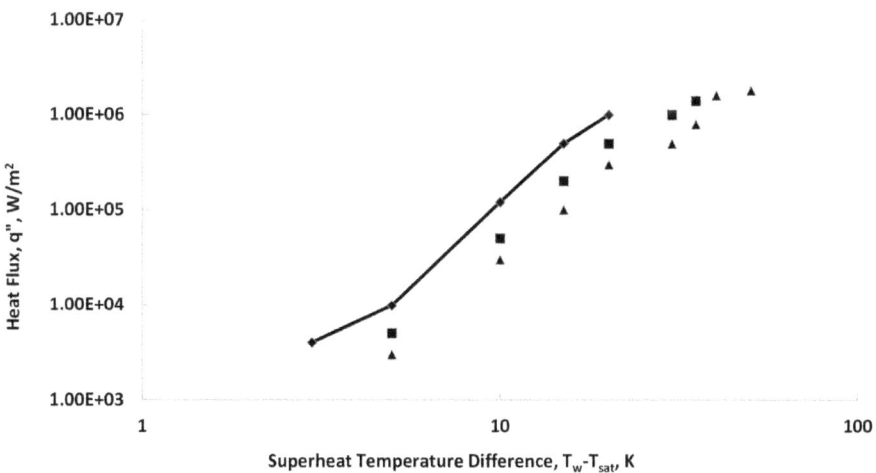

FIGURE 10.6   Typical experimental data of pool boiling with nanofluids: The pool boiling heat transfer coefficient is lower than that obtained in single-phase experiments.

homogeneous-fluid boiling is also shown. The pool boiling heat transfer coefficient is defined as the ratio $q''/(T_w-T_{sat})$.

While a more detailed study on the pool boiling coefficients may be needed, a tentative conclusion is that the addition of oxide nanoparticles decreases the pool boiling heat transfer coefficient, while the addition of metallic and CNT nanoparticles enhances this coefficient.

10.7.2 CONVECTIVE BOILING

Convective boiling occurs in channels, where the nanofluid is pumped. The main differences between pool boiling and convective boiling are the following:

1. The formed vapor is advected downstream by the fluid and does not stay close to the nucleation site to be slowly detached by buoyancy.
2. Because the channel is bounded, the liquid and vapor phases arrange themselves in one of the flow regimes that characterize the vapor-liquid two-phase flow.
3. There is frictional pressure drop along the channel that affects the saturation temperature, especially in microchannels.

The preponderance of experimental evidence on the convective boiling heat transfer coefficients points to the enhancement of the convective boiling coefficient with the addition of nanoparticles. None of the experimental studies is known to have observed a reduction of this coefficient (Michaelides, 2014), and most of the experimental studies indicate a significant (more than 50%) enhancement (Boudouh *et al.*, 2010), or at least no deterioration (Peng *et al.*, 2009; Rana *et al.*, 2013). The following mechanisms are responsible for the enhancement of the convective boiling heat transfer coefficients:

1. Channel-boundary modification by the nanoparticles with increased density of nucleation sites.
2. Increase of bubble-to-surface contact angles.
3. Inhibition of the formation of long-lasting dry vapor patches by the higher concentrations of deposited particles at the bubble-surface interface.
4. Increased frequency of bubble departure (by advection) from the nucleation sites.
5. Formation of larger bubbles that locally agitate the evaporating liquid.

Because of the high boiling heat transfer coefficients, nanofluids have been touted as the ideal heat transfer media for convective boiling in microchannels. A cause of concern with such applications is the long-term reliability of systems: As the base fluid evaporates, nanoparticles deposit close to the exit of the channel and block the channel. One may guard against the partial or total blocking of the cooling channels by restricting the fraction of base fluid allowed to evaporate in the channel. Nanoparticles will be carried away from the channel by the non-evaporating fraction of the base fluid.

10.7.3 Critical Heat Flux

The onset of critical heat flux (CHF) occurs with the formation and lateral expansion of dry vapor patches on the heated surfaces. The expansion spreads the vapor patches sideways, joins two or more vapor patches, and creates thin vapor layers over large parts of the heating surface, including microchannels. Because particles do not evaporate, they are trapped inside the liquid layer at the heated surface, and this creates higher particle concentrations adjacent to the dry patches. The increased concentration of particles has three effects on the vapor phase:

1. Restricts the sideways expansion of the vapor patch by forming a solid residue that obstructs the spreading of vapor.
2. Because of interfacial surface tension, it maintains thin interstitial liquid layers between individual particles and between the particles and the heating surface.
3. Particles close to the heated surface disturb the shape of the meniscus at the vapor-liquid interface and create a "structural disjoining pressure," which increases the wettability and inhibits the spreading of the dry patches (Wen, 2008).

All three mechanisms inhibit the spreading of vapor pockets and bubbles. As a result, the addition of nanoparticles in an evaporating fluid enhances the CHF, sometimes by a factor of 2 or 3, as it has been observed in several experimental studies. General observations from the several experimental studies on the CHF are (Michaelides, 2014) as follows:

1. Carbon nanotube (CNT) nanoparticles cause the highest CHF enhancement, followed by metal nanofluids and oxide nanofluids.
2. CHF enhancement in convective boiling is lower than that of pool boiling.
3. The vapor bubbles departing from the heating surface are significantly larger in nanofluids than in the pure fluid experiments. Most likely this is due to the surfactants used for the stabilization of the nanofluids, which also modify the surface tension of the base fluid.
4. The heating surfaces were modified with the deposition of a layer of particles, which sometimes formed a thin, porous layer with interstitial fluid.

10.8 EFFECTIVE DIFFUSIVITY AND MASS TRANSFER

Diffusion is the movement of mass at the molecular length scales and involves the transfer of a species, i, within a continuous fluid (gas or liquid) or a solid. The effective diffusivity of nanofluids, $\mathcal{D}_e$, of a species, i, is defined in terms of a local elemental volume by Eq. (10.2). Experimental and numerical studies prove that the diffusivity of the heterogeneous mixtures is uniform in nanofluids where the particles are uniformly distributed and is independent of the size of the segment, δy, provided that the latter is greater than the equivalent of ten nanoparticle equivalent diameters. If the nanoparticles are non-uniformly distributed in the suspension, the effective diffusivity is also non-uniform, and in general, it is a monotonic function of the dispersed particle concentration, α_d.

A parameter of interest in mass transfer applications is the mass transfer coefficient of the nanofluid, h_m, which is analogous to the heat transfer coefficient, h, and is defined in terms of the mass transfer equation

$$\dot{m}_i = h_{mi} A_c \left(\Delta C_i \right), \tag{10.10}$$

where ΔC_i is the difference of the concentration of the species, i, and A_c is the cross-sectional area through which the mass is transferred. The mass transfer coefficient is calculated using the Sherwood number, which is analogous to the Nusselt number

$$Sh = \frac{D h_{mi}}{\mathcal{D}_{ei}}. \tag{10.11}$$

The Sherwood number is obtained from empirical correlations, which typically emanate from heat transfer analysis and experiments using the analogy of heat and mass transfer (Michaelides, 2014).

10.8.1 ANALYTICAL RESULTS

The diffusivity of solid materials is much lower than that of fluid materials. Solid particles are impediments to the diffusion of mass in solid-fluid mixtures, and hence, it is expected that the effective diffusivity of solid suspensions, including nanofluids, would be lower than the corresponding diffusivity of the carrier fluid. Several analytical and experimental studies corroborate this fact. Among these, Cussler's (2009) derived the following expression for the ratio of the effective diffusivity of a suspension to the diffusivity of the carrier fluid, in terms of the solid particles' volumetric fraction, α_d

$$\frac{\mathcal{D}_{ie}}{\mathcal{D}_i} = \frac{1 - \alpha_d}{1 + 0.5\alpha_d}. \tag{10.12}$$

For nanoparticles, where $\alpha_d \ll 1$, this ratio is close to and slightly less than 1.

10.8.2 EXPERIMENTAL METHODS AND RESULTS

Experimental studies on the mass transfer and the effective diffusivity of nanofluids used a variety of methods and generated markedly different results. Several of the experimental studies used methods that actually measure the mass transfer coefficient, h_{mi}, rather than the molecular diffusivity, $\mathcal{D}_{ie}$. These studies concluded that the mass transfer coefficient of nanofluids is very high in comparison to that of the carrier fluid. For example, Olle et al. (2006) observed that the mass transfer coefficient of oxygen is higher than a factor of 6 in aqueous ferric (Fe_3O_4) nanofluids. Krishnamurti et al. (2006) observed a more dramatic increase (up to a factor of 14) of the mass transport coefficient in water-alumina (Al_2O_3) nanofluids. However, several other experimental studies observed lower mass transfer enhancement (e.g. Komati and Suresh, 2008), while a few studies (Turanov and Tolmachev, 2009; Feng and Johnson, 2012) observed a significant decrease (up to 33%) of the effective diffusivity.

A careful examination of the experimental results reveals that almost all the experimental studies, which have observed significant mass transfer enhancement, used a liquid reactor or simple optical methods, where the nanofluid flows freely, and there is ample evidence of local advection (Michaelides, 2014). Such studies actually measured the convective mass transfer coefficient, h_{mi} and some wrongly attributed the measurements to the effective diffusivity, $\mathcal{D}_{ie}$. The majority of the experimental studies, where zero (or almost zero) enhancement was observed determined the actual effective diffusivity, $\mathcal{D}_{ie}$, or the self-diffusion coefficient of a species. Experiments that used permeable membranes for the determination of the diffusivity in nanofluids strongly underestimated the effective diffusivity coefficients because the nanoparticles significantly increased the local viscosity of the heterogeneous nanofluid inside the membrane nanopores, partly blocked some of the nanopores, and thus, impeded the entire diffusion process (Michaelides, 2017).

A careful examination of the experimental data, the instruments used, and the methods the data was obtained indicate that the molecular diffusivity of the nanofluids remains almost constant and may actually decrease when nanoparticles are added to the base fluid, as predicted by Eq. (10.12). The slight decrease is most likely due to the obstruction by the particles and the higher local viscosity of the fluid. The experimental data also show that the mass transfer coefficients, h_m, which account for both the diffusion and the advection of species, are invariably increased when nanoparticles are added to the carrier fluid and the measuring instrument allows for the unhindered particle motion as well as the micro-convection process that follows the Brownian movement. The addition of nanoparticles causes locally mass advection, and this has a significant impact on the mass transfer of a species within the nanofluid. Because the molecular diffusion process is a very slow molecular process, in comparison to the other transport processes, any type of fluid advection, however slow or weak, significantly contributes to the mass transfer. All the relevant experimental data and the pertinent analytical studies show that if the Brownian movement and the associated unsteady micro-convection of the carrier fluid are not suppressed, the mass transfer process is significantly enhanced.

It must be noted that the lower diffusivities do not detract from the usefulness of nanofluids as effective mass transfer media. The objective of all the relevant technological applications is the enhancement of the overall mass transfer, not necessarily the molecular transfer (diffusivity). It is the enhancement of the convective mass transfer coefficient, h_{mi}, of a species that matters in engineering systems and not the effective diffusivity, $\mathcal{D}_{ie}$, of the species. If the overall mass transfer of the carrier fluid increases by the addition of small amounts of nanoparticles, nanofluids may prove to be inexpensive media for the development and improvement of microfluidic devices, such as the "lab on a chip" type of systems.

10.9 SPECIFIC HEAT CAPACITY

The specific heat capacity is an equilibrium thermodynamic property and represents the potential of a material to store heat. The conventional model of heterogeneous mixtures, which is based on the thermal equilibrium of the constitutive materials (Gibbs, 1878), offers the analytical framework for this property as derived in Eqs. (2.53) and (2.54). In the case of nanofluids that are composed of incompressible liquids and solids, the enthalpy is a function of the temperature alone, the specific heats at constant volume and constant pressure are equal, and the two equations are combined to yield

$$c_m = \frac{dh_m}{dT} = (1-Y)c_c + Yc_d. \tag{10.13}$$

Since most solids have lower specific heat capacities than water ($c_c > c_d$) it follows that the specific heat capacity of aqueous nanofluids is less than that of water. The experimental studies by Zhou and Ni (2008) with alumina nanoparticles and the molecular modelling study by Rajabpour *et al.,* (2013) confirm this conclusion. However, experiments with eutectic salts have shown different trends and observed specific heat capacity enhancements in the range 14–24% (Shin and Banerjee, 2011). The authors suggested the following three mechanisms to explain the observed enhancement:

1. The activation of surface energy of nanoparticles and the atomic energy bonds at the particle-fluid interface.
2. An increase in the interfacial thermal resistance between nanoparticles and the carrier fluid.
3. The possible development of a semi-solid layer at the interface that has higher heat capacity than the bulk carrier fluid.

While these effects have not been validated, a preliminary observation/conclusion may be drawn from the limited experimental and analytical data on the heat capacity of nanofluids: Aqueous nanofluids show reduced specific heat capacity, whereas some molten salt based nanofluids may exhibit heat capacity enhancement.

SUMMARY

The addition of nanoparticles increases the effective viscosity of nanofluids and, oftentimes, transforms the base fluid to non-Newtonian. In general, the thermal conductivity of the carrier fluids increases, sometimes by a factor of 2 or 3. Significant enhancements were also observed for the mass transfer coefficients of several nanofluids—on the order of 10—even though the effective diffusivity may slightly decrease. Reliable analysis and experimental data indicate that there is nothing "anomalous" with these suspensions of nanoparticles. The higher transport coefficients of these heterogeneous suspensions are readily explained with known effects related to the shape of nanoparticles, the material of the nanoparticles, the distribution and the configuration of nanoparticles inside the carrier fluid and the Brownian movement.

REFERENCES

Batchelor, G., 1977, The effect of Brownian motion on the bulk stress in a suspension of spherical particles, *J. Fluid Mech.*, **83**, 97–117.

Biercuk, M., Llaguno, M., Radosavljevic, M., Hyun, J., Johnson, A. and Fischer, J., 2002, Carbon nanotube composites for thermal management, *Appl. Phys. Lett.*, **80** (15), 2767.

Boudouh, M., Gualous, H.L. and De Labachelerie, M., 2010, Local convective boiling heat transfer and pressure drop of nanofluid in narrow rectangular channels, *Appl. Thermal Eng.*, **30** (17–18), 2619.

Brinkman, H., 1952, The viscosity of concentrated suspensions in solutions, *J. Chem Phys.*, **20**, 571–582.

Buongiorno, J., 2006, Convective transport in nanofluids, *J. Heat Transfer*, **128**, 240–250.

Choi, S.U.S., Zhang, Z.G., Yu, W., Lockwood, F.E. and Grulke, E.A., 2001, Anomalous thermal conductivity enhancement in nanotube suspensions, *Appl. Phys. Lett.*, **79**, 2252.

Cussler, E.L., 2009, *Diffusion: Mass Transfer in Fluid Systems*, 3rd edition, Cambridge University Press, New York.

Das, S.K., Nandy, P., Thiesen, P. and Roetzel, W., 2003, Temperature dependence of thermal conductivity enhancement for nanofluids, *J. Heat Transfer*, **125**, 567–572.

Ding, Y., Alias, H., Wen, D. and Williams, R.A., 2006, Heat transfer of aqueous suspensions of carbon nanotubes (CNT nanofluids), *Int. J. Heat Mass Transfer*, **49**, 240.

Einstein, A., 1906, Eine neue bestimung der molekuldimensionen, *Annalen der Physik*, **19**, 289–306.

Einstein, A., 1911, Berichtigung zu meiner Arbeit: "Eine neue bestimung der molekuldimensionen", *Annalen der Physik*, **34**, 591–592.

Feng, X. and Johnson, D.W., 2012, Mass transfer in SiO_2 nanofluids: A case against purported nanoparticle convection effects, *Int. J. Heat Mass Transf.*, **55**, 3447–3453.

Gibbs, J.W., 1878, On the equilibrium of heterogeneous substances, in *The Collective Works of J. Willard Gibbs*, Longmans, New York, NY.

Hojjat, M., Etemad, S.G. and Bagheri, R., 2010, Laminar heat transfer of non-Newtonian nanofluids in a circular tube, *Korean J. Chem. Eng*, **27**, 1391.

Incropera, F.P. and DeWitt, D.P., 2000, *Fundamentals of Heat and Mass Transfer*, 4th edition, Wiley, New York.

Kim, S.J., Bang, I.C., Buongiorno, J. and Hu, L.W., 2007, Surface wettability change during pool boiling of nanofluids and its effect on critical heat flux, *Int J Heat Mass Transf*, **50**, 4105.

Kim, S.J., Kim, C., Lee, W.H. and Park, S.R., 2011, Rheological properties of alumina nanofluids and their implication to the heat transfer enhancement mechanism, *J. Appl. Phys.*, **110**, 034316–1034316–6.

Komati, S. and Suresh, A.K., 2008, CO_2 absorption into amine solutions: A novel strategy for intensification, based on the addition of ferrofluids, *J. Chem. Techn. and Biotechnol.*, **86**, 1094–1100.

Krieger, I.M. and Dougherty, T.J., 1959, A mechanism for non-Newtonian flow in suspensions of rigid spheres, *Trans. Soc. Rheol.*, **3**, 137–152.

Krishnamurti, S., Bhattacharya, P., Phelan, P.E. and Prasher, R.S., 2006, Enhanced mass transport in nanofluids, *Nanoletters*, **6**, 419–423.

Li, Q. and Xuan, Y., 2002, Convective heat transfer and flow characteristics of Cu-water nanofluid, *Chinese, Technol. Sci. Series E.*, **45**, 408.

Liu, M.S., Lin, M.C.-C., Huang, I.T. and Wang, C.-C., 2005, Enhancement of thermal conductivity with carbon nanotube for nanofluids, *Int. Comm. in Heat and Mass Transfer*, **32**, 1202.

Maiga, E.S., Palm, J., Nguyen, C.T., Roy, G. and Galanis, N., 2005, Heat transfer enhancement by using nanofluids in forced convection flows, *Int. J. Heat Fluid Flow*, **26**, 530–546.

Maxwell, J.C., 1881, *A Treatise on Electricity and Magnetism*, 2nd edition, Clarendon Press, Oxford, England.

Michaelides, E.E., 1986, Heat transfer in particulate flows, *Int. J. Heat Mass Transfer*, **29**, 265.

Michaelides, E.E., 2006, *Particles, Bubbles and Drops: Their Motion, Heat and Mass Transfer*, World Scientific Publ., NJ.

Michaelides, E.E., 2013, Transport properties of nanofluids: A critical review, *J. Non-Equilibrium Thermodynamics*, **38**, 1–78.

Michaelides, E.E., 2014, *Nanofluidics: Thermodynamic and Transport Properties*, Springer, New York.

Michaelides, E.E., 2015, Brownian movement and thermophoresis of nanoparticles in liquids, *Int. J. Heat Mass Transf.*, **81**, 179–187.

Michaelides, E.E., 2017, Nanoparticle diffusivity in narrow cylindrical pores, *Int. J. of Heat and Mass Transfer*, **114**, 607–612.

Nan, C.W., Shi, Z. and Lin, Y., 2003, A simple model for thermal conductivity of carbon nanotube-based composites, *Chemical Physics Lett.*, **375**, 666–669.

Nguyen, C.T., Desgranges, F., Roy, G., Galanis, N., Marie, T., Boucher, S. and Mintsa, H.A., 2007, Temperature and particle-size dependent viscosity data for water based nanofluids-hysteresis phenomenon, *Int. J. Heat Fluid Flow*, **28**, 1492–1506.

Okada, M. and Suzuki, T., 1997, Natural convection of water-fine particle suspension in a rectangular cell, *Int. J. Heat Mass Transfer*, **40**, 3201.

Olle, B., Bucak, S., Holmes, T.C., Bromberg, L., Hatton, T.L. and Wang, D.I.C., 2006, Enhancement of oxygen mass transfer using functionalized magnetic nanoparticles, *Ind. and Eng. Chemistry Res.*, **45** (12), 4355–4363.

Onsager, L., 1931, Reciprocal relations in irreversible processes, I, *Phys. Review*, **37**, 405–426.

Pak, B.C. and Cho, Y., 1998, Hydrodynamic and heat transfer study of dispersed fluids with submicron metallic oxide particles, *Exp. Heat Transfer*, **11**, 151–170.

Park, K.J. and Jung, D., 2007, Enhancement of nucleate boiling heat transfer using carbon nanotubes, *Int J Heat Mass Transf.*, **5**, 4499.

Peng, H., Ding, G., Jiang, W., Hu, H. and Gao, Y., 2009, Heat transfer characteristics of refrigerant-based nanofluid flow boiling inside a horizontal smooth tube, *Int J Refrigeration*, **32** (6), 1259.

Putra, N., Roetzel, W. and Das, S.K., 2003, Natural convection of nano-fluids, *Heat and Mass Transfer*, **39**, 775.

Rajabpour, A., Akizi, F.Y., Heyhat, M.M. and Gordiz, K., 2013, Molecular dynamics simulation of the specific heat capacity of water-Cu nanofluids, *Intern. Nano Letters*, **3**, 58.

Rana, K.B., Rajvanshi, A.K. and Agrawal, G.D., 2013, A visualization study of flow boiling heat transfer with nanofluids, *J. Visualization*, **16**, 133.

Shin, D.H. and Banerjee, D., 2011, Enhancement of specific heat capacity of high-temperature silica-nanofluids synthesized in alkali chloride salt eutectics for solar thermal-energy storage applications, *Int. J. Heat Mass Transfer*, **54**, 2014.

Shukla, R. and Dhir, V., 2008, Effect of Brownian motion on thermal conductivity of nanofluids, *J. Heat Transfer*, **130**, 042406.

Tseng, W.J. and Lin, K.C., 2003, Rheology and colloidal structure of aqueous TiO2 nanoparticle suspensions, *Mater. Sci. Eng.*, **A355**, 186–192.

Turanov, A. and Tolmachev, Y., 2009, Heat- and mass-transport in aqueous silica nanofluids, *Heat and Mass Transfer*, **45**, 12, 1583–588.

Tzou, D.Y., 2008, Thermal instability of nanofluids in natural convection, *J. Heat Transfer*, **130**, 072401.

Wang, X., Xu, X. and Choi, S.U.S., 1999, Thermal conductivity of nanoparticle fluid mixture, *J. Thermophysics and Heat Transfer*, **13**, 474–480.

Wen, D., 2008, Mechanisms of thermal nanofluids on enhanced critical heat flux (CHF), *Int. J. Heat Mass Transfer*, **51**, 4958–4965.

Wen, D. and Ding, Y., 2004, Experimental investigation into convective heat transfer of nanofluids at the entrance region under laminar flow conditions, *Int. J. Heat Mass Transf.*, **47**, 5181.

Wen, D. and Ding, Y., 2005, Formulation of nanofluids for natural convective heat transfer applications, *Int. J. Heat Fluid Flow*, **26**, 855.

Xie, H., Lee, H., Youn, W. and Choi, M., 2003, Nanofluids containing multiwalled carbon nanotubes and their enhanced thermal conductivities, *J. Appl. Phys.*, **94** (8), 4967.

Zhou, S.Q. and Ni, R., 2008, Measurement of the specific heat capacity of water-based Al_2O_3 nanofluid, *Appl Phys Lett.*, **92**, 093123.

Index

For Product Safety Concerns and Information please contact our EU
representative GPSR@taylorandfrancis.com
Taylor & Francis Verlag GmbH, Kaufingerstraße 24, 80331 München, Germany

9 780367 544348